to accompany

Elementary Statistics

A Step by Step Approach

Fifth Edition

Allan G. Bluman
Community College of Allegheny County

Prepared by

Sally H. Robinson
South Plains College

Boston Burr Ridge, IL Dubuque, IA Madison, WI New York San Francisco St. Louis
Bangkok Bogotá Caracas Kuala Lumpur Lisbon London Madrid Mexico City
Milan Montreal New Delhi Santiago Seoul Singapore Sydney Taipei Toronto

The McGraw-Hill Companies

Instructor's Solutions Manual to accompany
ELEMENTARY STATISTICS: A STEP BY STEP APPROACH, FIFTH EDITION
ALLAN G. BLUMAN

Published by McGraw-Hill Higher Education, an imprint of The McGraw-Hill Companies, Inc., 1221 Avenue of the Americas, New York, NY 10020.

RECYCLED This book is printed on recycled, acid-free paper containing 10% postconsumer waste.

2 3 4 5 6 7 8 9 0 QPD/QPD 0 9 8 7 6 5 4

ISBN 0-07-254914-9

www.mhhe.com

Preface

This Instructor's Solutions Manual provides solutions to all odd and even exercises plus answers to all quiz questions in *Elementary Statistics: A Step by Step Approach, 5e* by Allan G. Bluman. Solutions are worked out step by step where appropriate and generally follow the same procedures used in the examples in the textbook. Answers may be carried to several decimal places to increase accuracy and to facilitate checking. Graphs are included with the solutions when appropriate or required. They are intended to convey a general idea and may not be to scale.

Caution: Answers generated using graphing calculators such as the TI-83 may vary from those shown in this manual.

Sally H. Robinson

Contents

Solutions to the Exercises

REVIEW EXERCISES - CHAPTER 1

1. Descriptive statistics describes a set of data. Inferential statistics uses a set of data to make predictions about a population.

2. Probability deals with events that occur by chance. It is used in gambling and insurance.

3. Answers will vary.

4. A population is the totality of all subjects under study. A sample is a subgroup of the population.

5. When the population is large, the researcher saves time and money using samples. Samples are used when the units must be destroyed.

6.
a. inferential e. inferential
b. descriptive f. inferential
c. descriptive g. descriptive
d. descriptive h. inferential

7.
a. ratio f. nominal
b. ordinal g. ratio
c. interval h. ratio
d. ratio i. ordinal
e. ratio j. ratio

8.
a. qualitative e. quantitative
b. quantitative f. quantitative
c. qualitative g. quantitative
d. quantitative

9.
a. discrete e. continuous
b. continuous f. continuous
c. discrete g. discrete
d. continuous

10.
a. 84.45-84.55 d. 16.5-17.5
b. 5.5-6.5 e. 8.715-8.725
c. 0.145-0.155 f. 96.5-97.5

11. Random samples are selected by using chance methods or random numbers.

11. continued
Systematic samples are selected by numbering each subject and selecting every *k*th number. Stratified samples are selected by dividing the population into groups and selecting from each group. Cluster samples are selected by using intact groups called clusters.

12.
a. cluster d. systematic
b. systematic e. stratified
c. random

13. Answers will vary.

14. Answers will vary.

15. Answers will vary.

16. Answers will vary.

17.
a. experimental c. observational
b. observational d. experimental

18.
a. Independent variable - type of pill taken
Dependent variable - number of infections
b. Independent variable - color of car
Dependent variable - running red lights
c. Independent variable - level of hostility
Dependent variable - cholesterol level
d. Independent variable - type of diet
Dependent variable - blood pressure

19. Answers will vary. Possible answers include:
(a) overall health of participants, amount of exposure to infected individuals through the workplace or home
(b) gender and/or age of driver, time of day
(c) diet, general health, heredity factors
(d) amount of exercise, heredity factors

20. Only twenty people were used in the study.

21. Claims can be proven only if the entire population is used.

22. The statement is meaningless since there is no definition of "the road less traveled." Also, there is no way to know that for <u>every</u>

22. continued
100 women, 91 would say that they have taken "the road less traveled."

23. Since the results are not typical, the advertisers selected only a few people for whom the product worked extremely well.

24. There is no mention of how this conclusion was obtained.

25. "74% more calories" than what? No comparison group is stated.

26. Since the word "may" is used, there is no guarantee that the product will help fight cancer.

27. What is meant by "24 hours of acid control"?

28. No. There are many other factors that contribute to criminal behavior.

29. Possible reasons for conflicting results: The amount of caffeine in the coffee or tea or the brewing method.

30. Answers will vary.

31. Answers will vary.

32. Answers will vary.

CHAPTER QUIZ
1. True
2. False, it is a data value.
3. False, the highest level is ratio.
4. False, it is stratified sampling.
5. False, it is a quantitative variable.
6. True
7. False, it is 5.5-6.5 inches.
8. c.
9. b.
10. d.
11. a.
12. c.
13. a.
14. descriptive, inferential
15. gambling, insurance
16. population
17. sample

18.
a. saves time

18. continued
b. saves money
c. use when population is infinite

19.
a. random
b. systematic
c. cluster
d. stratified

20. quasi-experimental

21. random

22.
a. inferential
b. descriptive
c. inferential
d. descriptive
e. inferential

23.
a. ratio
b. ordinal
c. interval
d. ratio
e. nominal

24.
a. continuous
b. discrete
c. discrete
d. continuous
e. continuous
f. discrete

25.
a. 3.15-3.25
b. 17.5-18.5
c. 8.5-9.5
d. 0.265-0.275
e. 35.5-36.5

EXERCISE SET 2-2

1. Frequency distributions are used to:
   1. organize data in a meaningful way
   2. determine the shape of the distribution
   3. facilitate computation procedures for finding descriptive measures such as the mean
   4. draw charts and graphs
   5. make comparisons between data sets

2. Categorical distributions are used with nominal or ordinal data, ungrouped distributions are used with data having a small range, and grouped distributions are used when the range of the data is large.

3.
a. $10.5 - 15.5$, $\frac{11+15}{2} = \frac{26}{2} = 13$, $15.5 - 10.5 = 5$
b. $16.5 - 39.5$, $\frac{17+39}{2} = \frac{56}{2} = 28$, $39.5 - 16.5 = 23$
c. $292.5 - 353.5$, $\frac{292+353}{2} = \frac{646}{2} = 323$, $353.5 - 292.5 = 61$
d. $11.75 - 14.75$, $\frac{11.75+14.75}{2} = \frac{26.5}{2} = 13.25$, $14.75 - 11.75 = 3$
e. $3.125 - 3.935$, $\frac{3.13+3.93}{2} = \frac{7.06}{2} = 3.53$, $3.935 - 3.125 = 0.81$

4. Five to twenty classes. Width should be an odd number so that the midpoint will have the same place value as the data.

5.
a. Class width is not uniform.
b. Class limits overlap, and class width is not uniform.
c. A class has been omitted.
d. Class width is not uniform.

6. An open-ended frequency distribution has either a first class with no lower limit or a last class with no upper limit. They are necessary to accomodate all the data.

7.

Class	*Tally*	*f*	*Percent*
W	𝍸 𝍸 𝍸 |	16	32%
BL	𝍸 𝍸 |||	13	26%
BR	𝍸 ||||	9	18%
Y	𝍸 |	6	12%
G	𝍸 |	6	12%
		50	100%

8. $H = 36 \quad L = 7$
Range $= 36 - 7 = 29$
Width $= 29 \div 6 = 4.83$ or 5

Limits	*Boundaries*	*f*	*cf*
7 - 11	6.5 - 11.5	2	2
12 - 16	11.5 - 16.5	5	7
17 - 21	16.5 - 21.5	9	16
22 - 26	21.5 - 26.5	2	18
27 - 31	26.5 - 31.5	0	18
32 - 36	31.5 - 36.5	1	19
		19	

9.

Class	*Boundaries*	*f*	*cf*
0	- 0.5 - 0.5	5	5
1	0.5 - 1.5	8	13
2	1.5 - 2.5	10	23
3	2.5 - 3.5	2	25
4	3.5 - 4.5	3	28
5	4.5 - 5.5	2	30
		30	

10. $H = 718 \quad L = 636$
Range $= 718 - 636 = 82$
Width $= 82 \div 6 = 13.\overline{6}$ or 14
Use width $= 15$ for odd number (rule 2)

Limits	*Boundaries*	*f*	*cf*
636 - 650	635.5 - 650.5	6	6
651 - 665	650.5 - 665.5	5	11
666 - 680	665.5 - 680.5	5	16
681 - 695	680.5 - 695.5	8	24
696 - 710	695.5 - 710.5	5	29
711 - 725	710.5 - 725.5	1	30
		30	

11. $H = 780 \quad L = 746$
Range $= 780 - 746 = 34$
Width $= 34 \div 6 = 5.\overline{6}$ or 6

Limits	*Boundaries*	*f*	*cf*
746 - 751	745.5 - 751.5	4	4
752 - 757	751.5 - 757.5	4	8
758 - 763	757.5 - 763.5	7	15
764 - 769	763.5 - 769.5	6	21
770 - 775	769.5 - 775.5	6	27
776 - 781	775.5 - 781.5	3	30
		30	

12. $H = 93 \quad L = 0$
Range $= 93 - 0 = 93$
Width $= 93 \div 7 \approx 13.29$ or 14
Use $w = 15$ for odd number.

12. continued

Limits	*Boundaries*	*f*	*cf*
0 - 14	-0.5 - 14.5	14	14
15 - 29	14.5 - 29.5	10	24
30 - 44	29.5 - 44.5	4	28
45 - 59	44.5 - 59.5	1	29
60 - 74	59.5 - 74.5	1	30
75 - 89	74.5 - 89.5	2	32
90 - 104	89.5 - 104.5	1	33
		33	

13. H = 70 L = 27
Range = 70 − 27 = 43
Width = 43 ÷ 7 = 6.1 or 7

Limits	*Boundaries*	*f*	*cf*
27 - 33	26.5 - 33.5	7	7
34 - 40	33.5 - 40.5	14	21
41 - 47	40.5 - 47.5	15	36
48 - 54	47.5 - 54.5	11	47
55 - 61	54.5 - 61.5	3	50
62 - 68	61.5 - 68.5	3	53
69 - 75	68.5 - 75.5	2	55
		55	

14. H = 4040 L = 70
Range = 4040 − 70 = 3970
Width = 3970 ÷ 8 = 496.25 or 497

Limits	*Boundaries*	*f*	*cf*
70 - 566	69.5 - 566.5	14	14
567 - 1063	566.5 - 1063.5	5	19
1064 - 1560	1063.5 - 1560.5	5	24
1561 - 2057	1560.5 - 2057.5	0	24
2058 - 2554	2057.5 - 2554.5	0	24
2555 - 3051	2554.5 - 3051.5	1	25
3052 - 3548	3051.5 - 3548.5	0	25
3549 - 4045	3548.5 - 4045.5	2	27
		27	

15.

Limits	*Boundaries*	*f*	*cf*
0 - 19	-0.5 - 19.5	13	13
20 - 39	19.5 - 39.5	18	31
40 - 59	39.5 - 59.5	10	41
60 - 79	59.5 - 79.5	5	46
80 - 99	79.5 - 99.5	3	49
100 - 119	99.5 - 119.5	1	50
		50	

16. H = 775 L = 5
Width = 775 − 5 = 770
Range = 770 ÷ 8 = 96.25 or 97

16. continued

Limits	*Boundaries*	*f*	*cf*
5 - 101	4.5 - 101.5	17	17
102 - 198	101.5 - 198.5	6	23
199 - 295	198.5 - 295.5	6	29
296 - 392	295.5 - 392.5	2	31
393 - 489	392.5 - 489.5	2	33
490 - 586	489.5 - 586.5	3	36
587 - 683	586.5 - 683.5	1	37
684 - 780	683.5 - 780.5	2	39
		39	

17. H = 11,413 L = 150
Range = 11,413 − 150 = 11,263
Width = 11,263 ÷ 10 = 1126.3 or 1127

Limits	*Boundaries*	*f*	*cf*
150 - 1276	149.5 - 1276.5	2	2
1277 - 2403	1276.5 - 2403.5	2	4
2404 - 3530	2403.5 - 3530.5	5	9
3531 - 4657	3530.5 - 4657.5	8	17
4658 - 5784	4657.5 - 5784.5	7	24
5785 - 6911	5784.5 - 6911.5	3	27
6912 - 8038	6911.5 - 8038.5	7	34
8039 - 9165	8038.5 - 9165.5	3	37
9166 - 10,292	9165.5 - 10,292.5	3	40
10,293 - 11,419	10,292.5 - 11,419.5	2	42
		42	

18.
H = 550 L = 306
Range = 550 − 306 = 244
Width = 244 ÷ 8 = 30.5 or 31

f_M, cf_M = McGwire f_S, cf_S = Sosa

Limits	*Boundaries*	f_M	cf_M	f_S	cf_S
306 - 336	305.5 - 336.5	1	1	0	0
337 - 367	336.5 - 367.5	6	7	10	10
368 - 398	367.5 - 398.5	19	26	16	26
399 - 429	398.5 - 429.5	15	41	21	47
430 - 460	429.5 - 460.5	18	59	15	62
461 - 491	460.5 - 491.5	6	65	3	65
492 - 522	491.5 - 522.5	3	68	1	66
523 - 553	522.5 - 553.5	2	70	0	66
		70		66	

EXERCISE SET 2-3

1.

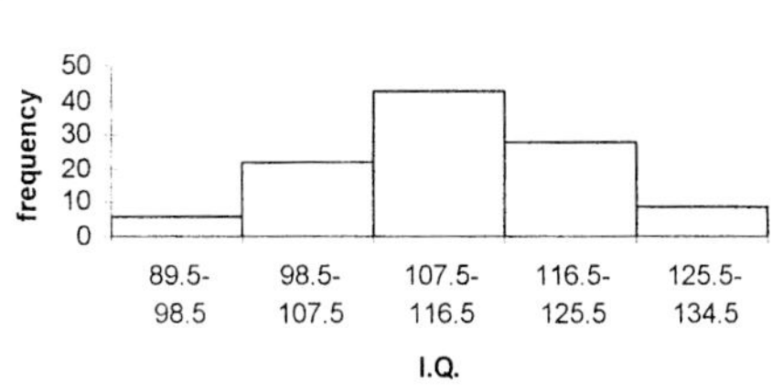

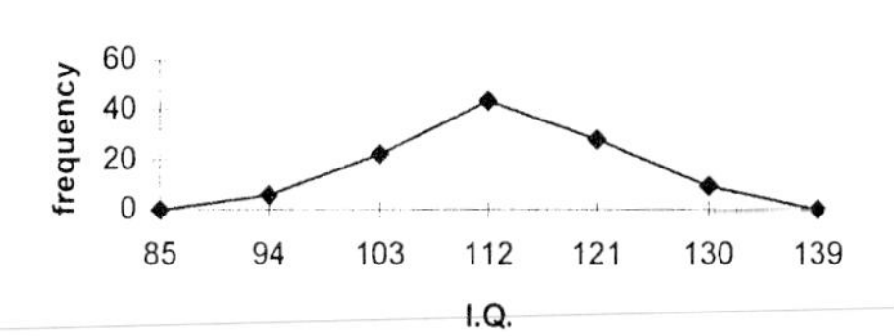

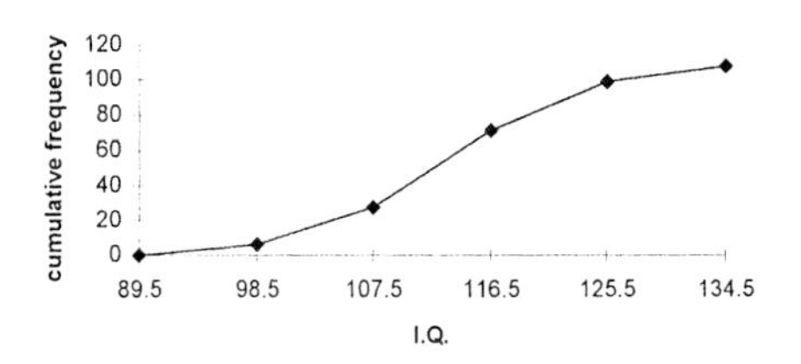

Eighty applicants do not need to enroll in the summer programs.

2.

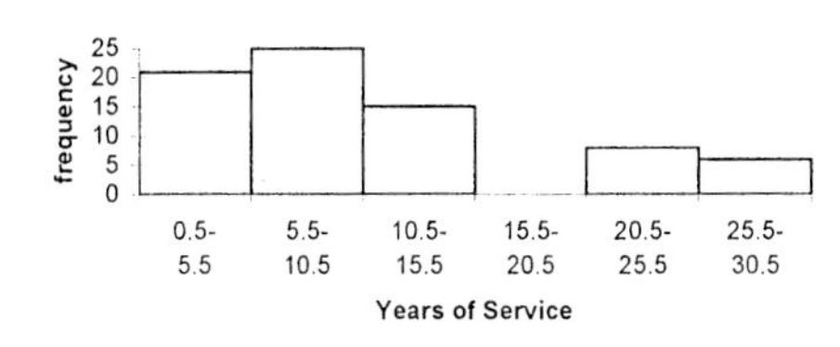

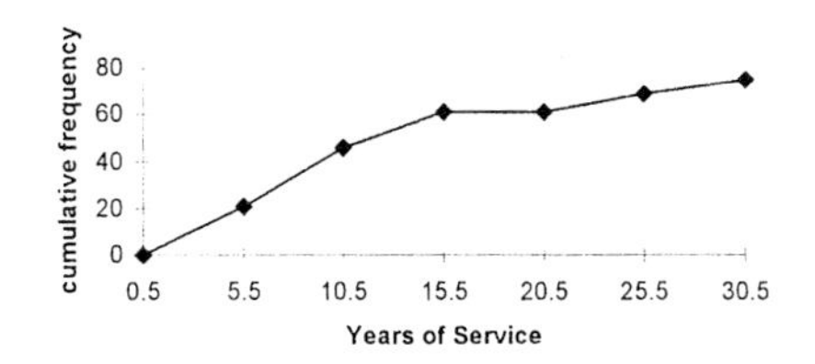

The majority of employees have worked for less than 11 years.

3.

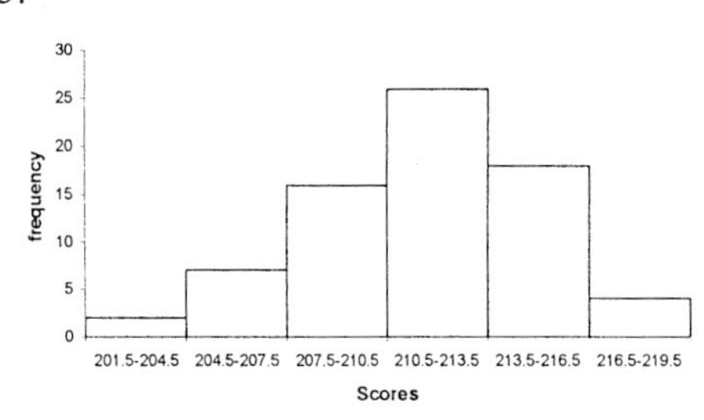

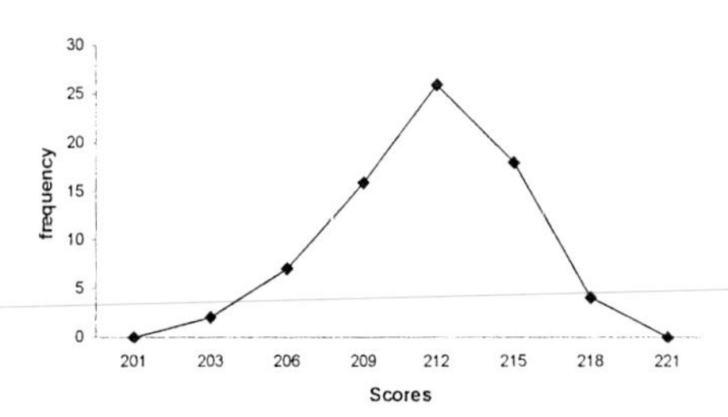

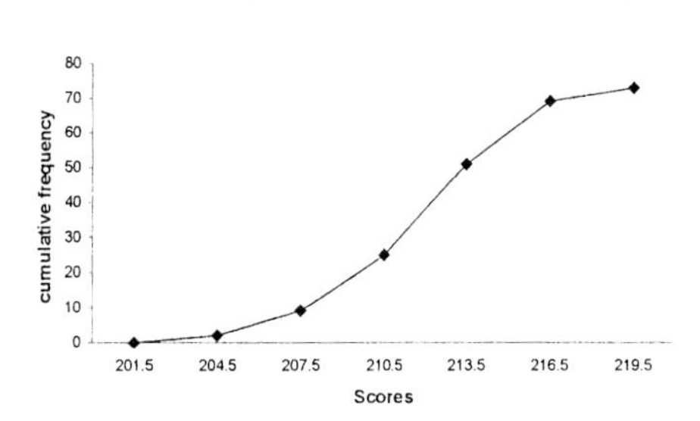

The distribution appears to be slightly left skewed.

4.

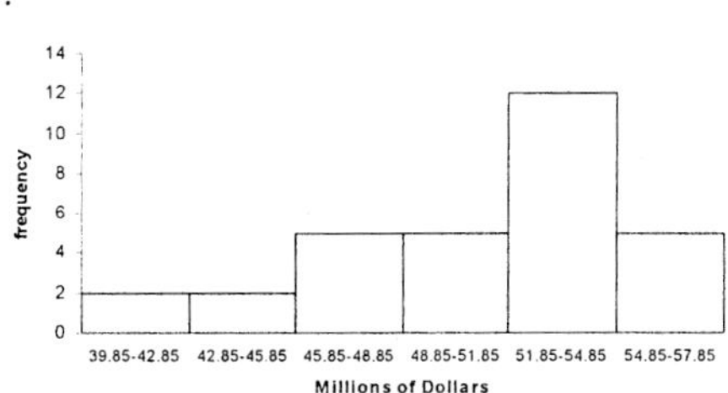

The distribution is left skewed or negatively skewed.

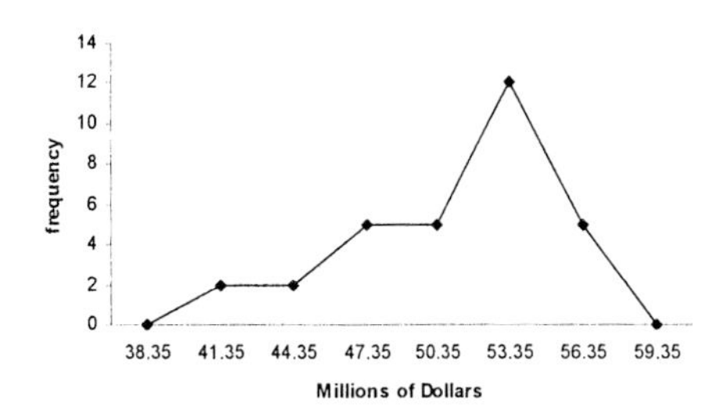

4. continued

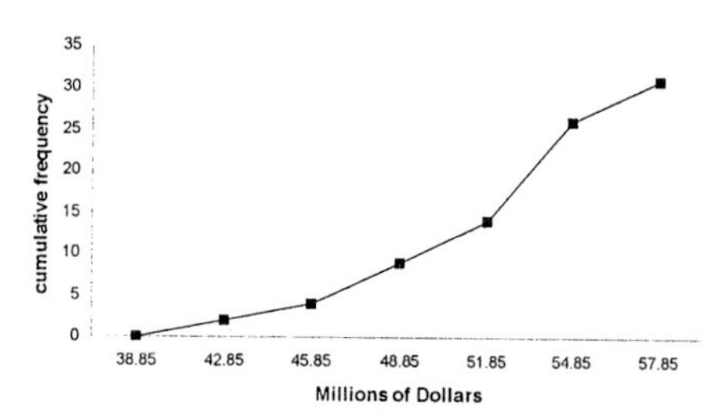

5.

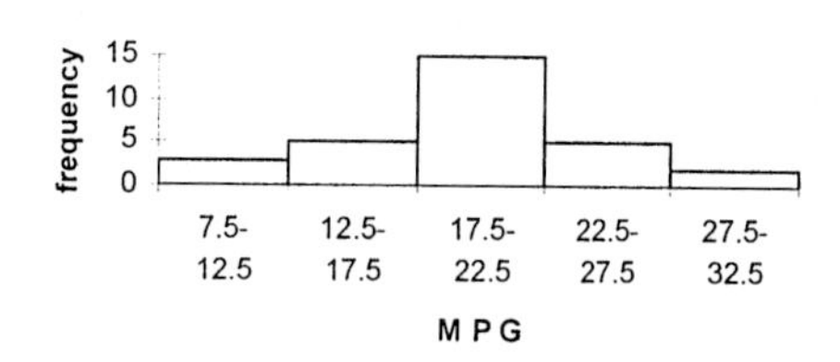

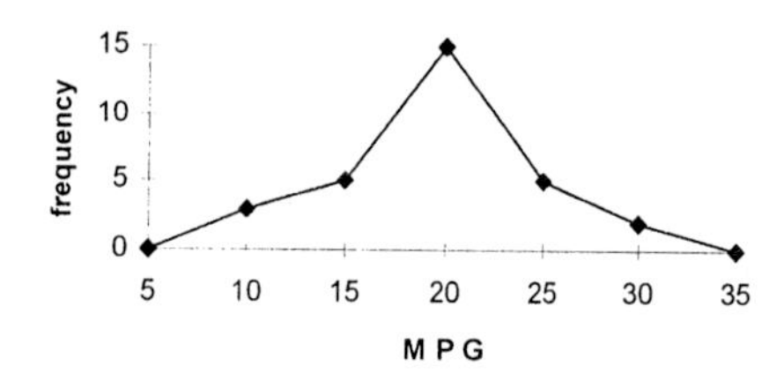

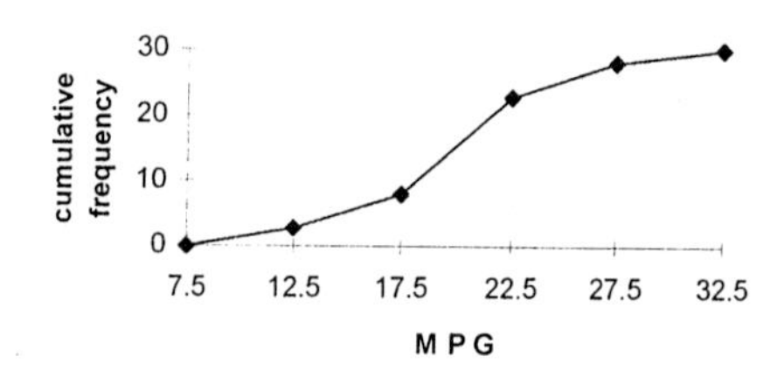

6.

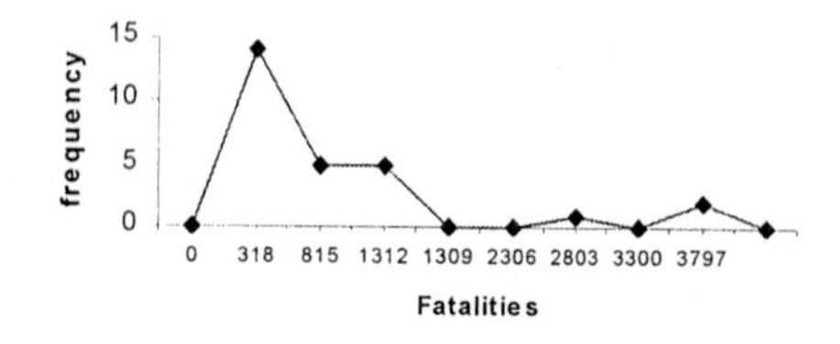

6. continued

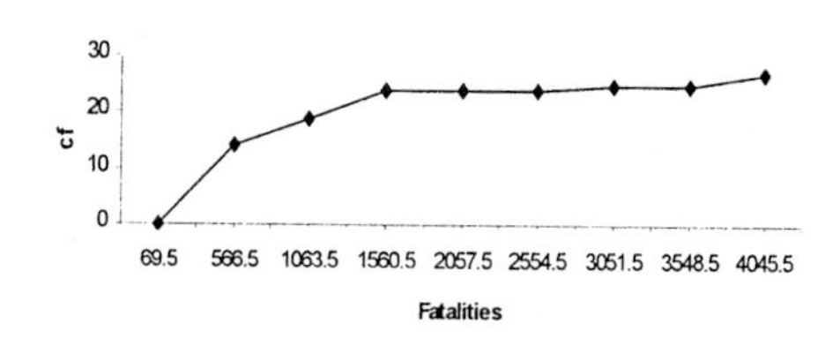

Nearly all of the states had between 69.5 and 1560.5 fatalities.

7.

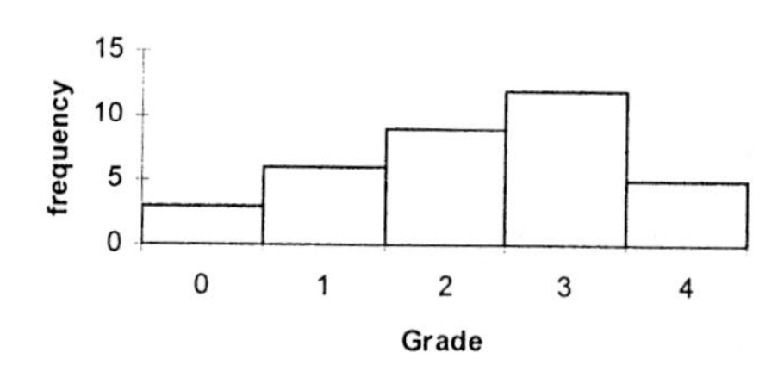

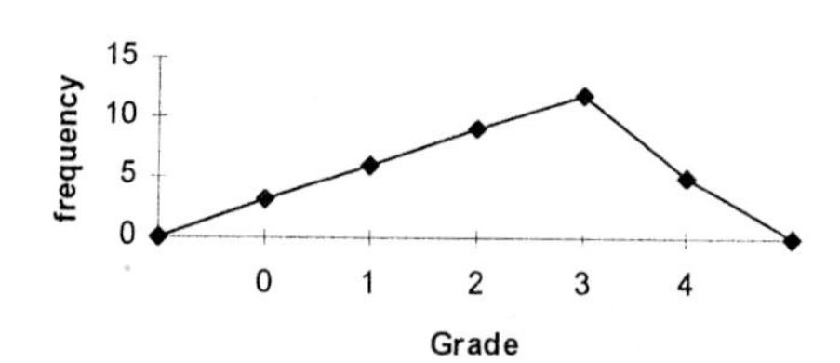

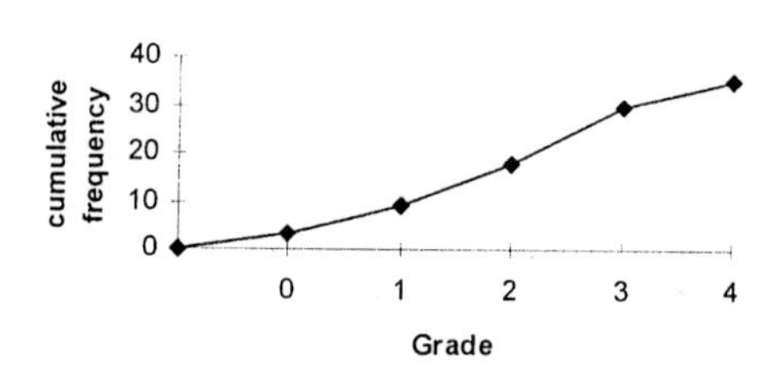

Yes, 26 out of the 35 students can enroll in the next course.

8.

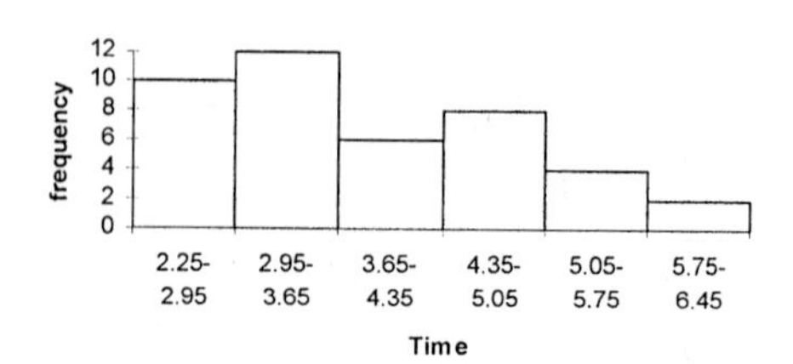

8. continued

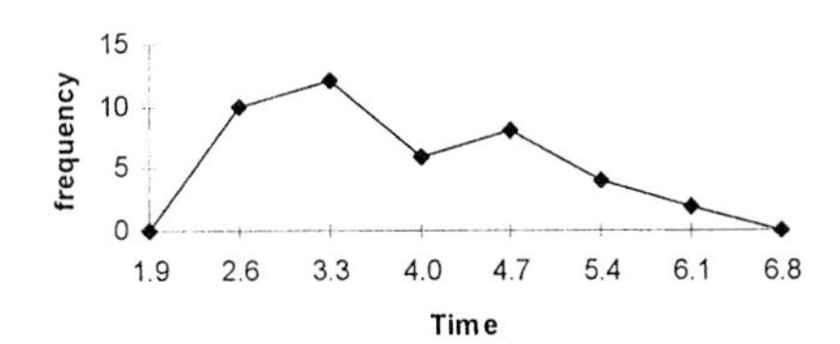

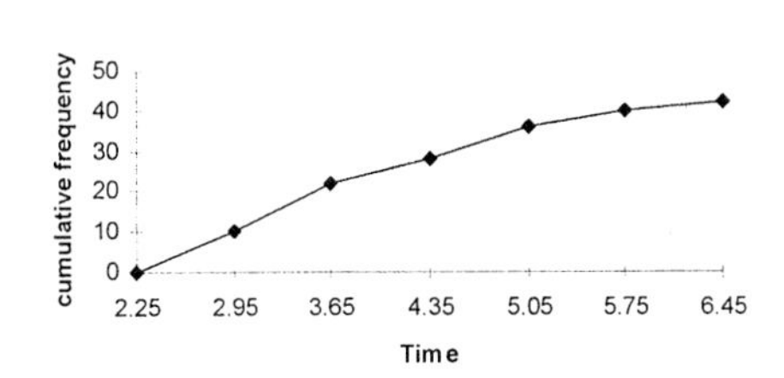

The most typical reaction times were 2.25 - 2.95 and 2.95 - 3.65.

9.

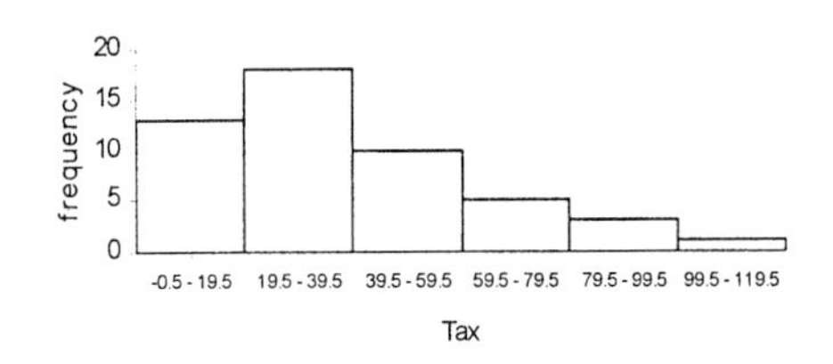

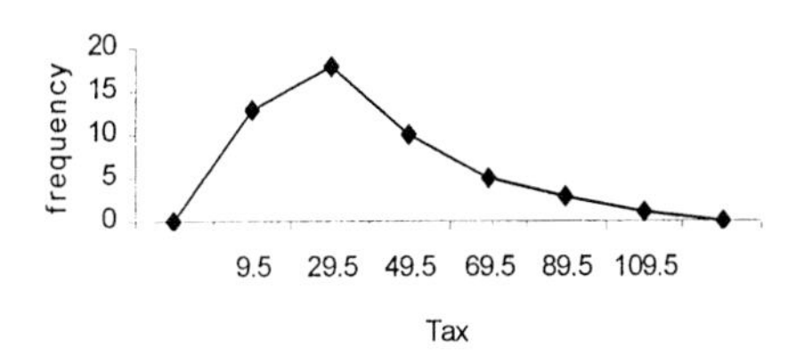

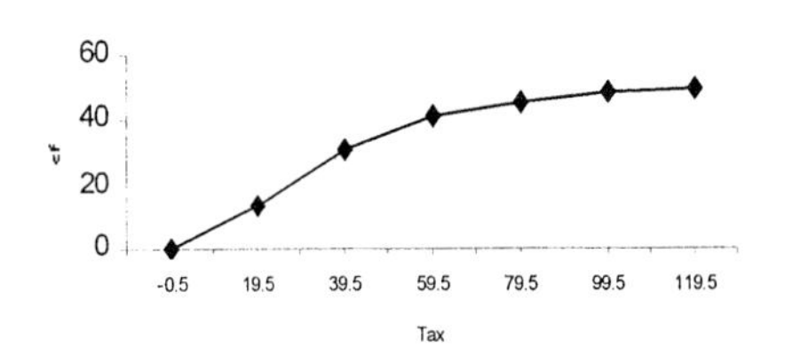

The majority of the states charge less than 40 cents per pack.

10.

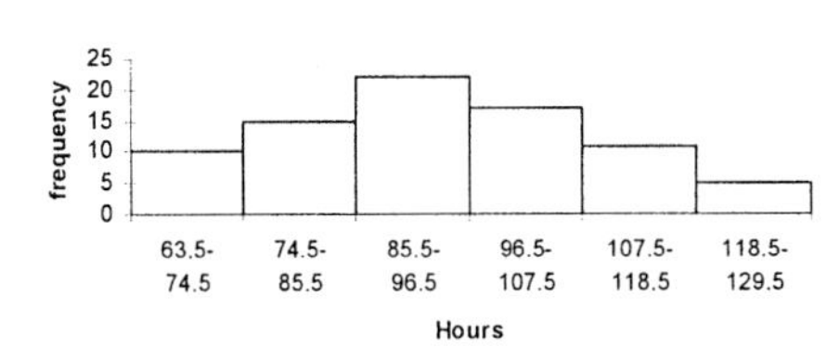

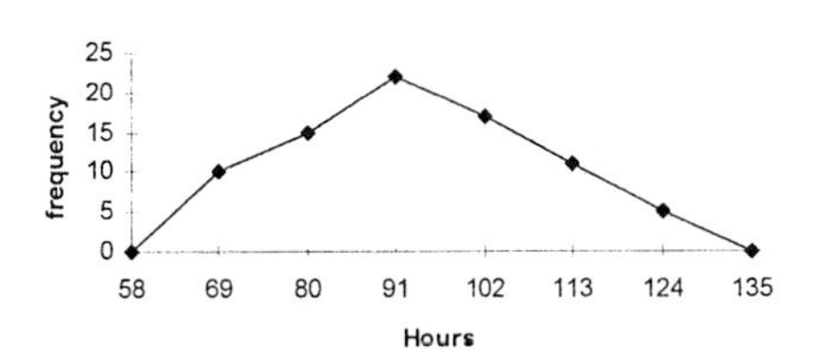

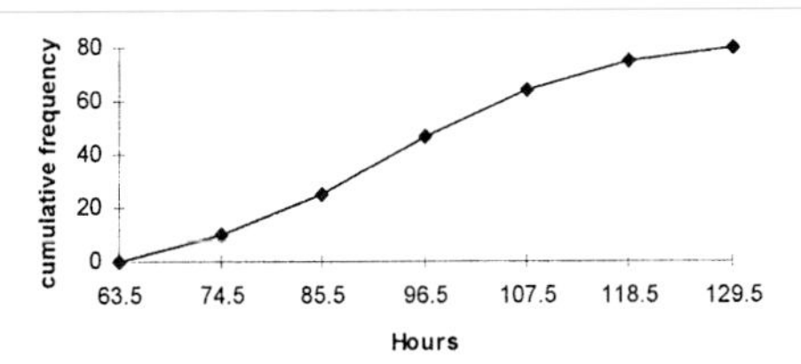

The histogram is approximately symmetrical about the center.

11.

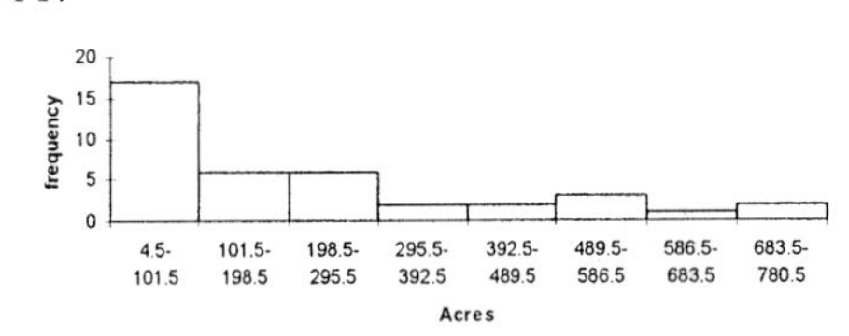

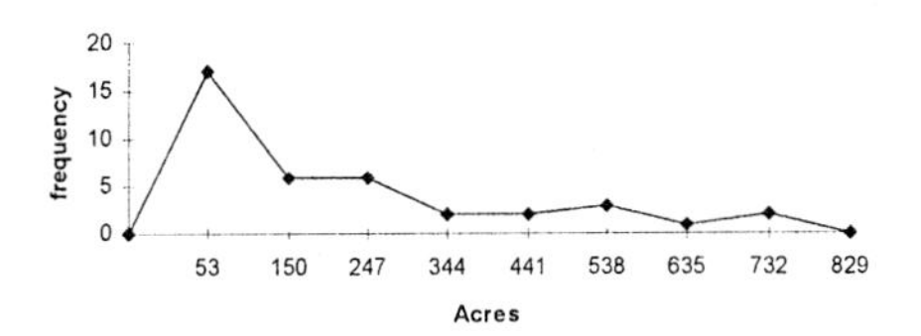

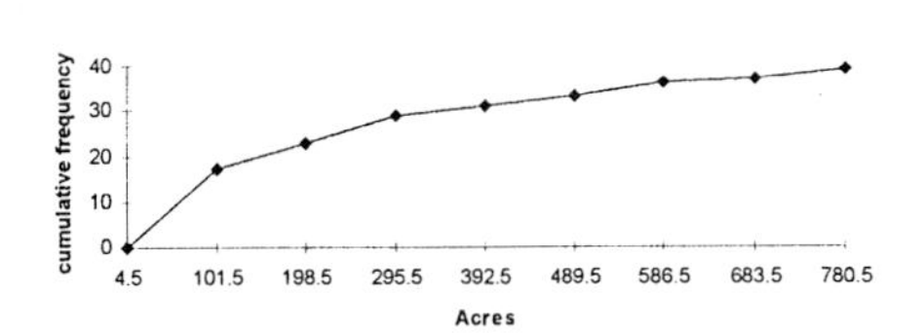

The majority of the parks had between 4.5 and 101.5 thousand acres.

12.

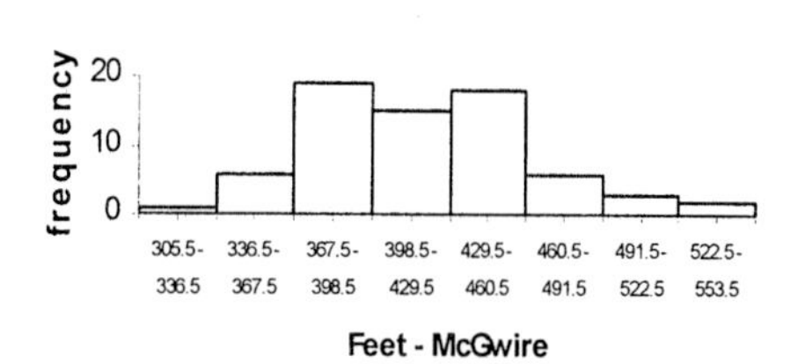

Feet - Sosa

The histograms show that the distances of McGwire's homeruns are more variable than Sosa's homerun distances.

13.

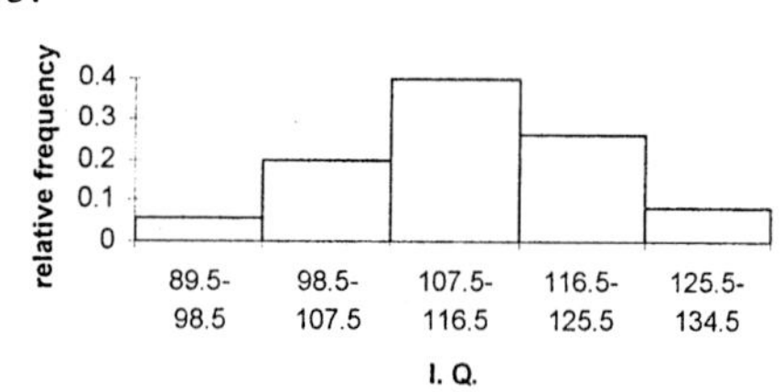

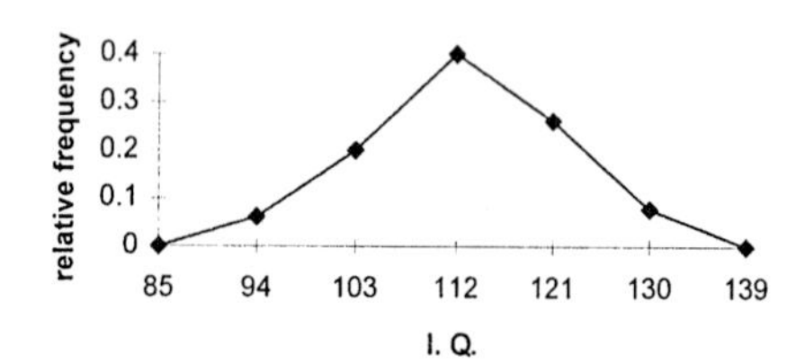

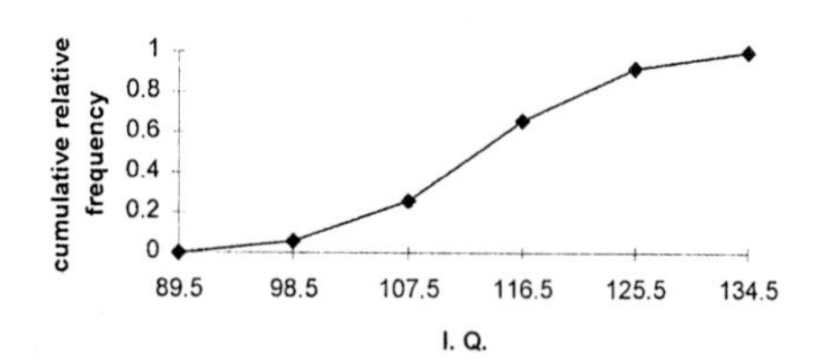

The proportion of applicants who need to enroll in a summer program is 0.26 or 26%.

14.

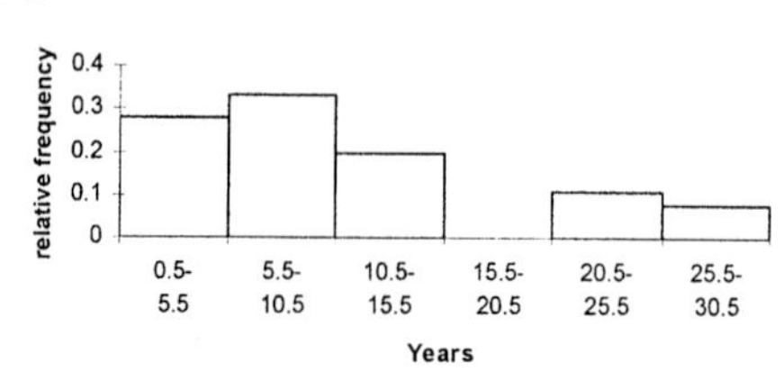

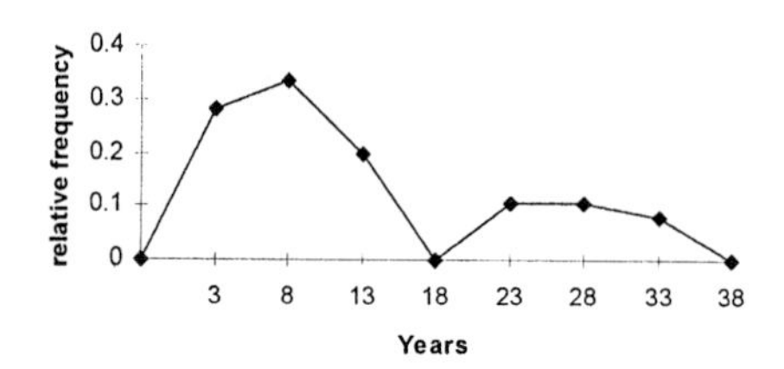

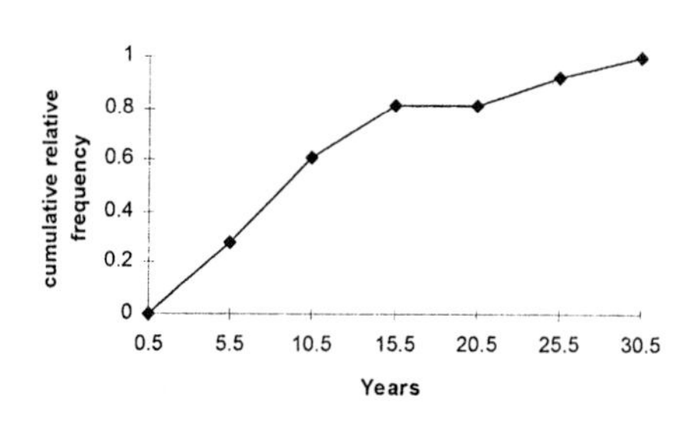

The proportion of employees who have been with the company longer than 20 years is 0.187 or 18.7%.

15. H = 270   L = 80
Range = 270 − 80 = 190
Width = 190 ÷ 7 = 27.1 or 28
Use width = 29 (rule 2)

Limits	*Boundaries*	*f*	*rf*	*crf*
80 - 108	79.5 - 108.5	8	0.17	0.17
109 - 137	108.5 - 137.5	13	0.28	0.45
138 - 166	137.5 - 166.5	2	0.04	0.49
167 - 195	166.5 - 195.5	9	0.20	0.69
196 - 224	195.5 - 224.5	10	0.22	0.91
225 - 253	224.5 - 253.5	2	0.04	0.95
254 - 282	253.5 - 282.5	2	0.04	0.99*
			0.99*	

*due to rounding

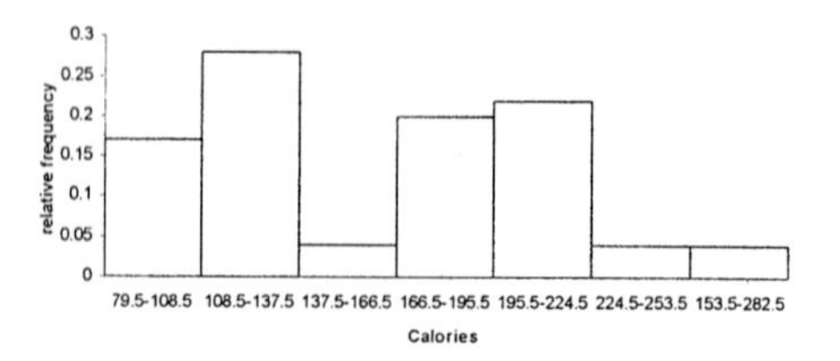

15. continued

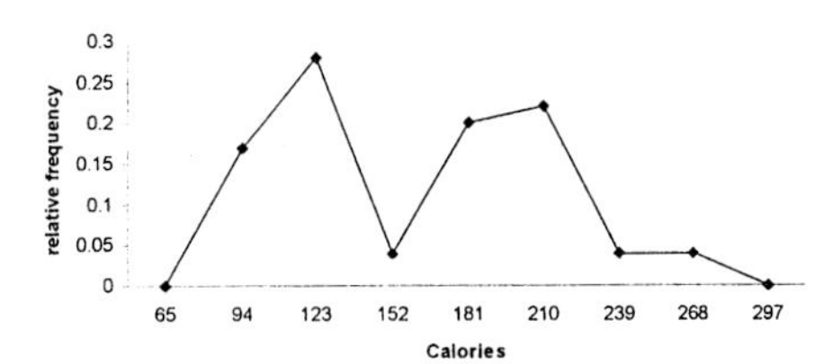

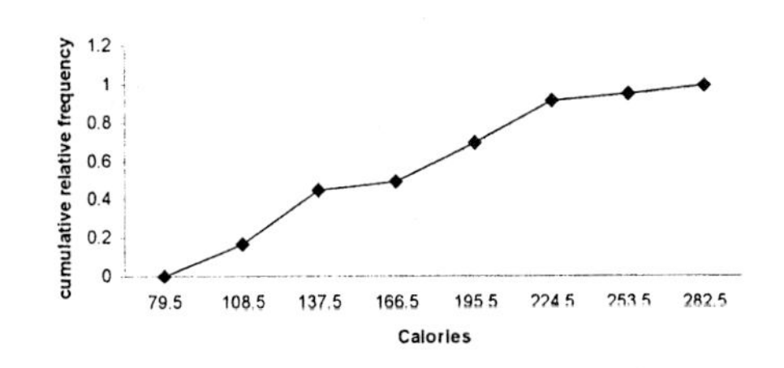

16.
H = 57   L = 12
Range = 57 − 12 = 45
Width = 45 ÷ 6 = 7.5 or 8

Limits	*Boundaries*	*f*	*rf*	*crf*
12 - 19	11.5 - 19.5	7	0.175	0.175
20 - 27	19.5 - 27.5	17	0.425	0.600
28 - 35	27.5 - 35.5	10	0.25	0.850
36 - 43	35.5 - 43.5	4	0.10	0.950
44 - 51	43.5 - 51.5	1	0.025	0.975
52 - 59	51.5 - 59.5	1	0.025	1.000
		40	1.000	

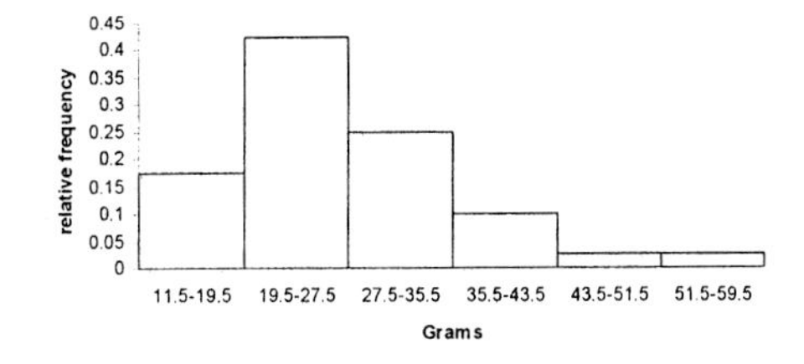

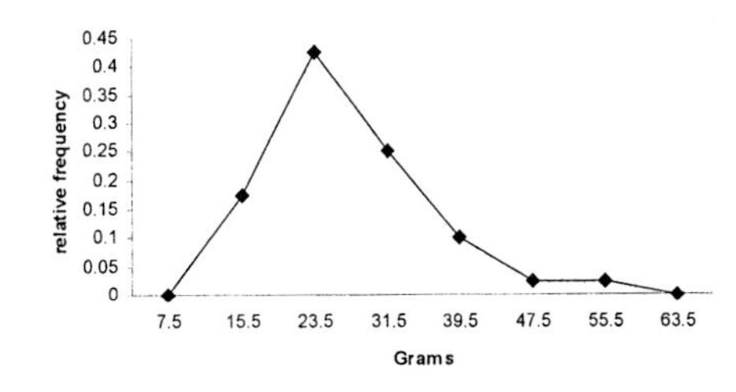

16. continued

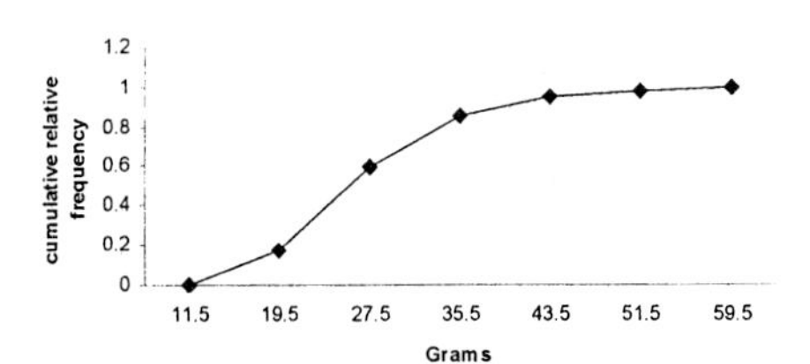

17.

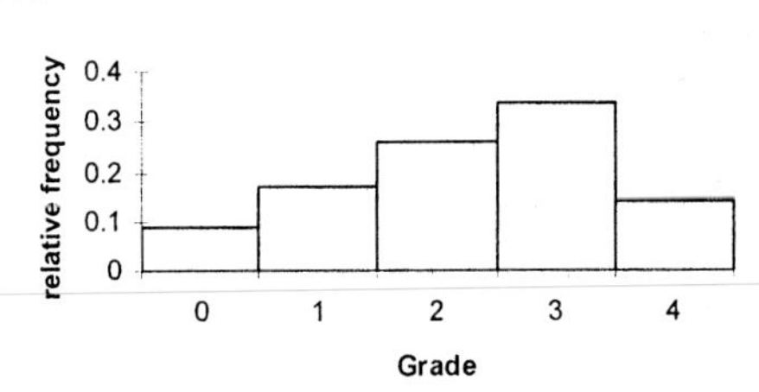

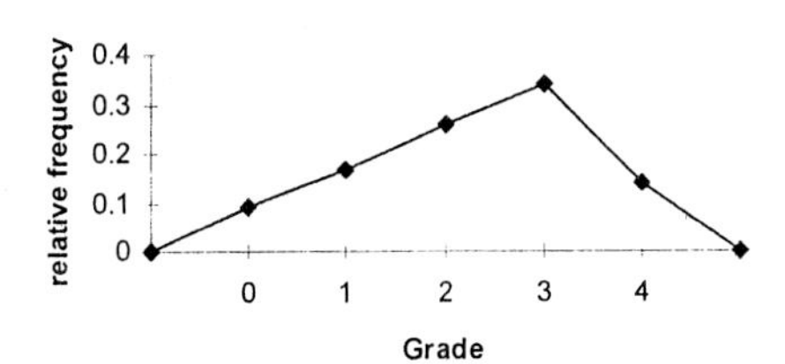

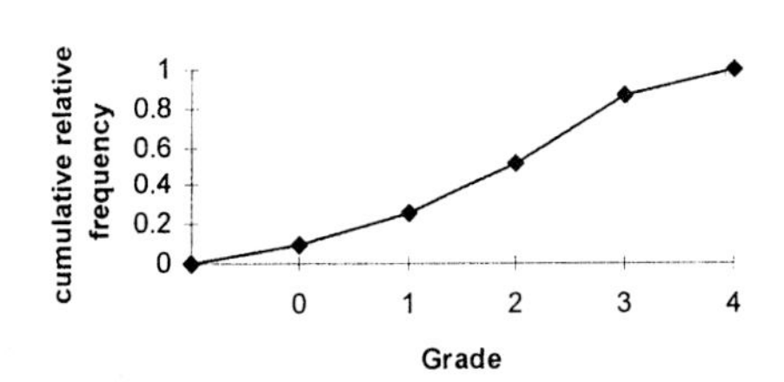

The proportion of students who cannot meet the requirement for the next course is 0.26 or 26%.

18.

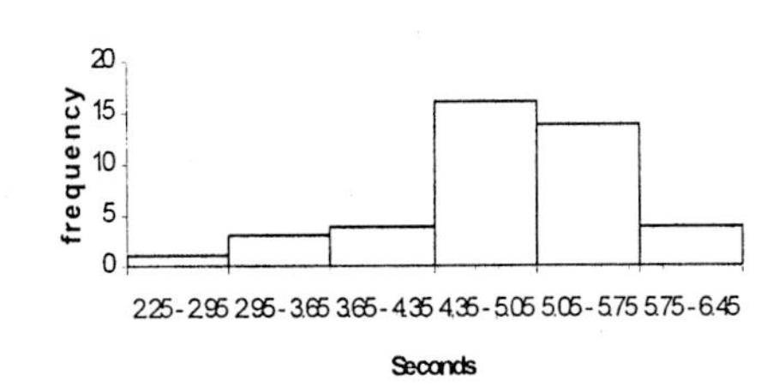

18. continued

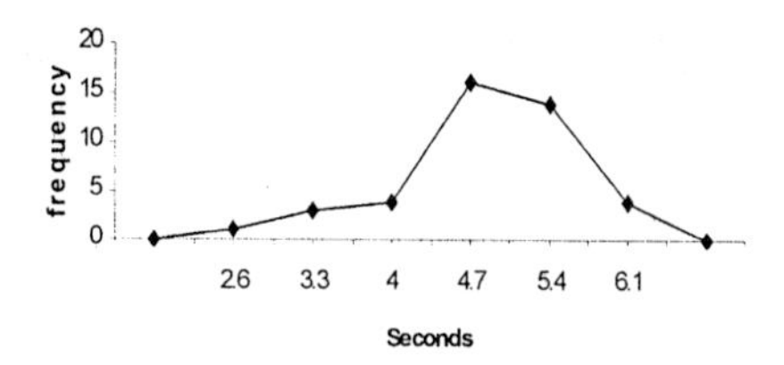

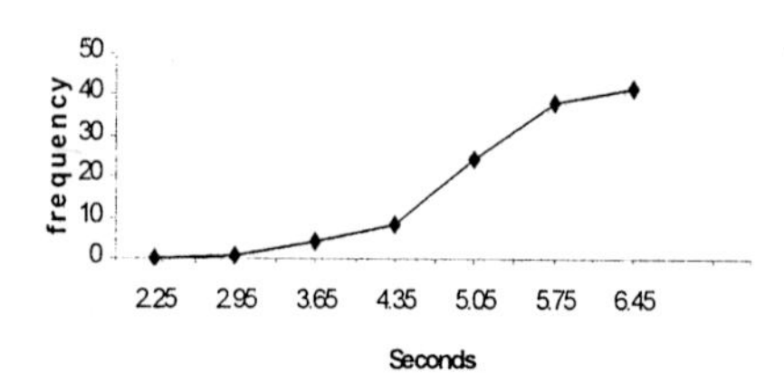

Based on the histograms, the older dogs have longer reaction times. Also, the reaction times for older dogs is more variable.

19.

Limits	*Boundaries*	X_m	f	cf
22 - 24	21.5 - 24.5	23	1	1
25 - 27	24.5 - 27.5	26	3	4
28 - 30	27.5 - 30.5	29	0	4
31 - 33	30.5 - 33.5	32	6	10
34 - 36	33.5 - 36.5	35	5	15
37 - 39	36.5 - 39.5	38	3	18
40 - 42	39.5 - 42.5	41	2	20
			20	

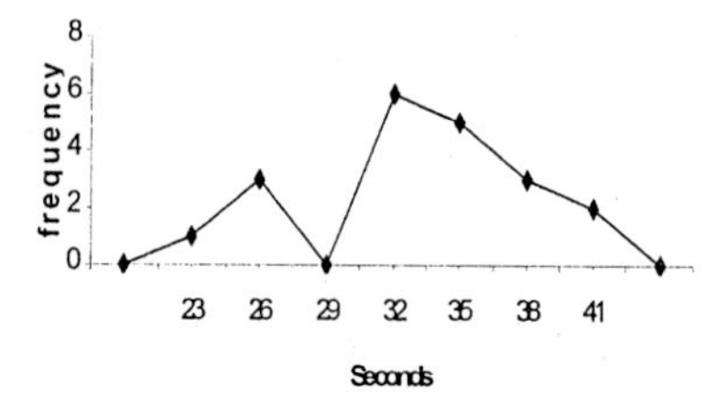

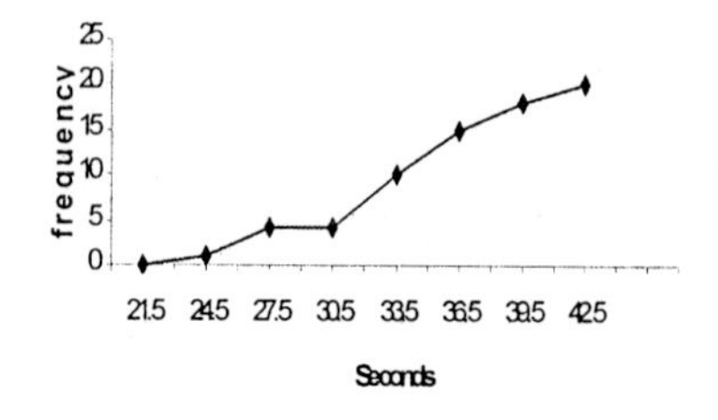

20.
a. 0
b. 14
c. 10
d. 16

EXERCISE SET 2-4

1.

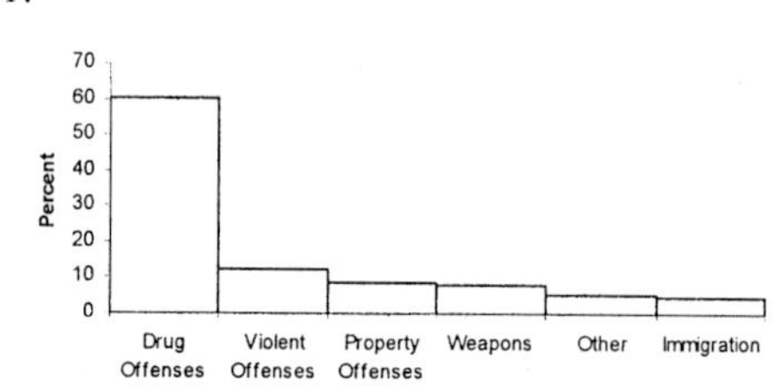

2.

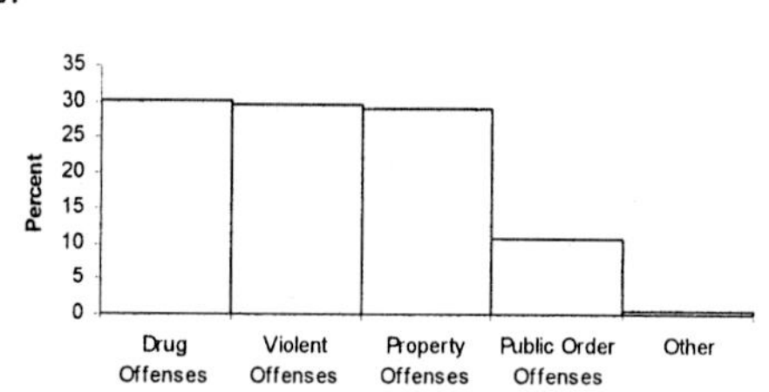

A higher percentage of people are in a federal prison for drug offenses than are in state prisons. One reason is that there are more federal drug offenses committed by bringing illegal drugs across the U. S. borders.

3.

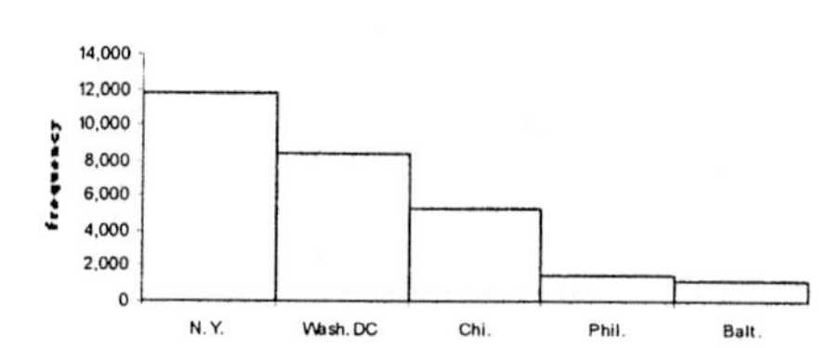

4.

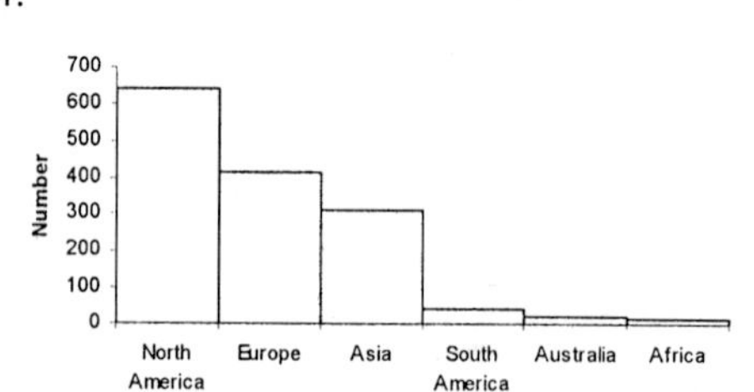

5.

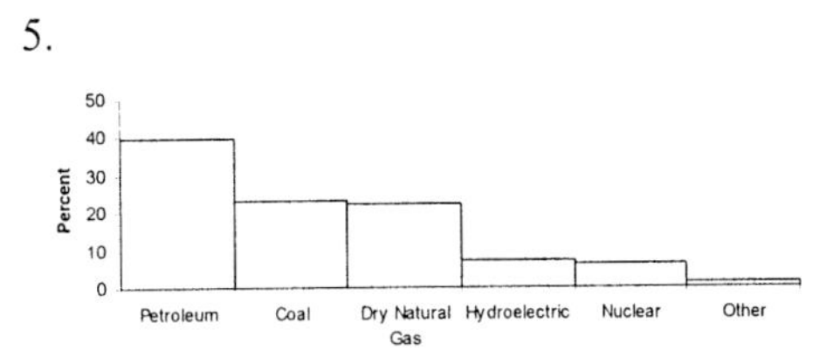

6.

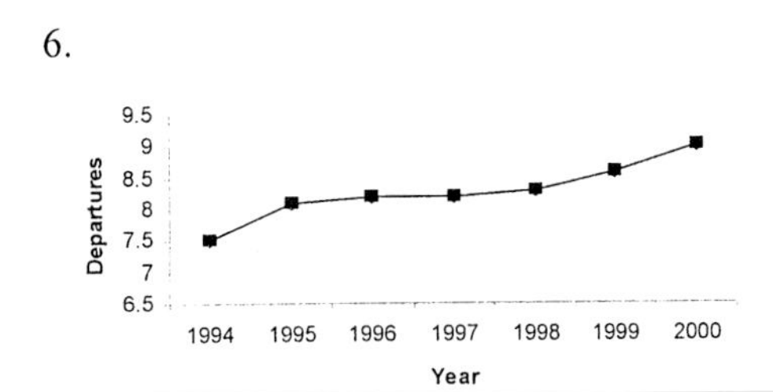

7.

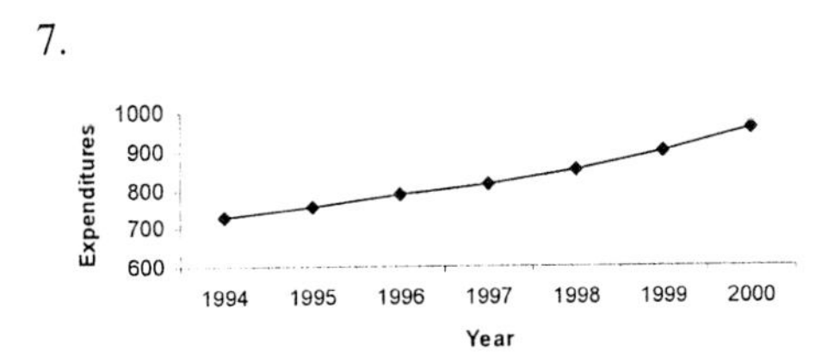

8.

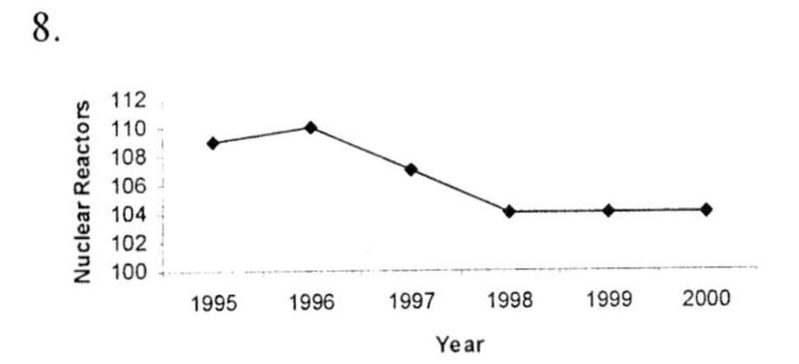

The highest number of reactors occurred in 1995 and 1996. The number of reactors declined between 1996 and 1998. Since 1998 the number of reactors has remained the same.

9.

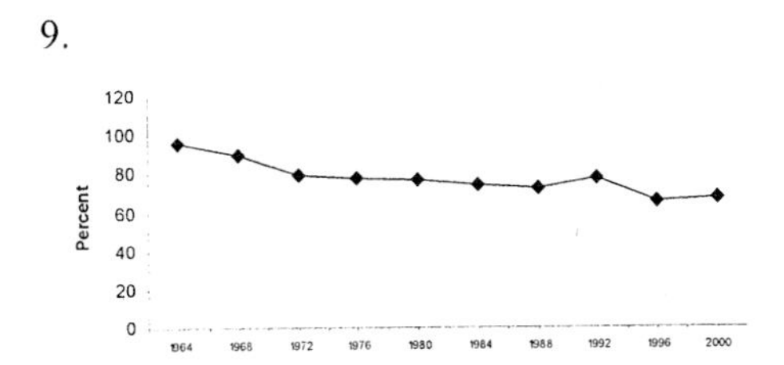

The graph shows a decline in the percents of registered voters voting in Presidential elections.

10.

Personal Business	146	14.6%	52.56°
Visit friends or family	330	33.0%	118.8°
Work-related	225	22.5%	81.0°
Leisure	299	29.9%	107.64°
	1000	100%	360°

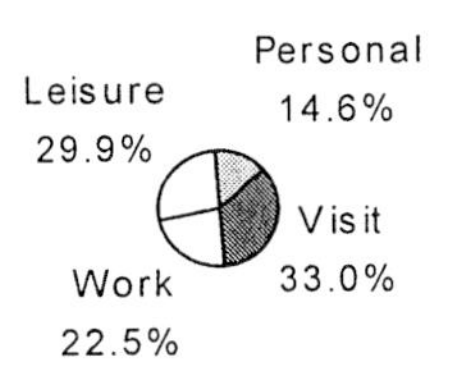

About $\frac{1}{3}$ of the travelers visit friends or relatives, with the fewest travelling for personal business.

11.

Personal Residence	7.8%	28.08°
Liquid Assets	5.0%	18.0°
Pension Accounts	6.9%	24.84°
Stocks, Funds, and Trusts	31.6%	113.76°
Business & Real Estate	46.9%	168.84°
Miscellaneous	1.8%	6.48°
	100.0%	360.00°

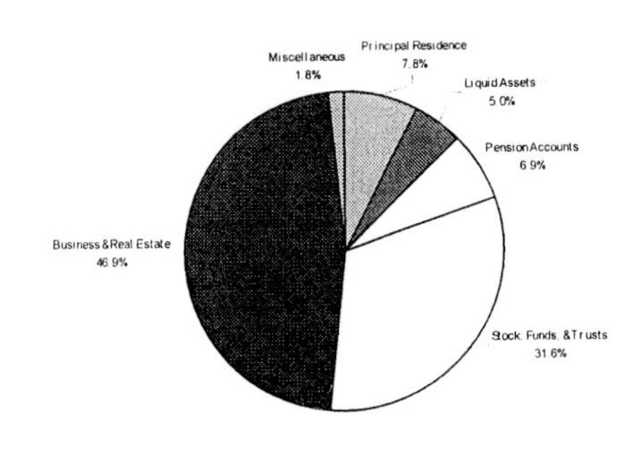

12.

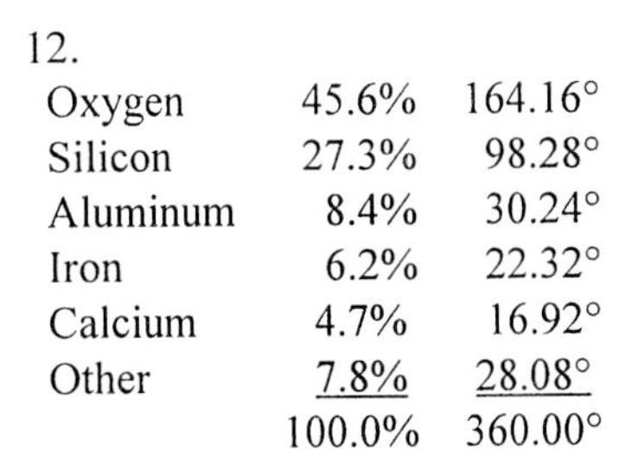

Oxygen	45.6%	164.16°
Silicon	27.3%	98.28°
Aluminum	8.4%	30.24°
Iron	6.2%	22.32°
Calcium	4.7%	16.92°
Other	7.8%	28.08°
	100.0%	360.00°

12. continued

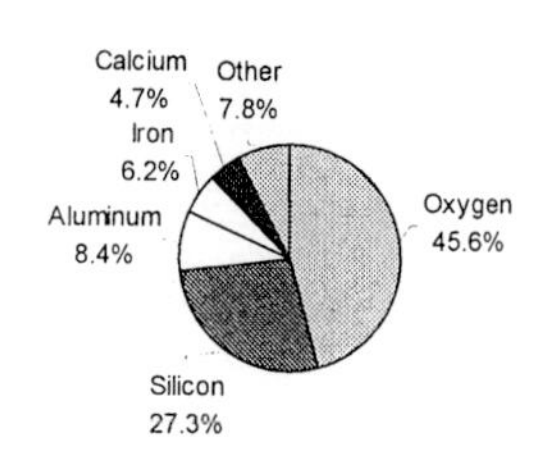

13.

Career change	34%	122.4°
New job	29%	104.4°
Start business	21%	75.6°
Retire	16%	57.6°
	100%	360.0°

Pie chart:

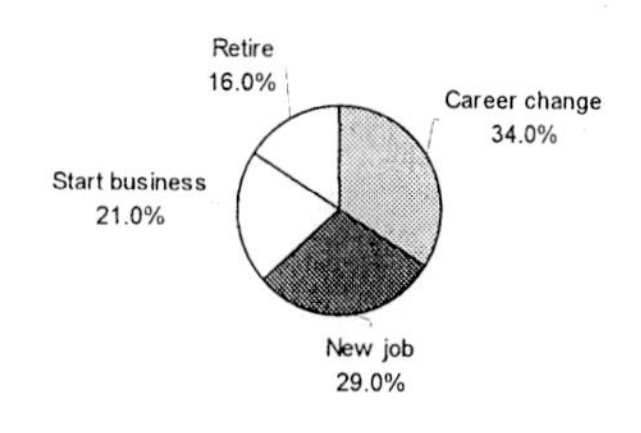

Pareto chart:

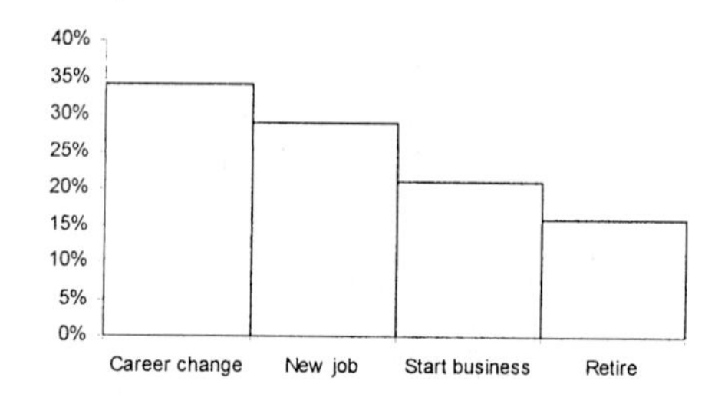

The Pareto chart is better at showing a comparison between categories.

14.
a. time series graph
b. pie graph
c. Pareto chart
d. pie graph
e. time series graph
f. Pareto chart

15.

Stem	Leaves
4	2 3
4	6 6 7 8 9 9
5	0 1 1 1 1 2 2 4 4 4 4 4
5	5 5 5 5 6 6 6 7 7 7 7 8
6	0 1 1 1 2 4 4
6	5 8 9

The majority of the Presidents were in their 50's at inauguration.

16.

Stem	Leaves
3	8
4	1
5	0 0 2 3 3 6 8 9
6	6 8 9 9
7	0 0 3 4 5 8
8	0 1 3 3 4 4 4 5 7 9 9 9
9	0 2 4

The majority of automobile thefts occurred in the 50's and 80's . The data is grouped towards the higher end of the distribution.

17.

Variety 1		Variety 2
2	1	3 8
3 0	2	5
9 8 8 5 2	3	6 8
3 3 1	4	1 2 5 5
9 9 8 5 3 3 2 1 0	5	0 3 5 5 6 7 9
	6	2 2

The distributions are similar but variety 2 seems to be more variable than variety 1.

18.

Females		Males
5	0	3
	1	5 9
	2	2
7 4 3 2 0	3	1 1
6	4	1 4 6 6
9 6 3 0	5	2 6 6 6 9
8 5	6	0 0 6 6
7 2 0	7	7
8 7 6 6 0 0	8	7 8
4 2	9	6 8

The distribution for unemployed males is more variable than the distribution for unemployed females. There are more unemployed females than males world-wide.

19.

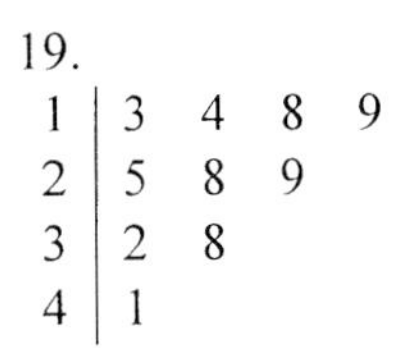

1	3 4 8 9
2	5 8 9
3	2 8
4	1

20.

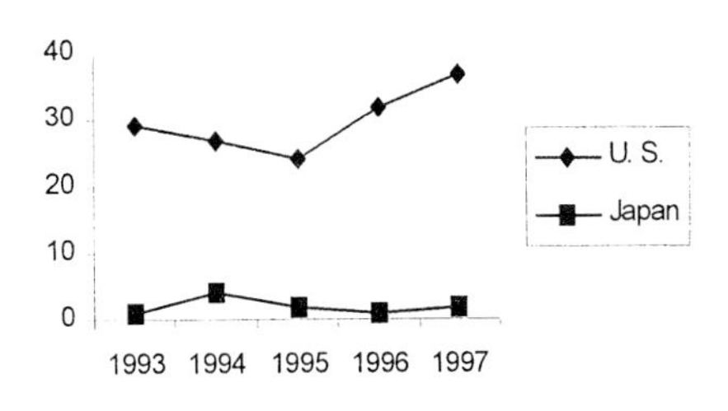

The United States has many more launches than Japan. The number of launches is relatively stable for Japan, while launches varied more for the U. S. The U. S. launches decreased slightly in 1995 and increased after that year.

21.

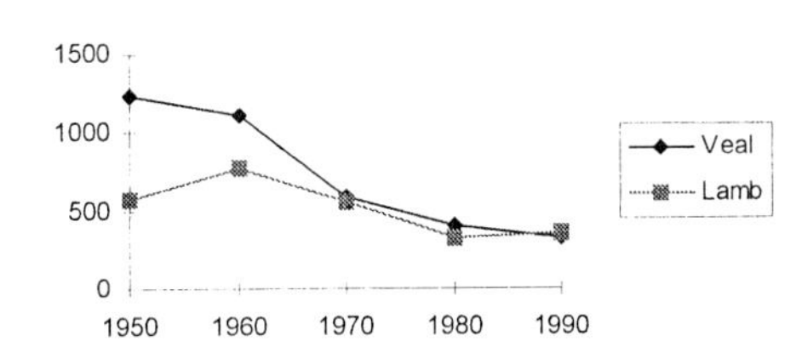

In 1950, veal production was considerably higher than lamb. By 1970, production was approximately the same for both.

22.

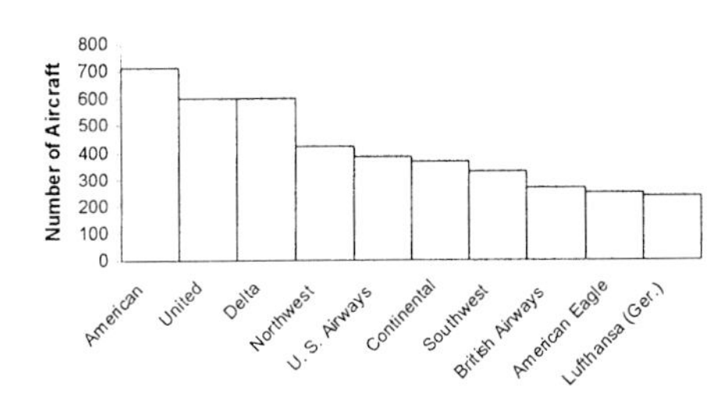

A Pareto chart is most appropriate.

23.

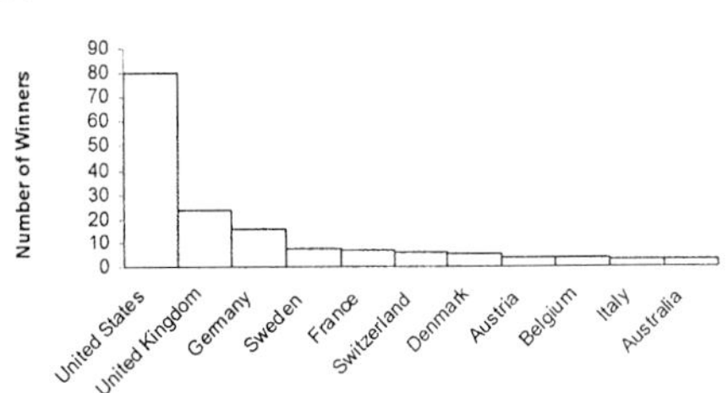

REVIEW EXERCISES - CHAPTER 2

1.

Class	f
Newspaper	7
Television	5
Radio	7
Magazine	6
	25

2.

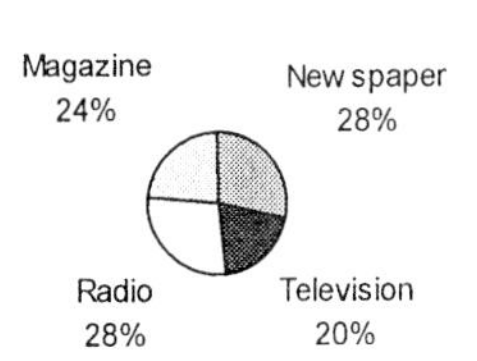

Most people get their information from newspapers and radio. However, there is not a large difference between any of the four groups.

3.

Class	f
baseball	4
golf ball	5
tennis ball	6
soccer ball	5
football	5
	25

4.

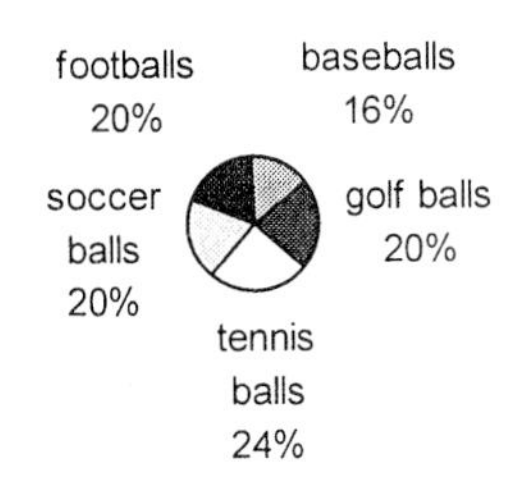

4. continued
More tennis balls were sold than any other type of ball.

5.

Class	f	cf
11	1	1
12	2	3
13	2	5
14	2	7
15	1	8
16	2	10
17	4	14
18	2	16
19	2	18
20	1	19
21	0	19
22	1	20
	20	

6.

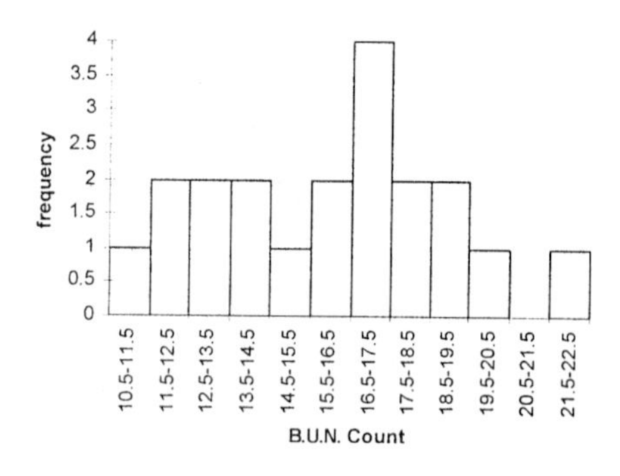

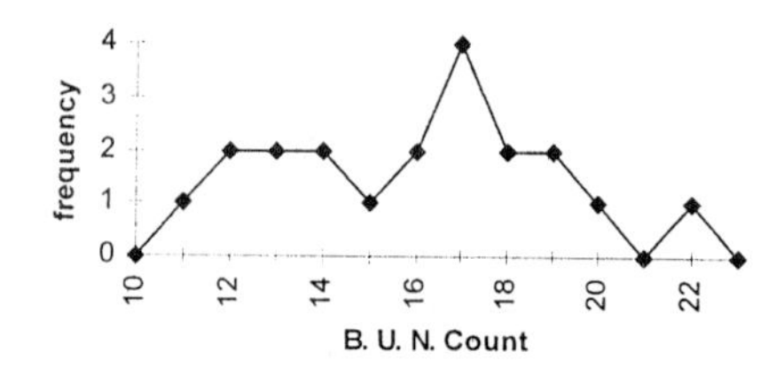

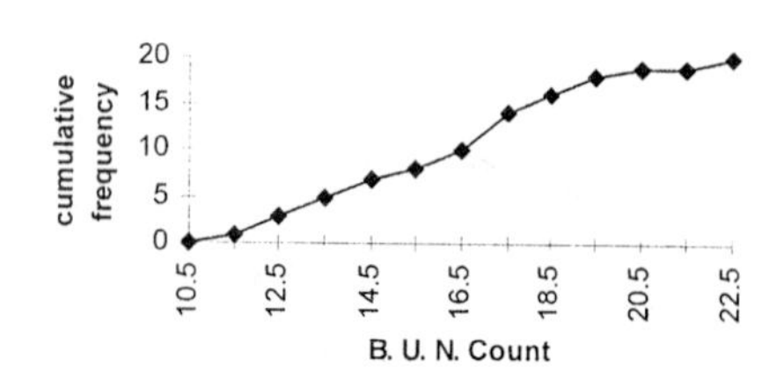

The distribution is somewhat uniform, with a slight peak in the 16.5 - 17.5 class. There is a gap in the 20.5 - 21.5 class.

7.

Limits	*Boundaries*	*f*	*cf*
1910 - 1919	1909.5 - 1919.5	1	1
1920 - 1929	1919.5 - 1929.5	2	3
1930 - 1939	1929.5 - 1939.5	15	18
1940 - 1949	1939.5 - 1949.5	12	30
1950 - 1959	1949.5 - 1959.5	20	50
1960 - 1969	1959.5 - 1969.5	18	68
1970 - 1979	1969.5 - 1979.5	18	86
1980 - 1989	1979.5 - 1989.5	6	92
1990 - 1999	1989.5 - 1999.5	8	100
		100	

8.

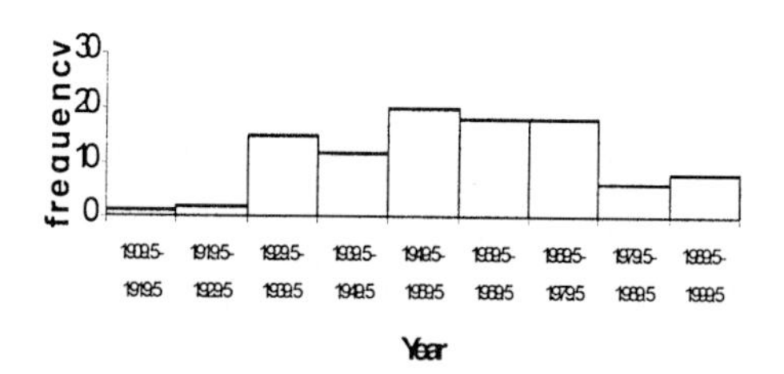

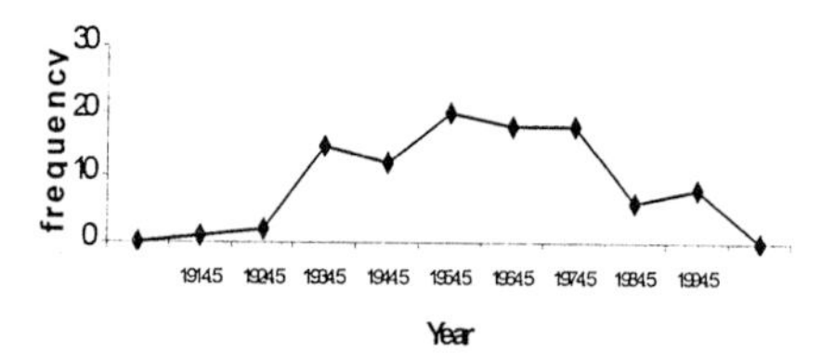

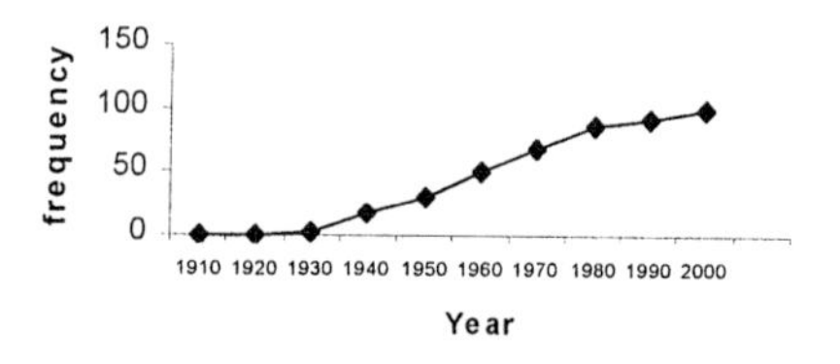

Most of the highest rated movies were made between 1930 and 1979. The decade with the largest number of top rated movies was 1950 - 1959.

9.

Limits	*Boundaries*	*f*	*cf*
170 - 188	169.5 - 188.5	11	11
189 - 207	188.5 - 207.5	9	20
208 - 226	207.5 - 226.5	4	24
227 - 245	226.5 - 245.5	5	29
246 - 264	245.5 - 264.5	0	29
265 - 283	264.5 - 283.5	0	29
284 - 302	283.5 - 302.5	0	29
303 - 321	302.5 - 321.5	1	30
		30	

10.

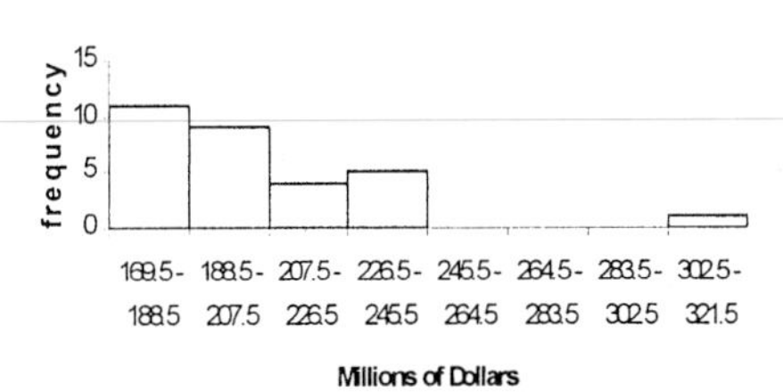

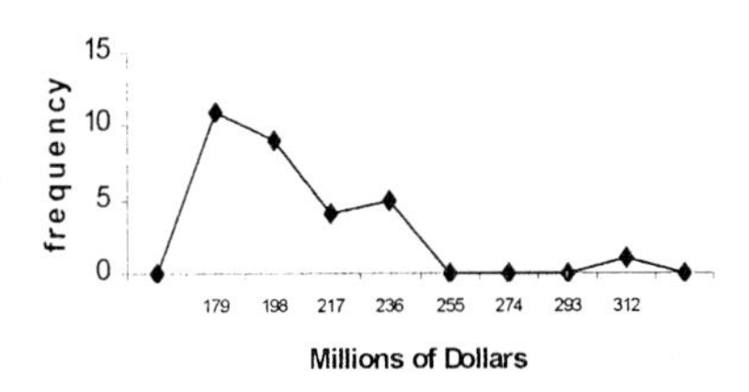

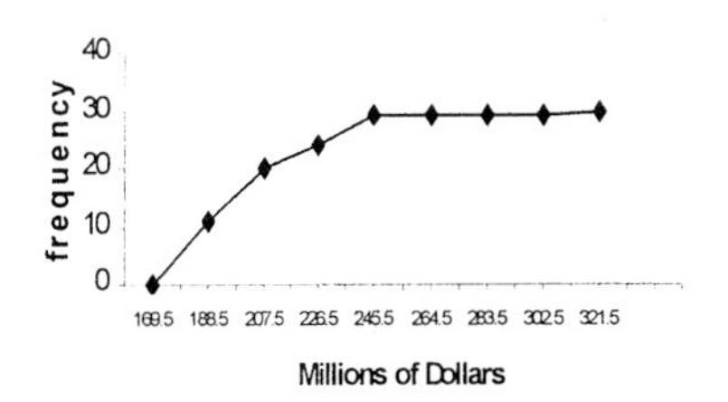

The typical value of the franchises is between \$169.5 - \$188.5 million. All but one of the franchises are valued between \$169.5 and \$245.5 million.

11.

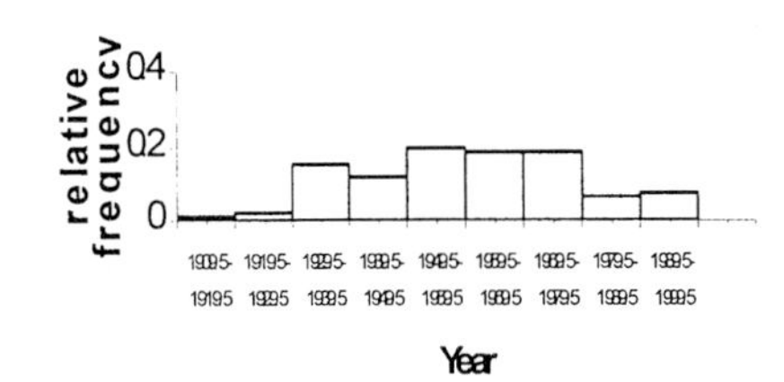

11. continued

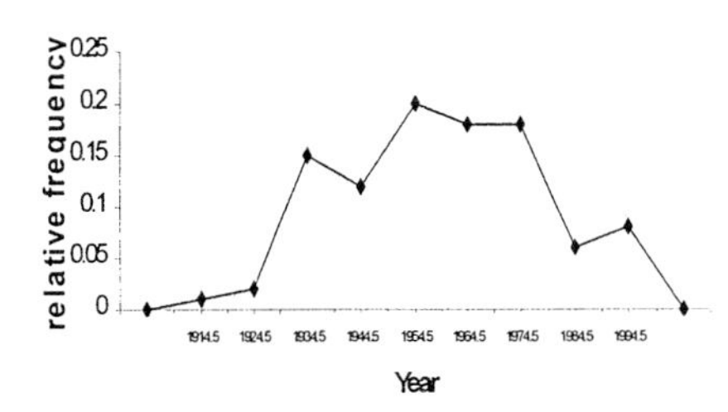

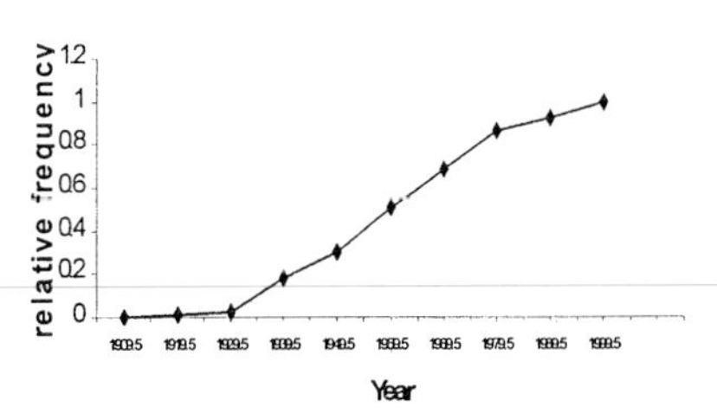

12.

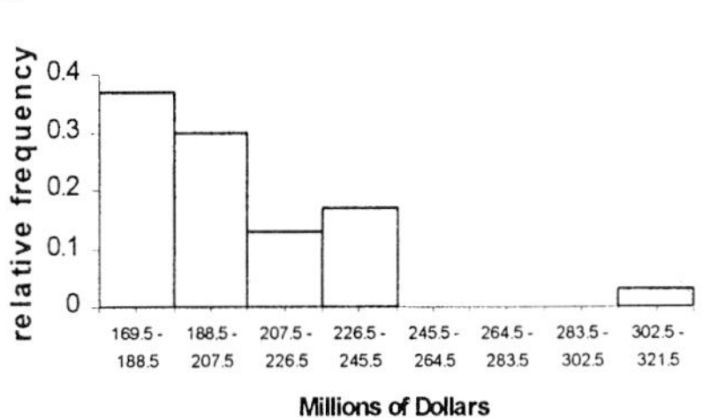

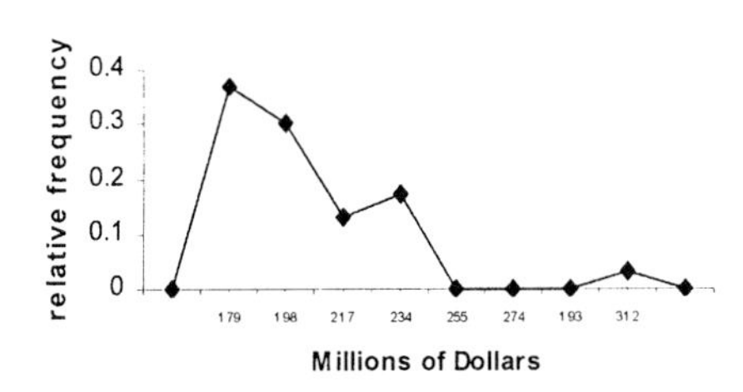

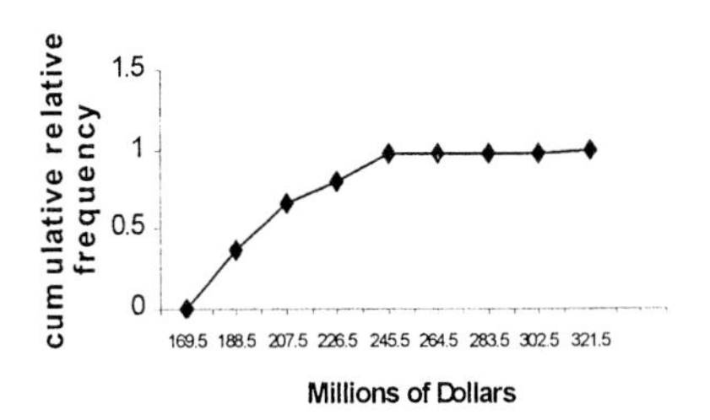

13.

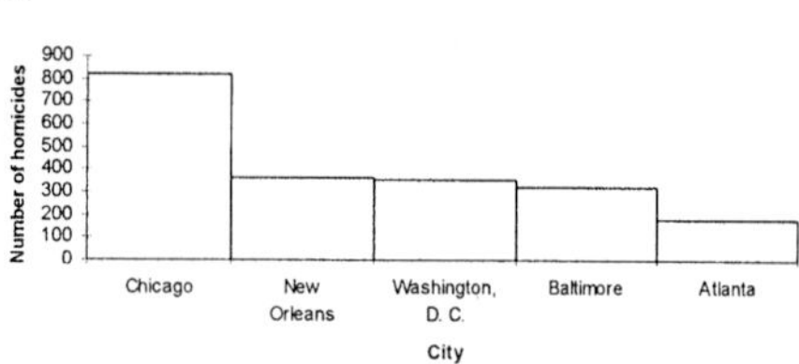

14.

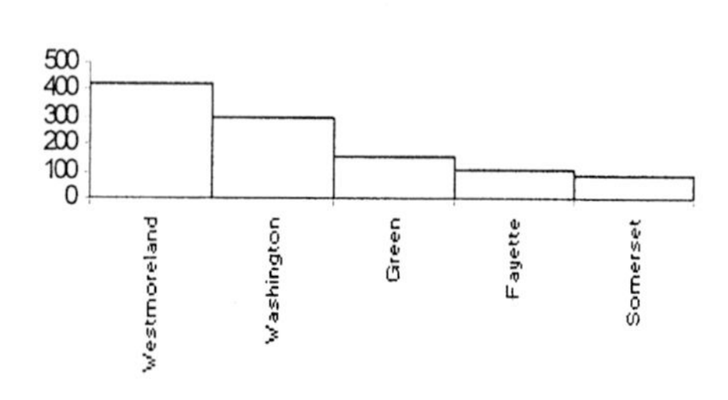

15.

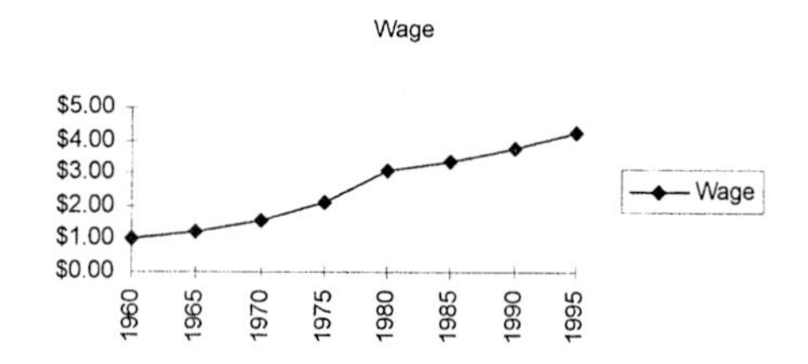

The minimum wage has increased over the years with the largest increase occurring between 1975 and 1980.

16.

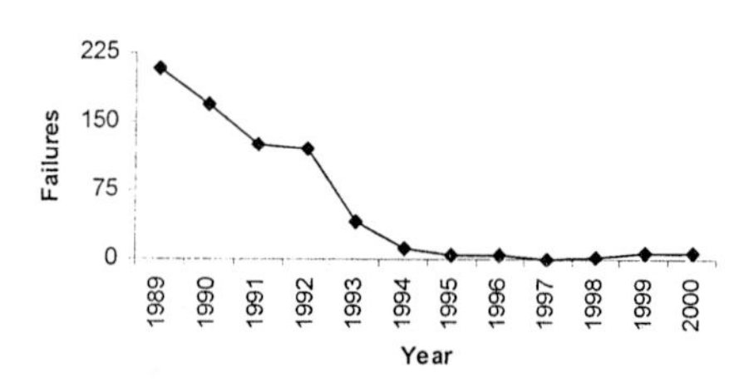

Failures decreased to only one failure in 1997, then increased slightly from 1998 to 1999.

17.

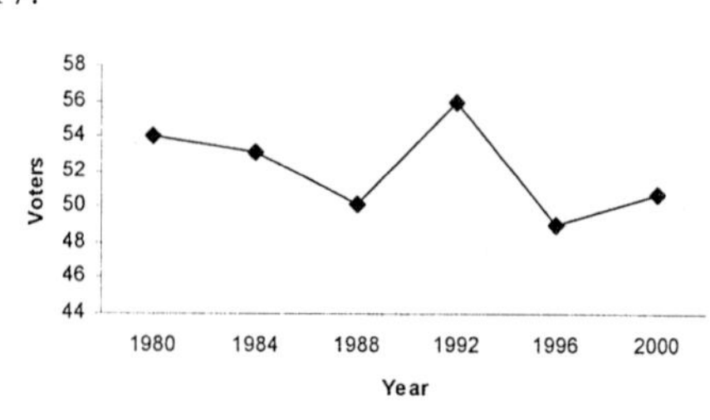

17. continued

The highest percent of voting-age population that voted in a Presidential election occurred in 1992, while the lowest percent occurred in 1996.

18.

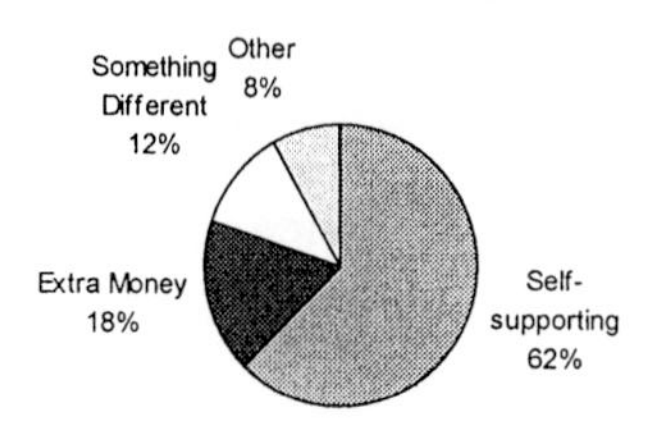

The majority of women worked to support themselves or their families.

19.

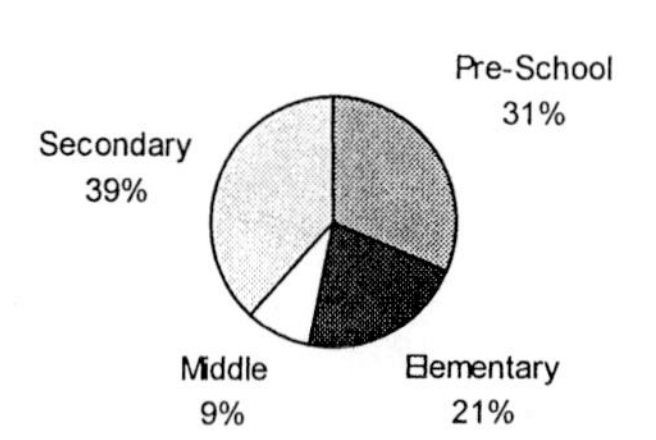

The fewest number of students were enrolled in the middle school field, and more students were in the secondary field than any other field.

20.

Stem	Leaves
2	9 9
3	2 4 5 6 8 8
4	1 2 3 7 7
5	1 3 5 8
6	2 2 2 3 7
7	2 3

21.

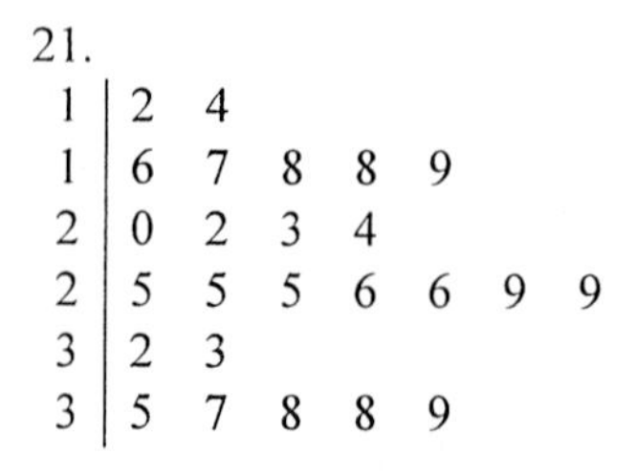

Stem	Leaves
1	2 4
1	6 7 8 8 9
2	0 2 3 4
2	5 5 5 6 6 9 9
3	2 3
3	5 7 8 8 9

22.

20	0	4	9			
21	0	1	2	7	8	8
22	2	7	7	7	8	
23	0	1	3	7	8	
24	1	2	2	3	7	
25	1	1	3	4	6	
26	0					

CHAPTER 2 QUIZ

1. False
2. False
3. False
4. True
5. True
6. False
7. False
8. c.
9. c.
10. b.
11. b.
12. Categorical, ungrouped, grouped
13. 5, 20
14. categorical
15. time series
16. stem and leaf plot
17. vertical or y
18.

	f	cf
H	6	6
A	5	11
M	6	17
C	8	25
	25	

19.

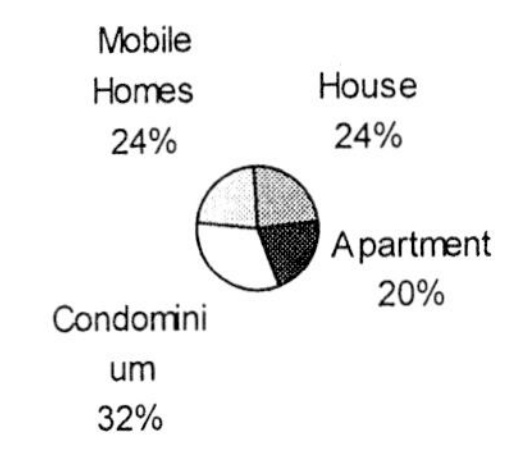

20.

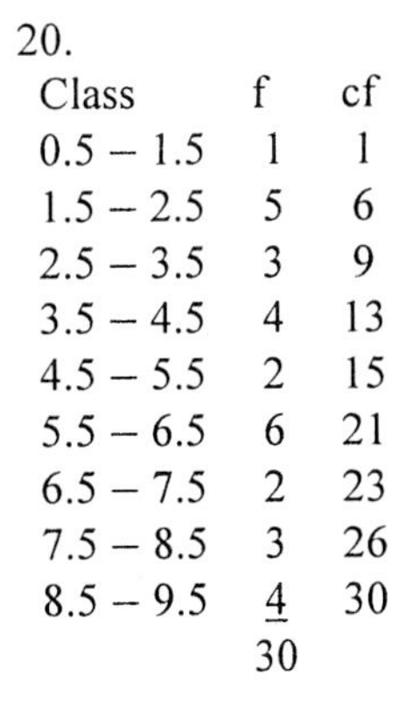

Class	f	cf
0.5 – 1.5	1	1
1.5 – 2.5	5	6
2.5 – 3.5	3	9
3.5 – 4.5	4	13
4.5 – 5.5	2	15
5.5 – 6.5	6	21
6.5 – 7.5	2	23
7.5 – 8.5	3	26
8.5 – 9.5	4	30
	30	

21.

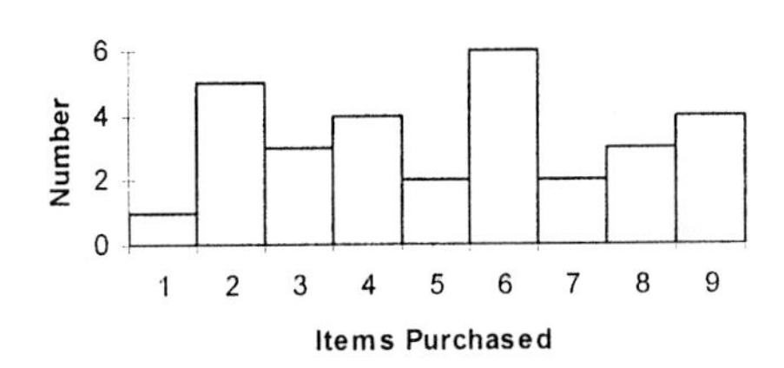

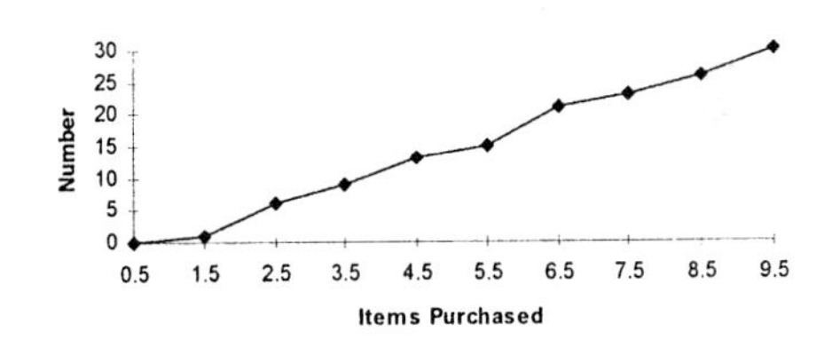

22.

Class	Boundaries	mp	f	cf
102 – 116	101.5 – 116.5	109	4	4
117 – 131	116.5 – 131.5	124	3	7
132 – 146	131.5 – 146.5	139	1	8
147 – 161	146.5 – 161.5	154	4	12
162 – 176	161.5 – 176.5	169	11	23
177 – 191	176.5 – 191.5	184	7	30
			30	

23.

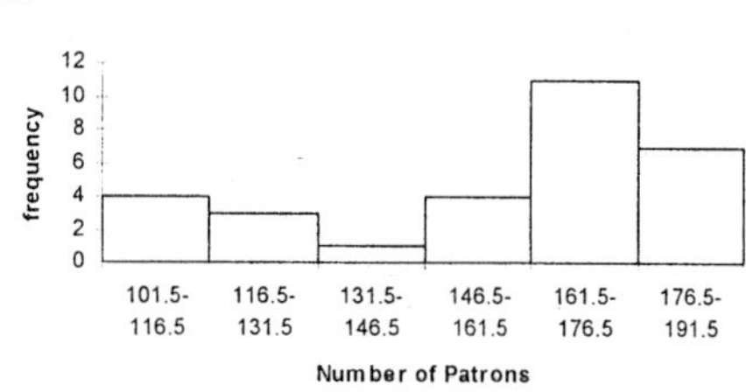

frequency
12
10
8
6
4
2
0
94 109 124 139 154 169 184 199
Number of Patrons

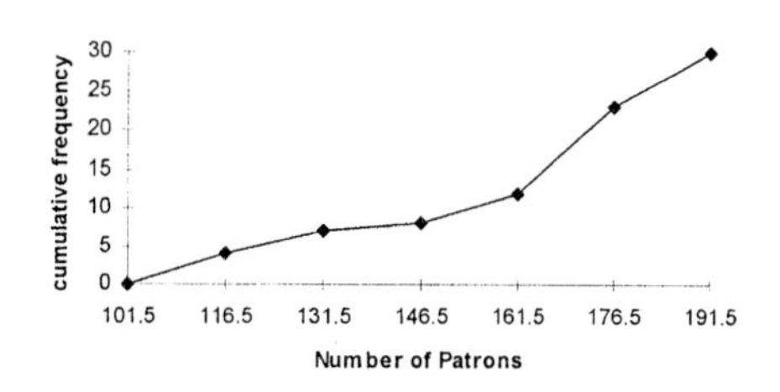

The distribution is somewhat U-shaped with a peak occurring in the 161.5 - 176.5 class.

24.

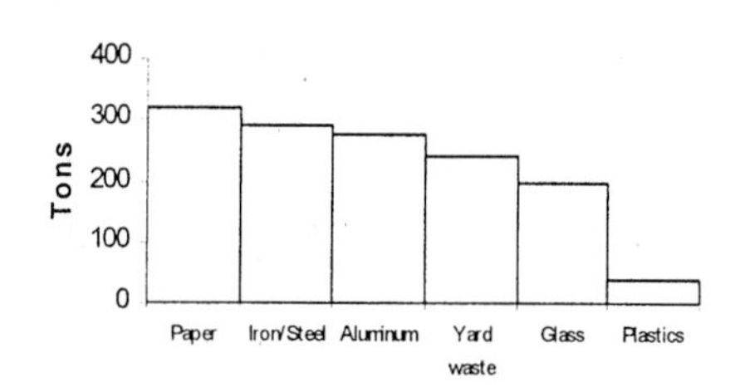

25.

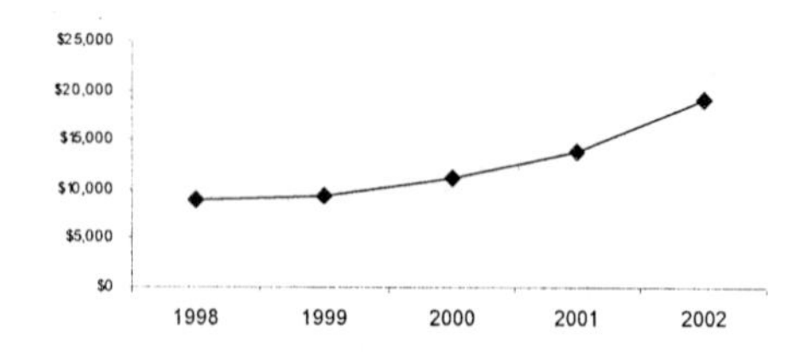

26.

Stem	Leaves
1	5 9
2	6 8
3	1 5 8 8 9
4	1 7 8
5	3 3 4
6	2 3 7 8
7	6 9
8	6 8 9
9	8

Note: Answers may vary due to rounding, TI 83's, or computer programs.

EXERCISE SET 3-2

1.
$\overline{X} = \frac{\Sigma X}{n} = \frac{93.09}{25} = 3.7236 \approx 3.724$

MD: 3.57, 3.64, 3.64, 3.65, 3.66, 3.67, 3.67, 3.68, 3.7, 3.7, 3.7, 3.73, **3.73**, 3.74, 3.74, 3.74, 3.75, 3.76, 3.77, 3.78, 3.78, 3.8, 3.8, 3.83, 3.86

Mode: 3.7 and 3.74 MR: $\frac{3.57+3.86}{2} = 3.715$

2.
$\overline{X} = \frac{\Sigma X}{n} = \frac{39.378}{20} = 1968.9$

MD: 1170, 1182, 1198, 1215, 1388, 1536, 1612, 1650, 1841, **1904, 2000**, 2123, 2151, 2307, 2425, 2499, 2540, 2625, 2800, 3212

$\text{MD} = \frac{1904+2000}{2} = 1952$

Mode: no mode MR: $\frac{1170+3212}{2} = 2191$

3.
$\overline{X} = \frac{\Sigma X}{n} = \frac{136}{9} = 15.1$

MD: 1, 2, 3, 3, **7**, 11, 18, 30, 61

Mode = 3 $\text{MR} = \frac{1+61}{2} = 31$

The median is probably the best measure of average because 61 is an extremely large data value and makes the mean artificially high.

4.
$\overline{X} = \frac{\Sigma X}{n} = \frac{948}{5} = 189.6$

MD: 75, 123, **151**, 259, 340

Mode: no mode $\text{MR} = \frac{75+340}{2} = 207.5$

5.
$\overline{X} = \frac{\Sigma X}{n} = \frac{218}{16} = 13.625$

MD: 1, 5, 6, 7, 8, 10, 10, **11, 12**, 14, 15, 16, 18, 22, 28, 35

$\text{MD} = \frac{11+12}{2} = 11.5$

Mode: 10 $\text{MR} = \frac{1+35}{2} = 18$

6.
$\overline{X} = \frac{\Sigma X}{n} = \frac{1,701,772}{12} = \$141,814.33$

MD: 115000, 115851, 125000, 125000, 127230, **146891, 147000**, 147000, 157300, 160500, 165000, 170000

6. continued

$\text{MD} = \frac{146891 + 147000}{2} = \$146,945.50$

Mode: \$125,000 and \$147,000 $\text{MR} = \frac{115,000 + 170,000}{2} = \$142,500$

7.
$\overline{X} = \frac{\Sigma X}{n} = \frac{79.6}{12} = 6.63$

MD: 5.4, 5.4, 6.2, 6.2, 6.4, **6.4, 6.5**, 7.0, 7.2, 7.2, 7.7, 8.0

$\text{MD} = \frac{6.4+6.5}{2} = 6.45$

Mode: no mode $\text{MR} = \frac{5.4+8.0}{2} = 6.7$

8.
$\overline{X} = \frac{\Sigma X}{n} = \frac{4,029,280}{50} = 80,585.6$

MD: 0, 0, 16, 155, ..., **18928, 22077**, ..., 484271, 489381, 1067227

$\text{MD} = \frac{18,928+22,077}{2} = 20{,}502.5$

Mode: 0 $\text{MR} = \frac{0+1,067,227}{2} = 533,613.5$

9.
$\overline{X} = \frac{\Sigma X}{n} = \frac{238,512}{42} = 5678.9$

MD: 150, 885, ..., **5315, 5370**, ..., 11070, 11413

$\text{MD} = \frac{5315+5370}{2} = 5342.5$

Mode: 4450 $\text{MR} = \frac{150+11,413}{2} = 5781.5$

The distribution is skewed to the right.

10.
McGwire:
$\overline{X} = \frac{\Sigma X}{n} = \frac{29,242}{70} = 417.7$

$\text{MD} = \frac{420+420}{2} = 420$

Mode: 430 $\text{MR} = \frac{306+550}{2} = 428$

Sosa:
$\overline{X} = \frac{\Sigma X}{n} = \frac{26,720}{66} = 404.8$

$\text{MD} = 410$

Mode: 420 $\text{MR} = \frac{340+500}{2} = 420$

The average of the distances of the homeruns hit by McGwire is larger than the average of the homerun distances hit by Sosa.

11.
For Year 1:

$\overline{X} = \frac{\Sigma X}{n} = \frac{24{,}911}{27} = 922.6$

MD = 527

Mode: no mode   MR $= \frac{69+4192}{2} = 2130.5$

For Year 2:

$\overline{X} = \frac{\Sigma X}{n} = \frac{24{,}615}{2} = 911.7$

MD = 485

Mode: 1430   MR $= \frac{70+4040}{2} = 2055$

The mean, median, and midrange of the traffic fatalities for Year 2 are somewhat less than those for the Year 1 fatalities, indicating that the number of fatalities has decreased.

12.

Class Limits	Boundaries	X_m	f	$f \cdot X_m$	cf
90 – 98	89.5 – 98.5	94	6	564	6
99 – 107	98.5 – 107.5	103	22	2266	28
108 – 116	107.5 – 116.5	112	43	4816	71
117 – 125	116.5 – 125.5	121	28	3388	99
126 – 134	125.5 – 134.5	130	9	1170	108
			108	12,204	

$\overline{X} = \frac{\sum f \cdot X_m}{n} = \frac{12204}{108} = 113$

modal class = 107.5 – 116.5

13.

Class Limits	*Boundaries*	X_m	f	$f \cdot X_m$
202 - 204	201.5 - 204.5	203	2	406
205 - 207	204.5 - 207.5	206	7	1442
208 - 210	207.5 - 210.5	209	16	3344
211 - 213	210.5 - 213.5	212	26	5512
214 - 216	213.5 - 216.5	215	18	3870
217 - 219	216.5 - 219.5	218	4	872
			73	15,446

$\overline{X} = \frac{\sum f \cdot X_m}{n} = \frac{15{,}446}{73} = 211.6$

modal class: 211 – 213

14.

Boundaries	X_m	f	$f \cdot X_m$
7.5 - 12.5	10	3	30
12.5 - 17.5	15	5	75
17.5 - 22.5	20	15	300
22.5 - 27.5	25	5	125
27.5 - 32.5	30	2	60
		30	590

$$\overline{X} = \frac{\sum f \cdot X_m}{n} = \frac{590}{30} = 19.7$$

modal class: 17.5 – 22.5

15.

Class	*Boundaries*	X_m	f	$f \cdot X_m$
0 - 2	-0.5 - 2.5	1	2	2
3 - 5	2.5 - 5.5	4	6	24
6 - 8	5.5 - 8.5	7	12	84
9 - 11	8.5 - 11.5	10	5	50
12 - 14	11.5 - 14.5	13	3	39
			28	199

$$\overline{X} = \frac{\sum f \cdot X_m}{n} = \frac{199}{28} = 7.1$$

modal class: 5.5 – 8.5

16.

Younger Dogs:

Class Limits	*Boundaries*	X_m	f	$f \cdot X_m$
2.3 - 2.9	2.25 - 2.95	2.6	10	26
3.0 - 3.6	2.95 - 3.65	3.3	12	39.6
3.7 - 4.3	3.65 - 4.35	4.0	6	24
4.4 - 5.0	4.35 - 5.05	4.7	8	37.6
5.1 - 5.7	5.05 - 5.75	5.4	4	21.6
5.8 - 6.4	5.75 - 6.45	6.1	2	12.2
			42	161

$$\overline{X} = \frac{\sum f \cdot X_m}{n} = \frac{161}{42} = 3.83$$

modal class: 2.95 – 3.65

Older Dogs:

Class Limits	*Boundaries*	X_m	f	$f \cdot X_m$
2.3 - 2.9	2.25 - 2.95	2.6	1	2.6
3.0 - 3.6	2.95 - 3.65	3.3	3	9.9
3.7 - 4.3	3.65 - 4.35	4.0	4	16.0
4.4 - 5.0	4.35 - 5.05	4.7	16	75.2
5.1 - 5.7	5.05 - 5.75	5.4	14	75.6
5.8 - 6.4	5.75 - 6.45	6.1	4	24.4
			42	203.7

16. continued

$\overline{X} = \frac{\sum f \cdot X_m}{n} = \frac{203.7}{42} = 4.85$

modal class: 4.35 – 5.05

No, the older dogs have a longer average reaction time than the younger dogs.

17.

Boundaries	X_m	f	$f \cdot X_m$
52.5 – 63.5	58	6	348
63.5 – 74.5	69	12	828
74.5 – 85.5	80	25	2000
85.5 – 96.5	91	18	1638
96.5 – 107.5	102	14	1428
107.5 – 118.5	113	5	565
		80	6807

$\overline{X} = \frac{\sum f \cdot X_m}{n} = \frac{6807}{80} = 85.1$

modal class: 74.5 – 85.5

18.

Class Limits	*Boundaries*	X_m	f	$f \cdot X_m$
10 – 20	9.5 – 20.5	15	2	30
21 – 31	20.5 – 31.5	26	8	208
32 – 42	31.5 – 42.5	37	15	555
43 – 53	42.5 – 53.5	48	7	336
54 – 64	53.5 – 64.5	59	10	590
65 – 75	64.5 – 75.5	70	3	210
			45	1929

$\overline{X} = \frac{\sum f \cdot X_m}{n} = \frac{1929}{45} = 42.9$

modal class: 32 – 42 or 31.5 – 42.5

19.

Class Limits	*Boundaries*	X_m	f	$f \cdot X_m$
13 – 19	12.5 – 19.5	16	2	32
20 – 26	19.5 – 26.5	23	7	161
27 – 33	26.5 – 33.5	30	12	360
34 – 40	33.5 – 40.5	37	5	185
41 – 47	40.5 – 47.5	44	6	264
48 – 54	47.5 – 54.5	51	1	51
55 – 61	54.5 – 61.5	58	0	0
62 – 68	61.5 – 68.5	65	2	130
			35	1183

$\overline{X} = \frac{\sum f \cdot X_m}{n} = \frac{1183}{35} = 33.8$

modal class: 26.5 – 33.5

20.

Class Limits	*Boundaries*	X_m	f	$f \cdot X_m$
150 – 158	149.5 – 158.5	154	5	770
159 – 167	158.5 – 167.5	163	16	2608
168 – 176	167.5 – 176.5	172	20	3440
177 – 185	176.5 – 185.5	181	21	3801
186 – 194	185.5 – 194.5	190	20	3800
195 – 203	194.5 – 203.5	199	15	2985
204 – 212	203.5 – 212.5	208	3	624
			100	18,028

$\overline{X} = \frac{\sum f \cdot X_m}{n} = \frac{18,028}{100} = 180.3$

modal class: 176.5 – 185.5

21.

Boundaries	X_m	f	$f \cdot X_m$
15.5 – 18.5	17	14	238
18.5 – 21.5	20	12	240
21.5 – 24.5	23	18	414
24.5 – 27.5	26	10	260
27.5 – 30.5	29	15	435
30.5 – 33.5	32	6	192
		75	1779

$\overline{X} = \frac{\sum f \cdot X_m}{n} = \frac{1779}{75} = 23.7$

modal class: 21.5 – 24.5

22.

Class Limits	*Boundaries*	X_m	f	$f \cdot X_m$
0.6 – 1.0	0.55 – 1.05	0.8	2	1.6
1.1 – 1.5	1.05 – 1.55	1.3	2	2.6
1.6 – 2.0	1.55 – 2.05	1.8	7	12.6
2.1 – 2.5	2.05 – 2.55	2.3	5	11.5
2.6 – 3.0	2.55 – 3.05	2.8	7	19.5
3.1 – 3.5	3.05 – 3.55	3.3	5	16.5
3.6 – 4.0	3.55 – 4.05	3.8	4	15.2
			32	79.6

$\overline{X} = \frac{\sum f \cdot X_m}{n} = \frac{79.6}{32} = 2.49$

modal class: 1.55 – 2.05 and 2.55 – 3.05

23.

Limits	*Boundaries*	X_m	f	$f \cdot X_m$
27 - 33	26.5 - 33.5	30	7	210
34 - 40	33.5 - 40.5	37	14	518
41 - 47	40.5 - 47.5	44	15	660
48 - 54	47.5 - 54.5	51	11	561
55 - 61	54.5 - 61.5	58	3	174
62 - 68	61.5 - 68.5	65	3	195
69 - 75	68.5 - 75.5	72	2	144
			55	2462

$$\overline{X} = \frac{\sum f \cdot X_m}{n} = \frac{2462}{55} = 44.8$$

modal class: 40.5 47.5

24.

Limits	*Boundaries*	X_m	f	$f \cdot X_m$
70 - 566	69.5 - 566.5	318	14	4452
567 - 1063	566.5 - 1063.5	815	5	4075
1064 - 1560	1063.5 - 1560.5	1312	5	6560
1561 - 2057	1560.5 - 2057.5	1809	0	0
2058 - 2554	2057.5 - 2554.5	2306	0	0
2555 - 3051	2554.5 - 3051.5	2803	1	2803
3052 - 3548	3051.5 - 3548.5	3300	0	0
3549 - 4045	3548.5 - 4045.5	3797	2	7594
			27	25,484

$$\overline{X} = \frac{\sum f \cdot X_m}{n} = \frac{25{,}484}{27} = 943.9$$

modal class: 69.5 – 566.5

25.

Limits	*Boundaries*	X_m	f	$f \cdot X_m$
0 - 19	-0.5 - 19.5	9.5	13	123.5
20 - 39	19.5 - 39.5	29.5	18	531.0
40 - 59	39.5 - 59.5	49.5	10	495.0
60 - 79	59.5 - 79.5	69.5	5	347.5
80 - 99	79.5 - 99.5	89.5	3	268.5
100 - 119	99.5 - 119.5	109.5	1	109.5
			50	1875.0

$$\overline{X} = \frac{\sum f \cdot X_m}{n} = \frac{1875}{50} = 37.5$$

modal class: 19.5 – 39.5

26.

$$\overline{X} = \frac{\sum w \cdot X}{\sum w} = \frac{8(10{,}000) + 10(12{,}000) + 12(8{,}000)}{8 + 10 + 12} = \frac{296{,}000}{8 + 10 + 12} = \frac{296{,}000}{30} = \$\,9866.67$$

27.

$$\overline{X} = \frac{\sum w \cdot X}{\sum w} = \frac{3(3.33)+3(3.00)+2(2.5)+2.5(4.4)+4(1.75)}{3+3+2+2.5+4} = \frac{41.99}{14.5} = 2.896$$

28.

$$\overline{X} = \frac{\sum w \cdot X}{\sum w} = \frac{40(1000)+30(3000)+50(800)}{1000+3000+800} = 35.4\%$$

29.

$$\overline{X} = \frac{\sum w \cdot X}{\sum w} = \frac{9(427000)+6(365000)+12(725000)}{9+6+12} = \frac{14,733,000}{27} = \$545,666.67$$

30.

$$\overline{X} = \frac{\sum w \cdot X}{\sum w} = \frac{20(83)+30(72)+50(90)}{100} = 83.2$$

31.

$$\overline{X} = \frac{\sum w \cdot X}{\sum w} = \frac{1(62)+1(83)+1(97)+1(90)+2(82)}{6} = \frac{496}{6} = 82.7$$

32.

a. Mode
b. Median
c. Median
d. Mode
e. Mean
f. Median

33.

a. Median
b. Mean
c. Mode
d. Mode
e. Mode
f. Mean

34.

Roman letters, $\overline{X}$
Greek letters, μ

35.

Both could be true since one could be using the mean for the average salary, and the other could be using the mode for the average.

36.

$5 \cdot 64 = 320$

37.

$5 \cdot 8.2 = 41$
$6 + 10 + 7 + 12 + x = 41$
$x = 6$

38.

The mean of the original data is 30.
The means will be:
a. 40
b. 20

38. continued
c. 300
d. 3
e. The results will be the same as adding, subtracting, multiplying, and dividing the mean by 10.

39.
a. $\frac{2}{\frac{1}{30}+\frac{1}{45}} = 36$ mph

b. $\frac{2}{\frac{1}{40}+\frac{1}{25}} = 30.77$ mph

c. $\frac{2}{\frac{1}{50}+\frac{1}{10}} = \$16.67$

40.
a. $\sqrt[3]{(1.35)(1.24)(1.18)} = 1.2547$

Average growth rate $1.25 - 1 = 0.25 = 25\%$

b. $\sqrt[4]{(1.08)(1.06)(1.04)(1.05)}$

$= 1.057397$

Average growth rate $= 1.057 - 1 = 0.057 = 5.7\%$

c. $\sqrt[5]{(1.1)(1.08)(1.12)(1.09)(1.03)} =$

$\sqrt[5]{1.4938197} = 1.08$

$1.08 - 1 = 0.08$ or 8% on average

d. $\sqrt[3]{(1.01)(1.03)(1.055)} = \sqrt[3]{1.0975165} = 1.03$

$1.03 - 1 = 0.03$ or 3%

41.

$$\sqrt{\frac{8^2+6^2+3^2+5^2+4^2}{5}} = \sqrt{30} = 5.48$$

EXERCISE SET 3-3

1.
The square root of the variance is equal to the standard deviation.

2.
One extremely high or low data value would influence the range.

3.
σ^2, σ

4.
s^2, s

5.
When the sample size is less than 30, the formula for the true standard deviation of the sample will underestimate the population standard deviation.

6.
a. s = 4.320
b. s = 5.066
c. s = 6.00
Data set A is least variable and data set C is the most variable.

7.
R = 15 − 6 = 9

$$s^2 = \frac{\sum X^2 - \frac{(\sum X)^2}{n}}{n-1} = \frac{1209 - \frac{(117)^2}{12}}{12-1} = \frac{68.25}{11} = 6.20$$

$$s = \sqrt{6.20} = 2.5$$

8.
R = 70 − 8 = 62

$$s^2 = \frac{\sum X^2 - \frac{(\sum X)^2}{n}}{n-1} = \frac{30{,}324 - \frac{(652)^2}{17}}{17-1} = 332.4$$

$$s = \sqrt{332.4} = 18.2$$

9.
For Temperature:
R = 61 − 29 = 32

$$s^2 = \frac{\sum X^2 - \frac{(\sum X)^2}{n}}{n-1} = \frac{20{,}777 - \frac{441^2}{10}}{10-1} = 147.66$$

$$s = \sqrt{147.66} = 12.15$$

For Precipitation:
R = 5.1 − 1.1 = 4.0

$$s^2 = \frac{\sum X^2 - \frac{(\sum X)^2}{n}}{n-1} = \frac{86.13 - \frac{26.3^2}{10}}{10-1} = 1.88$$

$$s = \sqrt{1.88} = 1.37$$

Temperature is more variable.

10.
Eastern states:
R = 37,741 − 20,966 = 16,775

$$s^2 = \frac{\sum X^2 - \frac{(\sum X)^2}{n}}{n-1} = \frac{5830685308 - \frac{183{,}684^2}{6}}{6-1} = 41{,}476{,}666.4$$

$$s = \sqrt{41{,}476{,}666.4} = 6440.24$$

10. continued
Western states:
$R = 101{,}510 - 54{,}339 = 47{,}171$

$$s^2 = \frac{\sum X^2 - \frac{(\sum X)^2}{n}}{n-1} = \frac{31{,}891{,}035{,}030 - \frac{428{,}362^2}{6}}{6-1} = 261{,}740{,}238.3$$

$$s = \sqrt{261{,}740{,}238.3} = 16{,}178.39$$

Western states are more variable.

11.
$R = 46 - 16 = 30$

$$s^2 = \frac{\sum X^2 - \frac{(\sum X)^2}{n}}{n-1} = \frac{9677 - \frac{313^2}{11}}{11-1} = \frac{770.727}{10} = 77.1$$

$$s = \sqrt{77.1} = 8.8$$

12.
$R = 3.80 - 3.08 = \$0.72$

$$s^2 = \frac{\sum X^2 - \frac{(\sum X)^2}{n}}{n-1} = \frac{87.6194 - \frac{25.4^2}{7}}{7-1} = \$0.08$$

$$s = \sqrt{0.08} = \$0.28$$

13.
$R = 22 - 1 = 21$

$$s^2 = \frac{\sum X^2 - \frac{(\sum X)^2}{n}}{n-1} = \frac{1061 - \frac{89^2}{15}}{15-1} = 38.1$$

$$s = \sqrt{38.1} = 6.2$$

14.
McGwire:
$R = 550 - 306 = 244$

$$s^2 = \frac{\sum X^2 - \frac{(\sum X)^2}{n}}{n-1} = \frac{12{,}367{,}642 - \frac{29{,}242^2}{70}}{70-1} = 2202.98$$

$$s = \sqrt{2202.98} = 46.9$$

Sosa:
$R = 500 - 340 = 160$

$$s^2 = \frac{\sum X^2 - \frac{(\sum X)^2}{n}}{n-1} = \frac{10{,}900{,}378 - \frac{26{,}720^2}{66}}{66-1} = 1274.25$$

$$s = \sqrt{1274.25} = 35.7$$

The distances of the homeruns are more variable for McGwire than for Sosa.

15.
For 1995:
$R = 4192 - 69 = 4123$

$$s^2 = \frac{\sum X^2 - \frac{(\sum X)^2}{n}}{n-1} = \frac{49{,}784{,}885 - \frac{24{,}911^2}{27}}{27-1} = 1{,}030{,}817.63$$

$s = \sqrt{1{,}030{,}817.63} = 1015.3$

For 1996:
$R = 4040 - 70 = 3970$

$$s^2 = \frac{\sum X^2 - \frac{(\sum X)^2}{n}}{n-1} = \frac{48{,}956{,}875 - \frac{24{,}615^2}{27}}{27-1} = 1{,}019{,}853.85$$

$s = \sqrt{1{,}019{,}853.85} = 1009.9$

The fatalities in 1995 are more variable.

16.
$R = 47196 - 734 = 46{,}462$

$$s^2 = \frac{\sum X^2 - \frac{(\sum X)^2}{n}}{n-1} = \frac{4{,}311{,}972{,}653}{50-1} = 87{,}999{,}441.9$$

$s = \sqrt{87{,}999{,}441.9} = 9380.8$

17.
$R = 11{,}413 - 150 = 11{,}263$

$$s^2 = \frac{\sum X^2 - \frac{(\sum X)^2}{n}}{n-1} = \frac{1{,}659{,}371{,}050 - \frac{238{,}512^2}{42}}{42-1} = \frac{304{,}895{,}475.1}{41} = 7{,}436{,}475.003$$

$s = \sqrt{7{,}436{,}475.003} = 2726.99$ or 2727

18.

X_m	f	$f \cdot X_m$	$f \cdot X_m^2$
94	6	564	53,016
103	22	2266	233,392
112	43	4816	539,392
121	28	3388	409,948
130	9	1170	152,100
	108	12,204	1,387,854

$$s^2 = \frac{1{,}387{,}854 - \frac{12{,}204^2}{108}}{108-1} = 82.26$$

$s = \sqrt{82.26} = 9.07$ or 9.1

19.

X_m	f	$f \cdot X_m$	$f \cdot X_m^2$
16	2	32	512
23	7	161	3703
30	12	360	10800
37	5	185	6845
44	6	264	11616
51	1	51	2601
58	0	0	0
65	2	130	8450
	35	1183	44527

$$s^2 = \frac{\sum f \cdot X_m^2 - \frac{(\sum f \cdot X_m)^2}{n}}{n-1} = \frac{44{,}527 - \frac{1183^2}{35}}{35-1} = \frac{4541.6}{34} = 133.58$$

$$s = \sqrt{133.58} = 11.6$$

20.

X_m	f	$f \cdot X_m$	$f \cdot X_m^2$
10	3	30	300
15	5	75	1125
20	15	300	6000
25	5	125	3125
30	2	60	1800
	30	590	12,350

$$s^2 = \frac{12350 - \frac{590^2}{30}}{30-1} = 25.7$$

$$s = \sqrt{25.7} = 5.07 \text{ or } 5.1$$

21.

Class	X_m	f	$f \cdot X_m$	$f \cdot X_m^2$
0 - 2	1	1	1	1
3 - 5	4	3	12	48
6 - 8	7	5	35	245
9 - 11	10	14	140	1400
12 - 14	13	6	78	1014
		29	266	2708

$$s^2 = \frac{\sum f \cdot X^2 - \frac{(\sum f \cdot X)^2}{n}}{n-1} = \frac{2708 - \frac{266^2}{29}}{29-1} = \frac{268.1379}{28} = 9.58$$

$$s = \sqrt{9.58} = 3.1$$

22.

X_m	f	$f \cdot X_m$	$f \cdot X_m^2$
2.4	12	28.8	69.12
3.1	13	40.3	124.93
3.8	7	26.6	101.08
4.5	5	22.5	101.25
5.2	2	10.4	54.08
5.9	1	5.9	34.81
	40	134.5	485.27

$$s^2 = \frac{485.27 - \frac{134.5^2}{40}}{40-1} = 0.8465$$

$$s = \sqrt{0.8465} = 0.92$$

23.

X_m	f	$f \cdot X_m$	$f \cdot X_m^2$
58	6	348	20184
69	12	828	57132
80	25	2000	160000
91	18	1638	148058
102	14	1428	145656
112	5	565	63845
	80	6807	595875

$$s^2 = \frac{\sum f \cdot X_m^2 - \frac{(\sum f \cdot X_m)^2}{n}}{n-1} = \frac{595875 - \frac{6807^2}{80}}{80-1} = \frac{16684.39}{79} = 211.2$$

$$s = \sqrt{211.2} = 14.5$$

24.

X_m	f	$f \cdot X_m$	$f \cdot X_m^2$
15	5	75	1125
26	10	260	6760
37	3	111	4107
48	7	336	16128
59	18	1062	62658
70	7	490	34300
	50	2334	125078

$$s^2 = \frac{125{,}078 - \frac{2334^2}{50}}{50-1} = 329.12$$

$$s = \sqrt{329.12} = 18.1$$

25.

X_m	f	$f \cdot X_m$	$f \cdot X_m^2$
56	2	112	6272
61	5	305	18605
66	8	528	34848
71	0	0	0
76	4	306	23104
81	5	405	32805
86	1	86	7396
	25	1740	123030

$$s^2 = \frac{\sum f \cdot X_m^2 - \frac{(\sum f \cdot X_m)^2}{n}}{n-1} = \frac{123030 - \frac{1740^2}{25}}{25-1} = \frac{1926}{24} = 80.3$$

$$s = \sqrt{80.25} = 9.0$$

26.

Younger Dogs:

X_m	f	$f \cdot X_m$	$f \cdot X_m^2$
2.6	10	26	67.6
3.3	12	39.6	130.68
4.0	6	24	96
4.7	8	37.6	176.72
5.4	4	21.6	116.64
6.1	2	12.2	74.42
	42	161	662.06

$$s^2 = \frac{662.06 - \frac{161^2}{42}}{42-1} = 1.1$$

$$s = \sqrt{1.1} = 1.0$$

Older Dogs:

X_m	f	$f \cdot X_m$	$f \cdot X_m^2$
2.6	1	2.6	6.76
3.3	3	9.9	32.67
4.0	4	16.0	64.0
4.7	16	75.2	353.44
5.4	14	75.6	408.54
6.1	4	24.4	148.84
	42	203.7	1014.25

$$s^2 = \frac{1014.25 - \frac{203.7^2}{42}}{42-1} = 0.6$$

$$s = \sqrt{0.6} = 0.8$$

The reaction times for the younger dogs are more variable than the reaction times for the older dogs.

27.

X_m	f	$f \cdot X_m$	$f \cdot X_m^2$
27	5	135	3645
30	9	270	8100
33	32	1056	34848
36	30	720	25920
39	12	468	18252
62	2	84	3528
	80	2733	94293

$$s^2 = \frac{\sum f \cdot X_m^2 - \frac{(\sum f \cdot X_m)^2}{n}}{n-1} = \frac{94293 - \frac{2733^2}{80}}{80-1} = \frac{926.89}{79} = 11.7$$

$s = \sqrt{11.7} = 3.4$

28.
C. Var $= \frac{s}{\overline{X}} = \frac{5}{110} = 0.045 = 4.5\%$
C. Var $= \frac{s}{\overline{X}} = \frac{4}{106} = 0.038 = 3.8\%$
The first class is more variable.

29.
C. Var $= \frac{s}{\overline{X}} = \frac{4{,}000}{40{,}000} = 0.10 = 10\%$

C. Var $= \frac{s}{\overline{X}} = \frac{2{,}000}{20{,}000} = 0.10 = 10\%$
They are equal.

30.
C. Var $= \frac{s}{\overline{X}} = \frac{5}{85} = 0.059 = 5.9\%$

C. Var $= \frac{s}{\overline{X}} = \frac{8}{110} = 0.073 = 7.3\%$
The history class is more variable.

31.
C. Var $= \frac{s}{\overline{X}} = \frac{6}{26} = 0.231 = 23.1\%$

C. Var $= \frac{s}{\overline{X}} = \frac{4000}{31{,}000} = 0.129 = 12.9\%$
The age is more variable.

32.
Formula: $1 - \frac{1}{k^2}$

a. $1 - \frac{1}{2^2} = \frac{3}{4}$ or 75%

b. $1 - \frac{1}{1.5^2} = 0.56$ or 56%

33.

a. $1 - \frac{1}{5^2} = 0.96$ or 96%

b. $1 - \frac{1}{4^2} = 0.9375$ or 93.75%

34.
a. $1 - \frac{1}{4^2} = 0.9375$

0.9375 (200) = 187.5

b. $1 - \frac{1}{2^2} = \frac{3}{4}$ or 0.75

0.75 (200) = 150

200 − 150 = 50

35.
$\overline{X} = 5.02 \quad s = 0.09$
At least 75% of the data values will fall withing two standard deviations of the mean; hence, 2($0.09) = $0.18 and $5.02 − $0.18 = $4.84 and $5.02 + $0.18 = $5.20. Hence at least 75% of the data values will fall between $4.84 and $5.20.

36.
$\overline{X}$ = $2.60   s = $0.15
$2.60 − 3($0.15) = $2.15; $2.60 + 3($0.15) = $3.05
At least 88.89% of the data values will fall between $2.15 and $3.05.

37.
$\overline{X} = 95 \quad s = 2$
At least 88.89% of the data values will fall within 3 standard deviations of the mean, hence 95 − 3(2) = 89 and 95 + 3(2) = 101. Therefore at least 88.89% of the data values will fall between 89 mg and 101 mg.

38.
$\overline{X} = 53 \quad x = 6$
53 − 2 (6) = 41 and 53 + 2 (6) = 65. At least 75% of the scores will fall between 41 and 65.

39.
$\overline{X} = 12 \quad s = 3$
20 − 12 = 8 and 8 ÷ 3 = 2.67
Hence, $1 - \frac{1}{k^2} = 1 - \frac{1}{2.67^2} = 1 - 0.14 = 0.86 = 86\%$
At least 86% of the data values will fall between 4 and 20.

40.
$\overline{X} = 4 \quad s = 0.10$
$4.18 - 4 = 0.18$ and $k = \frac{0.18}{0.10} = 1.8$
$1 - \frac{1}{k^2} = 1 - \frac{1}{1.8^2} = 0.69$ or 69%

41.
26.8 + 1(4.2) = 31
By the Empirical Rule, 68% of consumption is within 1 standard deviation of the mean. Then $\frac{1}{2}$ of 32%, or 16%, of consumption would be more than 31 pounds of citrus fruit per year.

42.
(a) 53 + 4.2K = 58.6
4.2K=5.6
K = 2
By Chebyshev's Theorem, $1 - \frac{1}{2^2} = .75$ or 75% of hours worked are within 2 standard deviations of the mean. Then $\frac{1}{2}$ of 25%, or 12.5%, work more than 58.6 hours per week.

42. continued
(b) By the Empirical Rule, $K = 2$ standard deviations of the mean is 95% of hours worked. Then $\frac{1}{2}$ of 5%, or 2.5%, worked more than 58.6 hours per week.

43.
$n = 30 \quad \overline{X} = 214.97 \quad s = 20.76$ At least 75% of the data values will fall between $\overline{X} \pm 2s$.
$\overline{X} - 2(20.76) = 214.97 - 41.52 = 173.45$ and $\overline{X} + 2(20.76) = 214.97 + 41.52 = 256.49$
In this case all 30 values fall within this range; hence Chebyshev's Theorem is correct for this example.

44.
$n = 30 \quad \overline{X} = 34.47 \quad s = 13.32$
$\overline{X} - 2s = 34.47 - 2(13.32) = 7.83$ and $\overline{X} + 2s = 34.47 + 2(13.32) = 61.11$
In this case 28 out of 30 data values fall withing the range of 7.83 to 61.11. This is 93.3% which is consistent with Chebyshev's Theorem.

45.
For $k = 1.5$, $1 - \frac{1}{1.5^2} = 1 - 0.44 = 0.56$ or 56%
For $k = 2$, $1 - \frac{1}{2^2} = 1 - 0.25 = 0.75$ or 75%
For $k = 2.5$, $1 - \frac{1}{2.5^2} = 1 - 0.16 = 0.84$ or 84%
For $k = 3$, $1 - \frac{1}{3^2} = 1 - 0.1111 = .8889$ or 88.89%
For $k = 3.5$, $1 - \frac{1}{3.5^2} = 1 - 0.08 = 0.92$ or 92%

46.
a. $s = 15.81$
b. $s = 15.81$
c. $s = 15.81$
d. $s = 79.06$
e. $s = 3.16$
f. The standard deviation is unchanged by adding or subtracting a specific number to each data value. If each data value is multiplied by a number the standard deviation increases by the number times the original standard deviation. For division the standard deviation is divided by the number.
g. When adding or subtracting the same number to each data value the mean will increase or decrease by that number, but the standard deviation will remain unchanged. When multiplying each data value by the same number the mean or standard deviation will be equal to that number times the original mean or standard deviation. When dividing each data value by the same number the mean or standard deviation will be equal to the original mean or standard deviation divided by that number.

47.
$\overline{X} = 13.3$

$$\text{Mean Dev} = \frac{|5-13.3|+|9-13.3|+|10-13.3|+|11-13.3|+|11-13.3|}{10}$$

$$+ \frac{|12-13.3|+|15-13.3|+|18-13.3|+|20-13.3|+|22-13.3|}{10} = 4.36$$

48.
a. $Sk = \frac{3(10-8)}{3} = 2$ positively skewed

b. $Sk = \frac{3(42-45)}{4} = -2.25$ negatively skewed

c. $Sk = \frac{3(18.6-18.6)}{1.5} = 0$ symmetrical

d. $Sk = \frac{3(98-97.6)}{4} = 0.3$ positively skewed

EXERCISE SET 3-4

1.
A z score tells how many standard deviations the data value is above or below the mean.

2.
A percentile rank indicates the percent of data values that fall below the specific rank.

3.
A percentile is a relative measure while a percent is an absolute measure of the part to the total.

4.
A quartile is a relative measure of position obtained by dividing the data set into quarters.

5.
$Q_1 = P_{25},\ Q_2 = P_{50},\ Q_3 = P_{75}$

6.
A decile is a relative measure of position obtained by dividing the data set into tenths.

7.
$D_1 = P_{10},\ D_2 = P_{20},\ D_3 = P_{30}$, etc

8.
$P_{50},\ Q_2,\ D_5$

9.
a. $z = \frac{X-\overline{X}}{s} = \frac{115-100}{10} = 1.5$

b. $z = \frac{124-100}{10} = 2.4$

c. $z = \frac{93-100}{10} = -0.7$

d. $z = \frac{100-100}{10} = 0$

e. $z = \frac{85-100}{10} = -1.5$

10.
a. $z = \frac{X-\overline{X}}{s} = \frac{2.7-2.5}{0.3} = 0.67$

b. $z = \frac{3.9-2.5}{0.3} = 4.67$

c. $z = \frac{2.8-2.5}{0.3} = 1$

d. $z = \frac{3.1-2.5}{0.3} = 2$

e. $z = \frac{2.2-2.5}{0.3} = -1$

11.
a. $z = \frac{X-\overline{X}}{s} = \frac{87-84}{4} = 0.75$

b. $z = \frac{79-84}{4} = -1.25$

11. continued

c. $z = \frac{93-84}{4} = 2.25$

d. $z = \frac{76-84}{4} = -2$

e. $z = \frac{82-84}{4} = -0.5$

12.

a. $z = \frac{X-\overline{X}}{s} = \frac{200-220}{10} = -2$

b. $z = \frac{232-220}{10} = 1.2$

c. $z = \frac{218-220}{10} = -0.2$

d. $z = \frac{212-220}{10} = -0.8$

e. $z = \frac{225-220}{10} = 0.5$

13.

a. $z = \frac{43-40}{3} = 1$

b. $z = \frac{75-72}{5} = 0.6$

The grade in part a is higher.

14.

For mathematics: $z = \frac{60-54}{3} = 2.0$ For history: $z = \frac{80-75}{2} = 2.5$

The student did better in history.

15.

a. $z = \frac{3.2-4.6}{1.5} = -0.93$ b. $z = \frac{630-800}{200} = -0.85$ c. $z = \frac{43-50}{5} = -1.4$

The score in part b is the highest.

16.

a. 58 b. 62.8 c. 64.5 d. 67.1

17.

a. 21st b. 58th c. 77th d. 29th

18.

a. 7 b. 25 c. 64 d. 76 e. 93

19.

a. a. 235 b. 255 c. 261 d. 275 e. 283

20.

a. 376 b. 389 c. 432 d. 473 e. 498

21.

a. 17th b. 39th c. 53rd d. 79th e. 91st

22.

Percentile $= \frac{\text{number of values below} + 0.5}{\text{total number of values}} \cdot 100\%$ Data: 78, 82, 86, 88, 92, 97

For 78, $\frac{0+.5}{6} \cdot 100\% = 8^{th}$ percentile For 82, $\frac{1+.5}{6} \cdot 100\% = 25^{th}$ percentile

For 86, $\frac{2+.5}{6} \cdot 100\% = 42^{nd}$ percentile For 88, $\frac{3+.5}{6} \cdot 100\% = 58^{th}$ percentile

For 92, $\frac{4+.5}{6} \cdot 100\% = 75^{th}$ percentile For 97, $\frac{5+.5}{6} \cdot 100\% = 92^{nd}$ percentile

23.

$c = \frac{6(30)}{100} = 1.8$ or 2 82

24.

For 12, $\frac{0+.5}{7} \cdot 100\% = 7^{th}$ percentile For 28, $\frac{1+.5}{7} \cdot 100\% = 21^{st}$ percentile

For 35, $\frac{2+.5}{7} \cdot 100\% = 36^{th}$ percentile For 42, $\frac{3+.5}{7} \cdot 100\% = 50^{th}$ percentile

For 47, $\frac{4+.5}{7} \cdot 100\% = 64^{th}$ percentile For 49, $\frac{5+.5}{7} \cdot 100\% = 79^{th}$ percentile

For 50, $\frac{6+.5}{7} \cdot 100\% = 93^{rd}$ percentile

25.

$c = \frac{n \cdot p}{100} = \frac{7(60)}{100} = 4.2$ or 5 Hence, 47 is the closest value to the 60th percentile.

26.

Percentile $= \frac{\text{number of values below} + 0.5}{\text{total number of values}} \cdot 100\%$ Data: 5, 12, 15, 16, 20, 21

For 5, $\frac{0+.5}{6} \cdot 100\% = 8^{th}$ percentile For 12, $\frac{1+.5}{6} \cdot 100\% = 25^{th}$ percentile

For 15, $\frac{3+.5}{6} \cdot 100\% = 42^{nd}$ percentile For 16, $\frac{4+.5}{6} \cdot 100\% = 58^{th}$ percentile

For 20, $\frac{5+.5}{6} \cdot 100\% = 75^{th}$ percentile For 21, $\frac{5+.5}{6} \cdot 100\% = 92^{nd}$ percentile

27.

$c = \frac{6(33)}{100} = 1.98$ or 2 5, 12, 15, 16, 20, 21

↑ P_{33}

28.

a. 3 16 17 18 19 20 21 22

↑ ↑

$Q_1 = 16.5$ $Q_3 = 20.5$

For Q_1 use P_{25}: $c = \frac{n \cdot p}{100} = \frac{8(25)}{100} = 2$. Hence, use the value between the 2nd and 3rd position.

$Q_1 = \frac{16+17}{2} = 16.5$

For Q_3 use P_{75}: $c = \frac{n \cdot p}{100} = \frac{8(75)}{100} = 6$. Hence use the value between the 6th and 7th position.

$Q_3 = \frac{20+21}{2} = 20.5$

$Q_3 - Q_1 = 20.5 - 16.5 = 4$ and $4(1.5) = 6$ $16.5 - 6 = 10.5$ and $20.5 + 6 = 26.5$

Only the value 3 lies outside the range of 10.5 to 26.5 and is a suspected outlier.

28. continued

b. 14 16 17 18 19 20 24 31 32 54

↑ (under 17) Q_1 ↑ (under 31) Q_3

For Q_1: $c = \frac{n \cdot p}{100} = \frac{10(25)}{100} = 2.5$. Round up to the 3rd position. $Q_1 = 17$.

For Q_3: $c = \frac{n \cdot p}{100} = \frac{10(75)}{100} = 7.5$. Round up to the 8th position. $Q_3 = 31$.

$Q_3 - Q_1$: $31 - 17 = 14$ and $14(1.5) = 21$ $17 - 21 = -4$ and $31 + 21 = 52$.
Only the value 54 lies outside the range of -4 to 52 and is a suspected outlier.

c. 200 321 327 343 350

↑ (under 321) Q_1 ↑ (under 343) Q_3

For Q_1: $c = \frac{n \cdot p}{100} = \frac{5(25)}{100} = 1.25$. Round up to the 2nd position. $Q_1 = 321$.

For Q_3: $c = \frac{n \cdot p}{100} = \frac{5(75)}{100} = 3.75$. Round up to the 4th position. $Q_3 = 343$.

$Q_3 - Q_1$: $343 - 321 = 22$ and $22(1.5) = 33$. $321 - 33 = 288$ and $343 + 33 = 376$.
Only the value 200 lies outside the range of 288 to 376 and is a suspected outlier.

d. 72 84 85 86 88 97 100

↑ (under 84) Q_1 ↑ (under 97) Q_3

For Q_1: $c = \frac{n \cdot p}{100} = \frac{7(25)}{100} = 1.75$. Round up to the 2nd position. $Q_1 = 84$.

For Q_3: $c = \frac{n \cdot p}{100} = \frac{7(75)}{100} = 5.25$. Round up to the 6th position. $Q_3 = 97$.

$Q_3 - Q_1$: $97 - 84 = 13$ and $13(1.5) = 19.5$. $84 - 19.5 = 64.5$ and $97 + 19.5 = 116.5$.
Since all values fall within the range of 64.5 to 116.5, there are no outliers.

e. 116 118 119 122 125 145

↑ (under 118) Q_1 ↑ (under 125) Q_3

For Q_1: $c = \frac{n \cdot p}{100} = \frac{6(25)}{100} = 1.5$. Round up to the 2nd position. $Q_1 = 118$.

For Q_3: $c = \frac{n \cdot p}{100} = \frac{6(75)}{100} = 4.5$. Round up to the 5th position. $Q_3 = 125$.

$Q_3 - Q_1$: $125 - 118 = 7$ and $7(1.5) = 10.5$. $118 - 10.5 = 107.5$ and $125 + 10.5 = 125.5$.

Only the value 145 is outside the range of 107.5 to 135.5 and is a suspected outlier.

f. 13 14 15 16 18 19 20 27 36

↑ (under 15) Q_1 ↑ (under 20) Q_3

For Q_1: $c = \frac{n \cdot p}{100} = \frac{9(25)}{100} = 2.25$. Round up to the 3rd position. $Q_1 = 15$.

28f. continued
For Q_3: $c = \frac{n \cdot p}{100} = \frac{9(75)}{100} = 6.75$. Round up to the 7th position. $Q_3 = 20$.

$Q_3 - Q_1$: $20 - 15 = 5$ and $5(1.5) = 7.5$. $15 - 7.5 = 7.5$ and $20 + 7.5 = 27.5$.

Only the value 36 lies outside the range of 7.5 to 27.5 and is a suspected outlier.

29.
a. 5, 12, 16, 25, 32, 38 $Q_1 = 12, Q_2 = 20.5, Q_3 = 32$

Midquartile $= \frac{12+32}{2} = 22$ Interquartile range: $32 - 12 = 20$

b. 53, 62, 78, 94, 96, 99, 103 $Q_1 = 62, Q_2 = 94, Q_3 = 99$

Midquartile $= \frac{62+99}{2} = 80.5$ Interquartile range: $99 - 62 = 37$

EXERCISE SET 3-5

1. Data arranged in order: 6, 8, 12, 19, 27, 32, 54

Minimum: 6
Q_1: 8
Median: 19
Q_3: 32
Maximum: 54
Interquartile Range: $32 - 8 = 24$

2. Data arranged in order: 7, 16, 19, 22, 48

Minimum: 7
Q_1: $\frac{7+16}{2} = 11.5$
Median: 19
Q_3: $\frac{22+48}{2} = 35$
Maximum: 48
Interquartile Range: $35 - 11.5 = 23.5$

3. Data arranged in order: 188, 192, 316, 362, 437, 589

Minimum: 188
Q_1: 192
Median: $\frac{316+362}{2} = 339$
Q_3: 437
Maximum: 589
Interquartile Range: $437 - 192 = 245$

4. Data arranged in order: 147, 156, 243, 303, 543, 632

Minimum: 147
Q_1: 156
Median: $\frac{243+303}{2} = 273$
Q_3: 543
Maximum: 632
Interquartile Range: $543 - 156 = 387$

5. Data arranged in order: 14.6, 15.5, 16.3, 18.2, 19.8

Minimum: 14.6
Q_1: $\frac{14.6+15.5}{2} = 15.05$
Median: 16.3
Q_3: $\frac{18.2+19.8}{2} = 19.0$
Maximum: 19.8
Interquartile Range: 3.95

6. Data arranged in order: 2.2, 3.7, 3.8, 4.6, 6.2, 9.4, 9.7

Minimum: 2.2
Q_1: 3.7
Median: 4.6
Q_3: 9.4
Maximum: 9.7
Interquartile Range: $9.4 - 3.7 = 5.7$

7. Minimum: 3
Q_1: 5
Median: 8
Q_3: 9
Maximum: 11
Interquartile Range: $9 - 5 = 4$

8. Minimum: 200
Q_1: 225
Median: 275
Q_3: 300
Maximum: 325
Interquartile Range: $300 - 225 = 75$

9. Minimum: 55
Q_1: 65
Median: 70
Q_3: 90
Maximum: 95
Interquartile Range: $90 - 65 = 25$

10. Minimum: 2000
Q_1: 3000
Median: 4000
Q_3: 5000
Maximum: 6000
Interquartile Range: $5000 - 3000 = 2000$

11.
$MD = \frac{3.9+4.7}{2} = 4.3$
$Q_1 = 2.0$ $Q_3 = 7.6$

11. continued

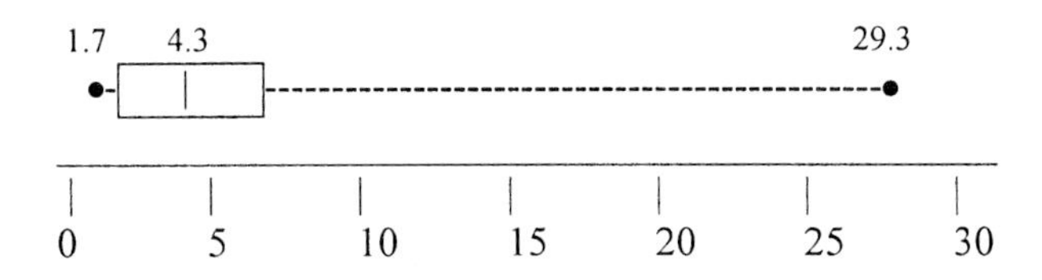

The distribution is positively skewed.

12.
MD = 9
$Q_1 = 5$ $Q_3 = 15$

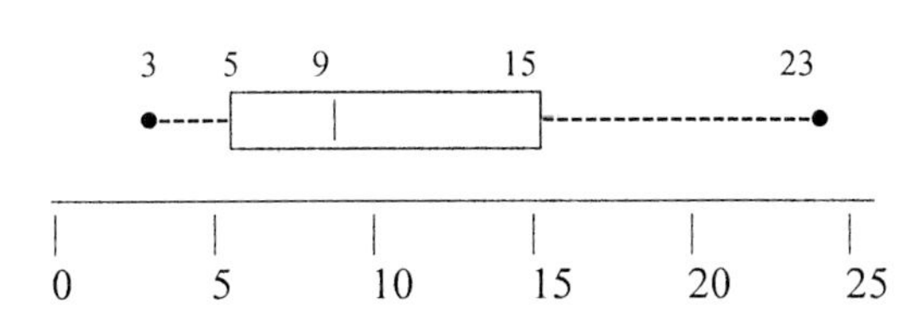

The distribution is positively skewed.

13. Data arranged in order: 13, 25, 25, 26, 28, 34, 35, 37, 42
Minimum: 13 Maximum: 42
MD = 28
$Q_1 = \frac{25+25}{2} = 25$ $Q_3 = \frac{35+37}{2} = 36$

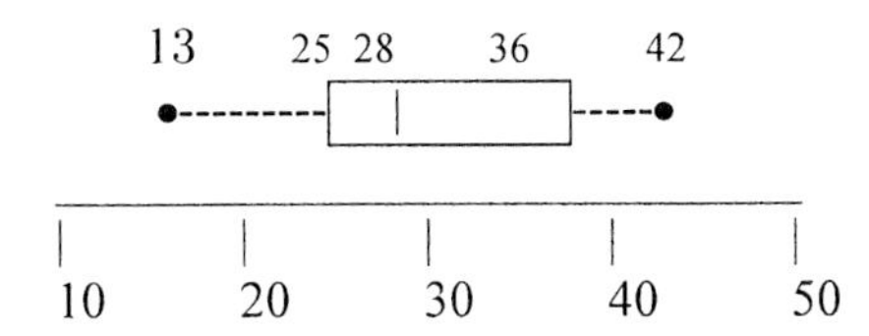

14. Data arranged in order: 28, 29, 30, 33, 34, 35, 37, 37, 37, 38, 39
Minimum: 28 Maximum: 39
MD: 35
Q_1: 30 Q_3: 37

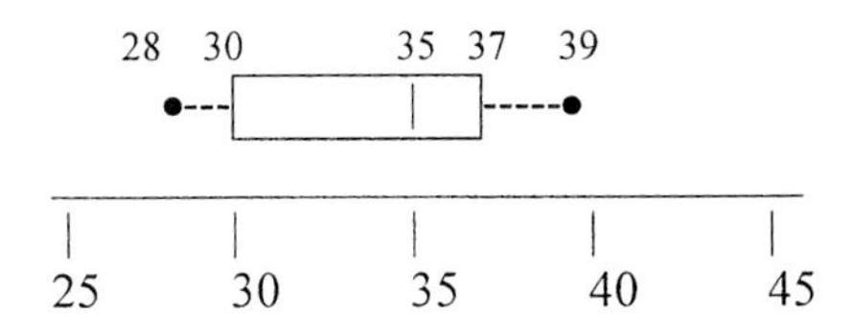

The distribution is negatively skewed.

15. Data arranged in order: 3.2, 3.9, 4.4, 8.0, 9.8, 11.7, 13.9, 15.9, 17.6, 21.7, 24.8, 34.1
Minimum: 3.2 Maximum: 34.1

MD: $\frac{11.7+13.9}{2} = 12.8$

Q_1: $\frac{4.4+8.0}{2} = 6.2$ Q_3: $\frac{17.6+21.7}{2} = 19.65$

15. continued

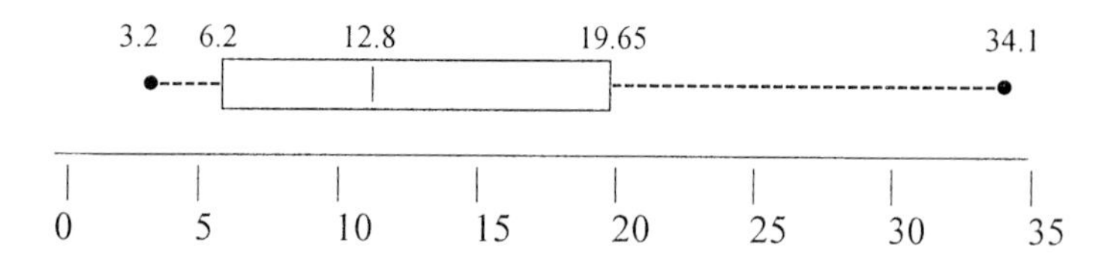

The distribution is positively skewed.

16.
For USA: min = 50,000, max = 425,628, MD = 72,100, Q_1 = 57,642.5, and Q_3 = 85,004

For South America: min = 46,563, max = 311,539, MD = 103,979, Q_1 = 56,242, and Q_3 = 274,026

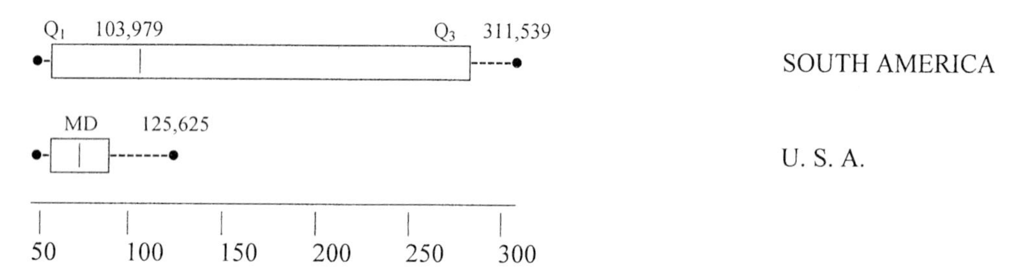

The range and variation of the capacity of the dams in South America is considerably larger than those of the United States.

17.
(a)
For April: $\overline{X} = 149.3$
For May: $\overline{X} = 264.3$
For June: $\overline{X} = 224.0$
For July: $\overline{X} = 123.3$

The month with the highest mean number of tornadoes is May.

(b)
For 2001: $\overline{X} = 186.0$
For 2000: $\overline{X} = 165.0$
For 1999: $\overline{X} = 219.75$

The year with the highest mean number of tornadoes is 1999.

(c) The 5-number summaries for each year are:

For 2001: 120, 127.5, 188, 244.5, 248
For 2000: 135, 135.5, 142, 194.5, 241
For 1999: 102, 139.5, 233, 300, 311

17c. continued

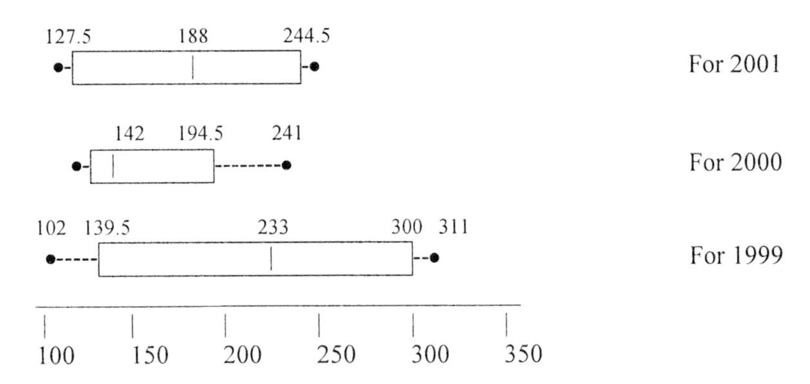

The distribution for 2001 is approximately symmetric while the distributions for 2000 and 1999 are skewed. The distribution for 2000 is positively skewed and the distribution for 1999 is negatively skewed. 2000 is the least variable and has the smallest median.

18. Data arranged in order: 39, 39, 42, 43, 43, 53, 54, 66, 91, 97

Minimum: 39
Q_1: 42
Median: $\frac{43+53}{2} = 48$
Q_3: 66
Maximum: 97
Interquartile Range: $66 - 42 = 24$
$1.5(24) = 36$ for mild outliers; $3(24) = 72$ for extreme outliers
There are no outliers.

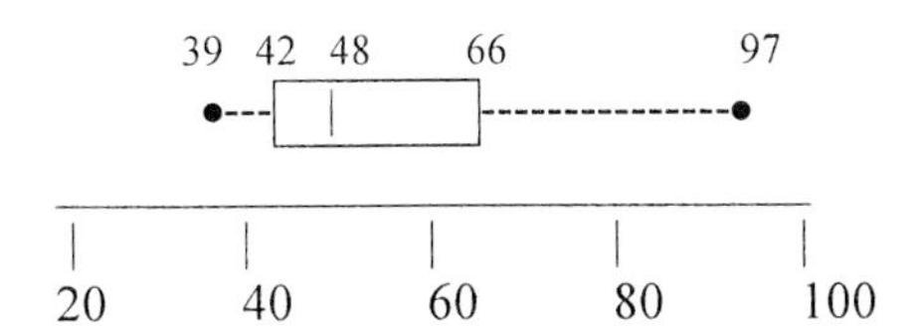

REVIEW EXERCISES - CHAPTER 3

1.

a. $\overline{X} = \frac{\sum X}{n} = \frac{2343+1240+1088+600+497+1925+1480+458}{8} = 1203.9$

b. 458 497 600 1088 1240 1480 1925 2343

↑

$MD = \frac{1088+1240}{2} = 1164$

c. no mode

d. $MR = \frac{458+2343}{2} = 1400.5$

e. $\text{Range} = 2343 - 458 = 1885$

f. $s^2 = \frac{\sum X^2 - \frac{(\sum X)^2}{n}}{n-1} = \frac{14,923,791 - \frac{9631^2}{8}}{8-1} = 475,610.1$

g. $s = \sqrt{475,610.1} = 689.6$

2.
Caribbean Sea:

a. $\overline{X} = \frac{108+75+\cdots+59+134}{19} = 4873.2$

b. 59, 75, 100, 108, 116, 134, 166, 171, 290, <u>436</u>, 687, 926, 1864, 2300, 3339, 4244, 5382, 29389, 42804
MD = 436

c. no mode

d. $MR = \frac{59+42,804}{2} = 21,431.5$

e. Range = 42,804 − 59 = 42,745

f. $s^2 = \frac{2,764,509,234 - \frac{92,590^2}{19}}{19-1} = 128,516,863.6$

g. $s = \sqrt{128,516,863.6} = 11,336.5$

Mediterranean Sea:

a. $\overline{X} = \frac{1927+229+\cdots+9301+9926}{10} = 3027.6$

b. 86, 95, 229, 540, 1411,/ 1927, 3189, 3572, 9301, 9926

$MD = \frac{1411+1927}{2} = 1669$

c. no mode

d. $MR = \frac{86+9926}{2} = 5006$

e. Range = 9926 − 86 = 9840

f. $s^2 = \frac{214,027,694 + \frac{30,276^2}{10}}{10-1} = 13,596,008.5$

g. $s = \sqrt{13,596,008.5} = 3687.3$

The Mediterranean islands are smaller and more variable than the Caribbean islands.

3.

Class	X_m	f	$f \cdot X_m$	$f \cdot X_m^2$	cf
1 - 3	2	1	2	4	1
4 - 6	5	4	20	100	5
7 - 9	8	5	40	320	10
10 - 12	11	1	11	121	11
13 - 15	14	1	14	196	12
		12	87	741	

a. $\overline{X} = \frac{\sum f \cdot X_m}{n} = \frac{87}{12} = 7.3$

b. Modal Class = 7 - 9 or 6.5 - 9.5

3. continued

c. $s^2 = \frac{741 - \frac{87^2}{12}}{11} = \frac{110.25}{11} = 10.0$

f. $s = \sqrt{10.0} = 3.2$

4.

Class	Boundaries	X_m	f	$f \cdot X_m$	$f \cdot X_m^2$	cf
11 - 15	10.5 - 15.5	13	3	39	507	3
16 - 20	15.5 - 20.5	18	5	90	1620	8
21 - 25	20.5 - 25.5	23	12	276	6348	20
26 - 30	25.5 - 30.5	28	9	252	7056	29
31 - 35	30.5 - 35.5	33	8	264	8712	37
36 - 40	35.5 - 40.5	38	3	114	4332	40
			40	1035	28,575	

a. $\overline{X} = \frac{\sum f \cdot X_m}{n} = \frac{1035}{40} = 25.9$

b. Modal Class = 20.5 - 25.5

c. $s^2 = \frac{28.575 - \frac{1035^2}{40}}{39} = \frac{1794.375}{39} = 46.0$

d. $s = \sqrt{46.0} = 6.8$

5.

Class Boundaries	X_m	f	$f \cdot X_m$	$f \cdot X_m^2$	cf
12.5 - 27.5	20	6	120	2400	6
27.5 - 42.5	35	3	105	3675	9
42.5 - 57.5	50	5	250	12,500	14
57.5 - 72.5	65	8	520	33,800	22
72.5 - 87.5	80	6	480	38,400	28
87.5 - 102.5	95	2	190	18,050	30
		30	1665	108,825	

a. $\overline{X} = \frac{\sum f \cdot X_m}{n} = \frac{1665}{30} = 55.5$

b. Modal class = 57.5 − 72.5

c. $s^2 = \frac{\sum f \cdot X_m^2 - \frac{(\sum f \cdot X_m)^2}{n}}{n - 1} = \frac{108825 - \frac{1665^2}{30}}{30 - 1} = \frac{16417.5}{29} = 566.1$

d. $s = \sqrt{566.1} = 23.8$

6.

Class	X_m	f	$f \cdot X_m$	$f \cdot X_m^2$	cf
10 - 12	11	6	66	726	6
13 - 15	14	4	56	784	10
16 - 18	17	14	238	4046	24
19 - 21	20	15	300	6000	39
22 - 24	23	8	184	4232	47
25 - 27	26	2	52	1352	49
28 - 30	29	1	29	841	50
		50	925	17981	

a. $\overline{X} = \frac{925}{50} = 18.5$

b. Modal Class $=$ 19 - 21 or 18.5 - 21.5

c. $s^2 = \frac{17981 - \frac{925^2}{50}}{50-1} = \frac{868.5}{49} = 17.7$

d. $s = \sqrt{17.7} = 4.2$

7.

$$\overline{X} = \frac{\sum w \cdot X}{\sum w} = \frac{12 \cdot 0 + 8 \cdot 1 + 5 \cdot 2 + 5 \cdot 3}{12 + 8 + 5 + 5} = \frac{33}{30} = 1.1$$

8.

$$\overline{X} = \frac{0.3(10{,}000) + 0.5(3000) + 0.2(1000)}{0.3 + 0.5 + 0.2} = \$4{,}700$$

9.

$$\overline{X} = \frac{\sum w \cdot X}{\sum w} = \frac{8 \cdot 3 + 1 \cdot 6 + 1 \cdot 30}{8 + 1 + 1} = \frac{60}{10} = 6$$

10.

C. Var $= \frac{5}{16} = 0.3125$

C. Var $= \frac{8}{43} = 0.186$

The number of books is more variable.

11.

Magazines: C. Var $= \frac{s}{\overline{X}} = \frac{12}{56} = 0.214$

Year: C. Var $= \frac{s}{\overline{X}} = \frac{2.5}{6} = 0.417$

The number of years is more variable.

12.

Percentile $= \frac{\text{number of values below} + 0.5}{\text{total number of values}} \cdot 100\%$

a. For 2, $\frac{0 + 0.5}{6} \cdot 100\% \Rightarrow 8^{\text{th}}$ percentile

For 4, $\frac{1 + 0.5}{6} \cdot 100\% \Rightarrow 25^{\text{th}}$ percentile

For 5, $\frac{2 + 0.5}{6} \cdot 100\% \Rightarrow 42^{\text{nd}}$ percentile

12. continued

For 6, $\frac{3+0.5}{6} \cdot 100\% \Rightarrow$ 58th percentile

For 8, $\frac{4+0.5}{6} \cdot 100\% \Rightarrow$ 75th percentile

For 9, $\frac{5+0.5}{6} \cdot 100\% \Rightarrow$ 92nd percentile

b. $c = \frac{np}{100} = \frac{6(30)}{100} = \frac{180}{100} = 1.8$ round up to 2

Hence, 4 is closest to the 30th percentile

c. 2 4 5 6 8 9

$\uparrow$ $\uparrow$ $\uparrow$

Q_1 MD Q_3 MD $= \frac{5+6}{2} = 5.5$

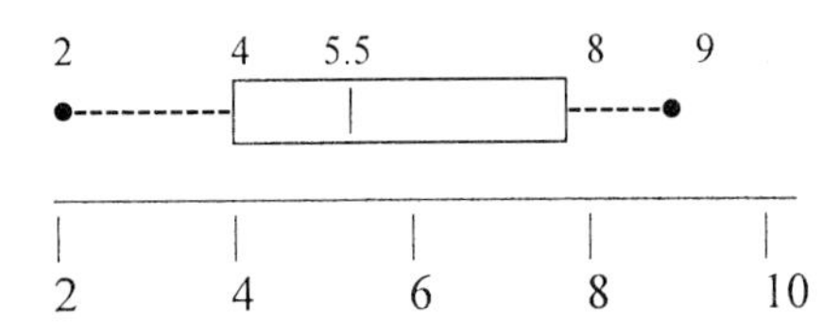

13.

a.

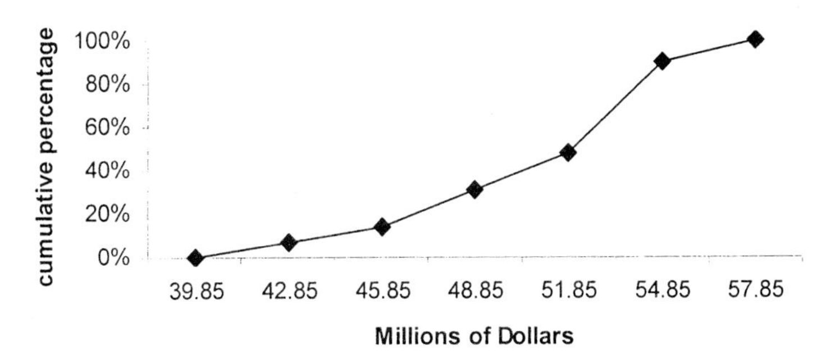

b. $P_{35} = 49$; $P_{65} = 52$; $P_{85} = 53$ (answers are approximate)

c. $44 \Rightarrow 15^{th}$ percentile; $48 \Rightarrow 33^{th}$ percentile; $54 \Rightarrow 91^{nd}$ percentile

14.

a. 400 506 511 514 517 521

$\uparrow$ $\uparrow$

Q_1 Q_3

For Q_1: $c = \frac{np}{100} = \frac{6(25)}{100} = 1.5$ round up to 2 $Q_1 = 506$

For Q_3: $c = \frac{np}{100} = \frac{6(75)}{100} = 4.5$ round up to 5 $Q_3 = 517$

$Q_3 - Q_1 = 517 - 506 = 11$; $11(1.5) = 16.5$; $506 - 16.5 = 489.5$ and $517 + 16.5 = 533.5$

Therefore, only the value 400 lies outside the range of 489.5 to 533.5 and is a suspected outlier.

b. 3 6 7 8 9 10 12 14 16 20

$\uparrow$ $\uparrow$

Q_1 Q_3

14b. continued

For Q_1: $c = \frac{np}{100} = \frac{10(25)}{100} = 2.5$ round up to 3 $\quad Q_1 = 7$

For Q_3: $c = \frac{np}{100} = \frac{10(75)}{100} = 7.5$ round up to 8 $\quad Q_3 = 14$

$Q_3 - Q_1 = 14 - 7 = 7$; $7(1.5) = 10.5$; $7 - 10.5 = -3.5$ and $14 + 10.5 = 24.5$
Since all values fall within the range of -3.5 to 24.5, there are no outliers.

c. 5 13 14 18 19 25 26 27

↑ ↑

$Q_1 = 13.5$ $\quad Q_3 = 25.5$

For Q_1: $c = \frac{np}{100} = \frac{8(25)}{100} = 2.0$ Use the value between the 2nd and 3rd position: $Q_1 = \frac{13+14}{2} = 13.5$

For Q_3: $c = \frac{np}{100} = \frac{8(75)}{100} = 6.0$ Use the value between the 6th and 7th position: $Q_3 = \frac{25+26}{2} = 25.5$

$Q_3 - Q_1 = 25.5 - 13.5 = 12$; $12(1.5) = 18$: $13.5 - 18 = -4.5$ and $25.5 + 18 = 43.5$
Since all values fall within the range of -4.5 to 43.5, there are no outliers.

d. 112 116 129 131 153 157 192

↑ ↑

Q_1 $\quad Q_3$

For Q_1: $c = \frac{np}{100} = \frac{7(25)}{100} = 1.75$ Round up to 2. $\quad Q_1 = 116$

For Q_3: $c = \frac{np}{100} = \frac{7(75)}{100} = 5.25$ Round up to 6. $\quad Q_3 = 157$

$Q_3 - Q_1 = 157 - 116 = 41$; $41(1.5) = 61.5$: $116 - 61.5 = 54.5$ and $157 + 61.5 = 218.5$
Since all values fall within the range of 54.5 to 218.5, there are no outliers.

15.
$\overline{X} = 0.32 \quad s = 0.03 \quad k = 2$
$0.32 - 2(0.03) = 0.26$ and $0.32 + 2(0.03) = 0.38$
At least 75% of the values will fall between \$0.26 and \$0.38.

16.
$\overline{X} = 42 \quad s = 3 \quad k = 3$
$42 - 3(3) = 33$ and $42 + 3(3) = 51$
At least 88.89% of the data values will fall between \$33 and \$51.

17.
$\overline{X} = 54 \quad s = 4 \quad 60 - 54 = 6 \quad k = \frac{6}{4} = 1.5 \quad 1 - \frac{1}{1.5^2} = 1 - 0.44 = 0.56$ or 56%

18.
$\overline{X} = 231 \quad s = 5 \quad 243 - 231 = 12 \quad k = \frac{12}{5} = 2.4 \quad 1 - \frac{1}{2.4^2} = 0.83$ or 83%

19.
$\overline{X} = 32 \quad s = 4 \quad 44 - 32 = 12 \quad k = \frac{12}{4} = 3 \quad 1 - \frac{1}{3^2} = 0.8889 = 88.89\%$

20.
a. $z = \frac{37-42}{5} = -1$ b. $z = \frac{72-80}{6} = -1.33$ Exam A has a better relative position.

21.
Before Christmas:
MD = 30 $Q_1 = 21$ $Q_3 = 33.5$

After Christmas:
MD = 18 $Q_1 = 14.5$ $Q_3 = 23$

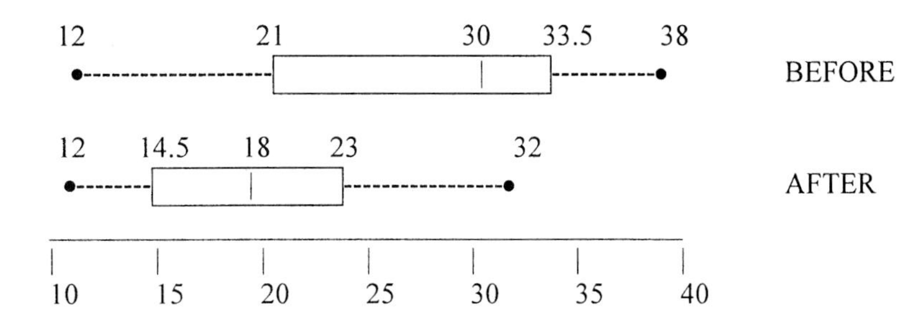

The employees worked more hours before Christmas than after Christmas. Also, the range and variability of the distribution of hours worked before Christmas is greater than that of hours worked after Christmas.

22. By the Empirical Rule, 68% of the scores will be within 1 standard deviation of the mean.
$1019 + 1(110) = 1129$
$1019 - 1(110) = 909$
Then 68% of scores will be 1019 ± 110 or between 909 and 1129.

CHAPTER 3 QUIZ

1. True
2. True
3. False
4. False
5. False
6. False
7. False
8. False
9. False
10. c.
11. c.
12. a. and b.
13. b.
14. d.
15. b.
16. statistic
17. parameters, statistics
18. standard deviation
19. σ
20. midrange
21. positively
22. outlier
23. a. 84.1   b. 85   c. none   d. 84   e. 12   f. 17.1   g. 4.1
24. a. 6.4   b. 5.5 - 8.5   c. 11.6   d. 3.4
25. a. 51.4   b. 35.5 - 50.5   c. 451.5   d. 21.2

26. a. 8.2 b. 6.5 - 9.5 c. 21.6 d. 4.6
27. 1.6
28. 4.46
29. 0.33; 0.162; newspapers
30. 0.3125; 0.229; brands
31. – 0.75; – 1.67; science
32. a. 0.5 b. 1.6 c. 15, c is higher
33. a. 6; 19; 31; 44; 56; 69; 81; 94 b. 27
c.

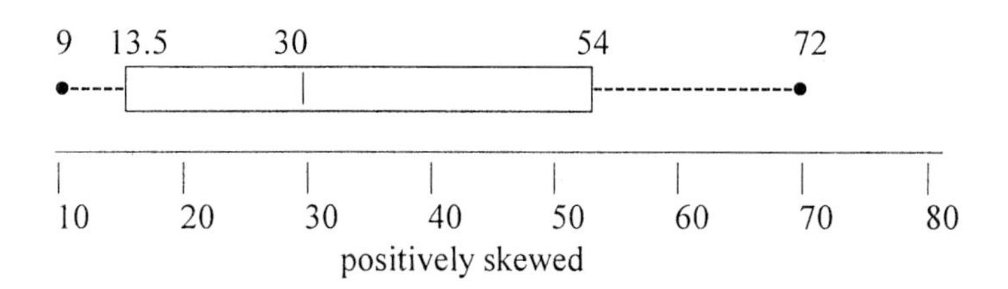

positively skewed

34.
a.

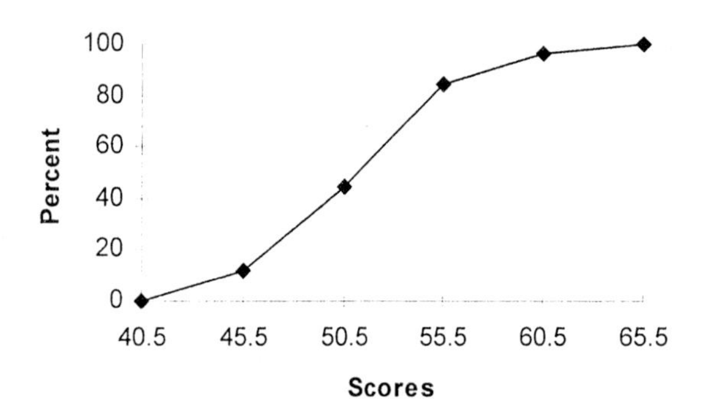

b. 47; 53; 65
c. 60^{th}percentile; 6th percentile; 98th percentile

35.
For Pre-buy:
MD = 1.625 Q_1 = 1.54 Q_3 = 1.65

For No Pre-buy:
MD = 3.95 Q_1 = 3.85 Q_3 = 3.99

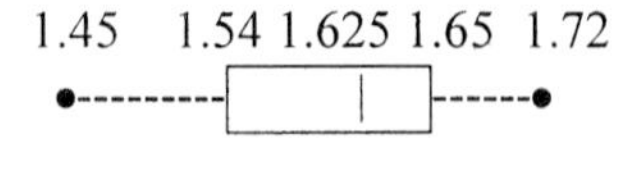

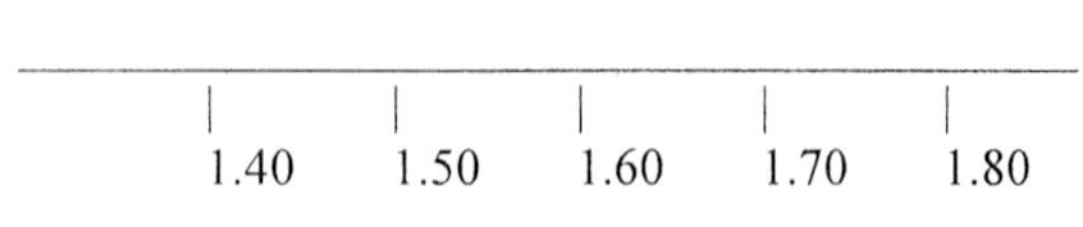

Pre-buy Cost

35. continued

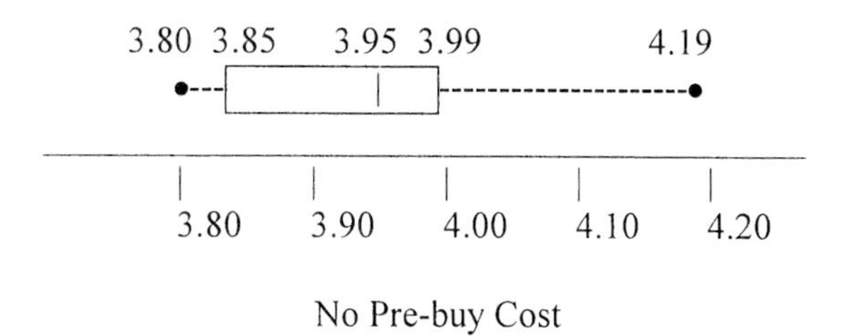

No Pre-buy Cost

The cost of pre-buy gas is much less than to return the car without filling it with gas. The variability of the return without filling with gas is larger than the variability of the pre-buy gas.

36.
For above 1129: 16%
For above 799: 97.5%

Note: Answers may vary due to rounding, TI-83's or computer programs.

EXERCISE SET 4-2

1.
A probability experiment is a chance process which leads to well-defined outcomes.

2.
The set of all possible outcomes of a probability experiment is called a sample space.

3.
An outcome is the result of a single trial of a probability experiment, whereas an event can consist of one or more outcomes.

4.
Equally likely events have the same probability of occurring.

5.
The range of values is $0 \leq P(E) \leq 1$.

6.
1

7.
0

8.
1

9.
$1 - 0.85 = 0.15$

10.
b, d, f, i

11.
a. empirical
b. classical
c. empirical
d. classical
e. empirical
f. empirical
g. subjective

12.
a. $\frac{1}{6}$
b. $\frac{1}{2}$
c. $\frac{1}{3}$
d. 1
e. 1
f. $\frac{5}{6}$
g. $\frac{1}{6}$

13.
There are 6^2 or 36 outcomes.
a. There are 5 ways to get a sum of 6. They are (1,5), (2,4), (3,3), (4,2), and (5,1). The probability then is $\frac{5}{36}$.

b. There are six ways to get doubles. They are (1,1), (2,2), (3,3), (4,4), (5,5), and (6,6). The probability then is $\frac{6}{36} = \frac{1}{6}$.

c. There are six ways to get a sum of 7. They are (1,6), (2,5), (3,4), (4,3), (5,2), and (6,1). There are two ways to get a sum of 11. They are (5,6) and (6,5). Hence, the total number of ways to get a 7 or 11 is eight. The probability then is $\frac{8}{36} = \frac{2}{9}$.

d. To get a sum greater than nine, one must roll a 10, 11, or 12. There are six ways to get a 10, 11, or 12. They are (4,6), (5,5), (6,4), (5,6), (6,5), and (6,6). The probability then is $\frac{6}{36} = \frac{1}{6}$.

e. To get a sum less than or equal to four, one must roll a 4, 3, or 2. There are six ways to do this. They are (3,1), (2,2), (1,3), (2,1), (1,2), and (1,1). The probability is $\frac{6}{36} = \frac{1}{6}$.

14.
a. $\frac{1}{13}$
b. $\frac{1}{4}$
c. $\frac{1}{52}$
d. $\frac{2}{13}$
e. $\frac{4}{13}$
f. $\frac{4}{13}$
g. $\frac{1}{2}$
h. $\frac{1}{26}$
i. $\frac{7}{13}$
j. $\frac{1}{26}$

15.
There are 24 possible outcomes.

(a) P(winning \$10) = P(rolling a 1)
P(rolling a 1) $= \frac{4}{24} = \frac{1}{6}$

(b) P(winning \$5 or \$10) = P(rolling either a 1 or 2)
P(1 or 2) $= \frac{12}{24} = \frac{1}{2}$

(c) P(winning a coupon) = P(rolling either a 3 or 4)
P(3 or 4) $= \frac{12}{24} = \frac{1}{2}$

16.
(a) P(begins with M) $= \frac{8}{50} = \frac{4}{25}$

(b) P(begins with a vowel) $= \frac{12}{50} = \frac{6}{25}$
P(not a vowel) $= 1 - \frac{6}{25} = \frac{19}{25}$

17.
(a) P(graduate school) $= \frac{110}{250} = \frac{11}{25}$ or 0.44

(b) P(medical school) $= \frac{10}{250} = \frac{1}{25}$ or 0.04

(c) P(not going to graduate school) $= 1 - \frac{110}{250} = \frac{140}{250}$ or 0.56

18.
P(doesn't believe in gun licensing) $= 1 - 0.69 = 0.31$

19.
(a) P(student) $= \frac{12}{18} = \frac{2}{3}$

(b) P(senior or junior) $= \frac{8}{18} = \frac{4}{9}$

(c) P(not a student) $= 1 - \frac{2}{3} = \frac{1}{3}$
or: P(not a student) = P(faculty or administrator) $= \frac{6}{18} = \frac{1}{3}$

20.
(a) P(education) $= \frac{106{,}000}{1{,}184{,}000} \approx 0.0895$

(b) P(not business) $= 1 - \frac{233{,}000}{1{,}184{,}000}$
P(not business) $= \frac{951{,}000}{1{,}184{,}000} \approx 0.8032$

21.
The sample space is BBB, BBG, BGB, GBB, GGB, GBG, BGG, and GGG.

a. All boys is the outcome BBB; hence P(all boys) $= \frac{1}{8}$.

b. All girls or all boys would be BBB and GGG; hence, P(all girls or all boys) $= \frac{1}{4}$.

c. Exactly two boys or two girls would be BBG, BGB, GBB, BBG, GBG, or BGG. The probability then is $\frac{6}{8} = \frac{3}{4}$.

d. At least one child of each gender means at least one boy or at least one girl. The outcomes are the same as those of part c, hence the probability is the same, $\frac{3}{4}$.

22.
There are 6 ways to get a 7 and 2 ways to get 11; hence, P(7 or 11) $= \frac{8}{36} = \frac{2}{9}$.

23.
The outcomes for 2, 3, or 12 are (1,1), (1,2), (2,1), and (6,6); hence P(2, 3, or 12) $= \frac{1+2+1}{36} = \frac{4}{36} = \frac{1}{9}$.

24.
P(Houston, Chicago, or Los Angeles) $= \frac{118}{2541}$ or 0.0464

P(some other city) $= 1 \quad \frac{118}{2541} = \frac{2423}{2541}$ or 0.9536

25.
a. There are 18 odd numbers; hence, P(odd) $= \frac{18}{36} = \frac{9}{19}$.

b. There are 11 numbers greater than 25 (26 through 36) hence, the probability is $\frac{11}{38}$.

c. There are 14 numbers less than 15 hence the probability is $\frac{14}{38} = \frac{7}{19}$.

26.
$\frac{39}{50} = 0.78$

27.
P(right amount or too little) $= 0.35 + 0.19$
P(right amount or too little) $= 0.54$

28.
$1 - 0.16 = 0.84$ or 84%

29.
(a)

	1	2	3	4	5	6
1	1	2	3	4	5	6
2	2	4	6	8	10	12
3	3	6	9	12	15	18
4	4	8	12	16	20	24
5	5	10	15	20	25	30
6	6	12	18	24	30	36

(b) P(multiple of 6) $= \frac{15}{36} = \frac{5}{12}$

(c) P(less than 10) $= \frac{17}{36}$

30.
P(individual or corporate taxes) $= 0.60$

31.
a. 0.08
b. 0.01
c. $0.08 + 0.27 = 0.35$
d. $0.01 + 0.24 + 0.11 = 0.36$

32.
Probably not.

33.
The statement is probably not based on empirical probability and probably not true.

34.
The outcomes will be:

0,0	0,1	0,2	0,3	0,4
1,0	1,1	1,2	1,3	1,4
2,0	2,1	2,2	2,3	2,4
3,0	3,1	3,2	3,3	3,4
4,0	4,1	4,2	4,3	4,4

a. $\frac{6}{25}$ b. $\frac{10}{25} = \frac{2}{5}$ c. $\frac{9}{25}$

d. $\frac{12}{25}$ e. $\frac{5}{25} = \frac{1}{5}$

35.
Actual outcomes will vary, however each number should occur approximately $\frac{1}{6}$ of the time.

36.
Actual outcomes will differ; however, the probabilities of 0, 1, and 2 heads will be approximately $\frac{1}{4}$, $\frac{1}{2}$, and $\frac{1}{4}$ respectively.

37.
a. 1:5, 5:1
b. 1:1, 1:1
c. 1:3, 3:1
d. 1:1, 1:1
e. 1:12, 12:1
f. 1:3, 3:1
g. 1:1, 1:1

EXERCISE SET 4-3

1.
Two events are mutually exclusive if they cannot occur at the same time. Examples will vary.

2.
a. no
b. no
c. yes
d. no
e. no
f. yes
g. yes

3.
$\frac{2}{12} = \frac{1}{6}$

4.
$\frac{16}{26} + \frac{2}{26} = \frac{18}{26} = \frac{9}{13}$

5.
$\frac{4}{19} + \frac{7}{19} = \frac{11}{19}$

6.
$\frac{10}{22} + \frac{7}{22} = \frac{17}{22}$

7.
a. $\frac{5}{17} + \frac{3}{17} = \frac{8}{17}$
b. $\frac{4}{17} + \frac{2}{17} = \frac{6}{17}$
c. $\frac{3}{17} + \frac{2}{17} + \frac{4}{17} = \frac{9}{17}$
d. $\frac{5}{17} + \frac{4}{17} + \frac{3}{17} = \frac{12}{17}$

8.
P(car or computer) =
$0.65 + 0.82 - 0.55 = 0.92$
P(neither) $= 1 - 0.92 = 0.08$

9.
P(football or basketball) =
$\frac{58 + 40 - 8}{200} = \frac{90}{200}$ or 0.45

P(neither) $= 1 - \frac{90}{200} = \frac{11}{20}$ or 0.55

10.
a. There are four 4's and 13 diamonds, but the 4 of diamonds is counted twice; hence, P(4 or diamond) = P(4) + P(diamonds) − P(4 of diamonds) $= \frac{4}{52} + \frac{13}{52} - \frac{1}{52} = \frac{16}{52} = \frac{4}{13}$.

b. P(club or diamond) $= \frac{13}{52} + \frac{13}{52} = \frac{26}{52}$ or $\frac{1}{2}$

c. P(jack or black) $= \frac{4}{52} + \frac{26}{52} - \frac{2}{52} = \frac{28}{52}$ or $\frac{7}{13}$

11.

	Junior	Senior	Total
Female	6	6	12
Male	12	4	16
Total	18	10	28

a. $\frac{18}{28} + \frac{12}{28} - \frac{6}{28} = \frac{24}{28} = \frac{6}{7}$

b. $\frac{10}{28} + \frac{12}{28} - \frac{6}{28} = \frac{16}{28} = \frac{4}{7}$

c. $\frac{18}{28} + \frac{10}{28} = \frac{28}{28} = 1$

12.

	Fiction	Non-Fiction	Total
Adult	30	70	100
Children	100	60	160
	130	130	260

(a) P(fiction) $= \frac{130}{260} = \frac{1}{2}$ or 0.5

(b) P(children's nonfiction) $= \frac{60}{260} = \frac{3}{13}$
P(not a children's nonfiction) $=$
$1 - \frac{3}{13} = \frac{10}{13}$ or 0.7692

(c) P(adult book or children's nonfiction) $=$
$\frac{100}{260} + \frac{60}{260} = \frac{160}{260}$ or $\frac{8}{13}$ or 0.6154

13.

	SUV	Compact	Mid-sized	Total
Foreign	20	50	20	90
Domestic	65	100	45	210
Total	85	150	65	300

(a) P(domestic) $= \frac{210}{300} = \frac{7}{10}$

(b) P(foreign and mid-sized) $= \frac{20}{300} = \frac{1}{15} = 0.0667$

(c) P(domestic or SUV) $= \frac{210}{300} + \frac{85}{300} - \frac{65}{300}$
$= \frac{230}{300} = \frac{23}{30}$ or 0.7667

14.

	Mammals	Birds	Reptiles	Amphibians	Total
U. S.	63	78	14	10	165
Foreign	251	175	64	8	498
Total	314	253	78	18	663

(a) P(found in U. S. and is a bird) $= \frac{78}{663} = \frac{2}{17}$ or 0.1176

(b) P(foreign or a mammal) $= \frac{498}{663} + \frac{314}{663} - \frac{251}{663} = \frac{561}{663}$ or 0.846

(c) P(warm-blooded) $=$
$\frac{314}{663} + \frac{253}{663} = \frac{567}{663}$ or 0.8552

15.

	Cashier	Clerk	Deli	Total
Married	8	12	3	23
Not Married	5	15	2	22
Total	13	27	5	45

a. P(stock clerk or married) = P(clerk) + P(married) − P(married stock clerk) $=$
$\frac{27}{45} + \frac{23}{45} - \frac{12}{45} = \frac{38}{45}$

15. continued
b. P(not married) $= \frac{22}{45}$

c. P(cashier or not married) = P(cashier) + P(not married) − P(unmarried cashier)
$= \frac{13}{45} + \frac{22}{45} - \frac{5}{45} = \frac{30}{45} = \frac{2}{3}$

16.

Comics	Morning	Evening	Weekly	Total
Yes	2	3	1	6
No	3	4	2	9
Total	5	7	3	15

a. P(weekly) $= \frac{3}{15}$ or $\frac{1}{5}$

b. P(morning or has comics) $=$
P(morning) + P(has comics) − P(morning with comics) $= \frac{5}{15} + \frac{6}{15} - \frac{2}{15} = \frac{9}{15} = \frac{3}{5}$

c. P(weekly or no comics) = P(weekly) + P(no comics) − P(weekly with no comics)
$= \frac{3}{15} + \frac{9}{15} - \frac{2}{15} = \frac{10}{15} = \frac{2}{3}$

17.

	Ch. 6	Ch. 8	Ch. 10	Total
Quiz	5	2	1	8
Comedy	3	2	8	13
Drama	4	4	2	10
Total	12	8	11	31

a. P(quiz show or channel 8) = P(quiz) + P(channel 8) − P(quiz show on ch. 8) $=$
$\frac{8}{31} + \frac{8}{31} - \frac{2}{31} = \frac{14}{31}$

b. P(drama or comedy) = P(drama) + P(comedy) $= \frac{13}{31} + \frac{10}{31} = \frac{23}{31}$

c. P(channel 10 or drama) = P(ch. 10) + P(drama) − P(drama on channel 10) $=$
$\frac{11}{31} + \frac{10}{31} - \frac{2}{31} = \frac{19}{31}$

18.

	1st Class	Ad	Magazine	Total
Home	325	406	203	934
Business	732	1021	97	1850
Total	1057	1427	300	2784

a. P(home) $= \frac{934}{2784} = \frac{467}{1392}$

b. P(advertisement or business) = P(ad) + P(business) − P(business and ad) $=$
$\frac{1427}{2784} + \frac{1850}{2784} - \frac{1021}{2784} = \frac{2256}{2784} = \frac{47}{58}$

18. continued

c. P(1st class or home) = P(1st class) + P(home) − P(1st class and home) = $\frac{1057}{2784} + \frac{934}{2784} - \frac{325}{2784} = \frac{1666}{2784} = \frac{833}{1392}$

19.
The total of the frequencies is 30.
a. $\frac{2}{30} = \frac{1}{15}$

b. $\frac{2+3+5}{30} = \frac{10}{30} = \frac{1}{3}$

c. $\frac{12+8+2+3}{30} = \frac{25}{30} = \frac{5}{6}$

d. $\frac{12+8+2+3}{30} = \frac{25}{30} = \frac{5}{6}$

e. $\frac{8+2}{30} = \frac{10}{30} = \frac{1}{3}$

20.
The total of the frequencies is 40.
a. $\frac{2}{40} = \frac{1}{20}$

b. $\frac{15+2}{40} = \frac{17}{40}$

c. $\frac{9}{40}$

d. $\frac{2+15+8}{40} = \frac{25}{40} = \frac{5}{8}$

21.
The total of the frequencies is 24.
a. $\frac{10}{24} = \frac{5}{12}$

b. $\frac{2+1}{24} = \frac{3}{24} = \frac{1}{8}$

c. $\frac{10+3+2+1}{24} = \frac{16}{24} = \frac{2}{3}$

d. $\frac{8+10+3+2}{24} = \frac{23}{24}$

22.

	High Chol.	Normal Chol.	Total
Alcoholic	87	13	100
Non-Alcoholic	43	157	200
Total	56	244	300

a. P(alcoholic with elevated cholesterol) = $\frac{87}{300} = \frac{29}{100}$

b. P(non-alcoholic) = $\frac{200}{300} = \frac{2}{3}$

c. P(non-alcoholic with normal cholesterol) = $\frac{157}{300}$

23.
a. There are 4 kings, 4 queens, and 4 jacks; hence P(king or queen or jack) = $\frac{12}{52} = \frac{3}{13}$

b. There are 13 clubs, 13 hearts, and 13 spades; hence, P(club or heart or spade) = $\frac{13+13+13}{52} = \frac{39}{52} = \frac{3}{4}$

c. There are 4 kings, 4 queens, and 13 diamonds but the king and queen of diamonds were counted twice, hence; P(king or queen or diamond) = P(king) + P(queen) + P(diamond) − P(king and queen of diamonds) = $\frac{4}{52} + \frac{4}{52} + \frac{13}{52} - \frac{2}{52} = \frac{19}{52}$

d. There are 4 aces, 13 diamonds, and 13 hearts. There is one ace of diamonds and one ace of hearts; hence, P(ace or diamond or heart) = P(ace) + P(diamond) + P(heart) − P(ace of hearts and ace of diamonds) = $\frac{4}{52} + \frac{13}{52} + \frac{13}{52} - \frac{2}{52} = \frac{28}{52} = \frac{7}{13}$

e. There are 4 nines, 4 tens, 13 spades, and 13 clubs. There is one nine of spades, one ten of spades, one nine of clubs and one ten of clubs. Hence, P(9 or 10 or spade or club) = P(9) + P(10) + P(spade) + P(club) − P(9 and 10 of clubs and spades) = $\frac{4}{52} + \frac{4}{52} + \frac{13}{52} + \frac{13}{52} - \frac{4}{52} = \frac{30}{52} = \frac{15}{26}$

24.
a. $\frac{5}{36} + \frac{6}{36} + \frac{5}{36} = \frac{16}{36} = \frac{4}{9}$

b. $\frac{6}{36} + \frac{3}{36} + \frac{5}{36} - \frac{2}{36} = \frac{12}{36} = \frac{1}{3}$

c. $\frac{6}{36} + \frac{3}{36} + \frac{6}{36} = \frac{15}{36} = \frac{5}{12}$

25.
P(red or white ball) = $\frac{7}{10}$

26.
There are $6^3 = 216$ possible outcomes.
a. $\frac{6}{216} = \frac{1}{36}$ since there are 6 triples, (1,1,1), (2,2,2), . . . , (6,6,6).

b. $\frac{6}{216} = \frac{1}{36}$ since there are six possible outcomes summing to 5. (1,2,2), (2,1,2), (2,2,1), (1,1,3), (1,3,1), and (3,1,1).

27.
P(mushrooms or pepperoni) =
P(mushrooms) + P(pepperoni) −
P(mushrooms and pepperoni)

27. continued
Let X = P(mushrooms and pepperoni)
Then $0.55 = 0.32 + 0.17 - X$
$X = 0.06$

28.
P(one or two car garage) =
$0.20 + 0.70 = 0.90$
Hence, P(no garage) $= 1 - 0.90 = 0.10$

29.
P(not a two-car garage) $= 1 - 0.70 = 0.30$

EXERCISE SET 4-4

1.
a. independent
b. dependent
c. dependent
d. dependent
e. independent
f. dependent
g. dependent
h. independent

2.
P(all 3 underweight) $= (0.18)^3 = 0.58\%$ or 0.6%

3.
P(two with elevated blood pressure) =
$(.68)^2 = 0.462$ or 46.2%

4.
P(all 4 used a seat belt) $= (.52)^4 = 7.3\%$

5.
P(5 are tippers) $= (0.73)^5 = 0.2073$

6.
P(two inmates are not citizens) $= (0.25)^2$
$= 0.0625$ or 6.3%

7.
(a) P(no computer) $= 1 - 0.543 = 0.457$
P(none of three has a computer) =
$(0.457)^3 = 0.0954$

(b) P(at least one has a computer) =
1 − P(none of three has a computer) =
$1 - 0.0954 = 0.9054$

(c) P(all three have computers) =
$(0.543)^3 = 0.1601$

8.
(a) P(both are spades) $= \frac{13}{52} \cdot \frac{12}{51} = \frac{1}{17}$

8. continued
(b) P(both are the same suit) =
$\frac{4}{4} \cdot \frac{12}{51} = \frac{4}{17}$

(c) P(both are kings) $= \frac{4}{52} \cdot \frac{3}{51} = \frac{1}{221}$

9.
P(all are citizens) $= (0.801)^3 = 0.5139$

10.
P(three agree to government's responsibility) $= (\frac{1}{2})^3 = \frac{1}{8}$

11.
P(all three have NFL apparel) =
$(0.31)^3 = 0.0298$

12.
P(both are defective) $= \frac{2}{6} \cdot \frac{1}{5} = \frac{1}{15}$

13.
P(no insurance) $= 0.12$
P(none are covered) $= (0.12)^4 = 0.0002$

14.
P(three murders with no weapon) $= (.06)^3 = 0.0002$ or 0.02%

15.
P(5 buy at least 1) $= (\frac{90}{120})^5 = \frac{243}{1024}$

16.
a. $\frac{4}{52} \cdot \frac{3}{51} \cdot \frac{2}{50} = \frac{1}{5525}$

b. $\frac{13}{52} \cdot \frac{12}{51} \cdot \frac{11}{50} = \frac{11}{850}$

c. $\frac{26}{52} \cdot \frac{25}{51} \cdot \frac{24}{50} = \frac{2}{17}$

17.
$\frac{5}{8} \cdot \frac{4}{7} \cdot \frac{3}{6} = \frac{5}{28}$

18.
$\frac{3}{8} \cdot \frac{2}{7} \cdot \frac{1}{6} = \frac{1}{56}$

19.
$\frac{18}{30} \cdot \frac{17}{29} = \frac{51}{145}$

20.
$\frac{12}{30} \cdot \frac{11}{29} = \frac{22}{145}$

21.

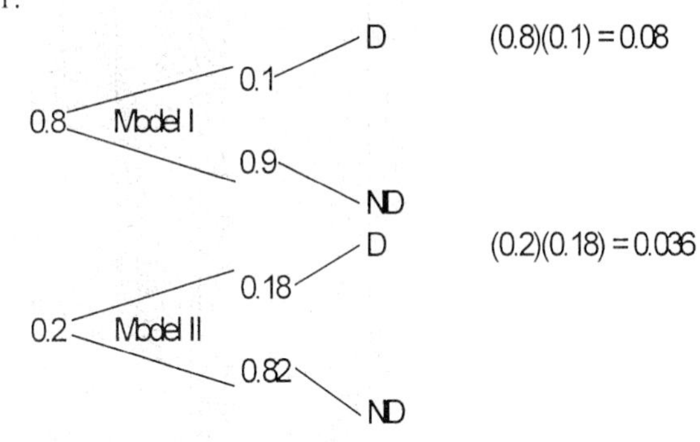

P(defective) $= 0.08 + 0.036 = 0.116$

22.

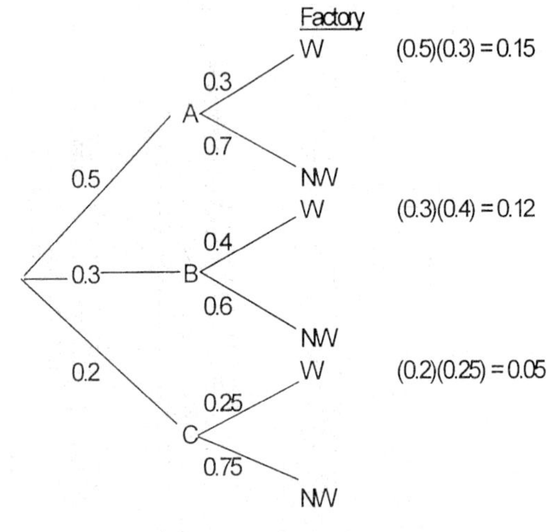

P(white) $= 0.15 + 0.12 + 0.05 = 0.32$

23.

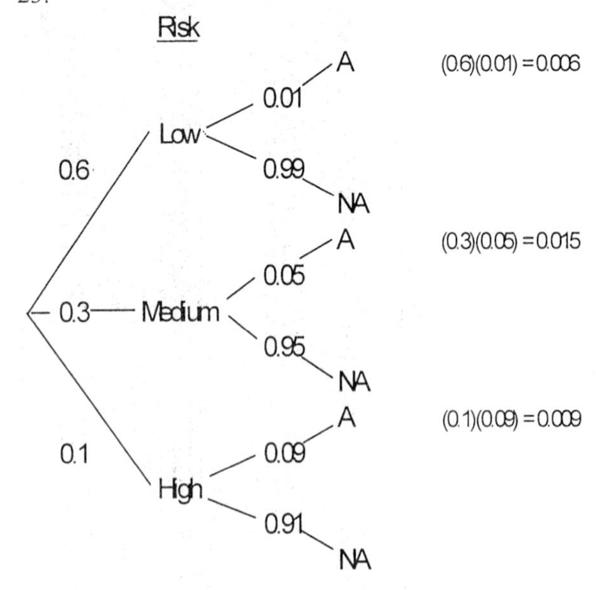

P(accident) $= .006 + .015 + .009 = 0.03$

24.
P(over \$100,000) $= (0.25)(0.05) + (0.75)$
$= 0.0275$ or 0.028

25.
P(red ball) $= \frac{1}{3} \cdot \frac{5}{8} + \frac{1}{3} \cdot \frac{3}{4} + \frac{1}{3} \cdot \frac{4}{6} = \frac{49}{72}$

26.
P(physics | sociology) $= \frac{0.092}{0.73} = 0.126$

27.
P(auto will be found within one week | it's been stolen) $= \frac{\text{P(stolen and found within 1 week)}}{\text{P(stolen)}}$
$= \frac{0.0009}{0.0015} = 0.6$

28.
P(2nd defective | 1st defective) =

$\frac{\text{P(1st and 2nd defective)}}{\text{P(1st defective)}} = \frac{\frac{2}{8} \cdot \frac{1}{7}}{\frac{2}{8}} = \frac{1}{7}$

29.
P(swim | bridge) $= \frac{\text{P(play bridge and swim)}}{\text{P(play bridge)}}$

$= \frac{0.73}{0.82} = 0.89$ or 89%

30.
P(calculus | dean's list) $= \frac{0.042}{0.21} = 0.2$

31.
P(garage | deck) $= \frac{0.42}{0.60} = 0.7$ or 70%

32.
P(salad | pizza) $= \frac{0.65}{0.95} = 0.684$ or 68.4%

33.
P(champagne | bridge) $= \frac{0.68}{0.83} = 0.82$ or 82%

34.

Class	Favor	Oppose	No Opinion	Total
Fr	15	27	8	50
Soph	23	5	2	30
Total	38	32	10	80

a. $\frac{27}{80} \div \frac{50}{80} = \frac{27}{50}$

b. $\frac{23}{80} \div \frac{38}{80} = \frac{23}{38}$

35.
(a) P(foreign patent | corporation) =

$\frac{\text{P(corporation and foreign patent)}}{\text{P(corporation)}} =$

35a. continued

$\frac{\frac{63,182}{147,497}}{\frac{134,076}{147,497}} = \frac{63,182}{134,076} = 0.4712$

(b) $P(\text{individual} \mid \text{U. S.}) = \frac{P(\text{U. S. \& individual})}{P(\text{U. S.})}$

$\frac{\frac{6129}{147,497}}{\frac{77,944}{147,497}} = \frac{6129}{77,944} = 0.0786$

36.

(a) $P(\text{gold} \mid \text{U. S.}) = \frac{P(\text{U. S. and gold})}{P(\text{U. S.})}$

$\frac{\frac{39}{928}}{\frac{97}{928}} = \frac{39}{97} = 0.4021$

(b) $P(\text{U. S.} \mid \text{gold}) = \frac{P(\text{gold and U. S.})}{P(\text{gold})}$

$\frac{\frac{39}{928}}{\frac{301}{928}} = \frac{39}{301} = 0.1296$

(c) No, because $P(\text{gold} \mid \text{U. S.}) \neq P(\text{gold})$.

37.

(a) P(none have been married) $= (0.703)^5 = 0.1717$

(b) P(at least one has been married) =
1 − P(none have been married)
$= 1 - 0.1717$
$= 0.8283$

38.

(a) P(all three caused by driver error) $= (0.54)^3 = 0.157$

(b) P(none caused by driver error) $= (0.46)^3 = 0.097$

(c) P(at least one caused by driver error) = 1 − P(none by driver error)
$= 1 - 0.0973 = 0.9027$

39.

P(at least one not immunized) = 1 − P(none of the six are not immunized)
= 1 − P(all six are immunized)
$= 1 - (0.76)^6 = 0.8073$

40.

P(at least one defective) =
1 − P(none defective) $= 1 - \frac{15}{18} \cdot \frac{14}{17} = \frac{16}{51}$

41.

If P(read to) = 0.58, then
P(not being read to) $= 1 - 0.58 = 0.42$

41. continued

P(at least one is read to) = 1 − P(none are read to)
= 1 − P(all five are not read to)
$= 1 - (0.42)^5 = 0.9869$

42.

(a) P(all three have assistantships) $= (0.6)^3 = 0.216$

(b) P (none have assistantships) $= (0.4)^3 = 0.064$

(c) P(at least one has an assistantship) = 1 − (none have assistantships)
$= 1 - 0.064 = 0.936$

43.

P(at least one club) = 1 − P(no clubs)

$1 - \frac{39}{52} \cdot \frac{38}{51} \cdot \frac{37}{50} \cdot \frac{36}{49} = 1 - \frac{6327}{20,825}$
$= \frac{14,498}{20,825}$

44.

P(at least one child) = 1 − P(no children)

$1 - \frac{13}{16} \cdot \frac{12}{15} \cdot \frac{11}{14} = 1 - \frac{143}{280} = \frac{137}{280}$

45.

P(at least one defective) = 1 − P(no defective) $= 1 - (.94)^5 = 0.266$ or 26.6%

46.

P(at least one will not improve) = 1 − P(all will improve) $= 1 - (0.75)^{12}$
$= 0.968$ or 96.8%

47.

P(at least one tail) = 1 − P(no tails)

$1 - (\frac{1}{2})^6 = 1 - \frac{1}{64} = \frac{63}{64}$

48.

P(at least one 7) = 1 − P(no 7's)

$1 - (\frac{9}{10})^3 = 1 - \frac{729}{1000} = \frac{271}{1000}$ or 0.271

49.

P(at least one 6) = 1 − P(no 6's)

$1 - (\frac{5}{6})^5 = 1 - \frac{3125}{7776} = \frac{4651}{7776}$

50.

P(at least one science) = 1 − P(no science)

$1 - \frac{7}{12} \cdot \frac{6}{11} \cdot \frac{5}{10} \cdot \frac{4}{9} = 1 - \frac{7}{99} = \frac{92}{99}$

51.
P(at least one even) = 1 − P(no evens)
$1 - (\frac{1}{2})^3 = 1 - \frac{1}{8} = \frac{7}{8}$

52.
P(at least one full professor) = 1 −
P(no full professors) = $1 - \frac{23}{30} \cdot \frac{22}{29} \cdot \frac{21}{28} \cdot \frac{20}{27}$
$1 - \frac{253}{783} = \frac{530}{783}$

53.
No, because $P(A \cap B) = 0$ therefore
$P(A \cap B) \neq P(A) \cdot P(B)$

54.
If independent, then P(compact | domestic) = P(compact)

$P(\text{compact}) = \frac{150}{300} = \frac{1}{2}$
$P(\text{compact} \mid \text{domestic}) = \frac{P(\text{domestic and compact})}{P(\text{domestic})}$
$= \frac{\frac{100}{300}}{\frac{210}{300}} = \frac{100}{210}$ or $\frac{10}{21}$

Thus, P(compact | domestic) ≠ P(compact) since $\frac{1}{2} \neq \frac{10}{21}$.

55.
P(enroll) = 0.55

P(enroll | DW) > P(enroll) which indicates that DW has a positive effect on enrollment.

P(enroll | LP) = P(enroll) which indicates that LP has no effect on enrollment.

P(enroll | MH) < P(enroll) which indicates that MH has a detrimental effect on enrollment.

Thus, all students should meet with DW.

56.
P(buy) = 0.35

If P(buy | ad) = 0.20, then the commercial adversely effects the probability of buying since the events are dependent and the probability that a person buys the product is less than 0.35.

If P(buy | ad) = 0.35, then the commercial has no effect on buying the product.

If P(buy | ad) = 0.55, then the commercial has an effect on buying the product.

EXERCISE SET 4-5

1.
$10^5 = 100{,}000$
$10 \cdot 9 \cdot 8 \cdot 7 \cdot 6 = 30{,}240$

2.
$9! = 9 \cdot 8 \cdot 7 \cdot 6 \cdot 5 \cdot 4 \cdot 3 \cdot 2 \cdot 1 = 362{,}880$

3.
$7! = 7 \cdot 6 \cdot 5 \cdot 4 \cdot 3 \cdot 2 \cdot 1 = 5040$

4.
$6! = 6 \cdot 5 \cdot 4 \cdot 3 \cdot 2 \cdot 1 = 720$

5.
$8! = 8 \cdot 7 \cdot 6 \cdot 5 \cdot 4 \cdot 3 \cdot 2 \cdot 1 = 40{,}320$

6.
$8 \cdot 6 \cdot 3 = 144$

7.
$5! = 5 \cdot 4 \cdot 3 \cdot 2 \cdot 1 = 120$

8.
$2 \cdot 25 \cdot 24 \cdot 23 = 27{,}600$
$2 \cdot 26 \cdot 26 \cdot 26 = 35{,}152$

9.
$10 \cdot 10 \cdot 10 = 1000$
$1 \cdot 9 \cdot 8 = 72$

10.
$9! = 9 \cdot 8 \cdot 7 \cdot 6 \cdot 5 \cdot 4 \cdot 3 \cdot 2 \cdot 1 = 362{,}880$

11.
$5 \cdot 2 = 10$

12.
$2 \cdot 4 = 8$

13.
(a) $8! = 8 \cdot 7 \cdot 6 \cdot 5 \cdot 4 \cdot 3 \cdot 2 \cdot 1 = 40{,}320$

(b) $10! = 10 \cdot 9 \cdot 8 \cdot 7 \cdot 6 \cdot 5 \cdot 4 \cdot 3 \cdot 2 \cdot 1$
$10! = 3{,}628{,}800$

(c) $0! = 1$

(d) $1! = 1$

(e) ${}_7P_5 = \frac{7!}{(7-5)!}$

$= \frac{7 \cdot 6 \cdot 5 \cdot 4 \cdot 3 \cdot 2 \cdot 1}{2 \cdot 1} = 2520$

13. continued

(f) ${}_{12}P_4 = \frac{12!}{(12-4)!}$

$= \frac{12 \cdot 11 \cdot 10 \cdot 9 \cdot 8 \cdot 7 \cdot 6 \cdot 5 \cdot 4 \cdot 3 \cdot 2 \cdot 1}{8 \cdot 7 \cdot 6 \cdot 5 \cdot 4 \cdot 3 \cdot 2 \cdot 1} = 11,880$

(g) ${}_5P_3 = \frac{5!}{(5-3)!}$

$= \frac{5 \cdot 4 \cdot 3 \cdot 2 \cdot 1}{2 \cdot 1} = 60$

(h) ${}_6P_0 = \frac{6!}{(6-0)!}$

$= \frac{6 \cdot 5 \cdot 4 \cdot 3 \cdot 2 \cdot 1}{6 \cdot 5 \cdot 4 \cdot 3 \cdot 2 \cdot 1} = 1$

(i) ${}_5P_5 = \frac{5!}{(5-5)!}$

$= \frac{5 \cdot 4 \cdot 3 \cdot 2 \cdot 1}{0!} = 120$

(j) ${}_6P_2 = \frac{6!}{(6-2)!}$

$= \frac{6 \cdot 5 \cdot 4 \cdot 3 \cdot 2 \cdot 1}{4 \cdot 3 \cdot 2 \cdot 1} = 30$

14.

${}_7P_7 = \frac{7!}{(7-7)!} = \frac{7!}{0!} = 5040$

15.

${}_4P_4 = \frac{4!}{(4-4)!} = \frac{4 \cdot 3 \cdot 2 \cdot 1}{0!} = 24$

16.

${}_7P_3 = \frac{7!}{(7-3)!} = \frac{7 \cdot 6 \cdot 5 \cdot 4!}{4!} = 210$

17.

${}_6P_3 = \frac{6!}{(6-3)!} = \frac{6!}{3!} = 120$

18.

${}_6P_6 = \frac{6!}{(6-6)!} = \frac{6!}{0!} = 720$

19.

${}_7P_4 = \frac{7!}{(7-4)!} = \frac{7 \cdot 6 \cdot 5 \cdot 4 \cdot 3 \cdot 2 \cdot 1}{3 \cdot 2 \cdot 1} = 840$

20.

${}_8P_3 = \frac{8!}{(8-3)!} = \frac{8 \cdot 7 \cdot 6 \cdot 5 \cdot 4 \cdot 3 \cdot 2 \cdot 1}{5 \cdot 4 \cdot 3 \cdot 2 \cdot 1} = 336$

21.

${}_{10}P_6 = \frac{10!}{(10-6)!} = \frac{10 \cdot 9 \cdot 8 \cdot 7 \cdot 6 \cdot 5 \cdot 4 \cdot 3 \cdot 2 \cdot 1}{4 \cdot 3 \cdot 2 \cdot 1} = 151,200$

22.

${}_5P_5 = \frac{5!}{(5-5)!} = \frac{5!}{0!} = 120$

23.

${}_{50}P_4 = \frac{50!}{(50-4)!} = \frac{50!}{46!} = 5,527,200$

24.

${}_{20}P_5 = \frac{20!}{(20-5)!} = \frac{20!}{15!}$

$= \frac{20 \cdot 19 \cdot 18 \cdot 17 \cdot 16 \cdot 15!}{15!} = 1,860,480$

25.

${}_5P_3 + {}_5P_4 + {}_5P_5 = \frac{5!}{2!} + \frac{5!}{1!} + \frac{5!}{0!}$

$= 60 + 120 + 120 = 300$

26.

${}_7P_5 = \frac{7!}{(7-5)!} = \frac{7!}{2!} = \frac{7 \cdot 6 \cdot 5 \cdot 4 \cdot 3 \cdot 2!}{2!} = 2520$

27.

a. $\frac{5!}{3!\,2!} = 10$

b. $\frac{8!}{5!\,3!} = 56$

c. $\frac{7!}{3!\,4!} = 35$

d. $\frac{6!}{4!\,2!} = 15$

e. $\frac{6!}{2!\,4!} = 15$

f. $\frac{3!}{3!\,0!} = 1$

g. $\frac{3!}{0!\,3!} = 1$

h. $\frac{9!}{2!\,7!} = 36$

i. $\frac{12!}{10!\,2!} = 66$

j. $\frac{4!}{1!\,3!} = 4$

28.

${}_{52}C_3 = \frac{52!}{49!\,3!} = \frac{52 \cdot 51 \cdot 50 \cdot 49!}{49! \cdot 3 \cdot 2 \cdot 1} = 22,100$

29.

${}_{10}C_3 = \frac{10!}{7!\,3!} = \frac{10 \cdot 9 \cdot 8 \cdot 7!}{7! \cdot 3 \cdot 2 \cdot 1} = 120$

30.

${}_{12}C_4 \cdot {}_9C_3 = \frac{12!}{8!\,4!} \cdot \frac{9!}{6!\,3!}$

$= \frac{12 \cdot 11 \cdot 10 \cdot 9 \cdot 8!}{8! \cdot 4 \cdot 3 \cdot 2 \cdot 1} \cdot \frac{9 \cdot 8 \cdot 7 \cdot 6!}{6! \cdot 3 \cdot 2 \cdot 1} = 41,580$

31.

${}_{10}C_4 = \frac{10!}{6!\,4!} = 210$

32.

${}_{10}C_3 = \frac{10!}{7!\,3!} = \frac{10 \cdot 9 \cdot 8 \cdot 7!}{7!\,3 \cdot 2 \cdot 1} = 120$

33.

${}_{20}C_5 = \frac{20!}{15!\,5!} = 15,504$

34.

${}_{11}C_6 = \frac{11!}{5!\,6!} = \frac{11 \cdot 10 \cdot 9 \cdot 8 \cdot 7 \cdot 6!}{6! \cdot 5 \cdot 4 \cdot 3 \cdot 2 \cdot 1} = 462$

35.

${}_8C_3 \cdot {}_{11}C_4 = 56 \cdot 330 = 18,480$

36.

$_4C_2 \cdot {_{12}C_5} \cdot {_7C_3} = \frac{4!}{2!\,2!} \cdot \frac{12!}{7!\,5!} \cdot \frac{7!}{4!\,3!}$

$= \frac{4\cdot3\cdot2!}{2!\cdot2\cdot1} \cdot \frac{12\cdot11\cdot10\cdot9\cdot8\cdot7!}{7!\cdot5\cdot4\cdot3\cdot2\cdot1} \cdot \frac{7\cdot6\cdot5\cdot4!}{4!\cdot3\cdot2\cdot1}$

$= 6 \cdot 792 \cdot 35 = 166{,}320$

37.

$_{12}C_4 = 495$

$_7C_2 \cdot {_5C_2} = 21 \cdot 10 = 210$

$_7C_2 \cdot {_5C_2} + {_7C_3} \cdot {_5C_1} + {_7C_4} =$

$21 \cdot 10 + 35 \cdot 5 + 35 =$

$210 + 175 + 35 = 420$

38.

$_{10}C_3 \cdot {_{10}C_3} = \frac{10!}{7!\,3!} \cdot \frac{10!}{7!\,3!}$

$= \frac{10\cdot9\cdot8\cdot7!}{7!\cdot3\cdot2\cdot1} \cdot \frac{10\cdot9\cdot8\cdot7!}{7!\cdot3\cdot2\cdot1} = 120 \cdot 120 = 14{,}400$

39.

$_6C_3 \cdot {_5C_2} = \frac{6!}{3!\,3!} \cdot \frac{5!}{3!\,2!}$

$= \frac{6\cdot5\cdot4\cdot3!}{3!\cdot3\cdot2\cdot1} \cdot \frac{5\cdot4\cdot3!}{3!\cdot2\cdot1} = 200$

40.

$_{12}C_6 \cdot {_{10}C_6} = \frac{12!}{6!\,6!} \cdot \frac{10!}{4!\,6!}$

$= \frac{12\cdot11\cdot10\cdot9\cdot8\cdot7\cdot6!}{6!\cdot6\cdot5\cdot4\cdot3\cdot2\cdot1} \cdot \frac{10\cdot9\cdot8\cdot7\cdot6!}{6!\cdot4\cdot3\cdot2\cdot1}$

$= 924 \cdot 210 = 194{,}040$

41.

$_{10}C_2 \cdot {_{12}C_2} = \frac{10!}{8!\,2!} \cdot \frac{12!}{10!\,2!}$

$= 45 \cdot 66 = 2{,}970$

42.

$_{25}C_5 = \frac{25!}{20!\,5!} = \frac{25\cdot24\cdot23\cdot22\cdot21\cdot20!}{20!\cdot5\cdot4\cdot3\cdot2\cdot1}$

$= 53{,}130$

43.

$_{17}C_2 = \frac{17!}{15!\,2!} = 136$

44.

$_9C_5 = \frac{9!}{4!\,5!} = \frac{9\cdot8\cdot7\cdot6\cdot5!}{5!\cdot4\cdot3\cdot2\cdot1} = 126$

45.

$_{11}C_7 = \frac{11!}{4!\,7!} = \frac{11\cdot10\cdot9\cdot8\cdot7!}{7!\cdot4\cdot3\cdot2\cdot1} = 330$

46.

$_{10}C_8 = \frac{10!}{2!\,8!} = \frac{10\cdot9\cdot8!}{8!\cdot2\cdot1} = 45$

47.

$_{20}C_8 = \frac{20!}{12!\,8!} = \frac{20\cdot19\cdot18\cdot17\cdot16\cdot15\cdot14\cdot13\cdot12!}{12!\cdot8\cdot7\cdot6\cdot5\cdot4\cdot3\cdot2\cdot1}$

$= 125{,}970$

48.

$_{17}C_8 = \frac{17!}{9!\,8!} = \frac{17\cdot16\cdot15\cdot14\cdot13\cdot12\cdot11\cdot10\cdot9!}{9!\cdot8\cdot7\cdot6\cdot5\cdot4\cdot3\cdot2\cdot1}$

$= 24{,}310$

49.

Selecting 1 coin there are 4 ways. Selecting 2 coins there are 6 ways. Selecting 3 coins there are 4 ways. Selecting 4 coins there is 1 way. Hence the total is $4 + 6 + 4 + 1 = 15$ ways. (List all possibilities.)

50.

X = number of chickens

2X = number of chicken legs

15 − X = number of cows

4(15 − X) = number of cow legs

$2X + 4(15 - X) = 46$

$2X + 60 - 4X = 46$

$-2X = -14$

$X = 7$

There are 7 chickens and 8 cows.

51.

a. $2 \cdot 4 \cdot 3 \cdot 2 \cdot 1 = 48$

b. $4 \cdot 6 + 3 \cdot 6 + 2 \cdot 6 + 1 \cdot 6 = 60$

c. $5! - 48 = 72$

52.

a. 4

b. $_4C_1 \cdot {_9C_1} \cdot 1 \cdot 1 \cdot 1 \cdot 1 = 36$

c. $48 \cdot 13 = 624$

d. $13 \cdot 12 \cdot {_4C_3} \cdot {_4C_2} = 3744$

EXERCISE SET 4-6

1.

P(2 face cards) $= \frac{12}{52} \cdot \frac{11}{51} = \frac{11}{221}$

2.

a. $\frac{_5C_4}{_{25}C_4} = \frac{\frac{5\cdot4!}{4!\cdot1}}{\frac{25\cdot24\cdot23\cdot22\cdot21!}{21!\cdot4\cdot3\cdot2\cdot1}} = \frac{1}{2530}$

b. $\frac{_5C_2\cdot{_{20}C_2}}{_{25}C_4} = \frac{\frac{5\cdot4\cdot3!}{3!\cdot2\cdot1}\cdot\frac{20\cdot19\cdot18!}{18!\cdot2\cdot1}}{25\cdot23\cdot22} = \frac{38}{253}$

c. $\frac{_{20}C_4}{_{25}C_4} = \frac{\frac{20\cdot19\cdot18\cdot17\cdot16!}{16!\cdot4\cdot3\cdot2\cdot1}}{25\cdot23\cdot22} = \frac{969}{2530}$

d. $\frac{_5C_1\cdot{_{20}C_3}}{_{25}C_4} = \frac{\frac{5\cdot4!}{4!\cdot1}\cdot\frac{20\cdot19\cdot18\cdot17!}{17!\cdot3\cdot2\cdot1}}{25\cdot23\cdot22} = \frac{114}{253}$

3.

a. There are $_4C_3$ ways of selecting 3 women and $_7C_3$ total ways to select 3 people; hence, P(all women) $= \frac{_4C_3}{_7C_3} = \frac{4}{35}$.

3. continued
b. There are $_3C_3$ ways of selecting 3 men; hence, P(all men) $= \frac{_3C_3}{_7C_3} = \frac{1}{35}$.

c. There are $_3C_2$ ways of selecting 2 men and $_4C_1$ ways of selecting one woman; hence, P(2 men and 1 woman) $= \frac{_3C_2 \cdot _4C_1}{_7C_3}$ $= \frac{12}{35}$.

d. There are $_3C_1$ ways to select one man and $_4C_2$ ways of selecting two women; hence, P(1 man and 2 women) $= \frac{_3C_1 \cdot _4C_2}{_7C_3}$ $= \frac{18}{35}$.

4. There are $_{49}C_3$ways to select 3 Republicans; hence, P(3 Republicans) = $\frac{_{49}C_3}{_{100}C_3} = \frac{18{,}424}{161{,}700} = 0.1139$

There are $_{50}C_3$ ways to select 3 Democrats; hence P(3 Democrats) $= \frac{_{50}C_3}{_{100}C_3}$
P(3 Democrats) $= \frac{19{,}600}{161{,}700} = 0.1212$

There are $_{49}C_1$ ways to select one Republican, $_1C_1$ ways to select one Independent, and $_{50}C_1$ ways to select one Democrat; hence P(one from each party) $= \frac{_{49}C_1 \cdot _1C_1 \cdot _{50}C_1}{_{100}C_3} = \frac{2450}{161{,}700} = 0.0152$

5.
(a) There are $_9C_4$ ways to select four from Pennsylvania; hence P(all four are from Pennsylvania) $= \frac{_9C_4}{_{56}C_4} = \frac{126}{367{,}290} = 0.0003$

(b) There are $_9C_2$ ways to select two from Pennsylvania and $_7C_2$ ways to select two from Virginia; hence P(two from Pennsylvania and two from Virginia) $= \frac{_9C_2 \cdot _7C_2}{_{56}C_4} = \frac{756}{367{,}290} = 0.0021$

6.
a. $\frac{_9C_4}{_{12}C_4} = \frac{126}{495} = \frac{14}{55}$

b. $\frac{_3C_1 \cdot _9C_3}{_{12}C_4} = \frac{252}{495} = \frac{28}{55}$

c. $\frac{_3C_3 \cdot _9C_1}{_{12}C_4} = \frac{1}{55}$

7.
$\frac{2}{50} \cdot \frac{1}{49} = \frac{1}{1225}$

8.
There are $_4C_3$ ways of getting 3 of a kind for

8. continued
one denomination and there are 13 denominations. There are $_4C_2$ways of getting two of a kind and 12 denominations left. There are $_{52}C_5$ ways to get five cards; hence,
P(full house) $= \frac{13 \cdot _4C_3 \cdot 12 \cdot _4C_2}{_{52}C_5} = \frac{6}{4165}$

9.
a. $\frac{_8C_4}{_{14}C_4} = \frac{70}{1001} = \frac{10}{143}$

b. $\frac{_6C_2 \cdot _8C_2}{_{14}C_4} = \frac{420}{1001} = \frac{60}{143}$

c. $\frac{_6C_4}{_{14}C_4} = \frac{15}{1001}$

d. $\frac{_6C_3 \cdot _8C_1}{_{14}C_4} = \frac{160}{1001}$

e. $\frac{_6C_1 \cdot _8C_3}{_{14}C_4} = \frac{336}{1001} = \frac{48}{143}$

10.
a. $\frac{_8C_3}{_{15}C_3} = \frac{56}{455} = \frac{8}{65}$

b. $\frac{_2C_2 \cdot _{13}C_1}{_{15}C_3} = \frac{13}{455} = \frac{1}{35}$

c. $\frac{_5C_3}{_{15}C_3} = \frac{10}{455} = \frac{2}{91}$

d. $\frac{_8C_1 \cdot _5C_1 \cdot _2C_1}{_{15}C_3} = \frac{80}{455} = \frac{16}{91}$

e. $\frac{_8C_2 \cdot _5C_1}{_{15}C_3} = \frac{140}{455} = \frac{4}{13}$

11.
(a) $\frac{_{11}C_2}{_{19}C_2} = \frac{55}{171} = 0.3216$

(b) $\frac{_8C_2}{_{19}C_2} = \frac{28}{171} = 0.1637$

(c) $\frac{_{11}C_1 \cdot _8C_1}{_{19}C_2} = \frac{88}{171} = 0.5146$

(d) It probably got lost in the wash!

12.
$\frac{_8C_3 \cdot _9C_4}{_{17}C_7} = \frac{56 \cdot 126}{19{,}448} = \frac{7056}{19{,}448} = \frac{882}{2431}$

13.
There are $6^3 = 216$ ways of tossing three dice, and there are 15 ways of getting a sum of 7; i.e., (1, 1, 5), (1, 5, 1), (5, 1, 1), (1, 2, 4), etc. Hence the probability of rolling a sum of 7 is $\frac{15}{216} = \frac{5}{72}$.

14.
$\frac{_4C_2 \cdot _8C_3}{_{12}C_5} = \frac{336}{792} = \frac{14}{33}$

15.
There are $5! = 120$ ways to arrange 5 washers in a row and 2 ways to have them in correct order, small to large or large to small; hence, the probability is $\frac{2}{120} = \frac{1}{60}$.

16.
There are ${}_{52}C_5 = \frac{52!}{47!\,5!} = 2{,}598{,}960$ possible hands.

a. $\frac{4}{2{,}598{,}960}$ b. $\frac{36}{2{,}598{,}960}$

c. $\frac{624}{2{,}598{,}960}$

REVIEW EXERCISES - CHAPTER 4

1.
a. $\frac{1}{6}$ b. $\frac{1}{6}$ c. $\frac{4}{6} = \frac{2}{3}$

2.
a. $\frac{13}{52} = \frac{1}{4}$ d. $\frac{4}{52} = \frac{1}{13}$

b. $\frac{11}{26}$ e. $\frac{26}{52} = \frac{1}{2}$

c. $\frac{1}{52}$

3.
$\frac{16}{45}$

4.
a. $\frac{1}{6}$ b. $\frac{3}{6} = \frac{1}{2}$ c. $\frac{2}{6} = \frac{1}{3}$

5.
$\frac{850}{1500} = \frac{17}{30}$

6.
a. $\frac{9}{35}$

b. $\frac{7}{35} + \frac{16}{35} = \frac{23}{35}$

c. $\frac{3}{35} + \frac{7}{35} + \frac{9}{35} = \frac{19}{35}$

d. $1 - \frac{16}{35} = \frac{19}{35}$

7.
a. $\frac{3}{30} = \frac{1}{10}$ c. $\frac{16+7+3}{30} = \frac{26}{30} = \frac{13}{15}$

b. $\frac{7+4}{30} = \frac{11}{30}$ d. $1 - \frac{4}{30} = \frac{26}{30} = \frac{13}{15}$

8.
Refer to the sample space for tossing two dice.

8. continued

a. There are 4 ways to roll a 5 and 5 ways to roll a 6; hence, $P(5 \text{ or } 6) = \frac{4}{36} + \frac{5}{36} = \frac{1}{4}$

b. There are 3 ways to get a 10, 2 ways to get an 11 and 1 way to get a 12; hence, $P(\text{sum greater than } 9) = \frac{3}{36} + \frac{2}{26} + \frac{1}{36} = \frac{1}{6}$

c. A sum less than 4 means 3 or 2, and greater than 9 means 10, 11, 12; hence, the probability is $\frac{2+1+3+2+1}{36} = \frac{9}{36} = \frac{1}{4}$.

d. Four, 8, and 12 are divisible by 4; hence, the probability of rolling a 4, 8, or 12 is $\frac{3+5+1}{36} = \frac{9}{36} = \frac{1}{4}$.

e. Since this is impossible, the answer is 0.

f. Since this is the entire sample space, the probability is $\frac{36}{36} = 1$.

9.
$0.80 + 0.30 - 0.12 = 0.98$

10.
P(John or Mary) =
P(John) + P(Mary) − P(John and Mary) =
$0.39 + 0.73 - 0.36 = 0.76$
The probability that neither purchases a new car is $1 - 0.76 = 0.24$.

11.
$(0.78)^5 = 0.289$ or 28.9%

12.
P(five borrowed books) $= (0.67)^5 = 0.1350$

P(none borrowed books) $= (0.33)^5 = 0.0039$

13.
a. $\frac{26}{52} \cdot \frac{25}{51} \cdot \frac{24}{50} = \frac{2}{17}$

b. $\frac{13}{52} \cdot \frac{12}{51} \cdot \frac{11}{50} = \frac{33}{2550} = \frac{11}{850}$

c. $\frac{4}{52} \cdot \frac{3}{51} \cdot \frac{2}{50} = \frac{1}{5525}$

14.
a. $\frac{1}{2} \cdot \frac{4}{52} = \frac{1}{26}$

b. $\frac{1}{2} \cdot \frac{26}{52} = \frac{1}{4}$

c. $\frac{1}{2} \cdot \frac{13}{52} = \frac{1}{8}$

15.
P(C or PP) = P(C) + P(PP) = $\frac{2+3}{13} = \frac{5}{13}$

16.

	X	Y	Z	Total
TV	18	32	15	65
Stereo	6	20	13	39
Total	24	52	28	104

a. $\frac{24}{104} + \frac{39}{104} - \frac{6}{104} = \frac{57}{104}$

b. $\frac{52}{104} + \frac{28}{104} = \frac{80}{104} = \frac{10}{13}$

c. $\frac{65}{104} + \frac{28}{104} - \frac{15}{104} = \frac{78}{104} = \frac{3}{4}$

17.

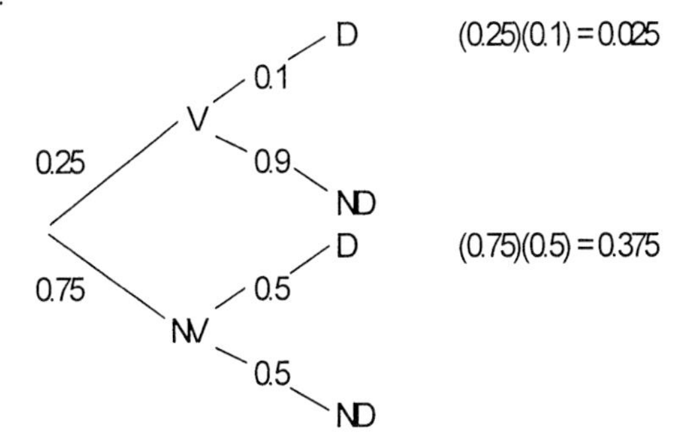

P(disease) = 0.025 + 0.375 = 0.4

18.

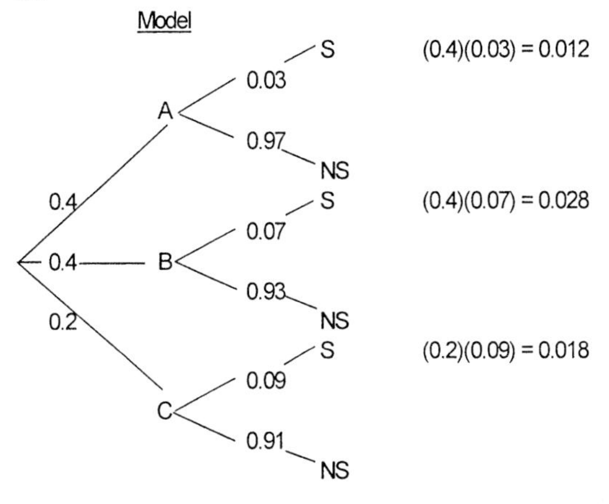

P(stereo) = 0.012 + 0.028 + 0.018 = 0.058 or 5.8%

19.
P(NC | C) = $\frac{\text{P(NC and C)}}{\text{P(C)}} = \frac{0.37}{0.73} = 0.51$

20.
P(warranty | TV) = $\frac{\text{P(warranty and TV)}}{\text{P(TV)}}$

20. continued
$= \frac{0.03}{0.11} = 0.273$

21.
$\frac{0.43}{0.75} = 0.573$ or 57.3%

22.
P(bus late | bad weather) =
$\frac{\text{P(bus late and bad weather)}}{\text{P(bad weather)}} = \frac{0.023}{0.40} = 0.058$

23.

	<4 yrs HS	HS	College	Total
Smoker	6	14	19	39
Non-Smoker	18	7	25	50
Total	24	21	44	89

a. There are 44 college graduates and 19 of them smoke; hence, the probability is $\frac{19}{44}$.

b. There are 24 people who did not graduate from high school, 6 of whom do not smoke; hence, the probability is $\frac{6}{24} = \frac{1}{4}$.

24.
P(at least one does not wear a helmet) =
1 − P(none do not wear a helmet)
= 1 − P(all 4 wear a helmet)
$= 1 - (0.23)^4 = 0.997$ or 99.7%

25.
P(at least one tail) = 1 − P(all heads)
$1 - (\frac{1}{2})^5 = 1 - \frac{1}{32} = \frac{31}{32}$

26.
P(at least one has chronic sinusitis) =
1 − P(none has chronic sinusitis)
$1 - (0.85)^5 = 0.556$ or 55.6%

27.
If repetitions are allowed:
$26 \cdot 26 \cdot 26 \cdot 10 \cdot 10 \cdot 10 = 175,760,000$

If repetitions are not allowed:
${}_{26}P_3 \cdot {}_{10}P_4 = \frac{26 \cdot 25 \cdot 24 \cdot 23!}{23!} \cdot \frac{10 \cdot 9 \cdot 8 \cdot 7 \cdot 6!}{6!}$
= 78,624,000

If repetitions are allowed in the letters but not in the digits:
$26 \cdot 26 \cdot 26 \cdot {}_{10}P_4 = 88,583,040$

28.
${}_5P_5 = \frac{5!}{(5-5)!} = \frac{5!}{0!} = 120$

29.
$_5C_3 \cdot {}_7C_4 = \frac{5!}{2!3!} \cdot \frac{7!}{3!4!} = 10 \cdot 35 = 350$

30.
$8! = {}_8P_8 = \frac{8!}{(8-8)!} = \frac{8!}{0!} = 40{,}320$

31.
$_{10}C_2 = \frac{10!}{8!2!} = 45$

32.
$_6C_3 \cdot {}_5C_2 \cdot {}_4C_1 = \frac{6!}{3!3!} \cdot \frac{5!}{3!2!} \cdot \frac{4!}{3!1!}$

$= 20 \cdot 10 \cdot 4 = 800$

33.
$26 \cdot 10 \cdot 10 \cdot 10 = 26{,}000$

34.
$5 \cdot 3 \cdot 2 = 30$

35.
$_{12}C_4 = \frac{12!}{8!4!} = \frac{12 \cdot 11 \cdot 10 \cdot 9 \cdot 8!}{4 \cdot 3 \cdot 2 \cdot 1 \cdot 8!} = 495$

36.
$_{13}C_3 = \frac{13!}{10!3!} = \frac{13 \cdot 12 \cdot 11 \cdot 10!}{10! \cdot 3 \cdot 2 \cdot 1} = 286$

37.
$_{20}C_5 = \frac{20!}{15!5!} = \frac{20 \cdot 19 \cdot 18 \cdot 17 \cdot 16 \cdot 15!}{15!5 \cdot 4 \cdot 3 \cdot 2 \cdot 1} = 15{,}504$

38.
$3 \cdot 5 \cdot 4 = 60$

39.
Total number of outcomes:
$26 \cdot 26 \cdot 26 \cdot 10 \cdot 10 \cdot 10 \cdot 10 = 175{,}760{,}000$

Total number of ways for USA followed by a number divisible by 5:
$1 \cdot 1 \cdot 1 \cdot 10 \cdot 10 \cdot 10 \cdot 2 = 2000$

Hence $P = \frac{2000}{175{,}760{,}000} = 0.000011$

40.
There are $_3C_2$ ways of attending two plays and $_5C_1$ ways of attending one movie, and a total of $_{10}C_3$ of attending 3 events; hence, the probability is:
$\frac{_3C_2 \cdot {}_5C_1}{_{10}C_3} = \frac{15}{120} = \frac{1}{8}$

41.
$\frac{_3C_1 \cdot {}_4C_1 \cdot {}_2C_1}{_9C_3} = \frac{2}{7}$

CHAPTER 4 QUIZ

1. False, subjective probability can be used when other types of probabilities cannot be found.
2. False, empirical probability uses frequency distributions.
3. True
4. False, P(A or B) = P(A) + P(B) − P(A and B)
5. False, the probabilities can be different.
6. False, complementary events cannot occur at the same time.
7. True
8. False, order does not matter in combinations.
9. b.
10. b. and d.
11. d.
12. b.
13. c.
14. b.
15. d.
16. b.
17. b.
18. sample space
19. zero and one
20. zero
21. one
22. mutually exclusive
23. a. $\frac{4}{52} = \frac{1}{13}$ c. $\frac{16}{52} = \frac{4}{13}$

    b. $\frac{4}{52} = \frac{1}{13}$
24. a. $\frac{13}{52} = \frac{1}{4}$ d. $\frac{4}{52} = \frac{1}{13}$

    b. $\frac{4+13-1}{52} = \frac{4}{13}$ e. $\frac{26}{52} = \frac{1}{2}$

    c. $\frac{1}{52}$
25. a. $\frac{12}{31}$ c. $\frac{27}{31}$

    b. $\frac{12}{31}$ d. $\frac{24}{31}$
26. a. $\frac{11}{36}$ d. $\frac{1}{3}$

    b. $\frac{5}{18}$ e. 0

    c. $\frac{11}{36}$ f. $\frac{11}{12}$
27. $0.75 + 0.25 - 0.16 = 0.84$
28. $(0.3)^5 = 0.002$

29. a. $\frac{26}{52} \cdot \frac{25}{51} \cdot \frac{24}{50} \cdot \frac{23}{49} \cdot \frac{22}{48} = \frac{253}{9996}$

b. $\frac{13}{52} \cdot \frac{12}{51} \cdot \frac{11}{50} \cdot \frac{10}{49} \cdot \frac{9}{48} = \frac{33}{66{,}640}$

c. 0

30. $\frac{0.35}{0.65} = 0.54$

31. $\frac{0.16}{0.3} = 0.53$

32. $\frac{0.57}{0.7} = 0.81$

33. $\frac{0.028}{0.5} = 0.056$

34. a. $\frac{1}{2}$ b. $\frac{3}{7}$

35. $1 - (0.45)^6 = 0.99$

36. $1 - (\frac{5}{6})^4 = 0.518$

37. $1 - (0.15)^6 = 0.9999886$

38. 2,646

39. 40,320

40. 1,365

41. 1,188,137,600; 710,424,000

42. 720

43. 33,554,432

44. 35

45. $\frac{1}{4}$

46. $\frac{3}{14}$

47. $\frac{12}{55}$

Note: Answers may vary due to rounding, TI-83's or computer programs.

EXERCISE SET 5-2

1.
A random variable is a variable whose values are determined by chance. Examples will vary.

2.
If the values a random variable can assume are countable, then the variable is called discrete; otherwise, it is called a continuous variable.

3.
The number of commercials a radio station plays during each hour.
The number of times a student uses his or her calculator during a mathematics exam.
The number of leaves on a specific type of tree.

4.
The weights of strawberries grown in a specific plot.
The heights of all seniors at a specific college.
The times it takes students to complete a mathematics exam.

5.
A probability distribution is a distribution which consists of the values a random variable can assume along with the corresponding probabilities of these values.

6.
Yes

7.
Yes

8.
No, probability values cannot be negative.

9.
Yes

10.
No, probability values cannot be greater than 1.

11.
No, probability values cannot be greater than 1.

12.
Continuous

13.
Discrete

14.
Discrete

15.
Continuous

16.
Continuous

17.
Discrete

18.
Continuous

19.

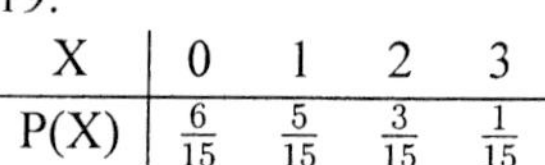

X	0	1	2	3
P(X)	$\frac{6}{15}$	$\frac{5}{15}$	$\frac{3}{15}$	$\frac{1}{15}$

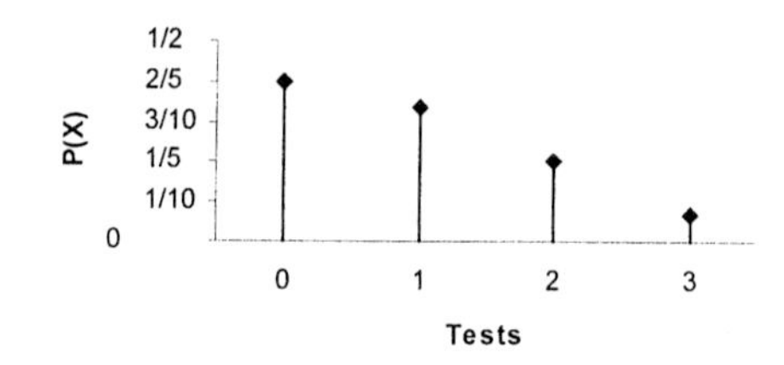

20.

X	\$1000	\$2000	\$3000
P(X)	$\frac{1}{2}$	$\frac{1}{4}$	$\frac{1}{4}$

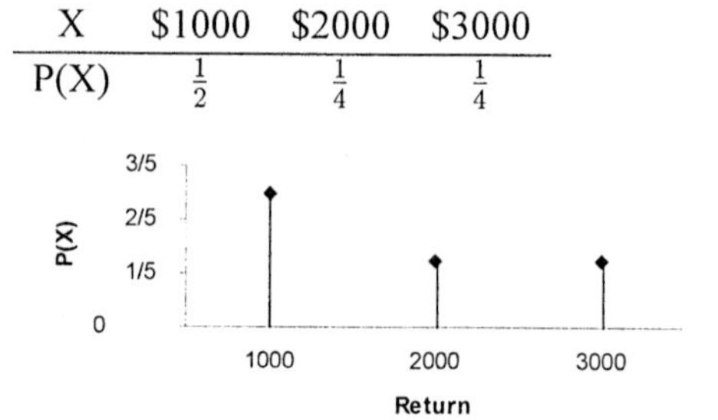

21.

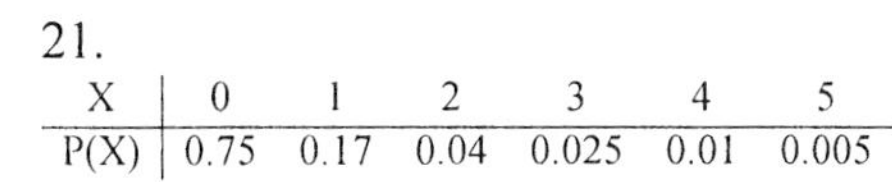

X	0	1	2	3	4	5
P(X)	0.75	0.17	0.04	0.025	0.01	0.005

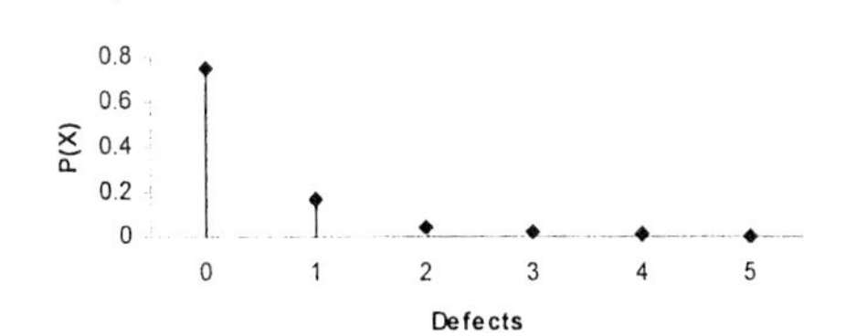

22.

X	0	1	2	3
P(X)	0.45	0.3	0.15	0.1

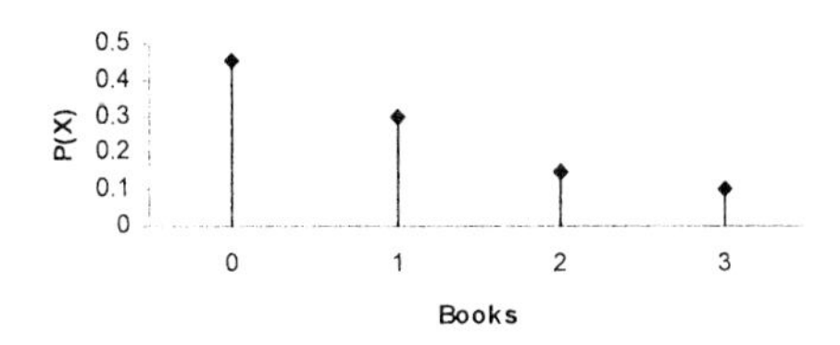

23.

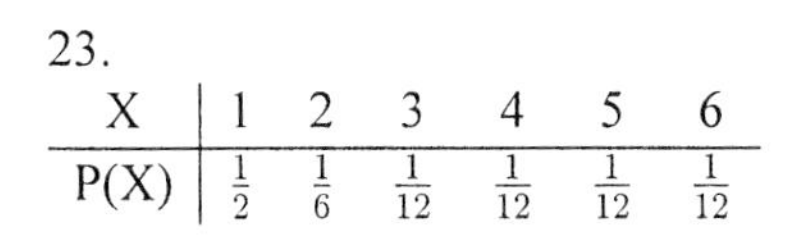

X	1	2	3	4	5	6
P(X)	$\frac{1}{2}$	$\frac{1}{6}$	$\frac{1}{12}$	$\frac{1}{12}$	$\frac{1}{12}$	$\frac{1}{12}$

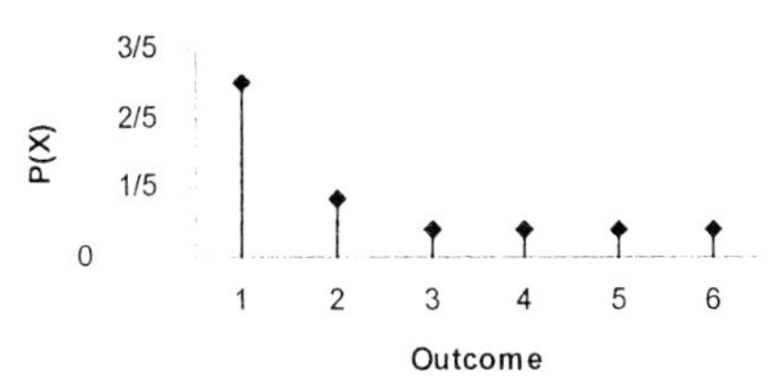

24.

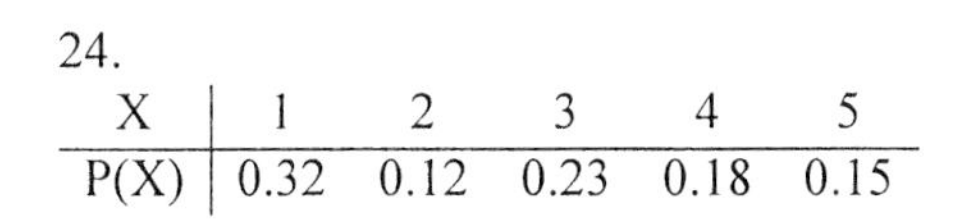

X	1	2	3	4	5
P(X)	0.32	0.12	0.23	0.18	0.15

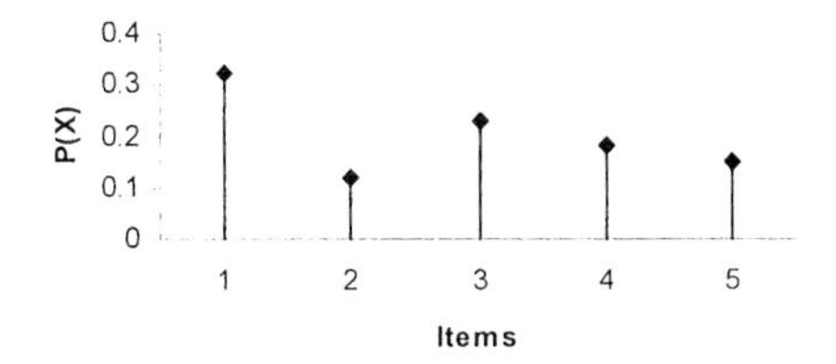

25.

X	1	2	3	4	5
P(X)	0.1	0.25	0.25	0.2	0.2

25. continued

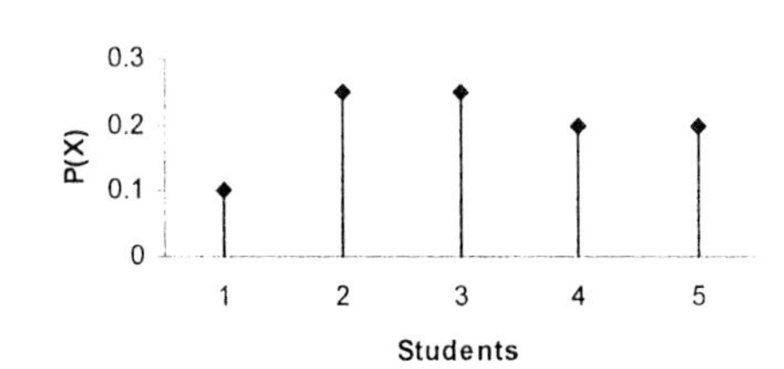

26.

X	0	1	2	3
P(X)	0.18	0.52	0.21	0.09

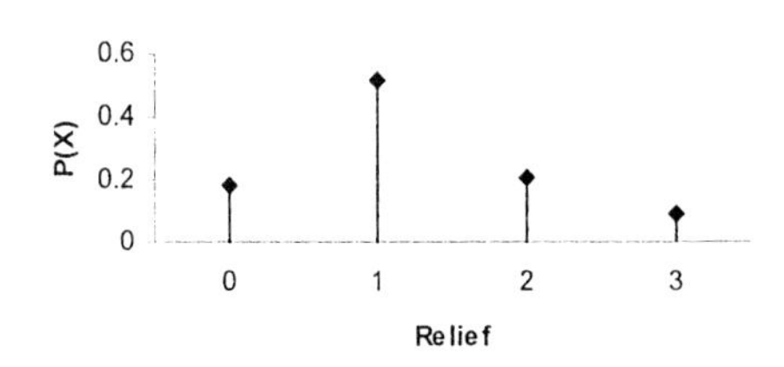

27.

X	\$1	\$5	\$10	\$20
P(X)	$\frac{3}{7}$	$\frac{2}{7}$	$\frac{1}{7}$	$\frac{1}{7}$

28.

X	0	1	2	3
P(X)	$\frac{1}{8}$	$\frac{3}{8}$	$\frac{3}{8}$	$\frac{1}{8}$

29.

X	1	2	3	4
P(X)	$\frac{1}{4}$	$\frac{1}{4}$	$\frac{3}{8}$	$\frac{1}{8}$

30.

X	2	3	4	5	6	7	8	9	10	11	12
P(X)	$\frac{1}{36}$	$\frac{1}{18}$	$\frac{1}{12}$	$\frac{1}{9}$	$\frac{5}{36}$	$\frac{1}{6}$	$\frac{5}{36}$	$\frac{1}{9}$	$\frac{1}{12}$	$\frac{1}{18}$	$\frac{1}{36}$

31.

X	1	2	3
P(X)	$\frac{1}{6}$	$\frac{1}{3}$	$\frac{1}{2}$

Yes.

32.

X	0.2	0.3	0.5
P(X)	0.2	0.3	0.5

Yes.

10. continued

X	P(X)	X · P(X)	$X^2 \cdot P(X)$
6	0.2	1.2	7.2
7	0.2	1.4	9.8
8	0.3	2.4	19.2
9	0.2	1.8	16.2
10	0.1	1.0	10.0
		$\mu = 7.8$	62.4

11.
$E(X) = \sum X \cdot P(X) = \$4995(\frac{1}{2500}) - \$5(\frac{2499}{2500}) = -\$3$

Alternate Solution:
$\$5000(\frac{1}{2500}) - \$5 = -\$3$

Yes, they will make $7500.

12.
$E(X) = \$1(\frac{10}{20}) + \$2(\frac{5}{20}) + \$5(\frac{3}{20}) + \$10(\frac{1}{20}) + \$100(\frac{1}{20}) - \$20.00 = -\$12.75$

Alternate Solution:
$E(X) = -19(\frac{10}{20}) - 18(\frac{5}{20}) - 15(\frac{3}{20}) - 10(\frac{1}{20}) + 80(\frac{1}{20}) = -\$12.75$

13.
$E(X) = \sum X \cdot P(X) = \$5.00(\frac{1}{6}) = \$0.83$
He should pay about $0.83.

14.
$E(X) = \$20.00(\frac{2}{36}) + \$5.00(\frac{6}{36}) - \$3.00$
$= 1.94 - \$3.00 = -\$1.06$

Alternate Solution:
$E(X) = 17.00(\frac{2}{36}) + 2.00(\frac{6}{36}) - 3.00(\frac{28}{36})$
$= -\$1.06$

15.
$E(X) = \sum X \cdot P(X) = \$1000(\frac{1}{1000}) + \$500(\frac{1}{1000}) + \$100(\frac{5}{1000}) - \$3.00$
$= -\$1.00$

Alternate Solution:
$E(X) = 997(\frac{1}{1000}) + 497(\frac{1}{1000}) + 97(\frac{5}{1000}) - 3(\frac{993}{1000}) = -\$1.00$

16.
$E(X) = 2(-1.00) = -\$2.00$

17.
$E(X) = \sum X \cdot P(X) = \$500(\frac{1}{1000}) - \$1.00$
$= -\$0.50$

17. continued
If 123 is boxed:
$E(X) = \$499(\frac{1}{1000}) - 1(\frac{999}{1000}) = -\$0.50$

There are 6 possibilities when a number with all different digits is boxed, $(3 \cdot 2 \cdot 1 = 6)$.
Hence,
$\$80.00 \cdot \frac{6}{1000} - \$1.00 = \$0.48 - \$1.00$
$= -\$0.52$

Alternate Solution:
$E(X) = 79(\frac{6}{1000}) - 1(\frac{994}{1000}) = -\$0.52$

18.
$\$940(0.028) - \$60.00(0.972) = \$26.32 - 58.32 = -\$32.00$

19.
The probabilities of each are:
Red: $\frac{18}{38}$ Black: $\frac{18}{38}$

1 – 18: $\frac{18}{38}$ 19 – 36: $\frac{18}{38}$

0: $\frac{1}{38}$ 00: $\frac{1}{38}$

Any single number: $\frac{1}{38}$

0 or 00: $\frac{2}{38}$

$E(X) = \sum X \cdot P(X)$

a. $\$1.00(\frac{18}{38}) - \$1.00(\frac{20}{38}) = -\$0.05$

b. $\$1.00(\frac{18}{38}) - \$1.00(\frac{20}{38}) = -\$0.05$

c. $\$35(\frac{1}{38}) - \$1.00(\frac{37}{38}) = -\$0.05$

d. $\$35(\frac{1}{38}) - \$1.00(\frac{37}{38}) = -\$0.05$

e. $\$17(\frac{2}{38}) - \$1.00(\frac{36}{38}) = -\$0.05$

20.

X	2	3	4	5	6	7	8	9	10	11	12
P(X)	$\frac{1}{36}$	$\frac{2}{36}$	$\frac{3}{36}$	$\frac{4}{36}$	$\frac{5}{36}$	$\frac{6}{36}$	$\frac{5}{36}$	$\frac{4}{36}$	$\frac{3}{36}$	$\frac{2}{36}$	$\frac{1}{36}$

$\mu = \sum X \cdot P(X) = 2(\frac{1}{36}) + 3(\frac{2}{36}) + 4(\frac{3}{36}) + 5(\frac{4}{36}) + 6(\frac{5}{36}) + 7(\frac{6}{36}) + 8(\frac{5}{36}) + 9(\frac{4}{36}) + 10(\frac{3}{36}) + 11(\frac{2}{36}) + 12(\frac{1}{36}) = 7$

20. continued
$\sigma^2 = \sum X^2 \cdot P(X) - \mu^2 = [2^2(\frac{1}{36}) + 3^2(\frac{2}{36}) + 4^2(\frac{3}{36}) + 5^2(\frac{4}{36}) + 6^2(\frac{5}{36}) + 7^2(\frac{6}{36}) + 8^2(\frac{5}{36}) + 9^2(\frac{4}{36}) + 10^2(\frac{3}{36}) + 11^2(\frac{2}{36}) + 12^2(\frac{1}{36})] - 7^2 = 5.83$ or 5.8

$\sigma = \sqrt{5.83} = 2.4$

21.
The expected value for a single die is 3.5, and since 3 die are rolled, the expected value is 3(3.5) = 10.5

22.
$\sigma^2 = \sum(X - \mu)^2 \cdot P(X)$
$\sigma^2 = \sum(X^2 - 2\mu X + \mu^2)P(X)$
$\sigma^2 = \sum[X^2P(X) - 2\mu XP(X) + \mu^2P(X)]$
$\sigma^2 = \sum x^2P(X) - 2\mu\sum XP(X) + \mu^2\sum P(X)$
$\sigma^2 = \sum X^2 \cdot P(X) - 2\mu \cdot \mu + \mu^2(1)$
$\sigma^2 = \sum X^2 \cdot P(X) - 2\mu^2 + \mu^2$
$\sigma^2 = \sum X^2 \cdot P(X) - \mu^2$

23.
Answers will vary.

24.
Answers will vary.

25.
Answers will vary.

26.
$E(X) = [\$100{,}000 \cdot \frac{1}{1{,}000{,}000} + 10{,}000 \cdot \frac{2}{50{,}000} + 1000 \cdot \frac{5}{10{,}000} + 100 \cdot \frac{10}{1000}] - \$0.30 =$
$\$0.10 + \$0.40 + \$0.50 + \$1.00 - \$0.37 = \$1.63$

EXERCISE SET 5-4

1.
a. Yes
b. Yes
c. Yes
d. No, there are more than two outcomes.
e. No
f. Yes
g. Yes
h. Yes
i. No, there are more than two outcomes.
j. Yes

2.
a. 0.420
b. 0.346

2. continued
c. 0.590
d. 0.251
e. 0.000
f. 0.250
g. 0.418
h. 0.176
i. 0.246

3.
a. $P(X) = \frac{n!}{(n-X)!\,X!} \cdot p^X \cdot q^{n-X}$

$P(X) = \frac{6!}{3! \cdot 3!} \cdot (0.03)^3(0.97)^3 = 0.0005$

b. $P(X) = \frac{4!}{2! \cdot 2!} \cdot (0.18)^2 \cdot (0.82)^2 = 0.131$

c. $P(X) = \frac{5!}{2! \cdot 3!} = (0.63)^3 \cdot (0.37)^2 = 0.342$

d. $P(X) = \frac{9!}{9! \cdot 0!} \cdot (0.42)^0 \cdot (0.58)^9 = 0.007$

e. $P(X) = \frac{10!}{5! \cdot 5!} \cdot (0.37)^5 \cdot (0.63)^5 = 0.173$

4.
a. $n = 6$, $p = 0.05$, $X = 3$
$P(X) = 0.002$
b. $n = 6$, $p = 0.05$, $X = 0, 1$
$P(X) = 0.735 + 0.232 = 0.967$
c. $n = 6$, $p = 0.05$, $X = 0$
$P(X) = 0.735$

5.
$n = 10$, $p = 0.5$, $X = 6, 7, 8, 9, 10$
$P(X) = 0.205 + 0.117 + 0.044 + 0.010 + 0.001 = 0.377$
No, because your score would be about 40%.

6.
$n = 10$, $p = 0.20$, $X = 6, 7, 8, 9, 10$
$P(X) = 0.006 + 0.001 + 0 + 0 + 0 = 0.007$

7.
$n = 9$, $p = 0.30$, $X = 3$
$P(X) = 0.267$

8.
$n = 15$, $p = 0.90$, $X = 9, 10, 11, \ldots, 15$
$P(X) = 1$

9.
$n = 7$, $p = 0.75$, $X = 0, 1, 2, 3$

9. continued
$P(X) = \frac{7!}{7!\,0!}(0.75)^0(0.25)^7 +$

$\frac{7!}{6!\,1!}(0.75)^1(0.25)^6 + \frac{7!}{5!\,2!}(0.75)^2(0.25)^5 +$

$\frac{7!}{4!\,3!}(0.75)^3(0.25)^4 = 0.071$

10.
n = 10, p = 0.6, X = 3
P(X) = 0.042

11.
n = 5, p = 0.40
a. X = 2, P(X) = 0.346
b. X = 0, 1, 2, or 3 people
P(X) = 0.078 + 0.259 + 0.346 + 0.230
= 0.913
c. X = 2, 3, 4, or 5 people
P(X) = 0.346 + 0.230 + 0.077 + 0.01
= 0.663
d. X = 0, 1, or 2 people
P(X) = 0.683

12.
a. n = 15, p = 0.3, X = 1, 2, 3, 4, 5, 6, 7
P(X) = 0.005 + 0.031 + 0.092 + 0.170 +
0.219 + 0.206 + 0.147 + 0.081 = 0.951
b. n = 15, p = 0.3, X = 7
P(X) = 0.081
c. n = 15, p = 0.3, X = 5, 6, 7, 8, 9, 10, 11
P(X) = 0.206 + 0.147 + 0.081 + 0.035 +
0.012 + 0.003 + 0.001 = 0.485

13.
a. n = 10, p = 0.2, X = 0, 1, 2, 3
P(X) = 0.107 + 0.268 + 0.302 + 0.201
= 0.878
b. n = 10, p = 0.2, X = 3, P(X) = 0.201
c. n = 10, p = 0.2, X = 5, 6, 7, 8, 9, 10
P(X) = 0.026 + 0.006 + 0.001 + 0 + 0 + 0
= 0.033

14.
a. $\mu = 100(0.75) = 75$
$\sigma^2 = 100(0.75)(0.25) = 18.75$ or 18.8
$\sigma = \sqrt{18.75} = 4.33$ or 4.3
b. $\mu = 300(0.3) = 90$
$\sigma^2 = 300(0.3)(0.7) = 63$
$\sigma = \sqrt{63} = 7.94$ or 7.9
c. $\mu = 20(0.5) = 10$
$\sigma^2 = 20(0.5)(0.5) = 5$
$\sigma = \sqrt{5} = 2.236$ or 2.2
d. $\mu = 10(0.8) = 8$
$\sigma^2 = 10(0.8)(0.2) = 1.6$

14d. continued
$\sigma = \sqrt{1.6} = 1.265$ or 1.3
e. $\mu = 1000(0.1) = 100$
$\sigma^2 = 1000(0.1)(0.9) = 90$
$\sigma = \sqrt{90} = 9.49$ or 9.5
f. $\mu = 500(0.25) = 125$
$\sigma^2 = 500(0.25)(0.75) = 93.75$
$\sigma = \sqrt{93.75} = 9.68$ or 9.7
g. $\mu = 50(\frac{2}{5}) = 20$
$\sigma^2 = 50(\frac{2}{5})(\frac{3}{5}) = 12$
$\sigma = \sqrt{12} = 3.464$ or 3.5
h. $\mu = 36(\frac{1}{6}) = 6$
$\sigma^2 = 36(\frac{1}{6})(\frac{5}{6}) = 5$
$\sigma = \sqrt{5} = 2.236$ or 2.2

15.
n = 800, p = 0.01
$\mu = 800(0.01) = 8$
$\sigma^2 = 800(0.01)(0.99) = 7.9$
$\sigma = \sqrt{7.92} = 2.8$

16.
n = 15, p = $\frac{1}{2}$
$\mu = 15(\frac{1}{2}) = 7.5$
$\sigma^2 = 15(\frac{1}{2})(\frac{1}{2}) = 3.75$ or 3.8
$\sigma = \sqrt{3.75} = 1.94$ or 1.9

17.
n = 500, p = 0.02
$\mu = 500(0.02) = 10$
$\sigma^2 = 500(0.02)(0.98) = 9.8$
$\sigma = \sqrt{9.8} = 3.1$

18.
n = 200, p = 0.83
$\mu = 200(0.83) = 166$
$\sigma^2 = 200(0.83)(0.17) = 28.2$
$\sigma = \sqrt{28.2} = 5.3$

19.
n = 1000, p = 0.21
$\mu = 1000(0.21) = 210$
$\sigma^2 = 1000(0.21)(0.79) = 165.9$
$\sigma = \sqrt{165.9} = 12.9$

20.
n = 80, p = 0.42
$\mu = 80(0.42) = 33.6$
$\sigma^2 = 80(0.42)(0.58) = 19.5$
$\sigma = \sqrt{19.5} = 4.4$

Using the mean as a guide, there should be about 34 seats.

21.
$n = 18, p = 0.25, X = 5$
$P(X) = \frac{18!}{13!\,5!}(0.25)^5(0.75)^{13} = 0.199$

22.
$n = 14, p = 0.63, X = 9$
$P(X) = \frac{14!}{5!\,9!}(0.63)^9(0.37)^5 = 0.217$

23.
$n = 10, p = \frac{1}{3}, X = 0, 1, 2, 3$
$P(X) = \frac{10!}{10!\,0!}(\frac{1}{3})^0(\frac{2}{3})^{10} + \frac{10!}{9!\,1!}(\frac{1}{3})^1(\frac{2}{3})^9$
$+ \frac{10!}{8!\,2!}(\frac{1}{3})^2(\frac{2}{3})^8 + \frac{10!}{7!\,3!}(\frac{1}{3})^3(\frac{2}{3})^7 = 0.559$

24.
$n = 20, p = 0.58, X = 12$
$P(X) = \frac{20!}{8!\,12!}(0.58)^{12}(0.42)^8 = 0.177$

25.
$n = 5, p = 0.13, X = 3, 4, 5$
$P(X) = \frac{5!}{2!\,3!}(0.13)^3(0.87)^2 +$
$\frac{5!}{1!\,4!}(0.13)^4(0.87)^1 + \frac{5!}{0!\,5!}(0.13)^5(0.87)^0$
$= 0.018$

26.
$n = 7, p = 0.14, X = 2$ or 3
$P(X) = \frac{7!}{5!\,2!}(0.14)^2(0.86)^5 +$
$\frac{7!}{4!\,3!}(0.14)^3(0.86)^4 = 0.246$

27.
$n = 12, p = 0.86, X = 10, 11, 12$
$P(X) = \frac{12!}{2!\,10!}(0.86)^{10}(0.14)^2 +$
$\frac{12!}{1!\,11!}(0.86)^{11}(0.14)^1 + \frac{12!}{0!\,12!}(0.86)^{12}(0.14)^0$
$= 0.77$

Yes. The probability is high, 77%.

28.

X	0	1	2	3
P(X)	0.125	0.375	0.375	0.125

29.
$n = 5, p = 0.2, X = 0, 1, 2, 3, 4, 5$

X	0	1	2	3	4	5
P(X)	0.328	0.410	0.205	0.051	0.006	0

29. continued

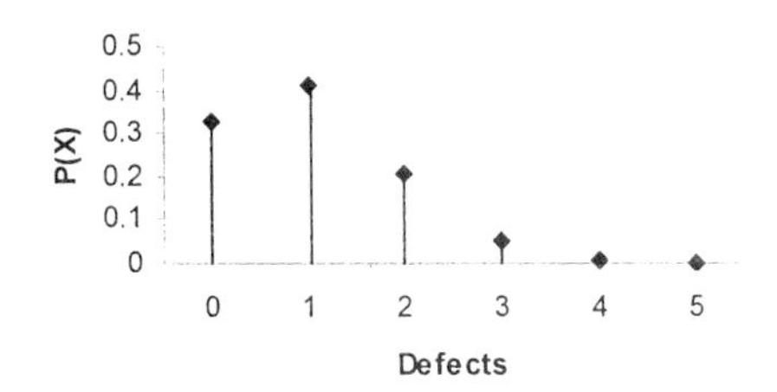

EXERCISE SET 5-5

1.
a. $P(M) = \frac{6!}{3!\,2!\,1!}(0.5)^3(0.3)^2(0.2)^1 = 0.135$

b. $P(M) = \frac{5!}{1!\,2!\,2!}(0.3)^1(0.6)^2(0.1)^2 = 0.0324$

c. $P(M) = \frac{4!}{1!\,1!\,2!}(0.8)^1(0.1)^1(0.1)^2 = 0.0096$

d. $P(M) = \frac{3!}{1!\,1!\,1!}(0.5)^1(0.3)^1(0.2)^1 = 0.18$

e. $P(M) = \frac{5!}{1!\,3!\,1!}(0.7)^1(0.2)^3(0.1)^1 = 0.0112$

2.
$P(M) = \frac{8!}{4!\,2!\,1!\,1!}(0.79)^4(0.12)^2(0.07)^1(0.02)^1$
$= 0.0066$

3.
$P(M) = \frac{8!}{3!\,2!\,3!}(0.25)^3(0.40)^2(0.35)^3 = 0.06$

4.
$P(M) = \frac{6!}{2!\,2!\,1!\,1!}(0.60)^2(0.25)^2(0.1)^1(0.05)^1$
$= 0.020$

5.
$P(M) = \frac{4!}{2!\,1!\,1!}(\frac{1}{6})^2(\frac{1}{6})^1(\frac{1}{6})^1 = \frac{1}{108}$

6.
$P(M) = \frac{8!}{1!\,3!\,3!\,1!}(\frac{9}{16})^1(\frac{3}{16})^3(\frac{3}{16})^3(\frac{1}{16})^1$
$= 0.0017$ or 0.002

7.
a. $P(5; 4) = 0.1563$
b. $P(2; 4) = 0.1465$
c. $P(6; 3) = 0.0504$
d. $P(10; 7) = 0.071$
e. $P(9; 8) = 0.1241$

8.
$\lambda = n \cdot p = (144)(0.02) = 2.88$
$P(3; 2.88) = \frac{e^{-2.88}(2.88)^3}{3!} = 0.2235$

9.
$p = \frac{1}{20,000} = 0.00005$
$\lambda = n \cdot p = 80,000(0.00005) = 4$
a. $P(0; 4) = 0.0183$
b. $P(1; 4) = 0.0733$
c. $P(2; 4) = 0.1465$
d. $P(3 \text{ or more}; 4) = 1 - [P(0; 4) + P(1; 4) + P(2; 4)]$
$= 1 - (0.0183 + 0.0733 + 0.1465)$
$= 0.7619$

10.
$\lambda = \frac{200}{400} = 0.5$
$P(1, 0.5) = 0.3033$

11.
$p = \frac{5}{1000} = \frac{1}{200}$
$\lambda = n \cdot p = (250) \cdot (\frac{1}{200}) = 1.25$
P(at least 2 orders) =
1 − [P(0 orders + P(1 order)]
$= 1 - [\frac{e^{-1.25}(1.25)^0}{0!} + \frac{e^{-1.25}(1.25)^1}{1!}]$

$= 1 - (0.2865 + 0.3581) = 0.3554$

12.
$p = \frac{5}{500} = 0.01$
$\lambda = n \cdot p = 100 \cdot (0.01) = 1$
$P(\text{at least } 2; 1) = 1 - [P(0; 1) + P(1; 1)]$
$= 1 - (0.3679 + 0.3679) = 0.2642$

13.
$\lambda = \frac{1}{1000} \cdot 3000 = 3$
$P(1 \text{ or more}; 3) = 0.1494 + 0.2240 + 0.2240 + 0.1680 + 0.1008 + 0.0504 + 0.0216 + 0.0081 + 0.0027 + 0.0008 + 0.0002 + 0.001 = 0.9502$

14.
$\lambda = 0.03(90) = 2.7$
$P(3{:}2.7) = 0.2205$

15.
$P(5; 4) = 0.1563$

16.
$p = 0.004$, $n = 150$, $\lambda = 0.004(150) = 0.6$
$P(5; 6) = 0.0004$

17.
a = 9 prefer hoods
b = 9 prefer hats
$X = 3, n = 6$
$P(A) = \frac{{}_9C_3 \cdot {}_9C_3}{{}_{18}C_6} = \frac{84}{221} = 0.38$

18.
P(at least 2 with defective pages) =
1 − [P(0 with defective pages) + P(1 with defective pages)]

$P(0) = \frac{{}_5C_0 \cdot {}_{20}C_5}{{}_{25}C_5} = \frac{2584}{8855} = 0.292$

$P(1) = \frac{{}_5C_1 \cdot {}_{20}C_4}{{}_{25}C_5} = \frac{1615}{3542} = 0.456$

P(at least 2 with defective pages) =
$1 - (0.292 + 0.456) = 0.252$

19.
$a = 15, b = 9, n = 4, X = 4$

$P(A) = \frac{{}_{15}C_4 \cdot {}_9C_0}{{}_{24}C_4} = \frac{65}{506} = 0.13$

20.
P(at least one defective) = 1 −
P(0 defective)
$a = 6, b = 18, n = 4, X = 0$

$P(0) = \frac{{}_6C_0 \cdot {}_{18}C_4}{{}_{24}C_4} = \frac{3060}{10,626} = 0.288$

P(at least one defective) =
$1 - 0.288 = 0.712$

21.
P(at least 1 defective) = 1 − P(0 defectives)
$a = 6, b = 18, n = 3, X = 0$

$P(0) = \frac{{}_6C_0 \cdot {}_{18}C_3}{{}_{24}C_3} = \frac{102}{253} = 0.403$
P(at least 1 defective) = $1 - 0.403 = 0.597$

REVIEW EXERCISES - CHAPTER 5

1.
No, the sum of the probabilities is greater than one.

2.
No, the sum of the probabilities is less than one.

3.
No, the sum of the probabilities is greater than one.

4.

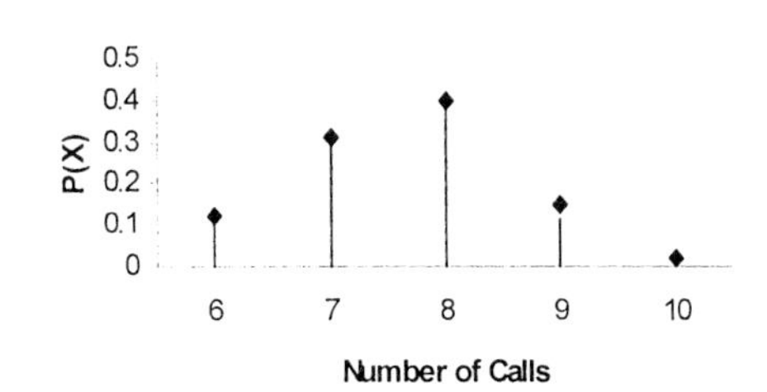

5.

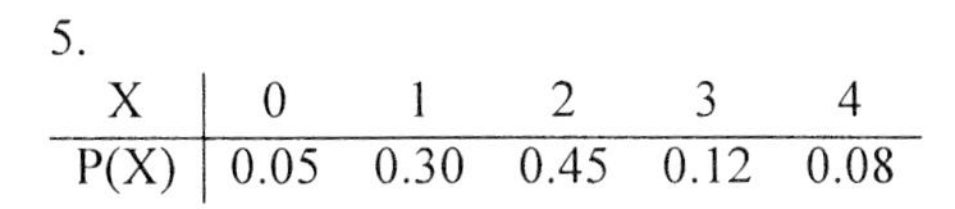

X	0	1	2	3	4
P(X)	0.05	0.30	0.45	0.12	0.08

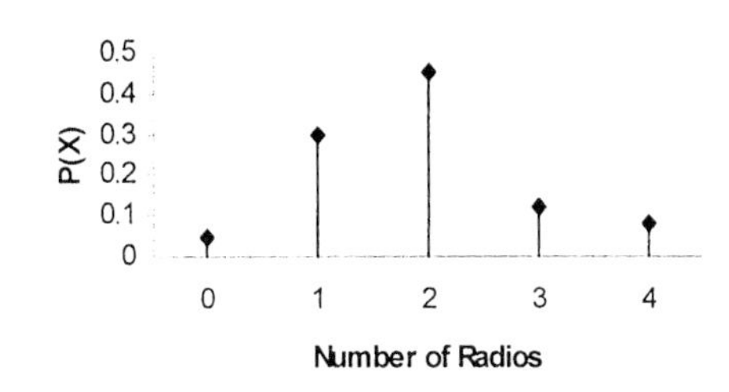

6.

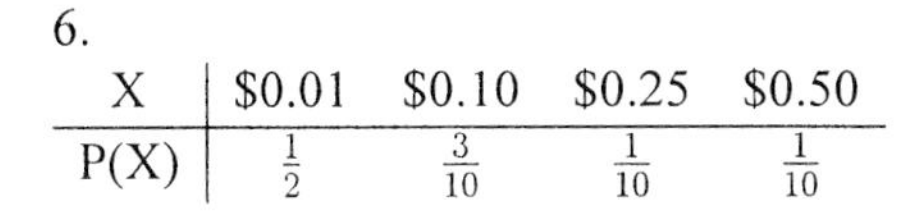

X	\$0.01	\$0.10	\$0.25	\$0.50
P(X)	$\frac{1}{2}$	$\frac{3}{10}$	$\frac{1}{10}$	$\frac{1}{10}$

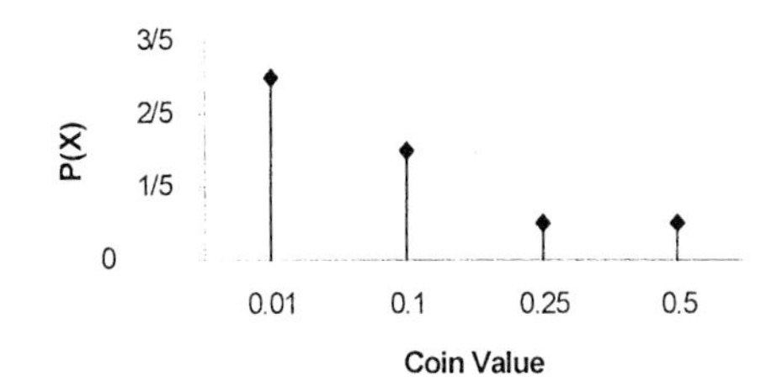

7.

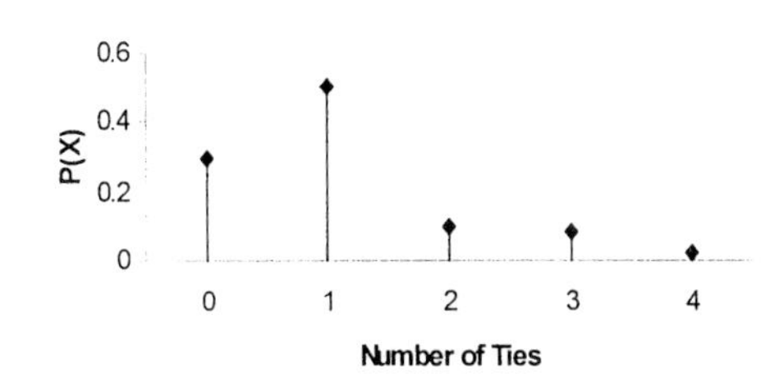

8.
$\mu = \sum X \cdot P(X) = 0(0.12) + 1(0.20) + 2(0.31) + 3(0.25) + 4(0.12) = 2.05$ or 2.1

8. continued
$\sigma^2 = \sum X^2 \cdot P(X) - \mu^2 = [0^2(0.12) + 1^2(0.2) + 2^2(0.31) + 3^2(0.25) + 4^2(0.12)] - 2.05^2 = 1.4075$ or 1.4

$\sigma = \sqrt{1.4075} = 1.186$ or 1.2

X	P(X)	$X \cdot P(X)$	$X^2 \cdot P(X)$
0	0.12	0.00	0.00
1	0.20	0.20	0.20
2	0.31	0.62	1.24
3	0.25	0.75	2.25
4	0.12	0.48	1.92
	$\mu =$	2.05	5.61

9.
$\mu = \sum X \cdot P(X) = 8(0.15) + 9(0.25) + 10(0.29) + 11(0.19) + 12(0.12) = 9.88$ or 9.9

$\sigma^2 = \sum X^2 \cdot P(X) - \mu^2 = [8^2(0.15) + 9^2(0.25) + 10^2(0.29) + 11^2(0.19) + 12^2(0.12)] - 9.88^2 = 1.5056$ or 1.5

$\sigma = \sqrt{1.5056} = 1.23$ or 1.2

X	P(X)	$X \cdot P(X)$	$X^2 \cdot P(X)$
8	0.15	1.20	9.60
9	0.25	2.25	20.25
10	0.29	2.90	29.00
11	0.19	2.09	22.99
12	0.12	1.44	17.28
	$\mu =$	9.88	99.12

10.
$\mu = \sum X \cdot P(X) = 1(0.42) + 2(0.27) + 3(0.15) + 4(0.1) + 5(0.06) = 2.11$ or 2.1

$\sigma^2 = \sum X^2 \cdot P(X) - \mu^2 = [1^2(0.42) + 2^2(0.27) + 3^2(0.15) + 4^2(0.1) + 5^2(0.06)] - 2.11^2 = 1.4979$ or 1.5

$\sigma = \sqrt{1.4979} = 1.22$ or 1.2

X	P(X)	$X \cdot P(X)$	$X^2 \cdot P(X)$
1	0.42	0.42	0.42
2	0.27	0.54	1.08
3	0.15	0.45	1.35
4	0.10	0.40	1.60
5	0.06	0.30	1.50
	$\mu =$	2.11	5.95

11.
$\mu = \sum X \cdot P(X) = 22(0.08) + 23(0.19) + 24(0.36) + 25(0.25) + 26(0.07) + 27(0.05) = 24.19$ or 24.2

$\sigma^2 = \sum X^2 \cdot P(X) - \mu^2 = [22^2(0.08) + 23^2(0.19) + 24^2(0.36) + 25^2(0.25) + 26^2(0.07) + 27^2(0.05)] - 24.19^2 = 1.4539$ or 1.5

$\sigma = \sqrt{1.4539} = 1.206$ or 1.2

X	P(X)	X · P(X)	X² · P(X)
22	0.08	1.76	38.72
23	0.19	4.37	100.51
24	0.36	8.64	207.36
25	0.25	6.25	156.25
26	0.07	1.82	47.32
27	0.05	1.35	36.45
	μ =	24.19	586.61

12.
$\mu = \sum X \cdot P(X)$
$= \frac{1}{5}(0.01) + \frac{1}{5}(0.05) + \frac{1}{5}(0.10) + \frac{1}{5}(0.25) + \frac{1}{5}(0.50) = \$0.182$ or $0.18

13.
$\mu = \sum X \cdot P(X)$
$= \frac{1}{2}(\$1.00) + \frac{18}{52}(\$5.00) + \frac{6}{52}(\$10.00) + \frac{2}{52}(\$100.00) = \$7.23$

To break even, a person should bet $7.23.

14.
$n = 10, p = 0.3, X = 5$
$P(X) = 0.103$

15.
a. 0.122
b. $1 - 0.002 + 0.009 = 0.989$
c. $0.002 + 0.009 + 0.032 = 0.043$

16.
a. In this case, it is easier to compute the probabilities for X = 0 and X = 1 and subtract the result from one.
$n = 15, p = 0.10, X > 2$
$P(X) = 1 - (0.206 + 0.343) = 0.451$
b. $n = 15, p = 0.10, X = 3$
$P(X) = 0.129$
c. $n = 15, p = 0.10, X < 4$
$P(X) = 0.206 + 0.343 + 0.267 + 0.129 + 0.043 = 0.988$

17.
$\mu = n \cdot p = 180(0.75) = 135$
$\sigma^2 = n \cdot p \cdot q = 180(0.75)(0.25) = 33.75$ or 33.8
$\sigma = \sqrt{33.75} = 5.809$ or 5.8

18.
$n = 50, p = 0.1$
$\mu = n \cdot p = 50(0.1) = 5$
$\sigma^2 = n \cdot p \cdot q = 50(0.1)(0.9) = 4.5$
$\sigma = \sqrt{4.5} = 2.12$ or 2.1

19.
$n = 8, p = 0.25$
$P(X \le 3) = \frac{8!}{8!\,0!}(0.25)^0(0.75)^8 + \frac{8!}{7!\,1!}(0.25)^1(0.75)^7 + \frac{8!}{6!\,2!}(0.25)^2(0.75)^6 + \frac{8!}{5!\,3!}(0.25)^3(0.75)^5 = 0.8862$ or 0.886

20.
$N = 500, p = 0.27$
$\mu = 500(0.27) = 135$
$\sigma^2 = 500(0.27)(0.73) = 98.55$ or 98.6
$\sigma = \sqrt{98.55} = 9.9$

21.
$n = 20, p = 0.75, X = 16$
P(16 have eaten pizza for breakfast) =

$\frac{20!}{4!\,16!}(0.75)^{16}(0.25)^4 = 0.1897$ or 0.190

22.
$n = 10, p = 0.25, X = 0, 1, 2, 3$
P(at most 3 have pizza for breakfast) =

$\frac{10!}{10!\,0!}(0.25)^0(0.75)^{10} + \frac{10!}{9!\,1!}(0.25)^1(0.75)^9 +$

$\frac{10!}{8!\,2!}(0.25)^2(0.75)^8 + \frac{10!}{7!\,3!}(0.25)^3(0.75)^7 =$

$= 0.776$

23.
$P(M) = \frac{20!}{12!\,4!\,3!\,1!}(0.7)^{12}(0.2)^4(0.08)^3(0.02)^1 = 0.008$

24.
$P(M) = \frac{12!}{8!\,3!\,1!}(0.9)^8(0.06)^3(0.04)^1 = 0.007$

25.
$P(M) = \frac{10!}{5!\,3!\,2!}(0.50)^5(0.40)^3(0.10)^2 = 0.050$

26.
$\lambda = n \cdot p = 0.04(100) = 4$
$P(8; 4) = 0.0298$ or 0.030

27.
a. P(6 or more; 6) = 1 − P(5 or less; 6)
= 1 − (0.0025 + 0.0149 + 0.0446 + 0.0892 + 0.1339 + 0.1606) = 0.5543
b. P(4 or more; 6) = 1 − P(3 or less; 6)
= 1 − (0.0025 + 0.0149 + 0.0446 + 0.0892) = 0.8488
c. P(5 or less; 6) = P(0; 6) + ... + P(6; 6)
= 0.4457

28.
$\lambda = n \cdot p = 1000(0.003) = 3$
P(6; 3) = 0.0504

29.
a = 13, b = 39, n = 5, X = 2
$P(2) = \frac{{}_{13}C_2 \cdot {}_{39}C_3}{{}_{52}C_5} = \frac{9,139}{33,320} = 0.27$

30.
a = 10, b = 40, n = 5, X = 2

$P(2) = \frac{{}_{10}C_2 \cdot {}_{40}C_3}{{}_{50}C_5} = \frac{22,230}{105,938} = 0.21$

31.
a. a = 7, b = 5
n = 3, X = 2 men, 1 woman
$P(2) = \frac{{}_7C_2 \cdot {}_5C_1}{{}_{12}C_3} = \frac{21}{44}$ or 0.477

b. a = 7, b = 5
n = 3, X = 0 men, 3 women

$P(0) = \frac{{}_7C_0 \cdot {}_5C_3}{{}_{12}C_3} = \frac{1}{22}$ or 0.045

c. a = 7, b = 5
n = 3, X = 1 man, 2 women

$P(1) = \frac{{}_7C_1 \cdot {}_5C_2}{{}_{12}C_3} = \frac{7}{22}$ or 0.318

CHAPTER 5 QUIZ

1. True
2. False, it is a discrete random variable.
3. False, the outcomes must be independent.
4. True
5. chance
6. $\mu = n \cdot p$
7. one
8. c.
9. c.
10. d.
11. No, the sum of the probabilities is greater than one.
12. Yes
13. Yes
14. Yes
15.

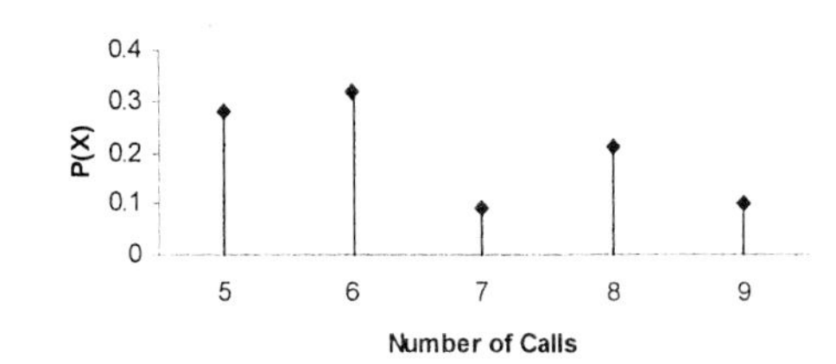

16.

X	0	1	2	3	4
P(X)	0.02	0.30	0.48	0.13	0.07

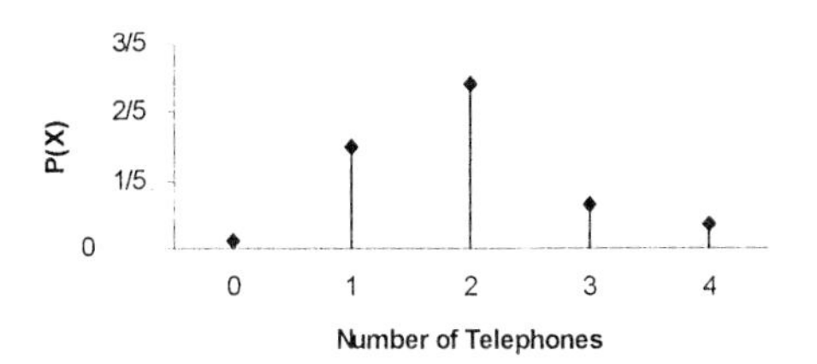

17.
$\mu = 0(0.10) + 1(0.23) + 2(0.31) + 3(0.27) + 4(0.09) = 2.02$
$\sigma^2 = [0^2(0.10) + 1^2(0.23) + 2^2(0.31) + 3^2(0.27) + 4^2(0.09)] - 2.02^2 = 1.3$
$\sigma = \sqrt{1.3} = 1.1$

18.
$\mu = 30(0.05) + 31(0.21) + 32(0.38) + 33(0.25) + 34(0.11) = 32.16$ or 32.2
$\sigma^2 = [30^2(0.05) + 31^2(0.21) + 32^2(0.38) + 33^2(0.25) + 34^2(0.11)] - 32.16^2 = 1.07$ or 1.1
$\sigma = \sqrt{1.07} = 1.0$

19.
$\mu = 4(\frac{1}{6}) + 5(\frac{1}{6}) + 2(\frac{1}{6}) + 10(\frac{1}{6}) + 3(\frac{1}{6}) + 7(\frac{1}{6}) = 5.17$ or 5.2

20.
$\mu = \$2(\frac{1}{2}) + \$10(\frac{5}{26}) + \$25(\frac{3}{26}) + \$100(\frac{1}{26}) = \$9.65$

21.
n = 20, p = 0.40, X = 5
P(5) = 0.124

22.
n = 20, p = 0.60
a. P(15) = 0.075
b. P(10, 11, ..., 20) = 0.117
c. P(0, 1, 2, 3, 4, 5) = 0.125

23.
n = 300, p = 0.80
$\mu = 300(0.80) = 240$
$\sigma^2 = 300(0.80)(0.20) = 48$
$\sigma = \sqrt{48} = 6.9$

24.
n = 75, p = 0.12
$\mu = 75(0.12) = 9$
$\sigma^2 = 75(0.12)(0.88) = 7.9$
$\sigma = \sqrt{7.9} = 2.8$

25.
$P(M) = \frac{30!}{15!\,8!\,5!\,2!}(0.5)^{15}(0.3)^8(0.15)^5(0.05)^2$

$= 0.0080$

26.
$P(M) = \frac{16!}{9!\,4!\,3!}(0.88)^9(0.08)^4(0.04)^3$

$= 0.0003$

27.
$P(M) = \frac{12!}{5!\,4!\,3!}(0.45)^5(0.35)^4(0.2)^3$

$= 0.061$

28.
$\lambda = 100(0.08) = 8,\ X = 6$
P(6; 8) = 0.122

29.
$\lambda = 8$
a. $P(X \geq 8; 8) = 0.1396 + \ldots + 0.0001$
$= 0.5471$
b. $P(X \geq 3; 8) = 1 - P(0, 1, \text{or } 2 \text{ calls})$
$= 1 - (0.0003 + 0.0027 + 0.0107)$
$= 1 - 0.0137 = 0.9863$
c. $P(X \leq 7; 8) = 0.0003 + \ldots + 0.1396$
$= 0.4529$

30.
a = 12, b = 36, n = 6, X = 3

$P(A) = \frac{{}_{12}C_3 \cdot {}_{36}C_3}{{}_{48}C_6} = \frac{\frac{12!}{9!\,3!} \cdot \frac{36!}{33!\,3!}}{\frac{48!}{42!\,6!}} = 0.128$

31.
a. $\frac{{}_6C_3 \cdot {}_8C_1}{{}_{14}C_4} = \frac{\frac{6!}{3!\,3!} \cdot \frac{8!}{7!\,1!}}{\frac{14!}{10!\,4!}} = 0.16$

b. $\frac{{}_6C_2 \cdot {}_8C_2}{{}_{14}C_4} = \frac{\frac{6!}{4!\,2!} \cdot \frac{8!}{6!\,2!}}{\frac{14!}{10!\,4!}} = 0.42$

c. $\frac{{}_6C_0 \cdot {}_8C_4}{{}_{14}C_4} = \frac{\frac{6!}{6!\,0!} \cdot \frac{8!}{4!\,4!}}{\frac{14!}{10!\,4!}} = 0.07$

Note to instructors: Graphs are not to scale and are intended to convey a general idea.

Answers are generated using Table E. Answers generated using the TI-83 will vary slightly.

EXERCISE SET 6-3

1.
The characteristics of the normal distribution are:
1. It is bell-shaped.
2. It is symmetric about the mean.
3. The mean, median, and mode are equal.
4. It is continuous.
5. It never touches the X-axis.
6. The area under the curve is equal to one.
7. It is unimodal.

2.
Many variables are normally distributed, and the distribution can be used to describe these variables.

3.
One or 100%.

4.
50% of the area lies below the mean, and 50% lies above the mean.

5.
68%, 95%, 99.7%

6.
The area is found by looking up z = 1.97 in Table E as shown in Block 1 of Procedure Table 6. Area = 0.4756

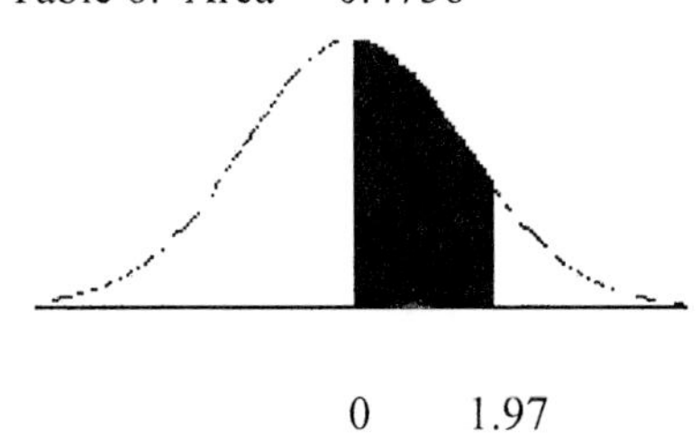

0 1.97

7.
The area is found by looking up z = 0.56 in Table E as shown in Block 1 of Procedure Table 6. Area = 0.2123

7. continued

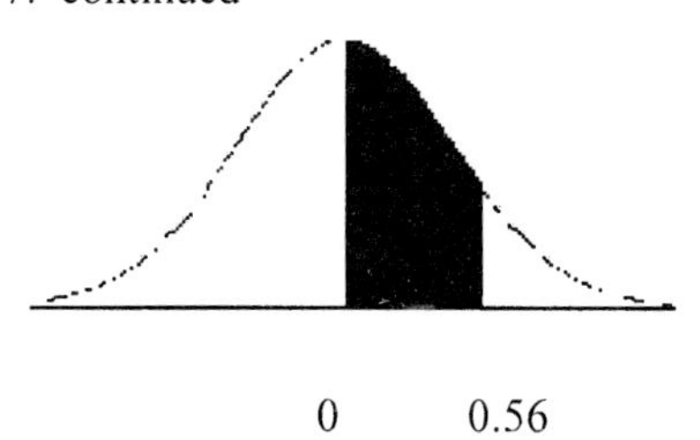

0 0.56

8.
The area is found by looking up z = 0.48 in Table E as shown in Block 1 of Procedure Table 6. Area = 0.1844

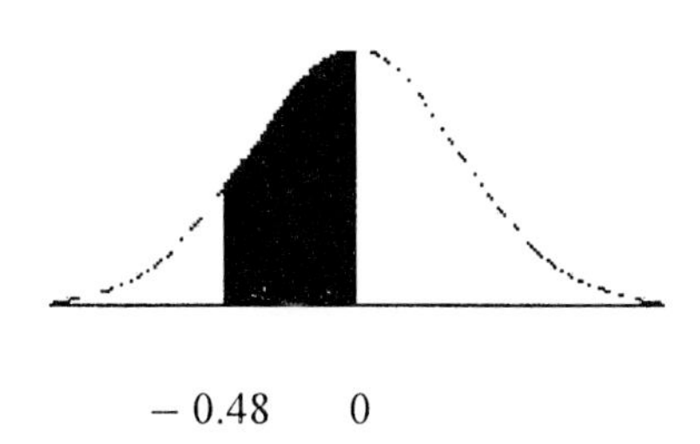

− 0.48 0

9.
The area is found by looking up z = 2.07 in Table E as shown in Block 1 of Procedure Table 6. Area = 0.4808

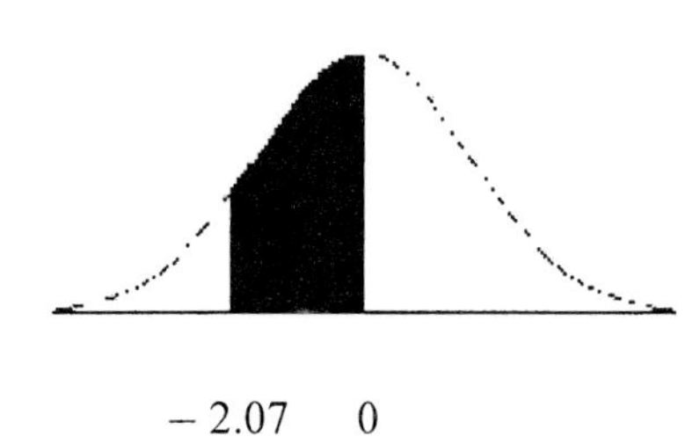

− 2.07 0

10.
The area is found by looking up z = 1.09 in Table E and subtracting the value from 0.5.
0.5 − 0.3621 = 0.1379

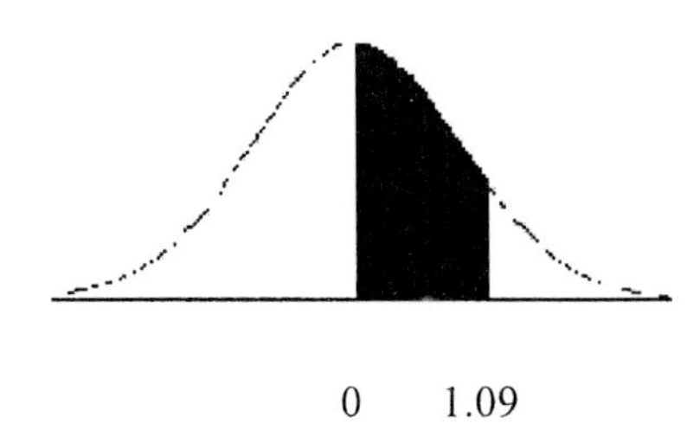

0 1.09

11.
The area is found by looking up z = 0.23 in Table E and subtracting it from 0.5 as shown in Block 2 of Procedure Table 6.
0.5 – 0.0910 = 0.4090

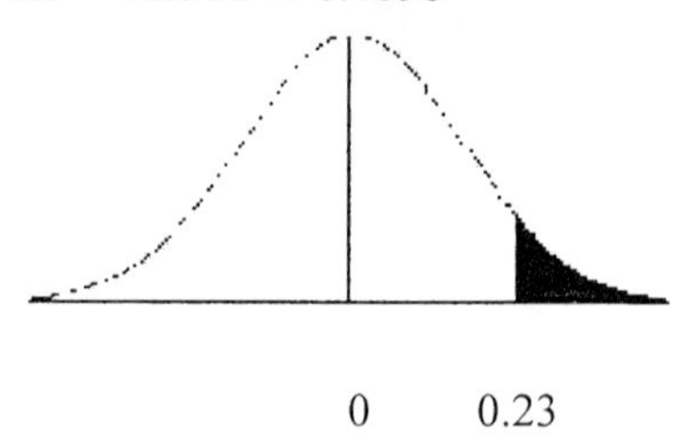

12.
The area is found by looking up z = 0.64 in Table E and subtracting the area from 0.5.
0.5 – 0.2389 = 0.2611

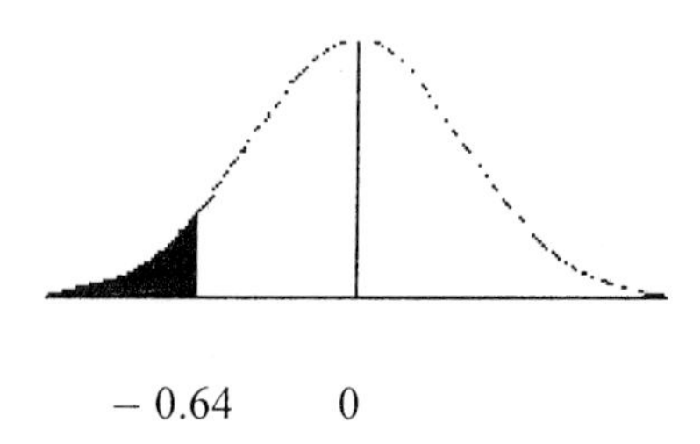

13.
The area is found by looking up z = 1.43 in Table E and subtracting it from 0.5 as shown in Block 2 of Procedure Table 6.
0.5 – 0.4236 = 0.0764

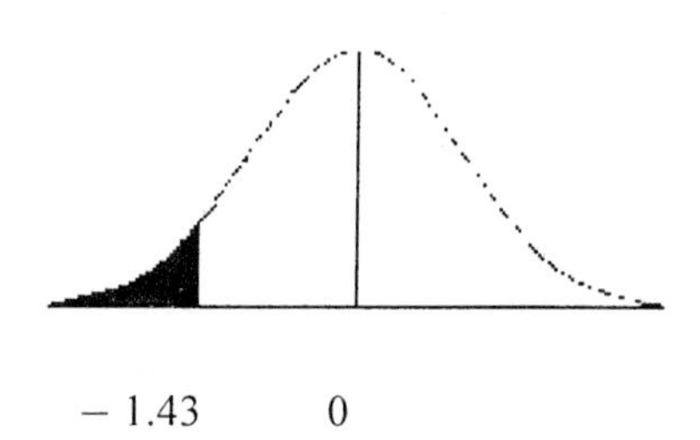

14.
The area is found by looking up the values 1.23 and 1.90 in Table E and subtracting the areas. 0.4713 – 0.3907 = 0.0806

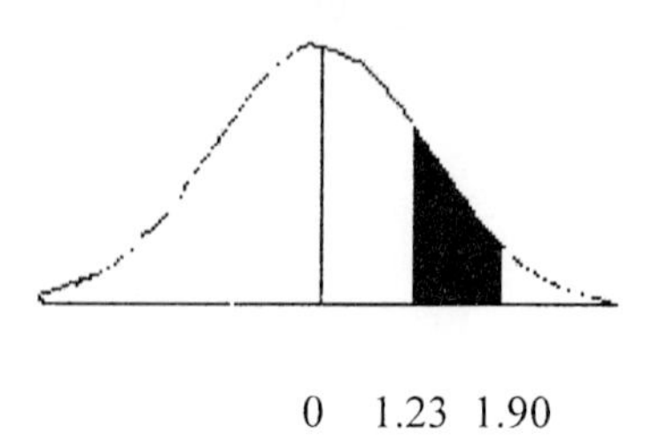

15.
The area is found by looking up the values 0.79 and 1.28 in Table E and subtracting the areas as shown in Block 3 of Procedure Table 6. 0.3997 – 0.2852 = 0.1145

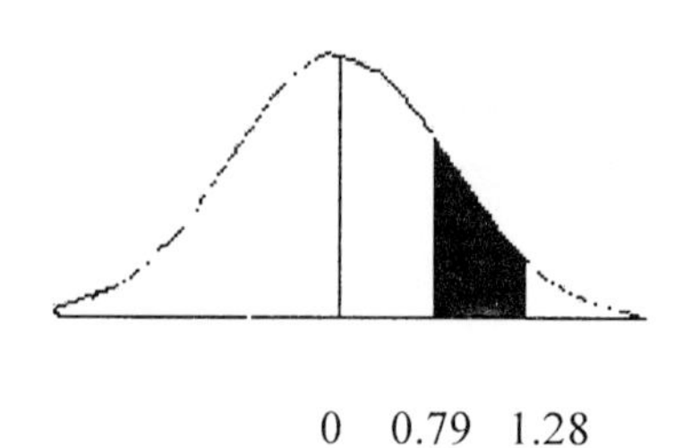

16.
The area is found by looking up the values 0.87 and 0.21 in Table E and subtracting the areas. 0.3078 – 0.0832 = 0.2246

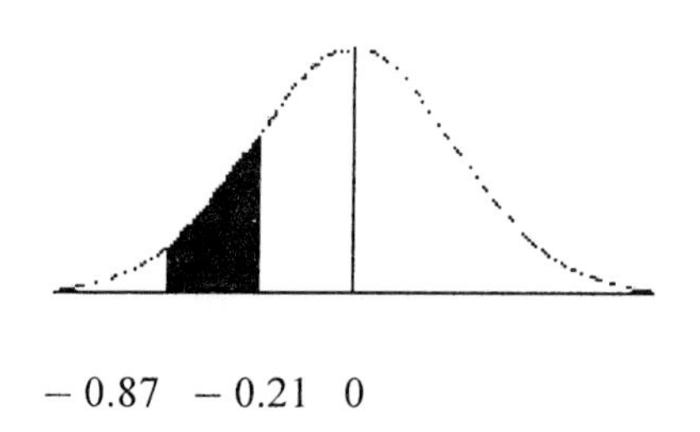

17.
The area is found by looking up the values 1.56 and 1.83 in Table E and subtracting the areas as shown in Block 3 of Procedure Table 6. 0.4664 – 0.4406 = 0.0258

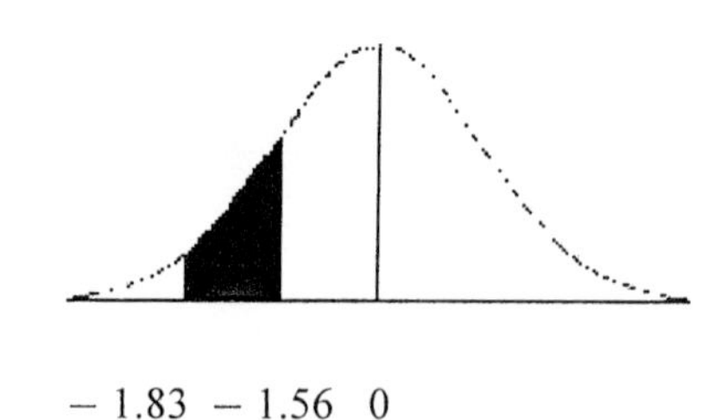

18.
0.3686 + 0.0948 = 0.4634

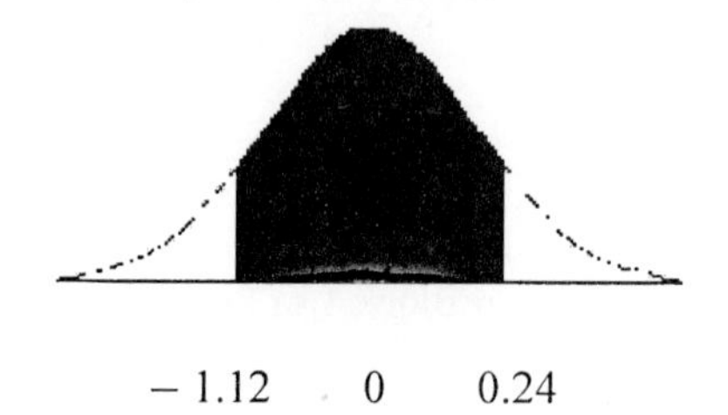

19.
The area is found by looking up the values 2.47 and 1.03 in Table E and adding them together as shown in Block 4 of Procedure Table 6. 0.3485 + 0.4932 = 0.8417

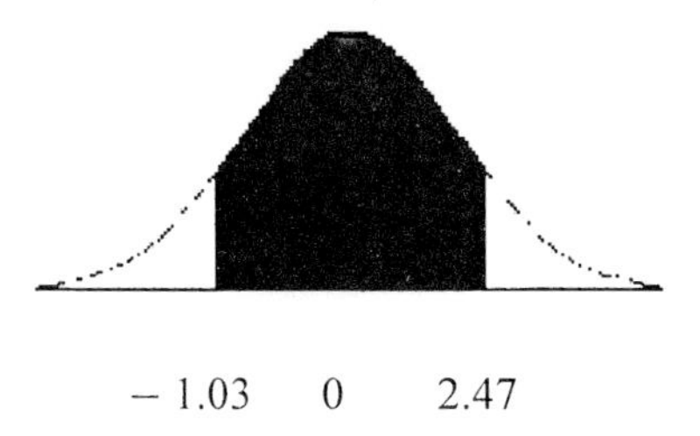

20.
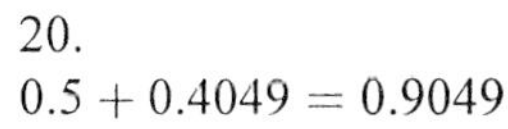
0.5 + 0.4049 = 0.9049

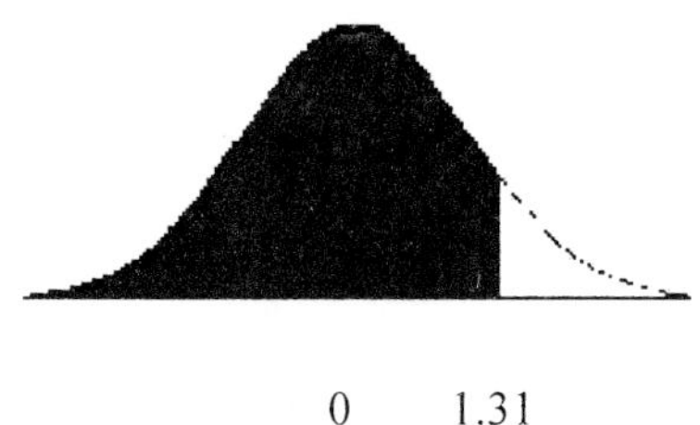

21.
The area is found by looking up z = 2.11 in Table E, then adding the area to 0.5 as shown in Block 5 of Procedure Table 6.
0.5 + 0.4826 = 0.9826

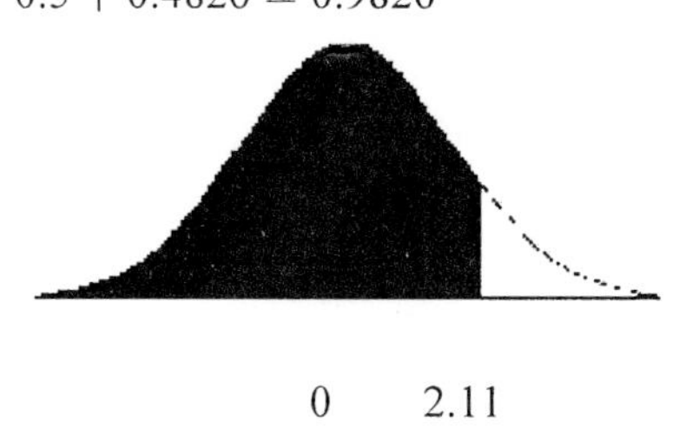

22.
0.5 + 0.4726 = 0.9726

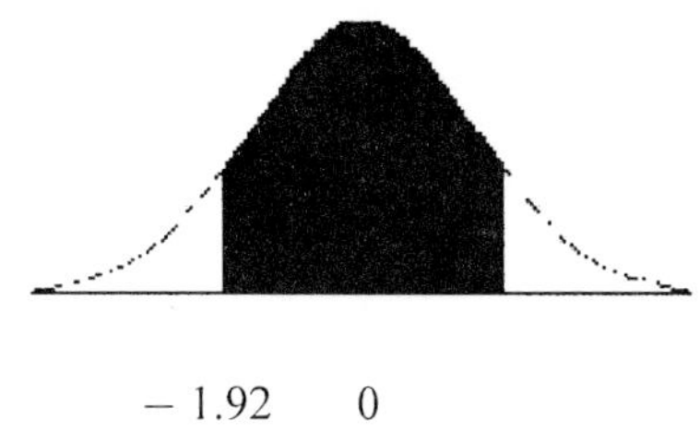

23.
The area is found by looking up z = 0.18 in Table E and adding it to 0.5 as shown in Block 6 of Procedure Table 6.
0.5 + 0.0714 = 0.5714

23. continued

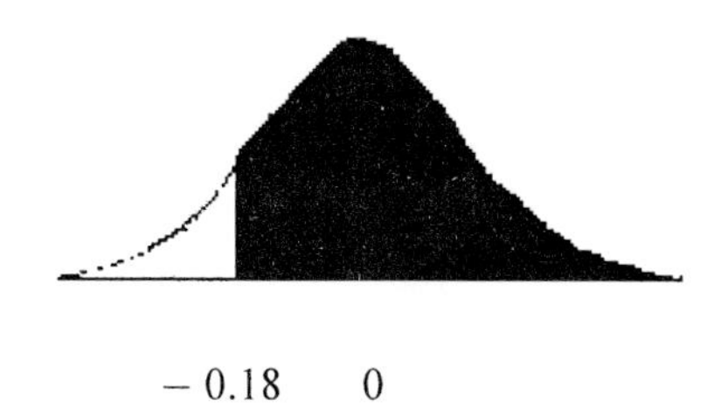

24.
0.5 – 0.4842 = 0.0158
0.5 – 0.4474 = 0.0526
0.0158 + 0.0526 = 0.0684

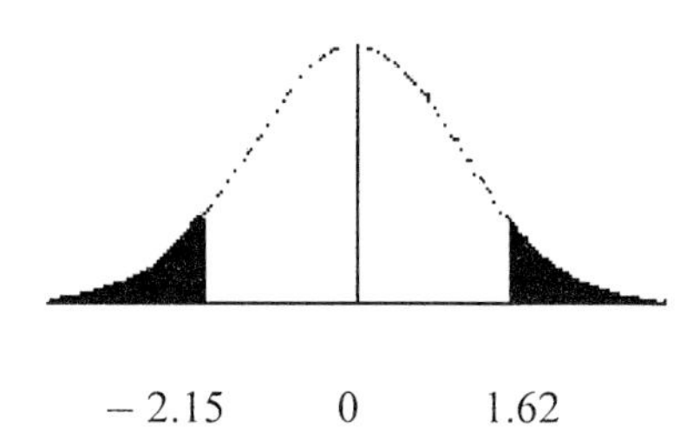

25.
The area is found by looking up the values 1.92 and – 0.44 in Table E, subtracting both areas from 0.5, and adding them together as shown in Block 7 of Procedure Table 6.
0.5 – 0.4726 = 0.0274
0.5 – 0.1700 = 0.3300
0.0274 + 0.3300 = 0.3574

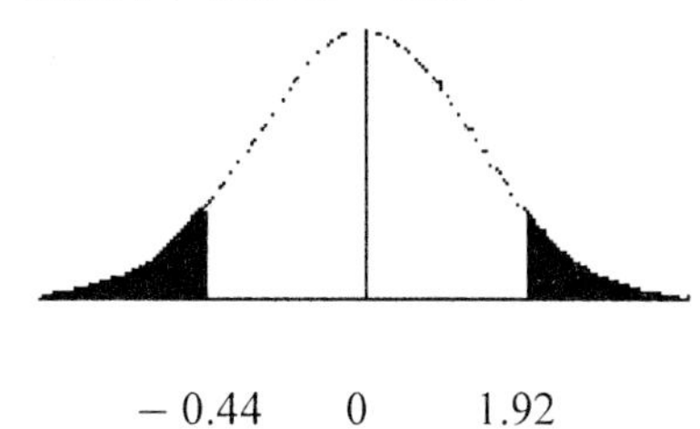

26.
0.4545

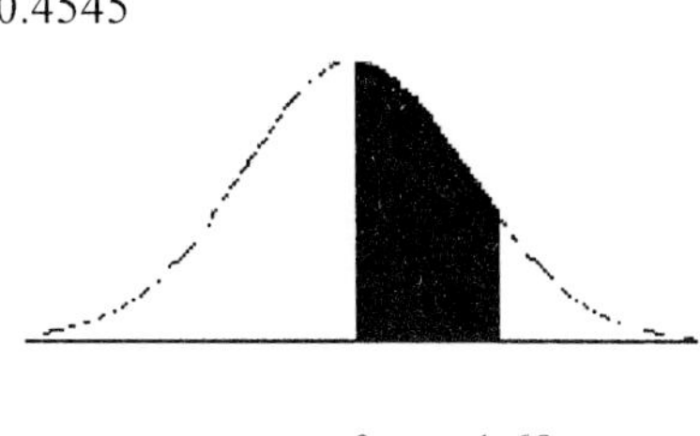

27.
The area is found by looking up z = 0.67 in Table E as shown in Block 1 of Procedure Table 6. Area = 0.2486

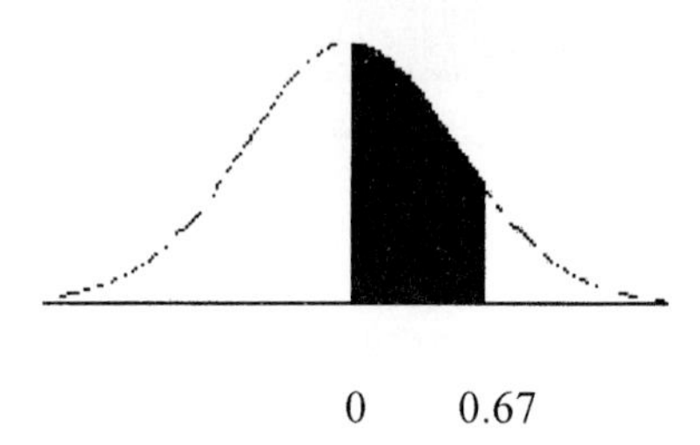

28.
0.3907

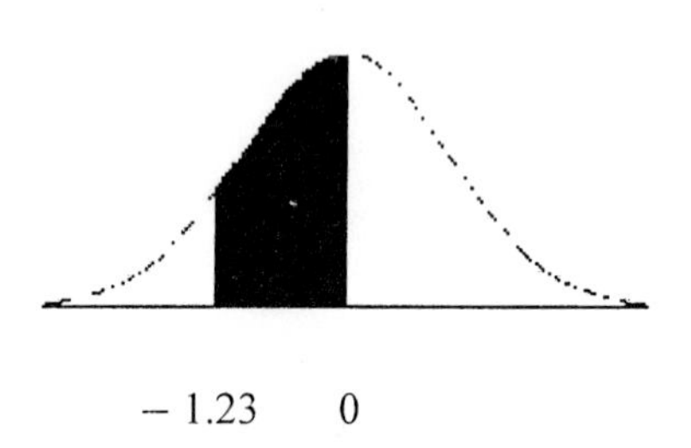

29.
The area is found by looking up z = 1.57 in Table E as shown in Block 1 of Procedure Table 6. Area = 0.4418

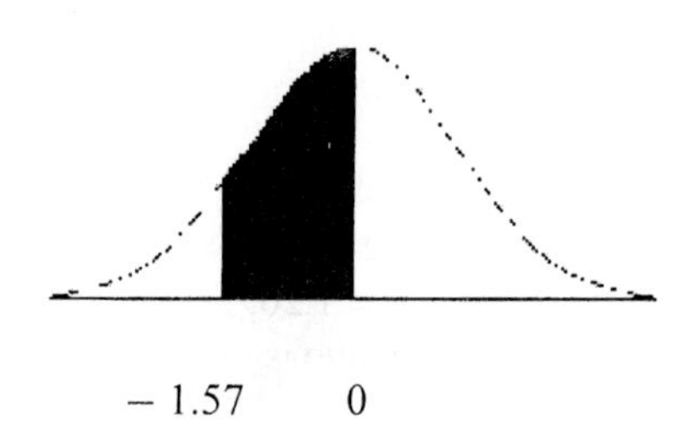

30.
0.5 – 0.4952 = 0.0048

31.
The area is found by looking up z = 2.83 in Table E then subtracting the area from 0.5 as shown in Block 2 of Procedure Table 6.
0.5 – 0.4977 = 0.0023

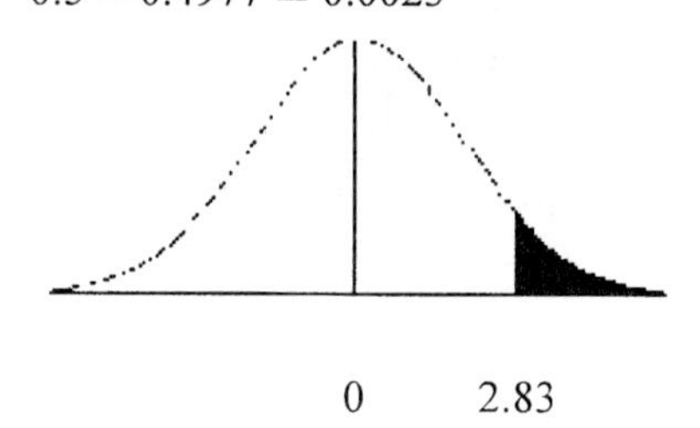

32.
0.5 – 0.4616 = 0.0384

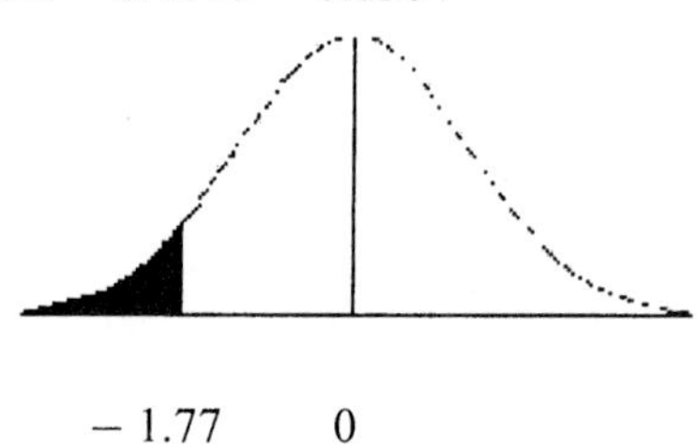

33.
The area is found by looking up z = 1.51 in Table E then subtracting the area from 0.5 as shown in Block 2 of Procedure Table 6.
0.5 – 0.4345 = 0.0655

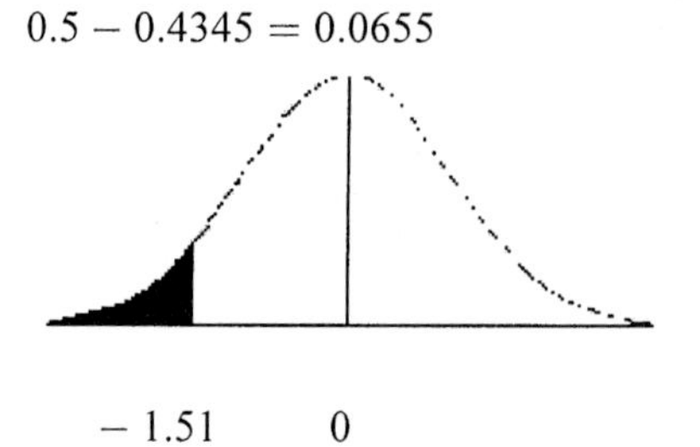

34.
0.0199 + 0.3643 = 0.3842

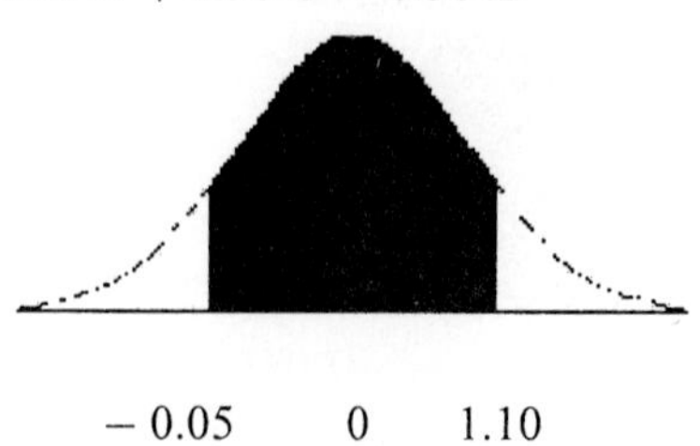

35.
The area is found by looking the values 2.46 and 1.74 in Table E and adding the areas together as shown in Block 4 of Procedure Table 6. 0.4931 + 0.4591 = 0.9522

35. continued

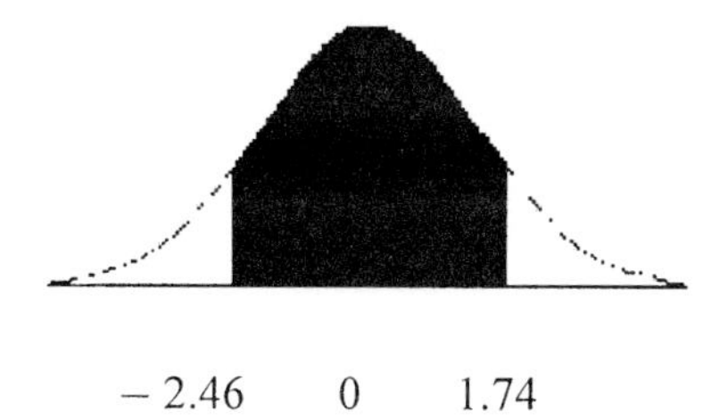

– 2.46 0 1.74

36.
0.4345 – 0.4066 = 0.0279

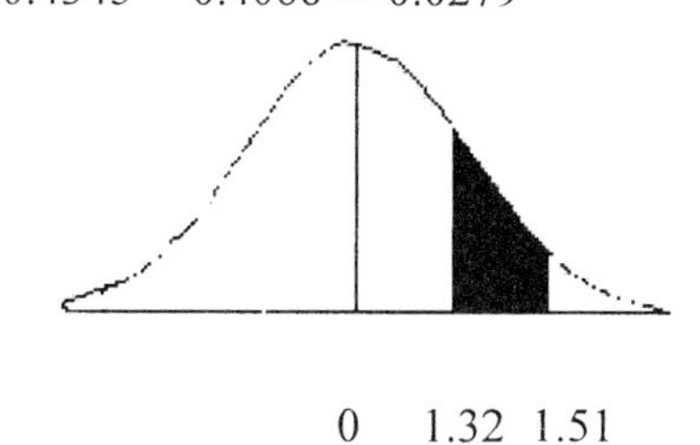

0 1.32 1.51

37.
The area is found by looking up the values 1.46 and 2.97 in Table E and subtracting the areas as shown in Block 3 of Procedure Table 6. 0.4985 – 0.4279 = 0.0706

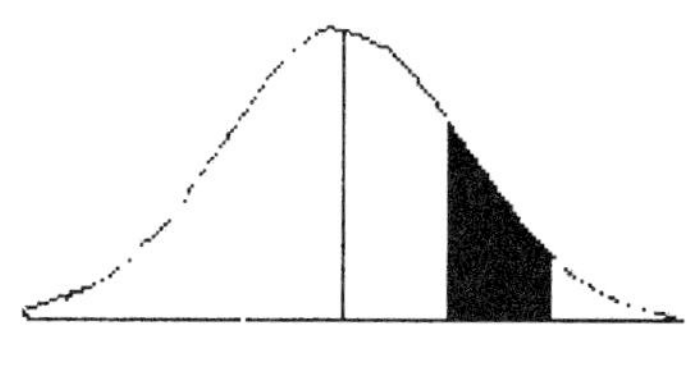

0 1.46 2.97

38.
0.5 + 0.4177 = 0.9177

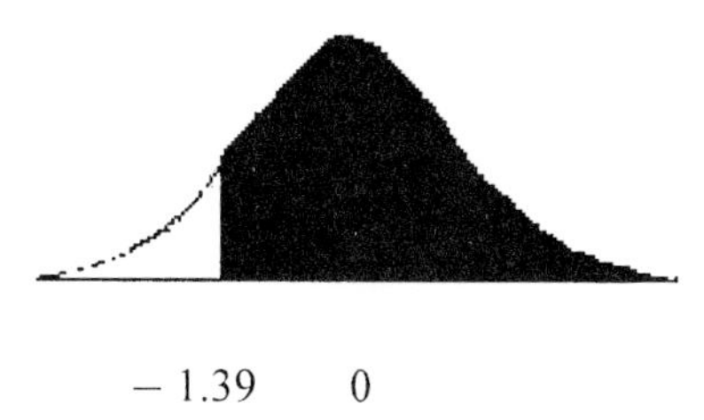

– 1.39 0

39.
The area is found by looking up z = 1.42 in Table E and adding 0.5 to it as shown in Block 5 of Procedure Table 6.
0.5 + 0.4222 = 0.9222

39. continued

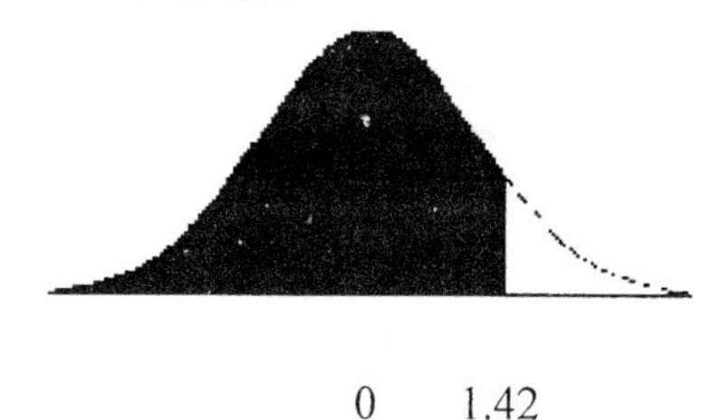

0 1.42

40.
+ 1.32

41.
z = – 1.94, found by looking up the area 0.4738 in Table E to get 1.94; it is negative because the z value is on the left side of 0.

42.
0.5 – 0.0239 = 0.4761
z = + 1.98

43.
z = – 2.13, found by subtracting 0.0166 from 0.5 to get 0.4834 then looking up the area to get z = 2.13; it is negative because the z value is on the left side of 0.

44.
0.9671 – 0.5 = 0.4671
z = + 1.84

45.
z = – 1.26, found by subtracting 0.5 from 0.8962 to get 0.3962, then looking up the area in Table E to get z = 1.26; it is negative because the z value is on the left side of 0.

46.
a. 0.5398 – 0.5 = 0.0398
z = 0.10

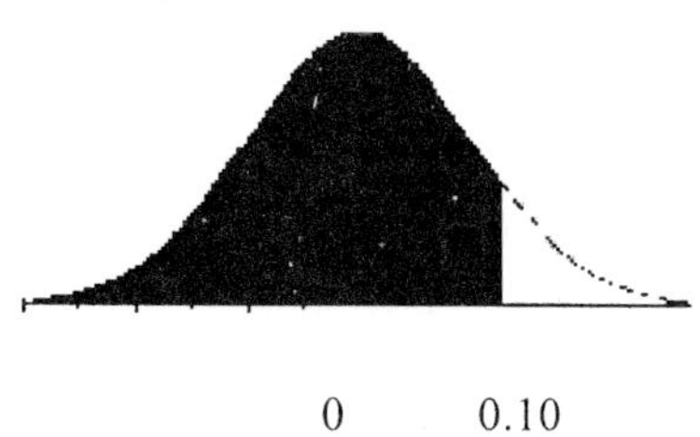

0 0.10

b. 0.7190 – 0.5 = 0.2190
z = 0.58

46b. continued

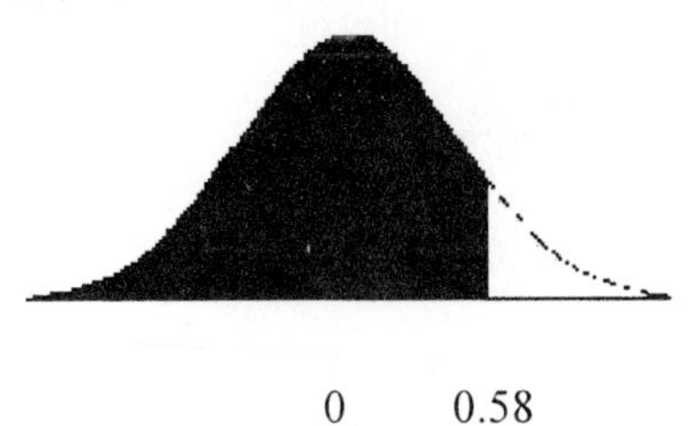

0 0.58

c. $0.9678 - 0.5 = 0.4678$
$z = 1.85$

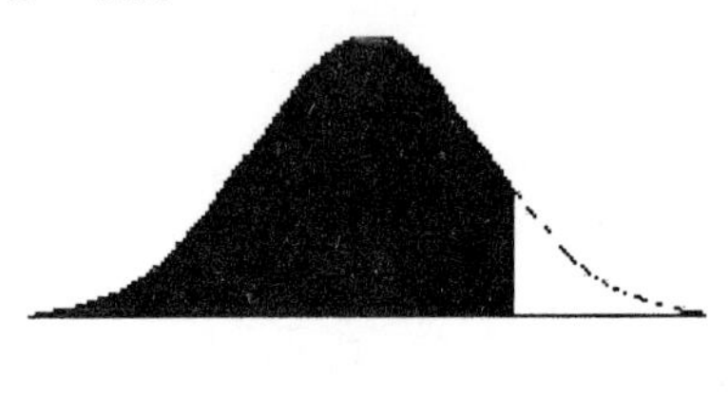

0 1.85

47.
a. $z = -2.28$, found by subtracting 0.5 from 0.9886 to get 0.4886. Find the area in Table E, then find z. It is negative since the z value falls to the left of 0.

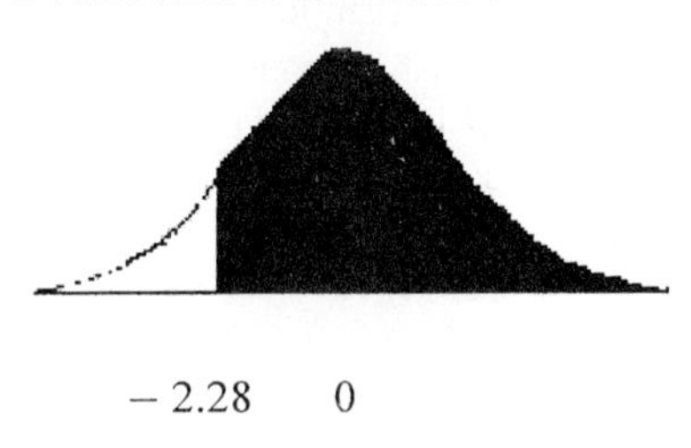

– 2.28 0

b. $z = -0.92$, found by subtracting 0.5 from 0.8212 to get 0.3212. Find the area in Table E, then find z. It is negative since the z value falls to the left of 0.

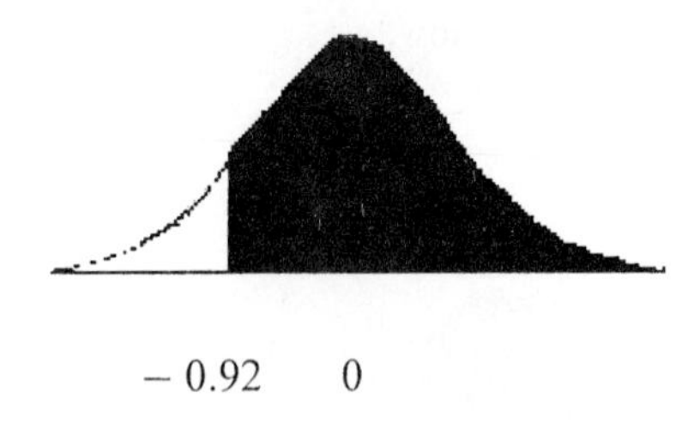

– 0.92 0

c. $z = -0.27$, found by subtracting 0.5 from 0.6064 to get 0.1064. Find the area in Table E, then find z. It is negative since the z value falls to the left of 0.

47c. continued

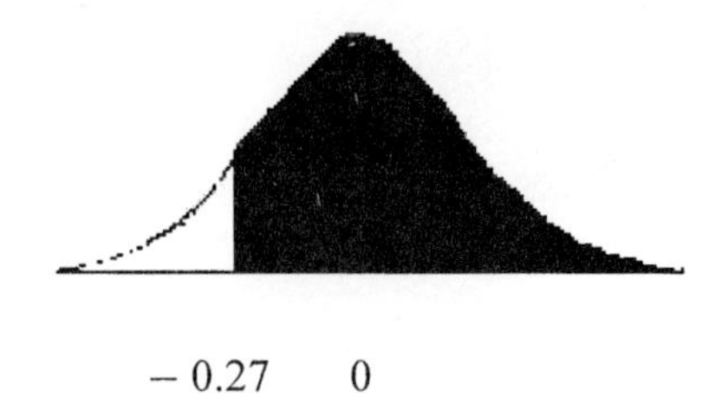

– 0.27 0

48.
$z = \pm 0.52$, approximately.

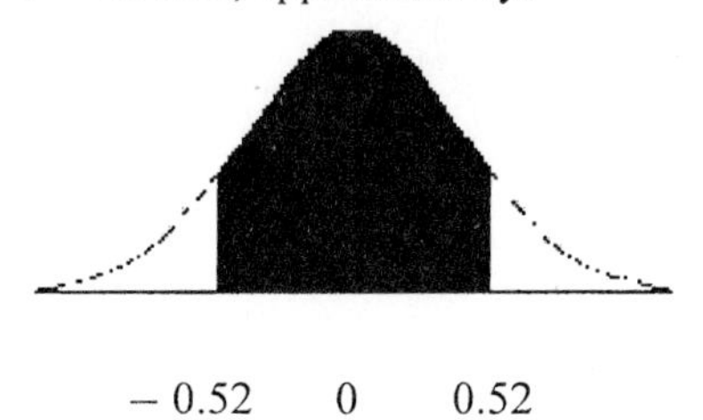

– 0.52 0 0.52

49.
a. $z = \pm 1.96$, found by:
$0.05 \div 2 = 0.025$ is the area in each tail.
$0.5 - 0.025 = 0.4750$ is the area needed to determine z.

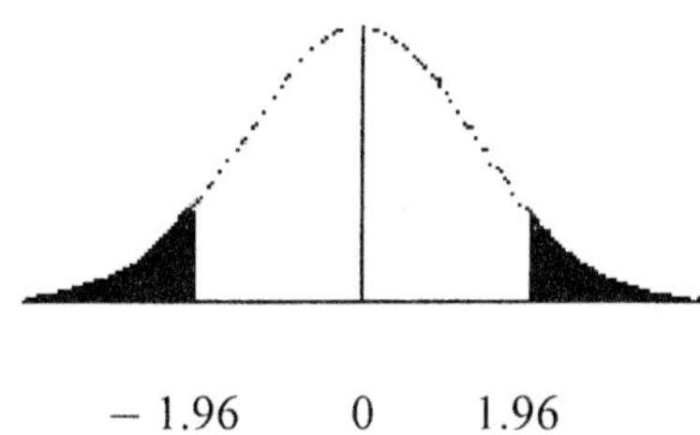

– 1.96 0 1.96

b. $z = \pm 1.65$, found by:
$0.10 \div 2 = 0.05$ is the area in each tail.
$0.5 - 0.05 = 0.4500$ is the area needed to determine z.

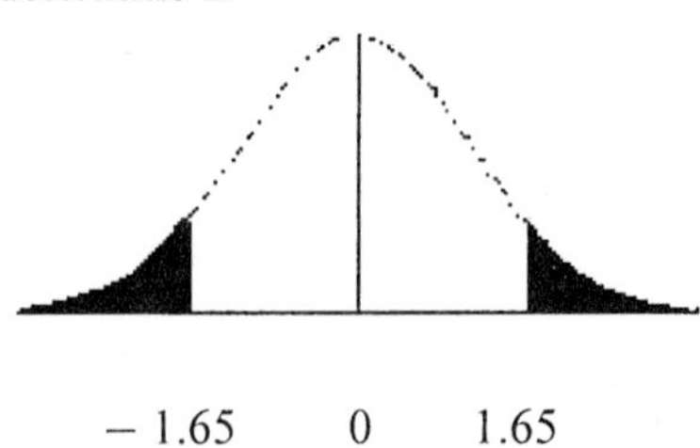

– 1.65 0 1.65

c. $z = \pm 2.58$, found by:
$0.01 \div 2 = 0.005$ is the area in each tail.
$0.5 - 0.005 = 0.4950$ is the area needed to determine z.

49c. continued

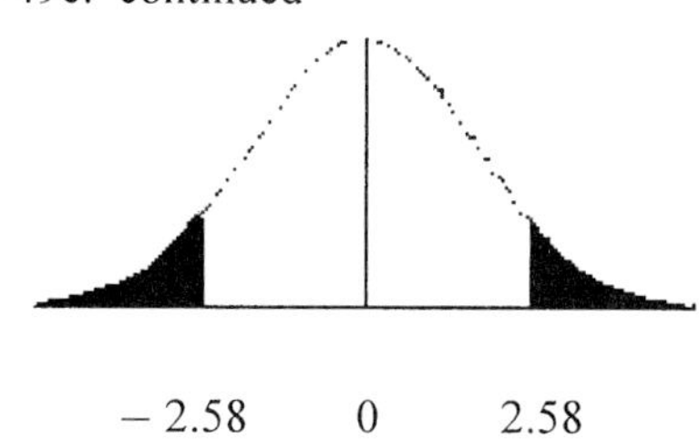

50.
For the 75th percentile z = 0.6745

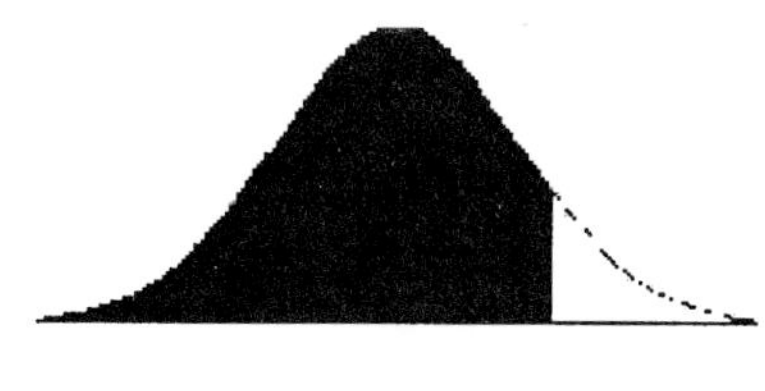

0 0.6745

For the 80th percentile z = 0.8416

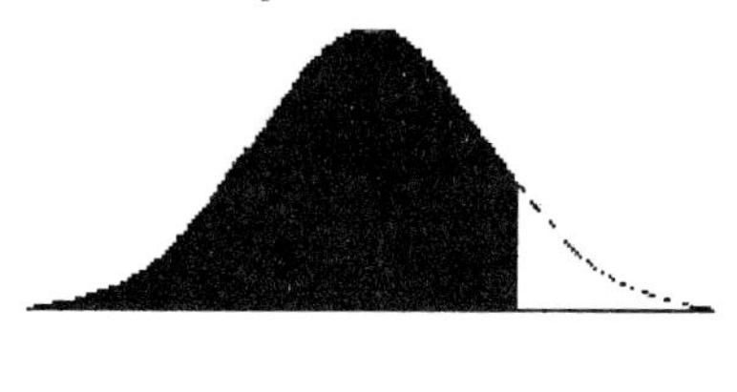

0 0.8416

For the 92th percentile z = 1.41

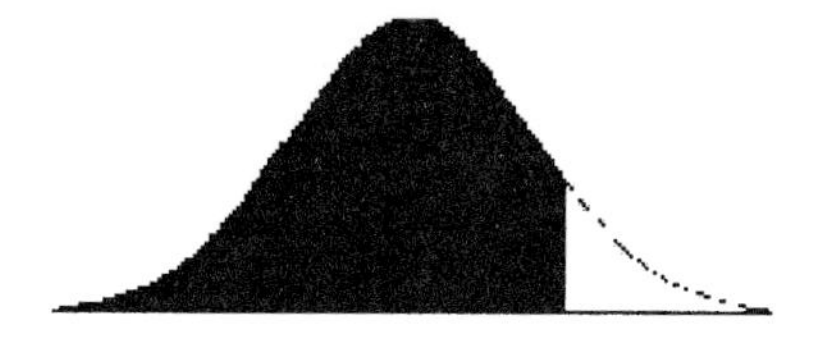

0 1.41

51.
$P(-1 < z < 1) = 2(0.3413) = 0.6826$

$P(-2 < z < 2) = 2(0.4772) = 0.9544$

$P(-3 < x < 3) = 2(0.4987) = 0.9974$

They are very close.

52.
$0.5 - 0.1234 = 0.3766$
For area = 0.3766, $z_0 = 1.16$
Thus, $P(z > 1.16) = 0.1234$

53.
For $z = -1.2$, area = 0.3849
$0.8671 - 0.3849 = 0.4822$
For area = 0.4822, z = 2.10
Thus, $P(-1.2 < z < 2.10) = 0.8671$

54.
For z = 2.5, area = 0.4938
$0.7672 - 0.4938 = 0.2734$
For area = 0.2734, z = 0.75
Thus, $P(-0.75 < z < 2.5) = 0.7672$

55.
For $z = -0.5$, area = 0.1915
$0.2345 - 0.1915 = 0.043$
For area = 0.043, z = 0.11
Thus, $P(-0.5 < z < 0.11) = 0.2345$

For $z = -0.5$, area = 0.1915
$0.2345 + 0.1915 = 0.4260$
For area = 0.426, $z = -1.45$
Thus, $P(-1.45 < z < -0.5) = 0.2345$

56.
$0.86 \div 2 = 0.43$
For area = 0.43, $z = \pm 1.48$
Thus, $P(-1.48 < z < 1.48) = 0.86$

57.

$$y = \frac{e^{-\frac{(X-0)^2}{2(1)^2}}}{1\sqrt{2\pi}} = \frac{e^{-\frac{X^2}{2}}}{\sqrt{2\pi}}$$

58.
Each x value (– 2, – 1.5, etc.) is substituted in the formula $y = \frac{e^{-\frac{X^2}{2}}}{\sqrt{2\pi}}$ to get the corresponding y value. The pairs are then plotted as shown below.

For $x = -2$, $y = \frac{e^{-\frac{(-2)^2}{2}}}{\sqrt{2\pi}} = \frac{e^{-2}}{\sqrt{6.28}}$
$= \frac{0.1353}{\sqrt{6.28}} = 0.05$

X	Y
-2.0	0.05
-1.5	0.13
-1.0	0.24
-0.5	0.35
0	0.40
0.5	0.35
1.0	0.24
1.5	0.13
2.0	0.05

58. continued

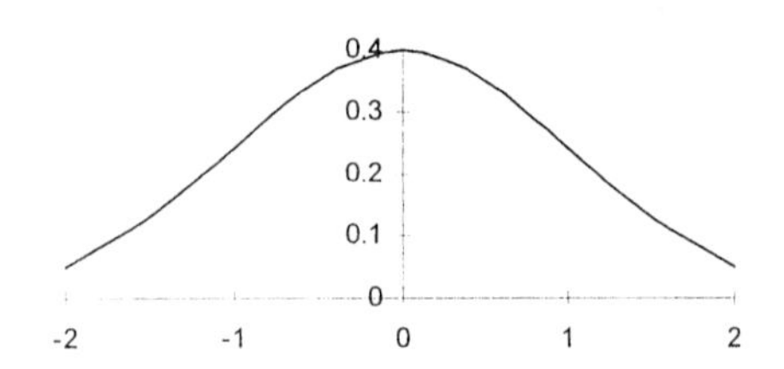

EXERCISE SET 6-4

1.
$z = \frac{\$3.00 - \$5.39}{\$0.79} = -3.03$
area = 0.4988
$P(z < -3.03) = 0.5 - 0.4988 = 0.0012$ or 0.12%

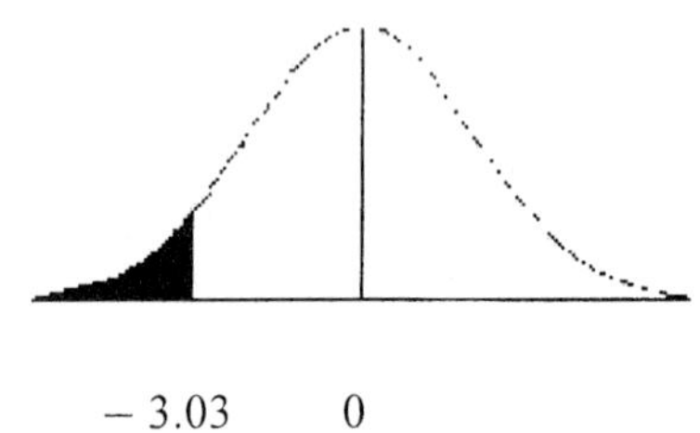

$-3.03 \quad 0$

2.
a. $z = \frac{20,000 - 27,989}{3250} = -2.46$
area = 0.4931

$z = \frac{30,000 - 27,989}{3250} = 0.62$
area = 0.2324

$P(-2.46 < z < 0.62) = 0.4931 + 0.2324$
$P = 0.7255$ or 72.55%

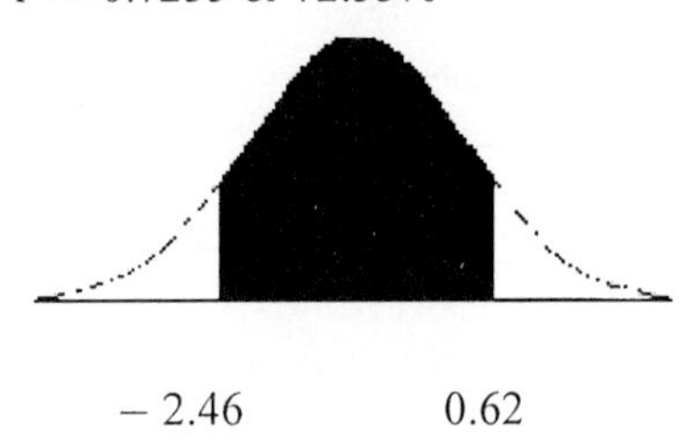

$-2.46 \quad 0.62$

b. $z = -2.46$
area = 0.4931

$P(z < -2.46) = 0.5 - 0.4931 = 0.0069$ or 0.007

2b. continued

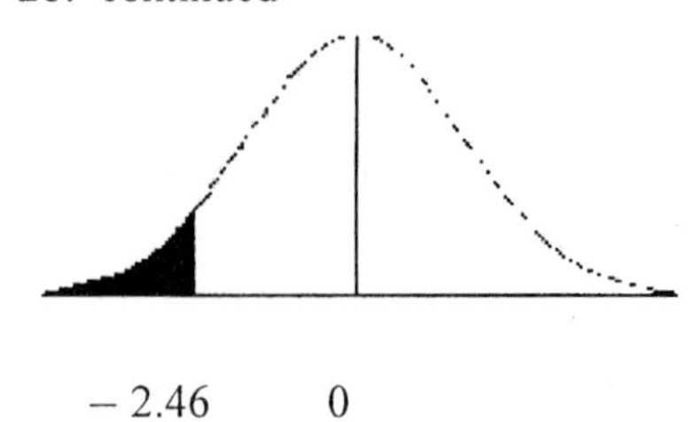

$-2.46 \quad 0$

3.
$z = \frac{X - \mu}{\sigma}$

a. $z = \frac{700,000 - 618,319}{50,200} = 1.63$
area = 0.4484

$P(z > 1.63) = 0.5 - 0.4484 = 0.0516$ or 5.16%

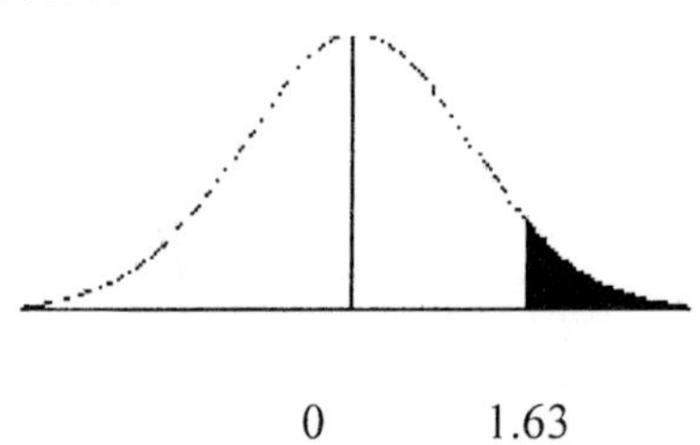

$0 \quad 1.63$

b. $z = \frac{500,000 - 618,319}{50,200} = -2.36$
area = 0.4909

$z = \frac{600,000 - 618,319}{50,200} = -0.36$
area = 0.1406

$P(-2.36 < z < -0.36) = 0.4909 - 0.1406$
$P = 0.3503$ or 35.03%

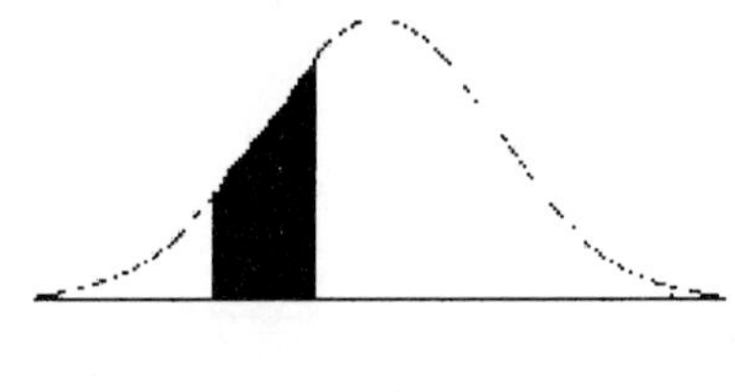

$-2.36 \quad -0.36$

4.
For the 90th percentile, area = 0.4 and $z = 1.28$
$x = 1.28(90) + 1019$
$x = 1134.2$ or 1134

For a score of 1200, $z = \frac{1200 - 1019}{90} = 2.01$
area = 0.4778

$P(z > 2.01) = 0.5 - 0.4778 = 0.0222$ or 2.22%

5.

$z = \frac{X - \mu}{\sigma}$

a. $z = \frac{200 - 225}{10} = -2.5$

area = 0.4938

$z = \frac{220 - 225}{10} = -0.5$

area = 0.1915

$P(-2.5 < z < -0.5) =$

$0.4938 - 0.1915 = 0.3023$ or 30.23%

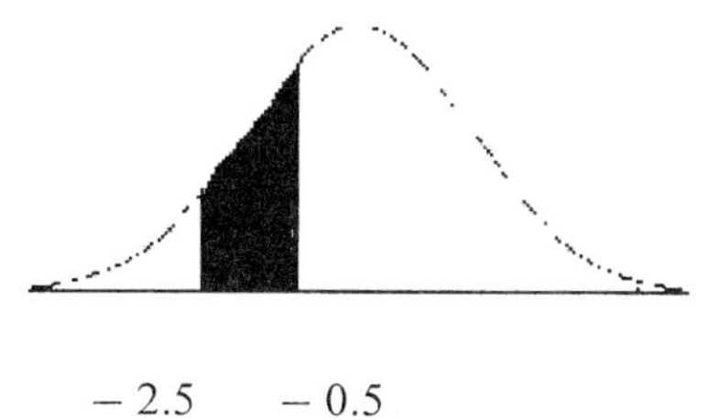

b. $z = -2.5$

area = 0.4938

$P(z < -2.5) = 0.5 - 0.4938 = 0.0062$ or 0.62%

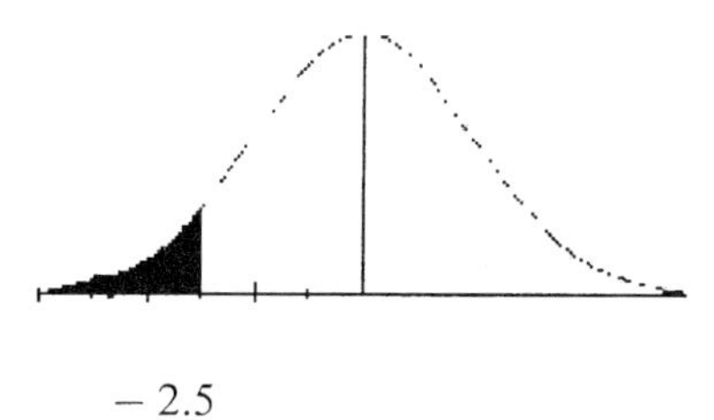

6.

a. $z = \frac{53 - 56}{4} = -0.75$

area = 0.2734

$z = \frac{59 - 56}{4} = 0.75$

area = 0.2734

$P(-0.75 < z < 0.75) = 0.2734 + 0.2734$

$= 0.5468$ or 54.68%

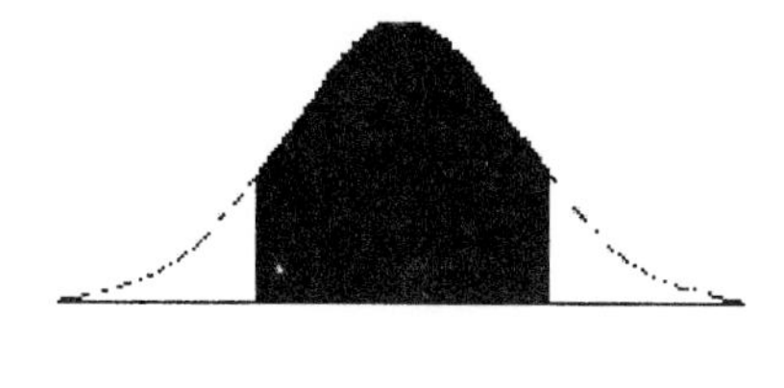

b. $z = \frac{58 - 56}{4} = 0.5$

area = 0.1915

6b. continued

$z = \frac{63 - 56}{4} = 1.75$

area = 0.4599

$P(0.5 < z < 1.75) = 0.4599 - 0.1915$

$= 0.2684$ or 26.84%

c. $z = \frac{50 - 56}{4} = -1.5$

area = 0.4332

$z = \frac{55 - 56}{4} = -0.25$

area = 0.0987

$P(-1.5 < z < -0.25) =$

$0.4332 - 0.0987 = 0.3345$ or 33.45%

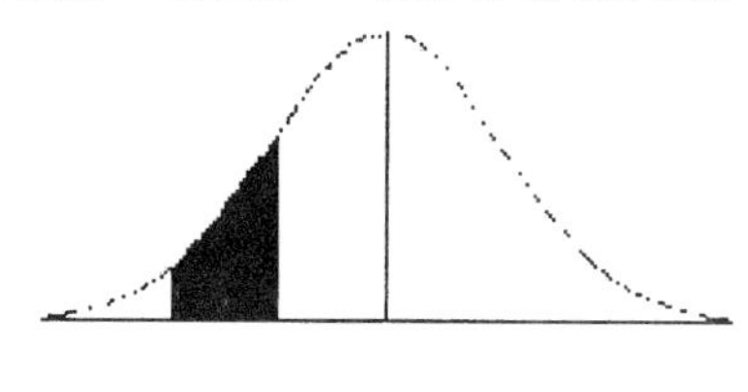

7.

$z = \frac{X - \mu}{\sigma}$

a. $z = \frac{\$90{,}000 - \$85{,}900}{\$11{,}000} = 0.37$

area = 0.1443

$P(z > 0.37) = 0.5 - 0.1443 = 0.3557$ or 35.57%

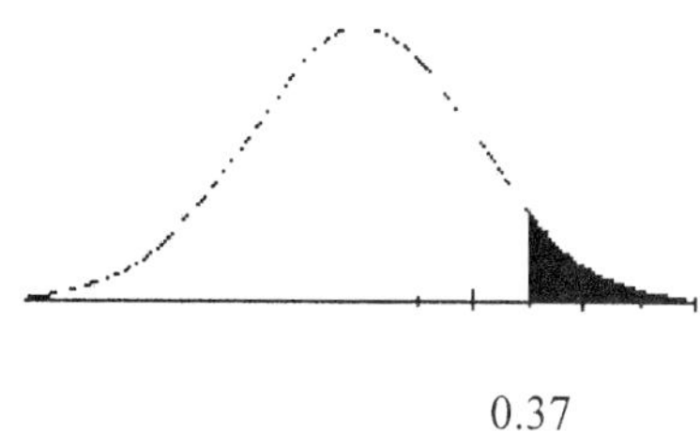

b. $z = \frac{\$75{,}000 - \$85{,}900}{\$11{,}000} = -0.99$

area = 0.3389

$P(z > -0.99) = 0.5 + 0.3389$

$= 0.8389$ or 83.89%

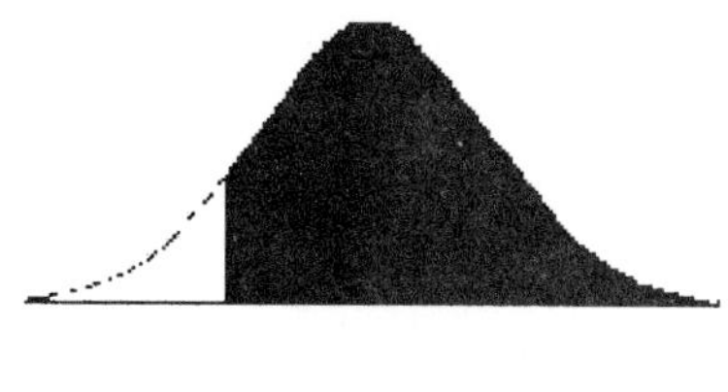

8.

a. $z = \frac{15{,}000 - 12{,}837}{1500} = 1.44$
area = 0.4251

$P(z > 1.44) = 0.5 - 0.4251 = 0.0749$ or 7.49%

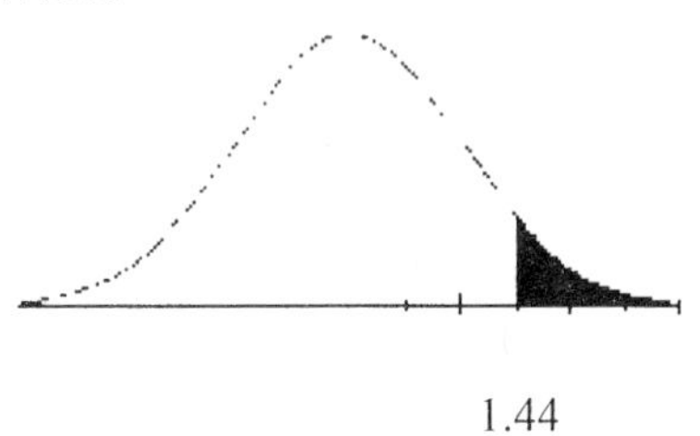

b. $z = \frac{13{,}000 - 12{,}837}{1500} = 0.11$
area = 0.0438

$z = \frac{14{,}000 - 12{,}837}{1500} = 0.78$
area = 0.2823

$P(0.11 < z < 0.78) =$
$0.2823 - 0.0438 = 0.2385$ or 23.85%

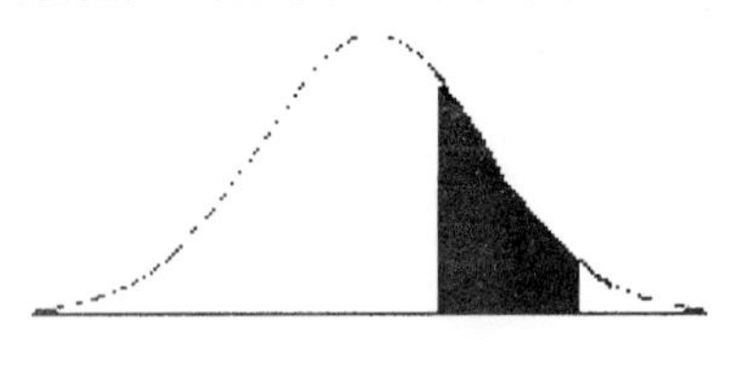

9.

$z = \frac{X - \mu}{\sigma}$

a. $z = \frac{180 - 200}{10} = -2.00$
area = 0.4772

$P(z \geq -2) = 0.5 + 0.4772 = 0.9772$ or 97.72%

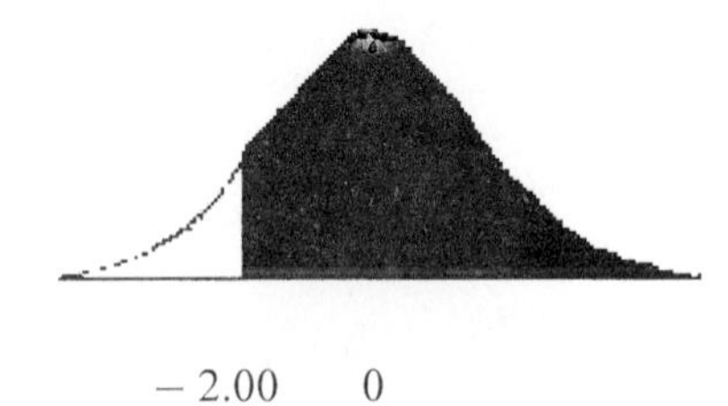

b. $z = \frac{205 - 200}{10} = 0.5$
area = 0.1915

$P(z \leq 0.5) = 0.5 + 0.1915 = 0.6915$ or 69.15%

9b. continued

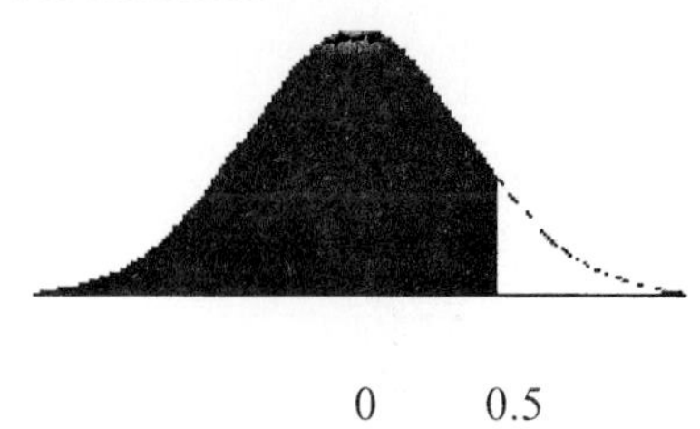

10.

$z = \frac{60 - 25.5}{6.1} = 5.66$
area = 0.4999

$P(z > 5.66) = 0.5 - 0.4999 = 0.0001$ or 0.01%

11.

$z = \frac{X - \mu}{\sigma}$

a. $z = \frac{1000 - 3262}{1100} = -2.06$
area = 0.4803

$P(z \geq -2.06) = 0.5 + 0.4803 = 0.9803$ or 98.03%

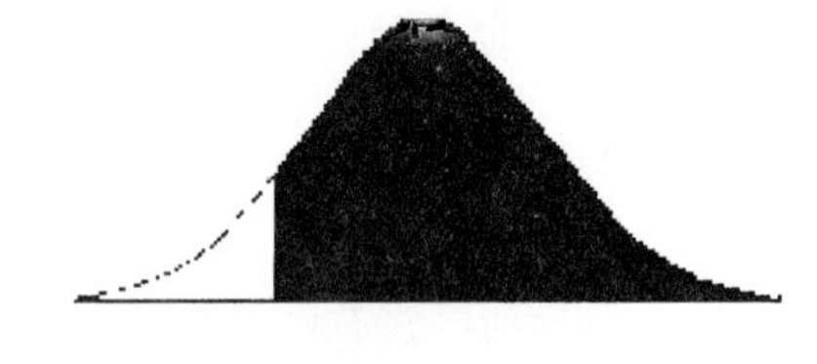

b. $z = \frac{4000 - 3262}{1100} = 0.67$
area = 0.2486

$P(z > 0.67) = 0.5 - 0.2486 = 0.2514$ or 25.14%

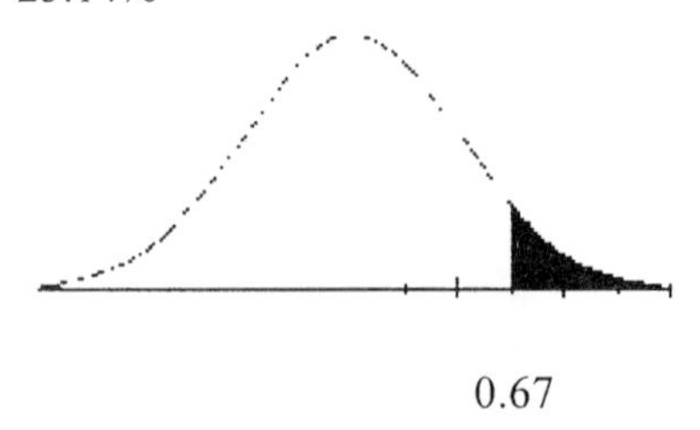

c. $z = \frac{3000 - 3262}{1100} = -0.24$
area = 0.0948

$P(-0.24 < z < 0.67) =$
$0.0948 + 0.2486 = 0.3434$ or 34.34%

11c. continued

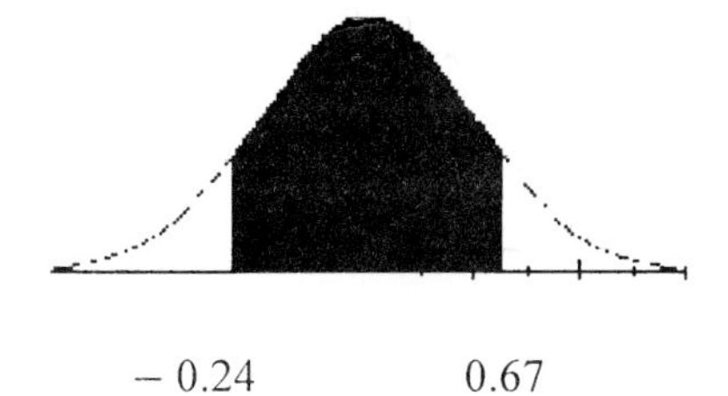

-0.24 $\quad$ 0.67

12.

a. $z = \frac{15 - 24.6}{5.8} = -1.66$
area = 0.4515

$z = \frac{30 - 24.6}{5.8} = 0.93$
area = 0.3238

$P(-1.66 < z < 0.93) = 0.4515 + 0.3238$
$= 0.7753$ or 77.53%

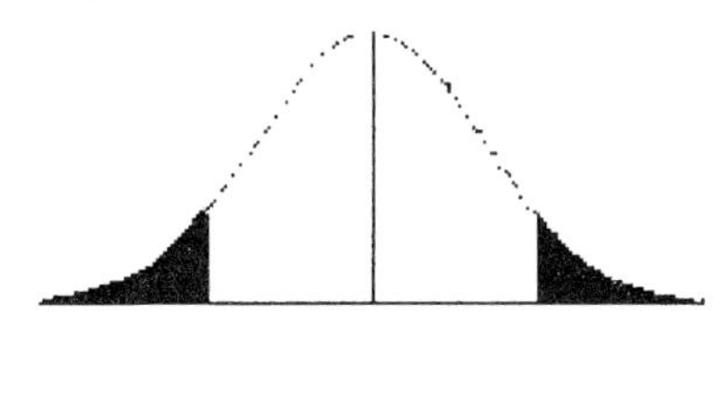

-1.66 $\quad$ 0 $\quad$ 0.93

b. $z = \frac{18 - 24.6}{5.8} = -1.14$
area = 0.3729

$z = \frac{28 - 24.6}{5.8} = 0.59$
area = 0.2224

$P(z < -1.14 \text{ or } z > 0.59) =$
$(0.5 - 0.3729) + (0.5 - 0.2224) = 0.1271$
$+ 0.2776 = 0.4047$ or 40.47%

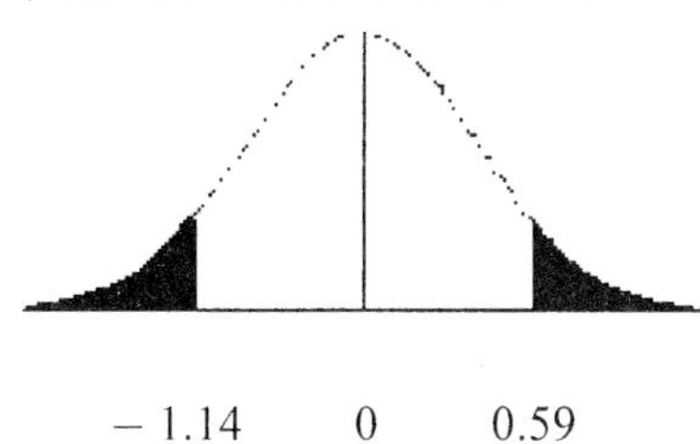

-1.14 $\quad$ 0 $\quad$ 0.59

13.

a. $z = \frac{280 - 300}{8} = -2.5$
area = 0.4938

$P(z > -2.5) = 0.5 + 0.4938 = 0.9938$ or
99.38%

13a. continued

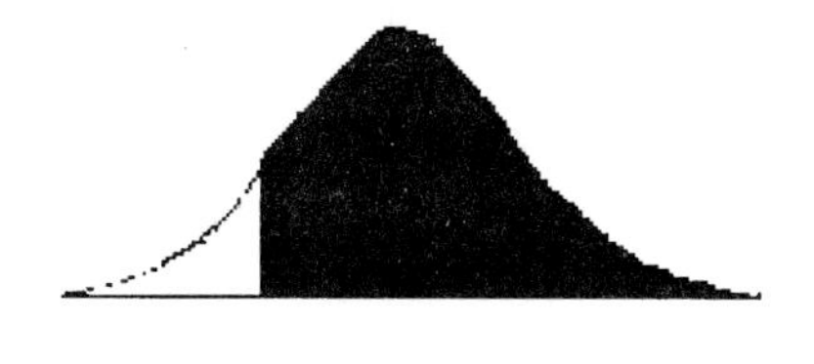

-2.5 $\quad$ 0

b. $z = \frac{293 - 300}{8} = -0.88$
area = 0.3106

$P(z < -0.88) = 0.5 - 0.3106 = 0.1894$ or
18.94%

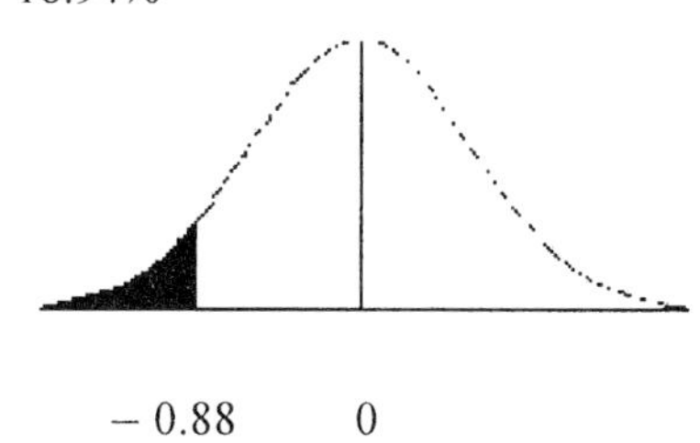

-0.88 $\quad$ 0

c. $z = \frac{285 - 300}{8} = -1.88$
area = 0.4699

$z = \frac{320 - 300}{8} = 2.5$
area = 0.4938

$P(-1.88 < z < 2.5) =$
$0.4699 + 0.4938 = 0.9637$ or 96.37%

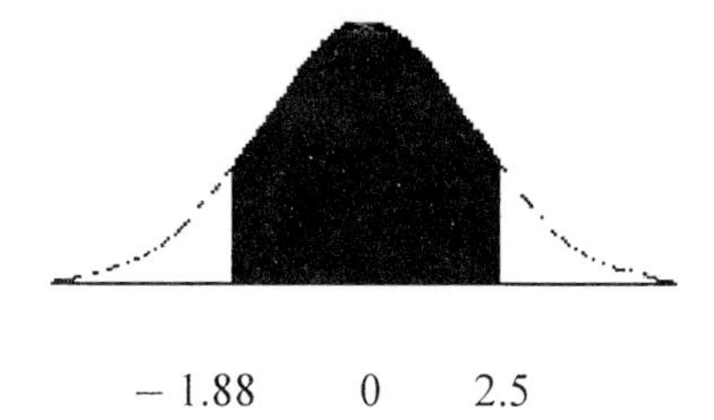

-1.88 $\quad$ 0 $\quad$ 2.5

14.

a. $z = \frac{62 - 64.2}{3.2} = -0.69$
area = 0.2549

$P(z > -0.69) = 0.5 + 0.2549 = 0.7549$ or
75.49%

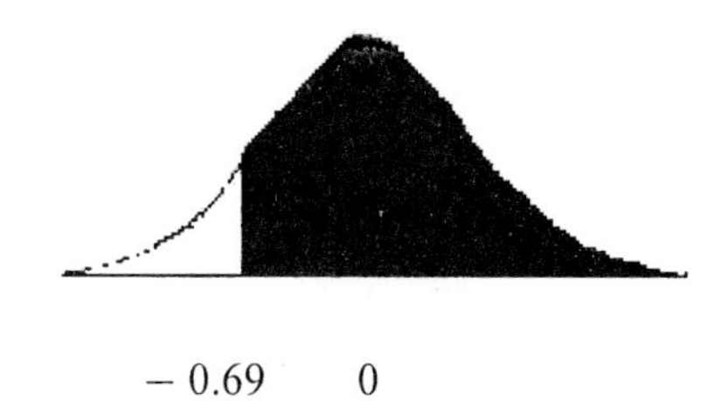

-0.69 $\quad$ 0

14. continued
b. $z = \frac{67-64.2}{3.2} = 0.88$
area = 0.3106
$P(z < 0.88) = 0.5 + 0.3106 = 0.8106$ or 81.06%

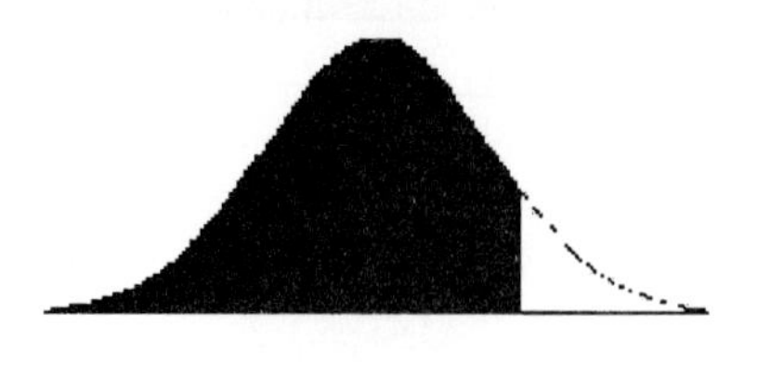

0 0.88

c. $z = \frac{65-64.2}{3.2} = 0.25$
area = 0.0987

$z = \frac{68-64.2}{3.2} = 1.19$
area = 0.3830

$P(0.25 < z < 1.19) = 0.3830 - 0.0987$
$= 0.2843$ or 28.43%

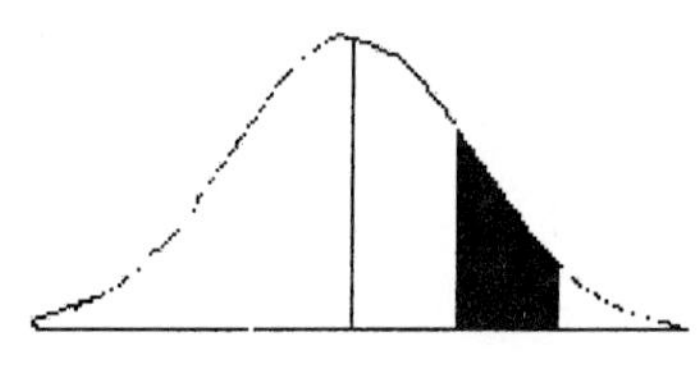

0 0.25 1.19

15.
a. $z = \frac{130-132}{8} = -0.25$
area = 0.0987

$P(z > -0.25) = 0.5 + 0.0987 = 0.5987$ or 59.87%

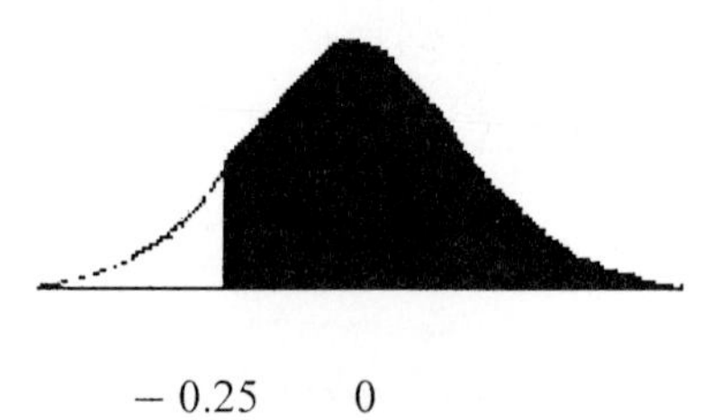

−0.25 0

b. $z = \frac{140-132}{8} = 1.00$
area = 0.3413

$P(z < 1) = 0.5 + 0.3413 = 0.8413$ or 84.13%

15b. continued

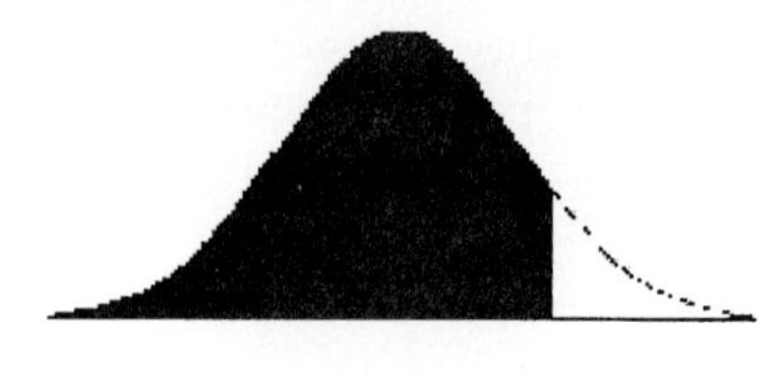

0 1.00

c. $z = \frac{131-132}{8} = -0.13$
area = 0.0517

$z = \frac{136-132}{8} = 0.50$
area = 0.1915

$P(-0.13 < z < 0.50) = 0.0517 + 0.1915$
$= 0.2432$ or 24.32%

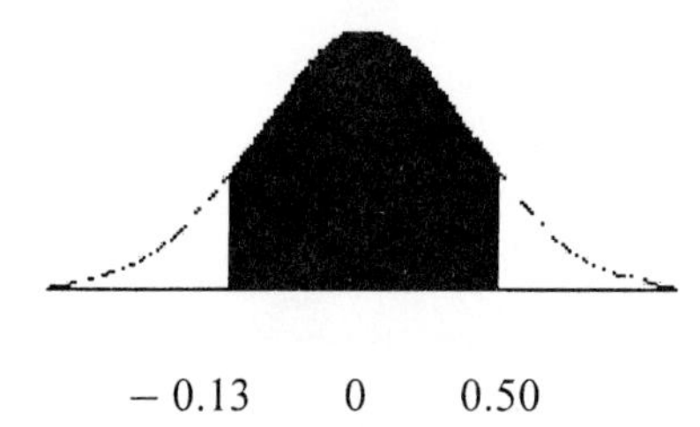

−0.13 0 0.50

16.
The top 15% (area) is in the right tail of the normal curve. The corresponding z score is found using area $= 0.5 - 0.15 = 0.35$.
Thus $z = 1.04$.
$x = 1.04(8) + 80 = 88.32$ points

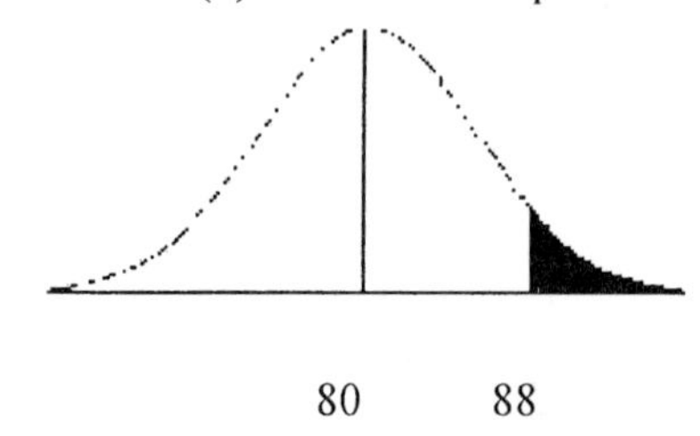

80 88

17.
The top 75% (area) includes all but the left 25% of the curve. The corresponding z score is -0.67.
$x = -0.67(15) + 100 = 89.95$ points

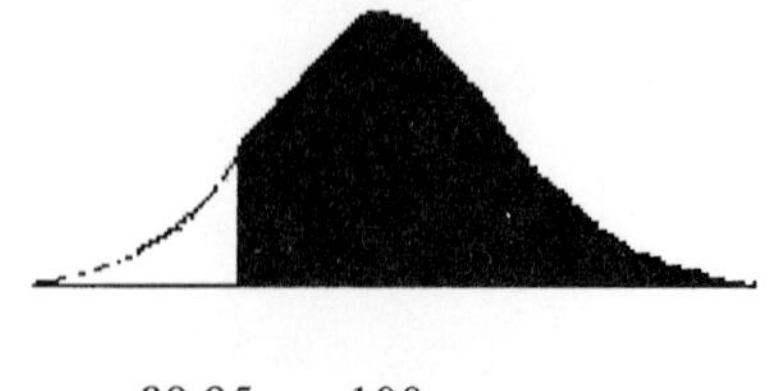

89.95 100

18.
The middle 50% means that 25% of the area will be on either side of the mean. Thus, area = 0.25 and z = ± 0.67.
$x = 0.67(103) + 792 = 861.01$
$x = -0.67(103) + 792 = 722.99$

The scores are between 723 and 861.

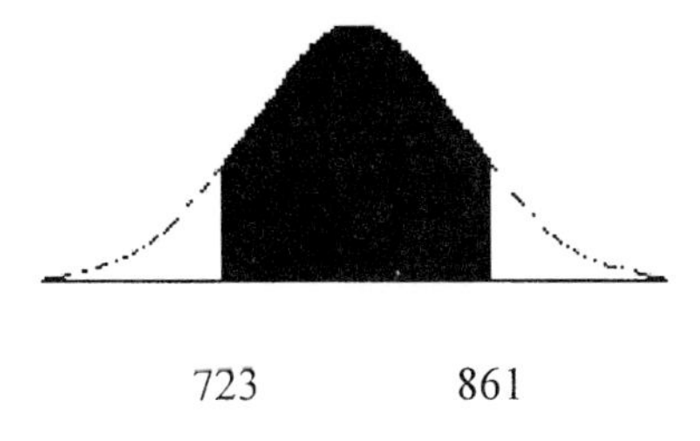

723 861

19.
The middle 80% means that 40% of the area will be on either side of the mean. The corresponding z scores will be ± 1.28.
$x = -1.28(92) + 1810 = 1694.24$ sq. ft.
$x = 1.28(92) + 1810 = 1927.76$ sq. ft.

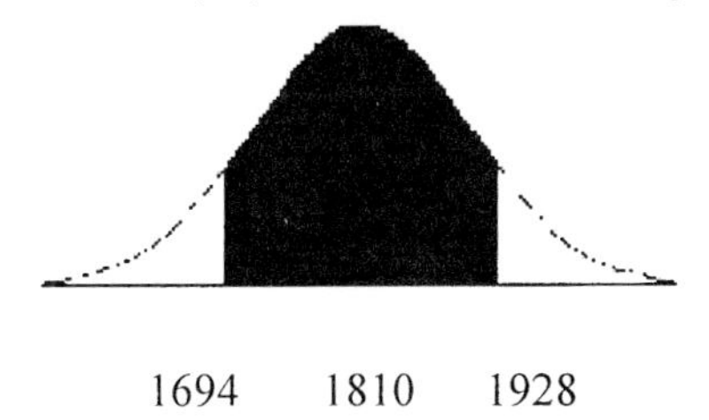

1694 1810 1928

20.
z = ± 1.28
$x = -1.28(1500) + 145{,}500 = \$143{,}580$
$x = 1.28(1500) + 145{,}500 = \$147{,}420$

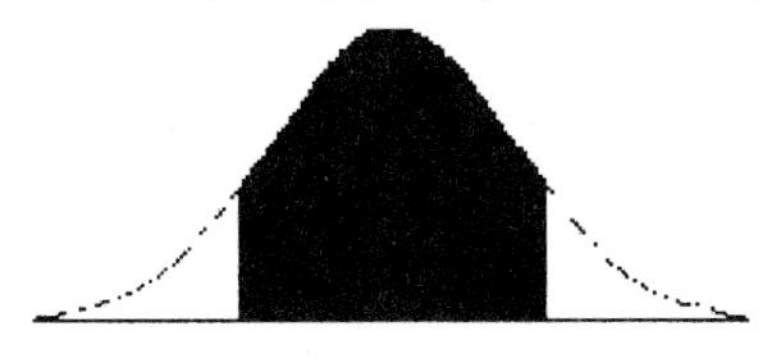

143,580 145,500 147,420

21.
$z = \frac{1200-949}{100} = 2.51$
area = 0.4940

$P(z > 2.51) = 0.5 - 0.4940 = 0.006$ or 0.6%

21. continued

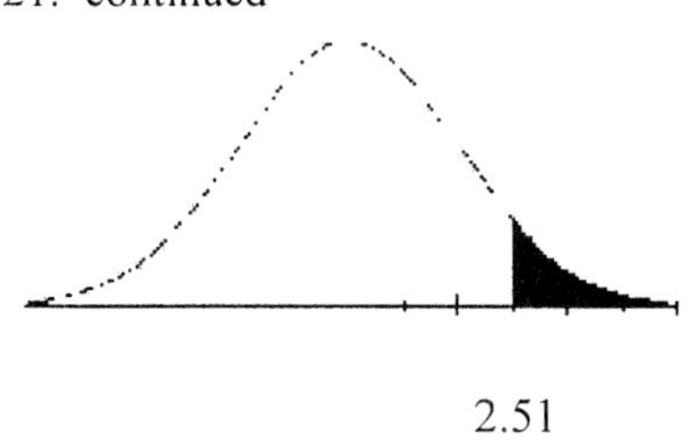

2.51

For the least expensive 10%, the area is 0.4 on the left side of the curve. Thus,
z = -1.28.
$x = -1.28(100) + 949 = \$821$

22.
The bottom 5% (area) is in the left tail of the normal curve. The corresponding z score is found using area = $0.5 - 0.05 = 0.45$.
Thus z = -1.65.
$x = -1.65(18) + 122.6 = 92.9$ points

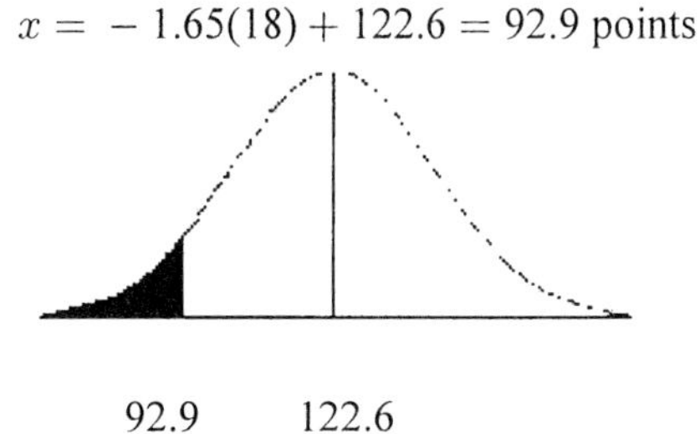

92.9 122.6

23.
The middle 60% means that 30% of the area will be on either side of the mean. The corresponding z scores will be ± 0.84.
$x = -0.84(1150) + 8256 = \$7290$
$x = 0.84(1150) + 8256 = \$9222$

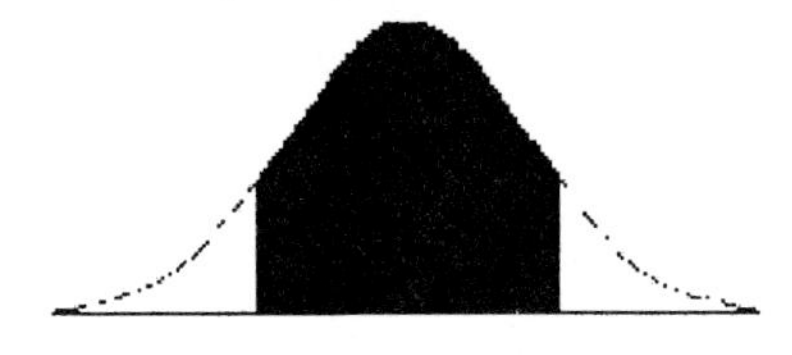

\$7290 \$8256 \$9222

24.
For the oldest 20%, the area is 0.3 on the right side of the curve. Thus, z = 0.84.
$22.8 = 0.84s + 19.4$
$s = 4.048$

25.
For the fewest 15%, the area is 0.35 on the left side of the curve. Thus, z = -1.04.
$x = -1.04(1.7) + 5.9$
$x = 4.132$ days

25. continued
For the longest 25%, the area is 0.25 on the right side of the curve. Thus, z = 0.67.
$x = 0.67(1.7) + 5.9$
$x = 7.039$ days

26.
a. For the top 3%, the area is 0.47 on the right side of the curve. Thus, z = 1.88.
$x = 1.88(100) + 400$
$x = 588$ minimum score to receive the award.

b. For the bottom 1.5%, the area is 0.485 on the left side of the curve. Thus, z = -2.17.
$x = -2.17(100) + 400$
$x = 183$ minimum score to avoid summer school.

27.
The bottom 18% means that 32% of the area is between 0 and $-z$. The corresponding z score will be -0.92.
$x = -0.92(6256) + 24{,}596 = \$18{,}840.48$

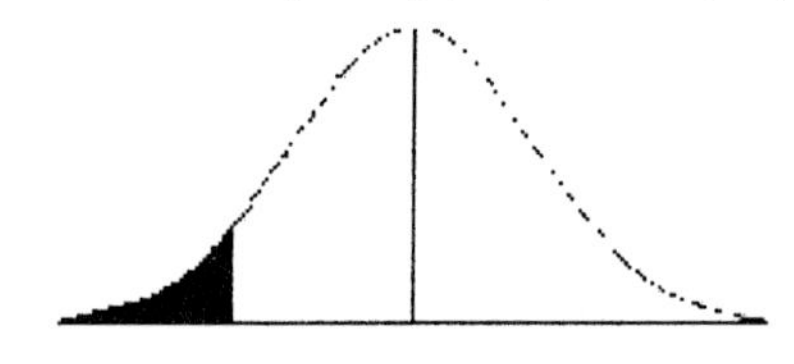

\$18,840.48 \$24,596

28.
The middle 50% means that 25% of the area is on either side of the mean. The corresponding z scores will be ± 0.67.
$x = -0.67(5) + 40 = \$36.65$
$x = 0.67(5) + 40 = \$43.35$

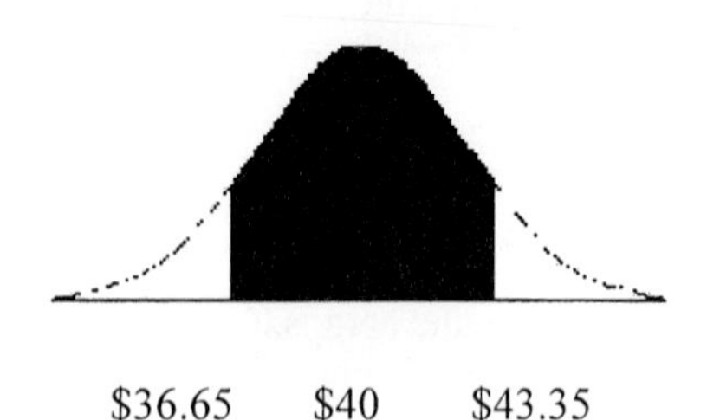

\$36.65 \$40 \$43.35

29.
The 10% to be exchanged would be at the left, or bottom, of the curve; therefore, 40% of the area is between 0 and $-z$. The corresponding z score will be -1.28.
$x = -1.28(5) + 25 = 18.6$ months.

29. continued

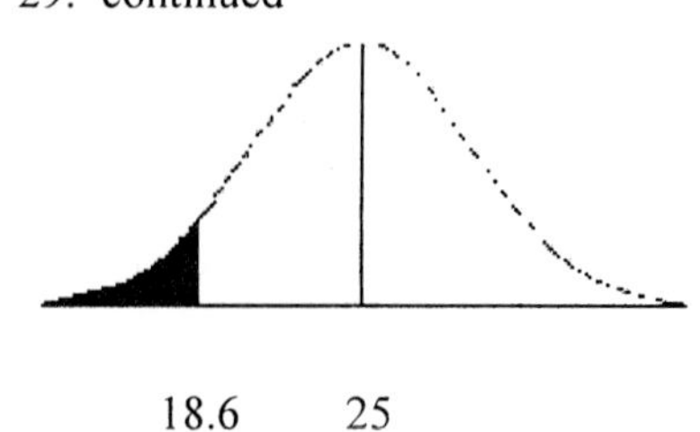

18.6 25

30.
The top 15% means that 35% of the area is between 0 and z. The corresponding z score is 1.04.
$x = 1.04(8) + 62 = 70.32 \approx 70$

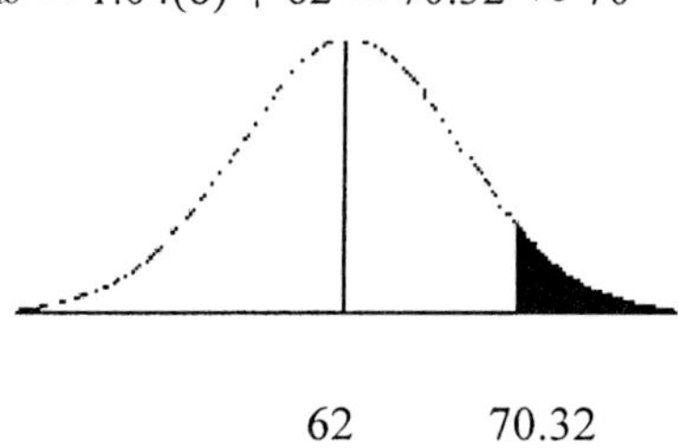

62 70.32

31.
a. $\mu = 120$ $\sigma = 20$
b. $\mu = 15$ $\sigma = 2.5$
c. $\mu = 30$ $\sigma = 5$

32.
No. Any subgroup would not be a perfect representation of the seniors; therefore, the mean and standard deviation would be different.

33.
There are several mathematical tests that can be used including drawing a histogram and calculating Pearson's index of skewness.

34.
No. The shape of the distributions would be the same, since z scores are raw scores scaled by the standard deviation.

35.
2.68% area in the right tail of the curve means that 47.32% of the area is between 0 and z, corresponding to a z score of 1.93.
$z = \frac{X-\mu}{\sigma}$
$1.93 = \frac{105-100}{\sigma}$
$1.93\sigma = 5$
$\sigma = 2.59$

36.
3.75% area in the left tail means that 46.25% of the area is between 0 and $-z$, corresponding to a z score of -1.78.
$-1.78 = \frac{85-\mu}{6}$
$-1.78(6) = 85 - \mu$
$\mu = 95.68$

37.
1.25% of the area in each tail means that 48.75% of the area is between 0 and $\pm z$. The corresponding z scores are ± 2.24.
Then $\mu = \frac{42+48}{2} = 45$ and $X = \mu + z\sigma$.
$48 = 45 + 2.24\sigma$
$\sigma = 1.34$

38.
The cutoff for the A's and F's would be:
$x = \mu + z\sigma$
$x = 60 + 1.65(10)$
$x = 76.5$ for the A's
$x = 60 + (-1.65)(10)$
$x = 43.5$ for the F's

For the B's and D's:
$x = 60 + (0.84)(10)$
$x = 68.4$ for the B's
$x = 60 + (-0.84)(10)$
$x = 51.6$ for the D's

The grading scale would be:

77 and up	A
68 – 76	B
53 – 67	C
44 – 52	D
0 – 43	F

39.
Histogram:

The histogram shows a positive skew.

$PI = \frac{3(970.2-853.5)}{376.5} = 0.93$

$IQR = Q_3 - Q_1 = 910 - 815 = 95$
$1.5(IQR) = 1.5(95) = 142.5$
$Q_1 - 142.5 = 672.5$
$Q_3 + 142.5 = 1052.5$

39. continued
There are several outliers.

Conclusion: The distribution is not normal.

40.
Histogram:

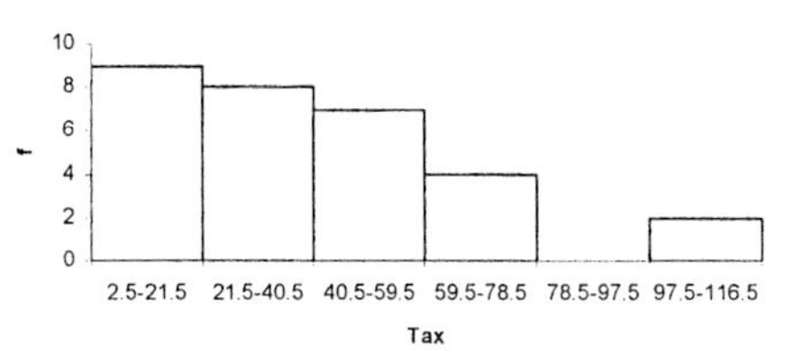

The histogram shows a positive skew.

$PI = \frac{3(39.9-33.5)}{27.18} = 0.71$

$IQR = Q_3 - Q_1 = 58 - 20 = 38$
$1.5(IQR) = 1.5(38) = 57$
$Q_1 - 57 = -37$
$Q_3 + 57 = 115$
There are no outliers.

Conclusion: The distribution is not normal.

41.
Histogram:

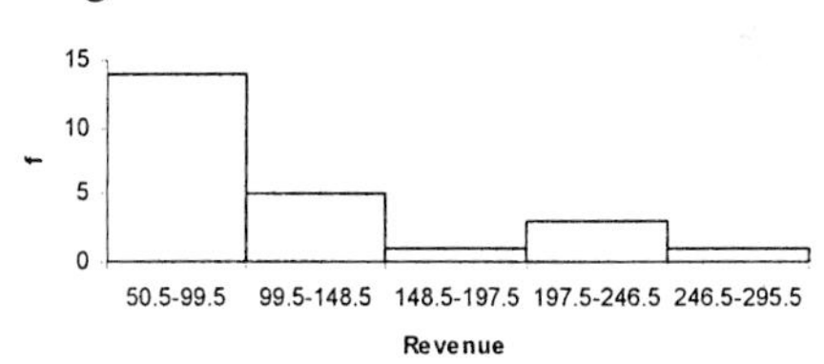

The histogram shows a positive skew.

$PI = \frac{3(115.3-92.5)}{66.32} = 1.03$

$IQR = Q_3 - Q_1 = 154.5 - 67 = 87.5$
$1.5(IQR) = 1.5(87.5) = 131.25$
$Q_1 - 131.25 = -64.25$
$Q_3 + 131.25 = 285.75$
There is one outlier.

Conclusion: The distribution is not normal.

42.
Histogram:

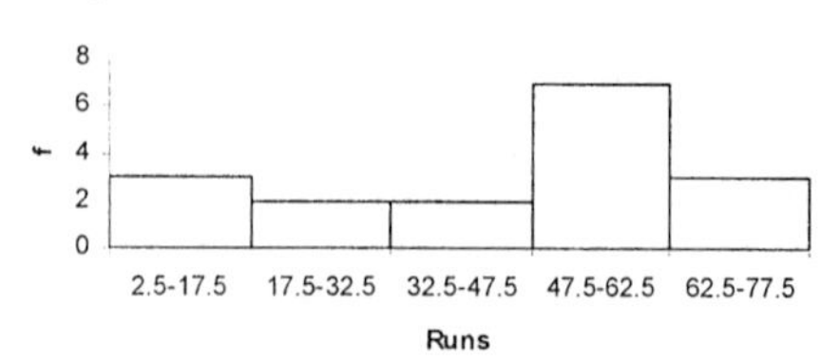

The histogram shows a negative skew.

$PI = \frac{3(45.2-52)}{20.58} = -0.99$

$IQR = Q_3 - Q_1 = 60.5 - 29.5 = 31$
$1.5(IQR) = 1.5(31) = 46.5$
$Q_1 - 46.5 = -17$
$Q_3 + 46.5 = 107$
There are no outliers.

Conclusion: The distribution is not normal.

EXERCISE SET 6-5

1.
The distribution is called the sampling distribution of sample means.

2.
The sample is not a perfect representation of the population. The difference is due to what is called sampling error.

3.
The mean of the sample means is equal to the population mean.

4.
The standard deviation of the sample means is called the standard error of the mean.
$\sigma_X = \frac{\sigma}{\sqrt{n}}$

5.
The distribution will be approximately normal when sample size is large.

6.
$z = \frac{X-\mu}{\sigma}$

7.
$z = \frac{\overline{X}-\mu}{\sigma/\sqrt{n}}$

8.
$z = \frac{\overline{X}-\mu}{\frac{\sigma}{\sqrt{n}}} = \frac{17-17.2}{\frac{2.5}{\sqrt{55}}} = -0.59$

8. continued

area = 0.2224

$z = \frac{18-17.2}{\frac{2.5}{\sqrt{55}}} = 2.37$ area = 0.4911
$P(-0.59 < z < 2.37) = 0.2224 + 0.4911$
$= 0.7135$ or 71.35%

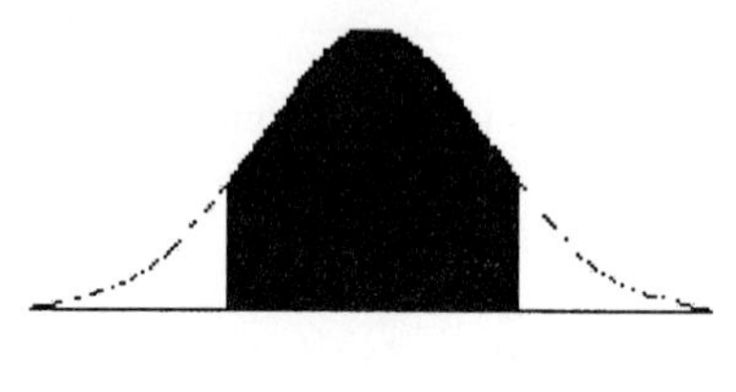

17 17.2 18

9.
$z = \frac{\overline{X}-\mu}{\frac{\sigma}{\sqrt{n}}} = \frac{\$175-\$186.80}{\frac{\$32}{\sqrt{50}}} = -2.61$
area = 0.4955
$P(z < -2.61) = 0.5 - 0.4955 = 0.0045$ or 0.45%

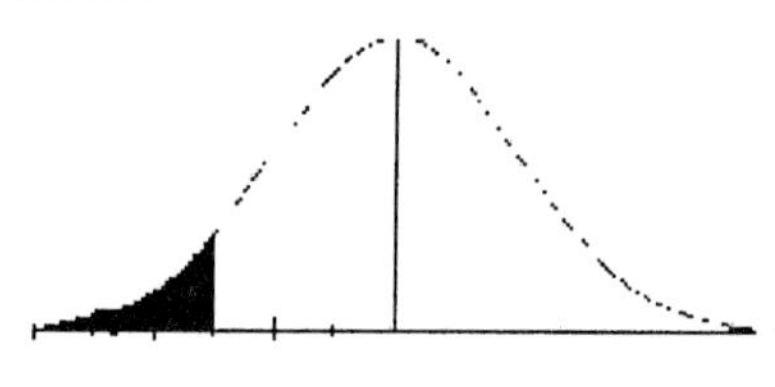

\$175 \$186.8

10.
a. $z = \frac{\$50{,}000-\$52{,}174}{\$7500} = -0.29$
area = 0.1141
$P(z < -0.29) = 0.5 - 0.1141 = 0.3859$ or 38.6%

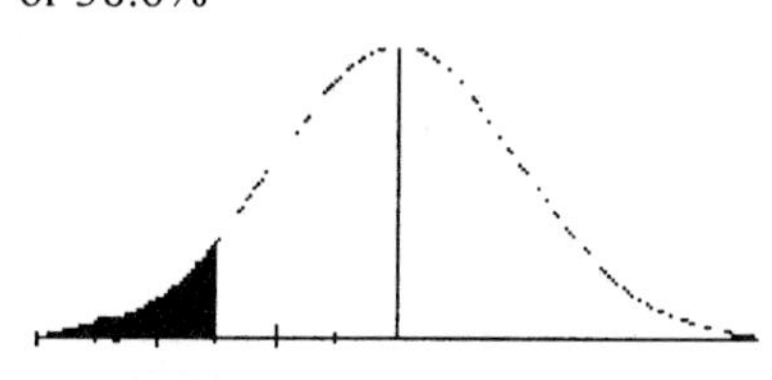

\$50,000 \$52,174

b. $z = \frac{\$50{,}000-\$52{,}174}{\frac{\$7500}{\sqrt{100}}} = -2.90$
area = 0.4981
$P(z < -2.90) = 0.5 - 0.4981 = 0.0019$ or 0.19%

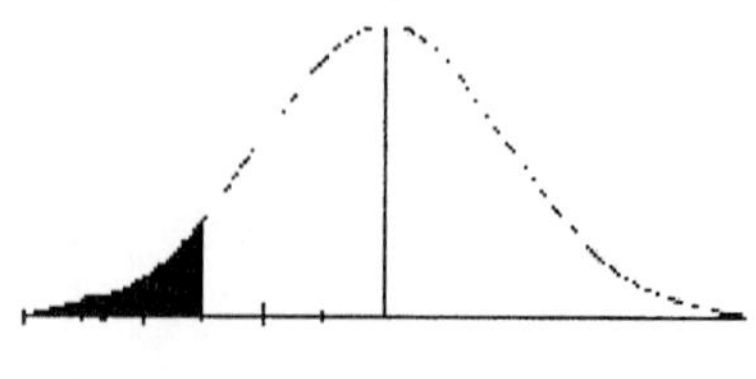

\$50,000 \$52,174

11.
$z = \frac{\overline{X} - \mu}{\frac{\sigma}{\sqrt{n}}} = \frac{128.3 - 126}{\frac{15.7}{\sqrt{25}}} = 0.73$
area = 0.2673
$P(z > 0.73) = 0.5 - 0.2673 = 0.2327$ or 23.27%

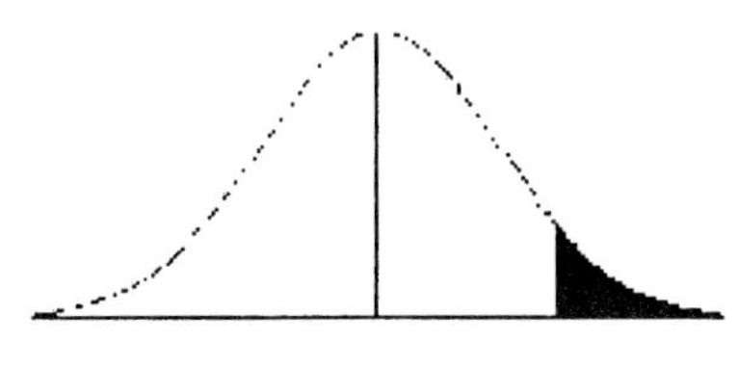

126 128.3

12.
a. $z = \frac{\$40{,}000 - \$29{,}863}{\$5100} = 1.99$
area = 0.4767
$P(z > 1.99) = 0.5 - 0.4767 = 0.0233$ or 2.33%

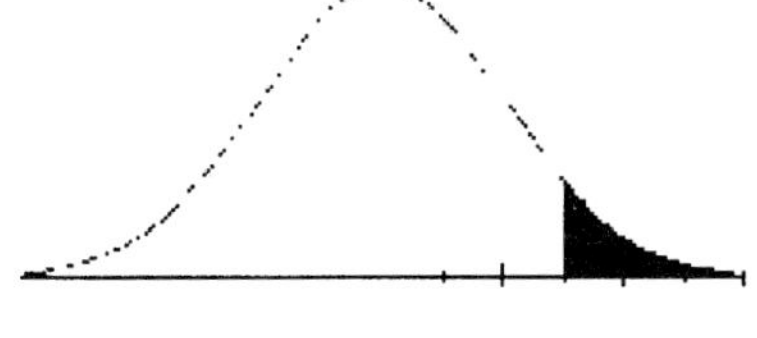

\$29,863 \$40,000

b. $z = \frac{\$30{,}000 - \$29{,}863}{\frac{\$5100}{\sqrt{80}}} = 0.24$
area = 0.0948
$P(z > 0.24) = 0.5 - 0.0948 = 0.4052$ or 40.52%

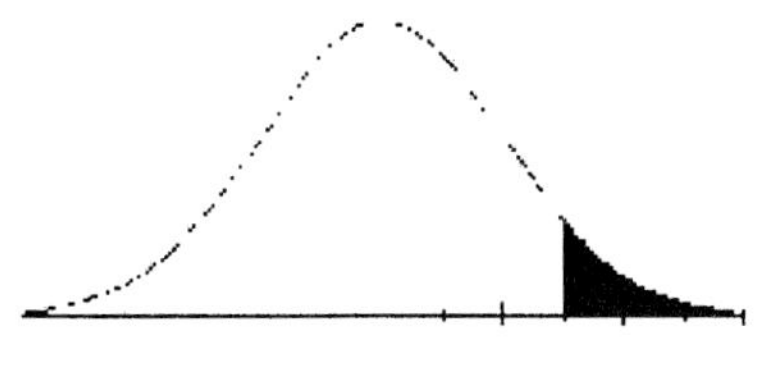

\$29,863 \$30,000

13.
$z = \frac{\overline{X} - \mu}{\frac{\sigma}{\sqrt{n}}} = \frac{\$2.00 - \$2.02}{\frac{\$0.08}{\sqrt{40}}} = -1.58$
area = 0.4429
$P(z < -1.58) = 0.5 - 0.4429 = 0.0571$ or 5.71%

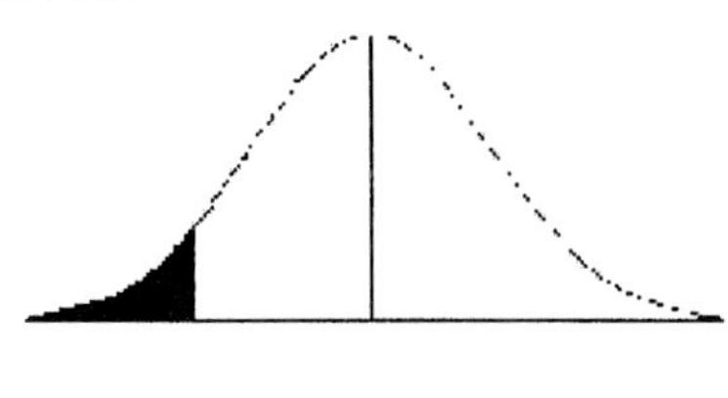

\$2.00 \$2.02

14.
$z = \frac{1050 - 1019}{\frac{100}{\sqrt{200}}} = 4.38$
area = 0.4999
$P(z \geq 4.38) = 0.5 - 0.4999 = 0.0001$ or 0.01%
(Note: Using the TI-83 Plus the answer is 0.000006)
Hence, we would be surprised to get a sample mean of 1050 since the probability is very small.

15.
$z = \frac{\overline{X} - \mu}{\frac{\sigma}{\sqrt{n}}} = \frac{27 - 30}{\frac{5}{\sqrt{22}}} = -2.81$
area = 0.4975
$z = \frac{\overline{X} - \mu}{\frac{\sigma}{\sqrt{n}}} = \frac{31 - 30}{\frac{5}{\sqrt{22}}} = 0.94$ area = 0.3264
$P(-2.81 < z < 0.94) = 0.4975 + 0.3264 = 0.8239$ or 82.39%

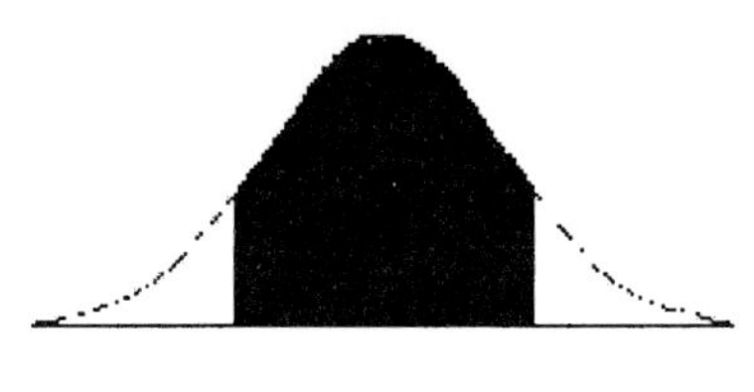

27 30 31

16.
$z = \frac{2.7 - 3.1}{\frac{0.9}{\sqrt{47}}} = -3.05$ area = 0.4989
$P(z < -3.05) = 0.5 - 0.4989 = 0.0011$ or 0.11%

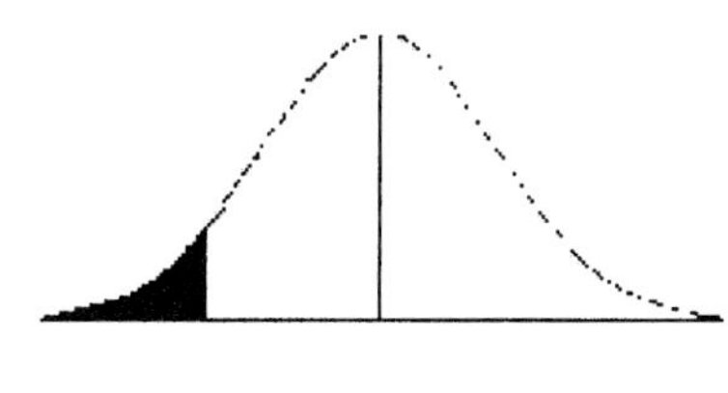

2.7 3.1

17.
$z = \frac{\overline{X} - \mu}{\frac{\sigma}{\sqrt{n}}} = \frac{44.2 - 43.6}{\frac{5.1}{\sqrt{50}}} = 0.83$
area = 0.2967
$P(z > 0.83) = 0.5 - 0.2967 = 0.2033$ or 20.33%

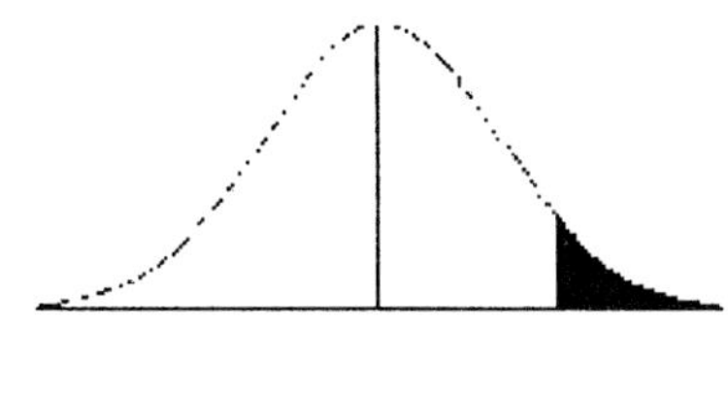

43.6 44.2

18.

$z = \frac{32-30.83}{\frac{5}{\sqrt{10}}} = 0.74$ area = 0.2704

$z = \frac{33-30.83}{\frac{5}{\sqrt{10}}} = 1.37$ area = 0.4147

$P(0.74 < z < 1.37) = 0.4147 - 0.2704$
$= 0.1443$ or 14.43%

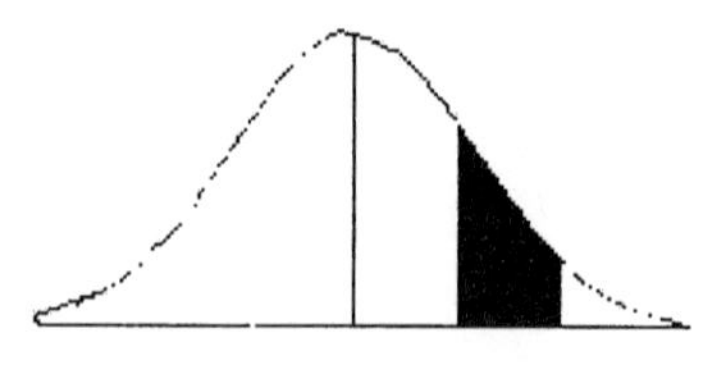

30.83 32 33

19.

$z = \frac{\overline{X}-\mu}{\frac{\sigma}{\sqrt{n}}} = \frac{1980-2000}{\frac{187.5}{\sqrt{50}}} = -0.75$

area = 0.2734

$z = \frac{\overline{X}-\mu}{\frac{\sigma}{\sqrt{n}}} = \frac{1990-2000}{\frac{187.5}{\sqrt{50}}} = -0.38$

area = 0.1480

$P(-0.75 < z < -0.38) = 0.2734 - 0.1480 = 0.1254$ or 12.54%

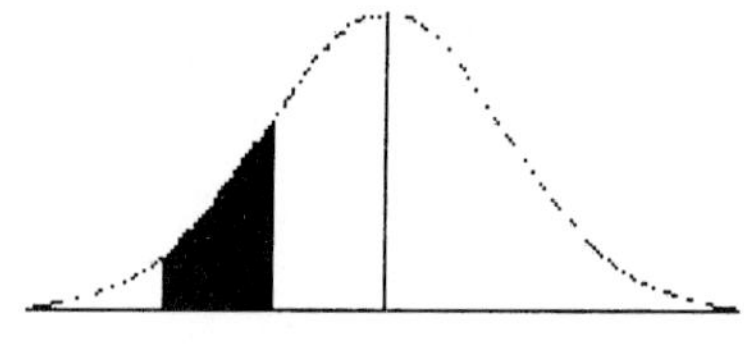

1980 1990 2000

20.

a. $z = \frac{X-\mu}{\sigma} = \frac{26{,}000-24{,}393}{4362} = 0.37$

area = 0.1443

$P(z < 0.37) = 0.5 + 0.1443 = 0.6443$ or 64.43%

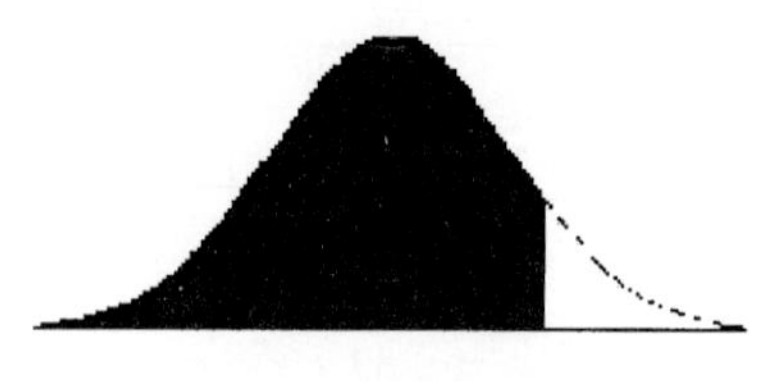

26,000

b. $z = \frac{26{,}000-24{,}393}{\frac{4362}{\sqrt{25}}} = 1.84$

area = 0.4671

$P(z < 1.84) = 0.5 + 0.4671 = 0.9671$ or 96.71%

20b. continued

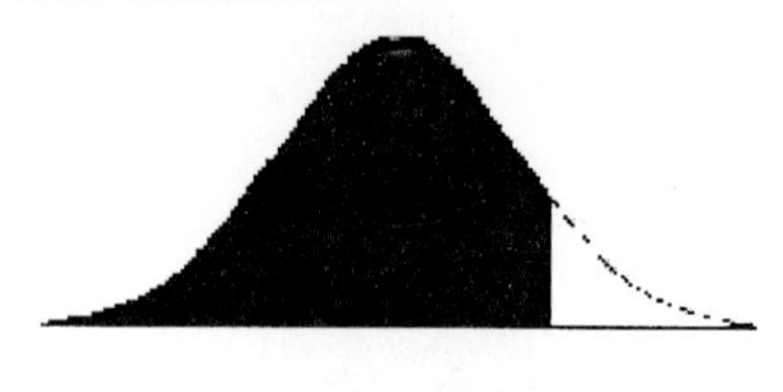

\$26,000

c. Sample means are less variable than individual means.

21.

a. $z = \frac{X-\mu}{\sigma} = \frac{43-46.2}{8} = -0.4$

area = 0.1554

$P(z < -0.4) = 0.5 - 0.1554 = 0.3446$ or 34.46%

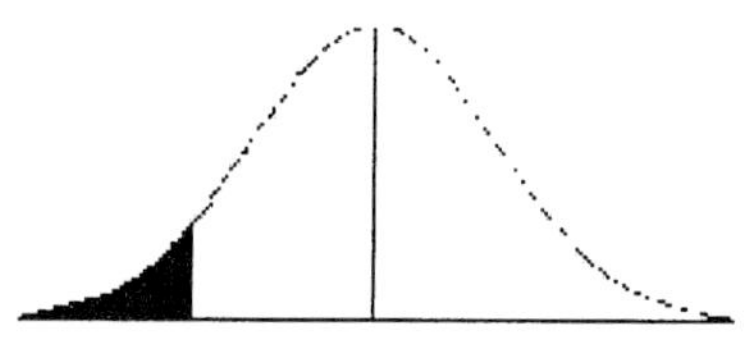

43 46.2

b. $z = \frac{43-46.2}{\frac{8}{\sqrt{50}}} = -2.83$ area = 0.4977

$P(z < -2.83) = 0.5 - 0.4977 = 0.0023$ or 0.23%

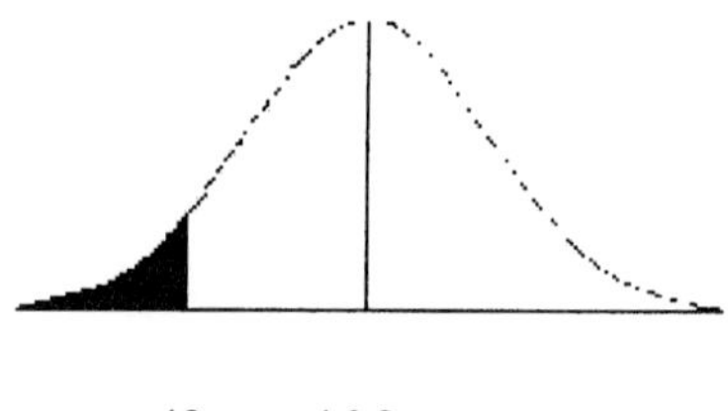

43 46.2

c. Yes, since it is withing one standard deviation of the mean.

d. Very unlikely, since the probability would be less than 1%.

22.

a. $z = \frac{121.8-120}{5.6} = 0.32$

$P(0 < z < 0.32) = 0.1255$ or 12.55%

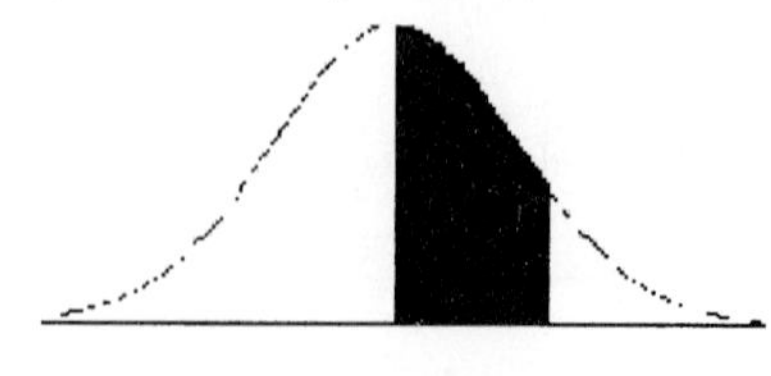

120 121.8

22. continued

b. $z = \frac{121.8-120}{\frac{5.6}{\sqrt{30}}} = 1.76$

$P(0 < z < 1.76) = 0.4608$

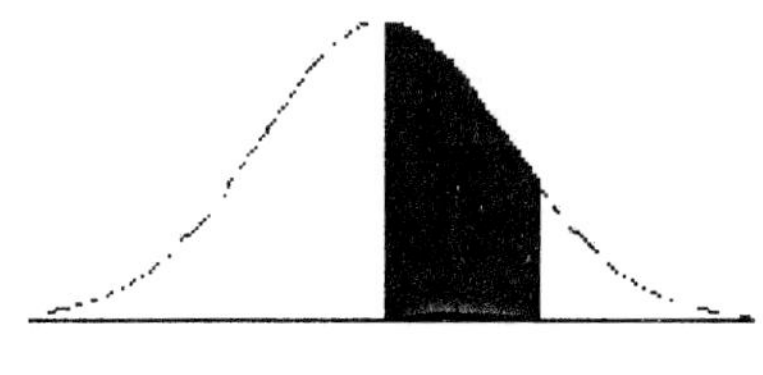

120 121.8

c. Sample means are less variable than individual data.

23.

a. $z = \frac{220-215}{15} = 0.33$ area = 0.1293

$P(z > 0.33) = 0.5 - 0.1293 = 0.3707$ or 37.07%

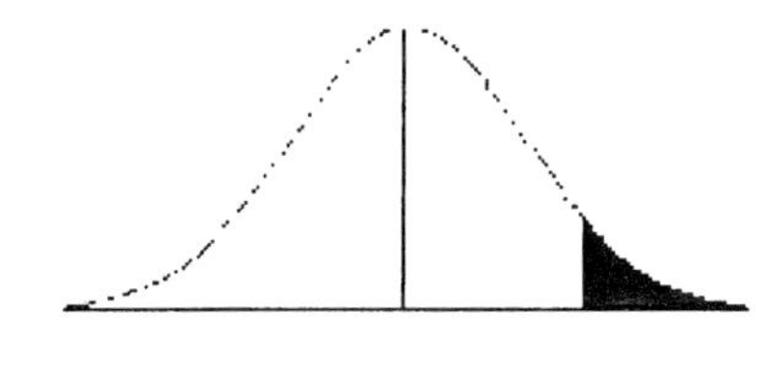

215 220

b. $z = \frac{220-215}{\frac{15}{\sqrt{25}}} = 1.67$ area = 0.4525

$P(z > 1.67) = 0.5 - 0.4525 = 0.0475$ or 4.75%

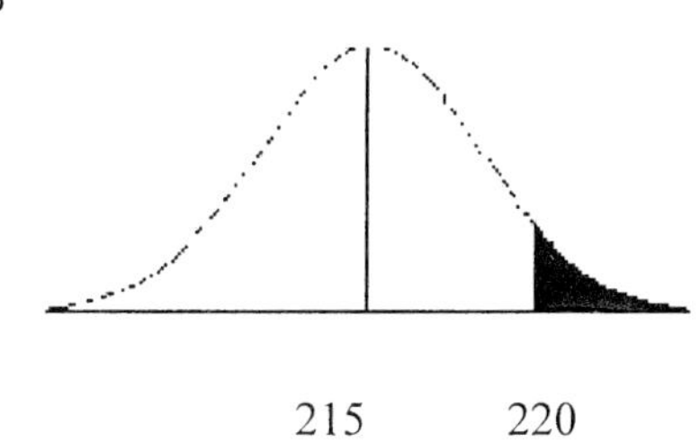

215 220

24.

a. $z_1 = \frac{36-36.2}{3.7} = -0.05$ area = 0.0199

$z_2 = \frac{37.5-36.2}{3.7} = 0.35$ area = 0.1368

$P(-0.05 < z < 0.35) = 0.0199 + 0.1368$
$= 0.1567$ or 15.67%

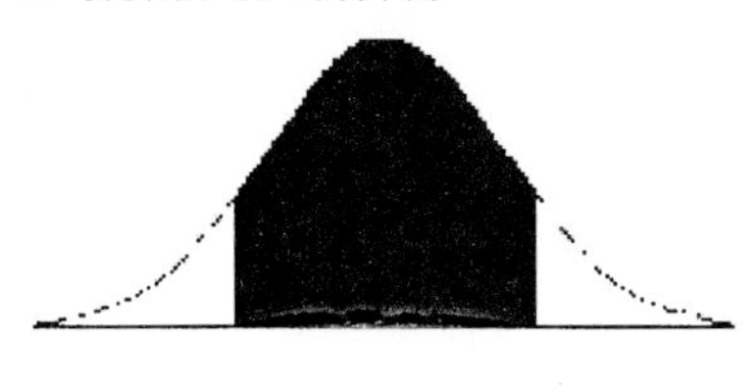

36 36.2 37.5

24. continued

b. $z_1 = \frac{36-36.2}{\frac{3.7}{\sqrt{15}}} = -0.21$ area = 0.0832

$z_2 = \frac{37.5-36.2}{\frac{3.7}{\sqrt{15}}} = 1.36$ area = 0.4131

$P(-0.21 < z < 1.36) = 0.0832 + 0.4131$
$= 0.4963$ or 49.63%

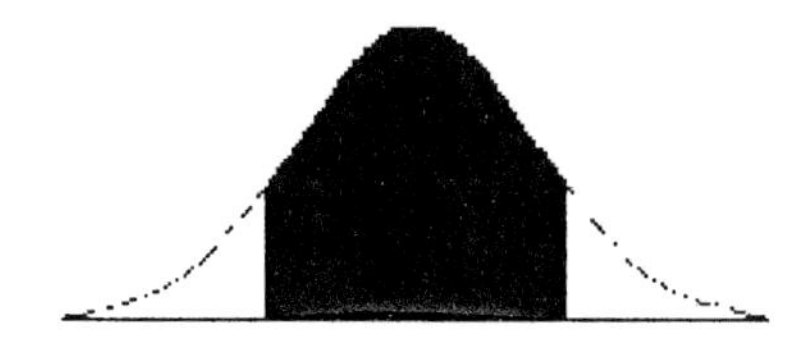

36 36.2 37.5

25.

a. $z_1 = \frac{46-48.25}{4.20} = -0.54$ area = 0.2054

$z_2 = \frac{48-48.25}{4.20} = -0.06$ area = 0.0239

$P(-0.54 < z < -0.06) = 0.2054 - 0.0239 = 0.1815$ or 18.15%

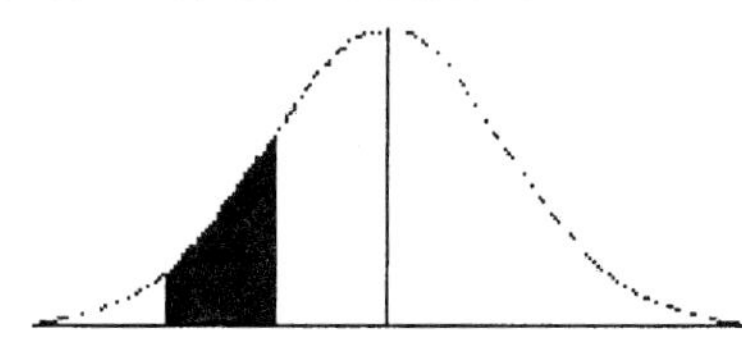

46 48 48.25

b. $z_1 = \frac{46-48.25}{\frac{4.20}{\sqrt{20}}} = -2.40$ area = 0.4918

$z_2 = \frac{48-48.25}{\frac{4.20}{\sqrt{20}}} = -0.27$ area = 0.1064

$P(-2.40 < z < -0.27) = 0.4918 - 0.1064 = 0.3854$ or 38.54%

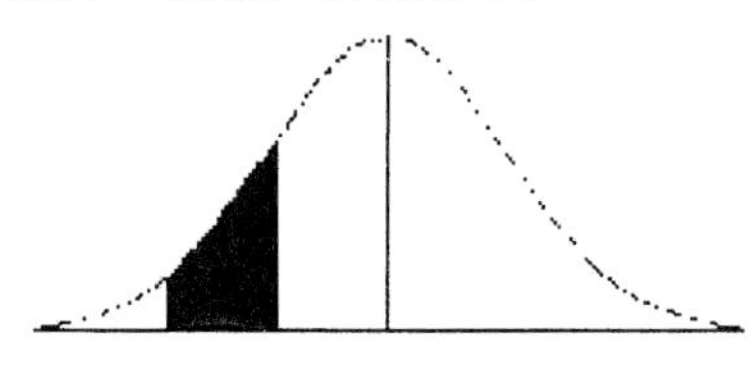

46 48 48.25

c. Means are less variable than individual data.

26.

Since $50 > 0.05(500)$ or 25, the correction factor must be used.

It is $\sqrt{\frac{500-50}{500-1}} = 0.950$

26. continued

$z = \frac{\overline{X}-\mu}{\frac{\sigma}{\sqrt{n}}\cdot\sqrt{\frac{N-n}{n-1}}} = \frac{70-72}{\frac{5.3}{\sqrt{50}}\cdot(0.95)} = -2.81$

area = 0.5 − 0.4975 = 0.0025

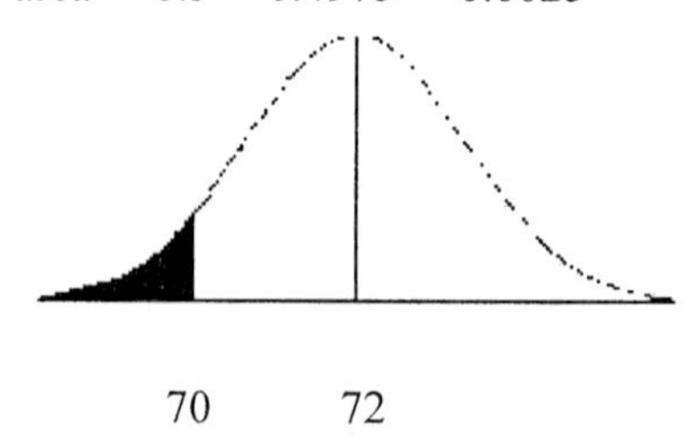

70 72

27.

Since 50 > 0.05(800) or 40, the correction factor is necessary.

It is $\sqrt{\frac{800-50}{800-1}} = 0.969$

$z = \frac{\overline{X}-\mu}{\frac{\sigma}{\sqrt{n}}\cdot\sqrt{\frac{N-n}{n-1}}} = \frac{83{,}500-82{,}000}{\frac{5000}{\sqrt{50}}(0.969)} = 2.19$

area = 0.4857

P(z > 2.19) = 0.5 − 0.4857 = 0.0143 or 1.43%

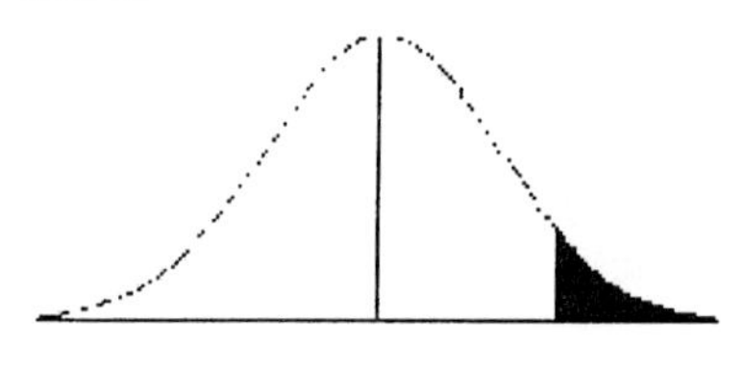

82,000 83,500

28.

The upper 95% is the same as 5% in the left tail; therefore, 45% of the area is between 0 and − z. The corresponding z score is − 1.65.

$-1.65 = \frac{\overline{X}-2000}{\frac{100}{\sqrt{20}}}$

$-1.65(\frac{100}{\sqrt{20}}) + 2000 = \overline{X}$

$\overline{X} = 1963.10$

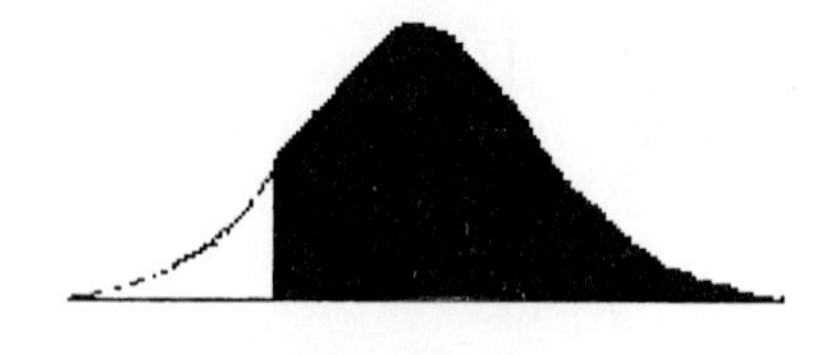

1963.10 2000

29.

$\sigma_{X} = \frac{\sigma}{\sqrt{n}} = \frac{15}{\sqrt{100}} = 1.5$

$2(1.5) = \frac{15}{\sqrt{n}}$

29. continued

$3 \cdot \sqrt{n} = 15$

$\sqrt{n} = 5$

n = 25, the sample size necessary to double the standard error.

30.

$\frac{1.5}{2} = \frac{15}{\sqrt{n}}$

$0.75 \cdot \sqrt{n} = 15$

$\sqrt{n} = \frac{15}{0.75} = 20$

n = 400, the sample size necessary to cut the standard error in half.

EXERCISE SET 6-6

1.

When p is approximately 0.5, and as *n* increases, the shape of the binomial distribution becomes similar to the normal distribution.The normal approximation should be used only when $n \cdot p$ and $n \cdot q$ are both greater than or equal to 5. The correction for continuity is necessary because the normal distribution is continuous and the binomial is discrete.

2.

For each problem use the following formulas:

$\mu = np \quad \sigma = \sqrt{npq} \quad z = \frac{\overline{X}-\mu}{\sigma}$

Be sure to correct each X for continuity.

a. $\mu = 0.5(30) = 15$

$\sigma = \sqrt{(0.5)(0.5)(30)} = 2.74$

$z = \frac{17.5-15}{2.74} = 0.91$ area = 0.3186

$z = \frac{18.5-15}{2.74} = 1.28$ area = 0.3997

P(17.5 < X < 18.5) = 0.3997 − 0.3186 = 0.0811 = 8.11%

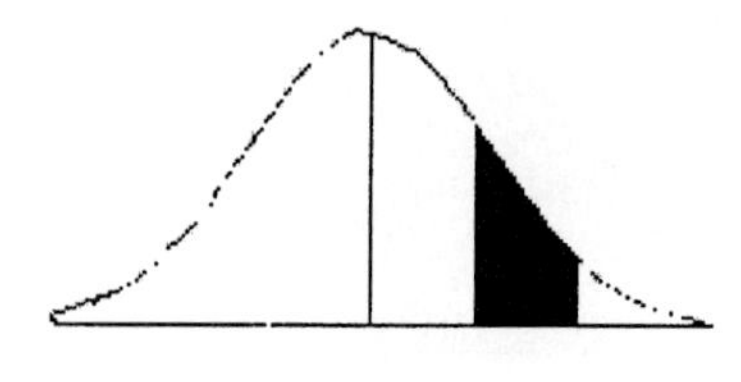

15 17.5 18.5

b. $\mu = 0.8(50) = 40$
$\sigma = \sqrt{(50)(0.8)(0.2)} = 2.83$

$z = \frac{43.5-40}{2.83} = 1.24$ area = 0.3925

$z = \frac{44.5-40}{2.83} = 1.59$ area = 0.4441

$P(43.5 < X < 44.5) = 0.4441 - 0.3925$
$= 0.0516$ or 5.16%

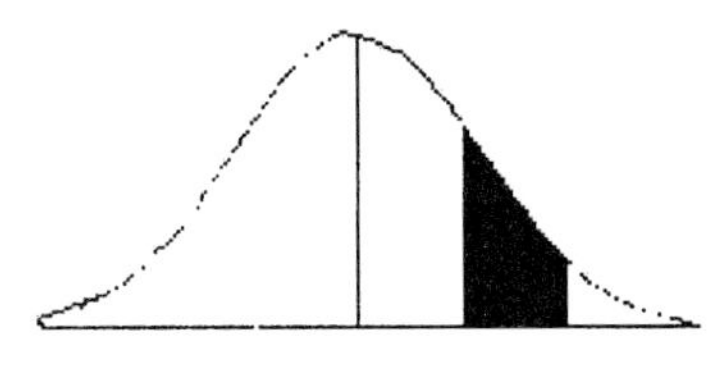

40 43.5 44.5

c. $\mu = 0.1(100) = 10$
$\sigma = \sqrt{(0.1)(0.9)(100)} = 3$

$z = \frac{11.5-10}{3} = 0.50$ area = 0.1915

$z = \frac{12.5-10}{3} = 0.83$ area = 0.2967

$P(11.5 < X < 12.5) = 0.2967 - 0.1915$
$= 0.1052$ or 10.52%

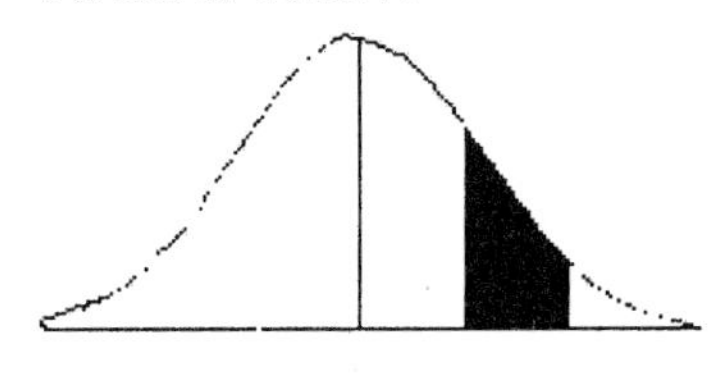

10 11.5 12.5

d. $\mu = 10(0.5) = 5$
$\sigma = \sqrt{(0.5)(0.5)(10)} = 1.58$

$z = \frac{6.5-5}{1.58} = 0.95$ area = 0.3289

$P(X \geq 6.5) = 0.5 - 0.3289 = 0.1711$ or
17.11%

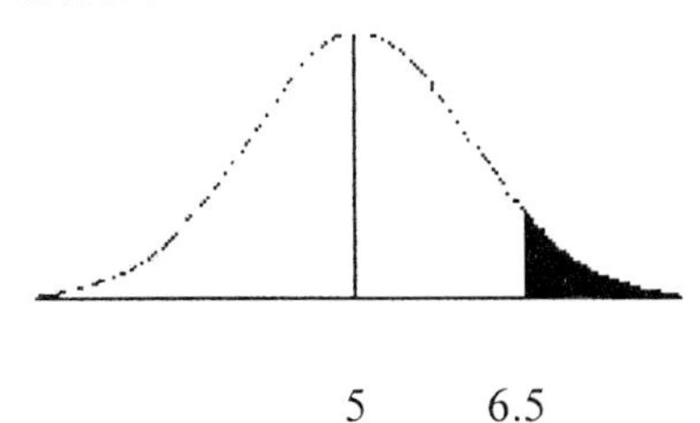

5 6.5

e. $\mu = 20(0.7) = 14$
$\sigma = \sqrt{(20)(0.7)(0.3)} = 2.05$

2e. continued
$z = \frac{12.5-14}{2.05} = -0.73$ area = 0.2673

$P(X \leq 12.5) = 0.5 - 0.2673 = 0.2327$ or
23.27%

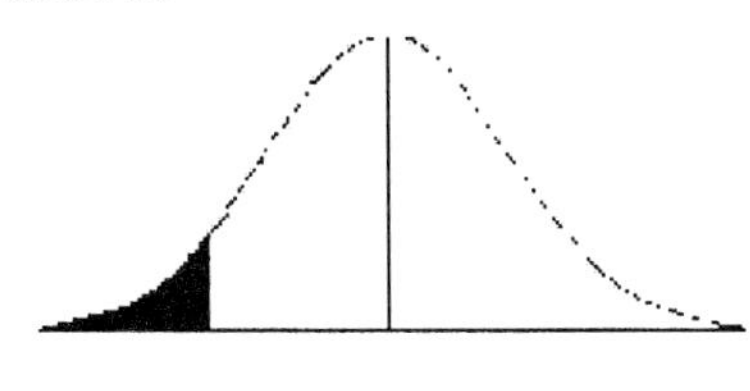

12.5 14

f. $\mu = 50(0.6) = 30$
$\sigma = \sqrt{(50)(0.6)(0.4)} = 3.46$

$z = \frac{40.5-30}{3.46} = 3.03$ area = 0.4988

$P(X \leq 40.5) = 0.5 + 0.4988 = 0.9988$ or
99.88%

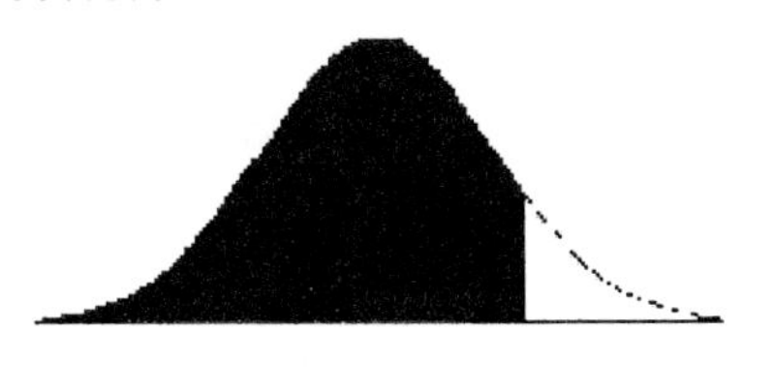

30 40.5

3.
a. $np = 20(0.50) = 10 \geq 5$ Yes
$nq = 20(0.50) = 10 \geq 5$
b. $np = 10(0.60) = 6 \geq 5$ No
$nq = 10(0.40) = 4 < 5$
c. $np = 40(0.90) = 36 \geq 5$ No
$nq = 40(0.10) = 4 < 5$
d. $np = 50(0.20) = 10 \geq 5$ Yes
$nq = 50(0.80) = 40 \geq 5$
e. $np = 30(0.80) = 24 \geq 5$ Yes
$nq = 30(0.20) = 6 \geq 5$
f. $np = 20(0.85) = 17 \geq 5$ No
$nq = 20(0.15) = 3 > 5$

4.
$\mu = 500(0.56) = 280$
$\sigma = \sqrt{(500)(0.56)(0.44)} = 11.1$

$z = \frac{249.5-280}{11.1} = -2.75$ area = 0.4970

$P(X > 249.5) = 0.5 + 0.4970 = 0.9970$ or
99.7%

4. continued

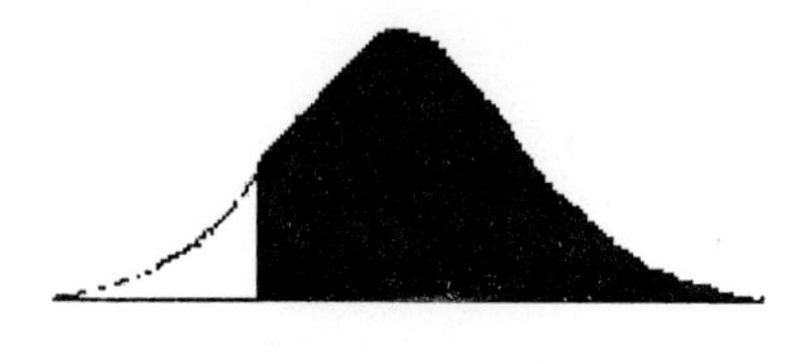

5.
$p = \frac{2}{5} = 0.4$ $\mu = 400(0.4) = 160$
$\sigma = \sqrt{(400)(0.4)(0.6)} = 9.8$

$z = \frac{169.5-160}{9.8} = 0.97$ area $= 0.3340$

$P(X > 169.5) = 0.5 - 0.3340 = 0.1660$ or 16.6%

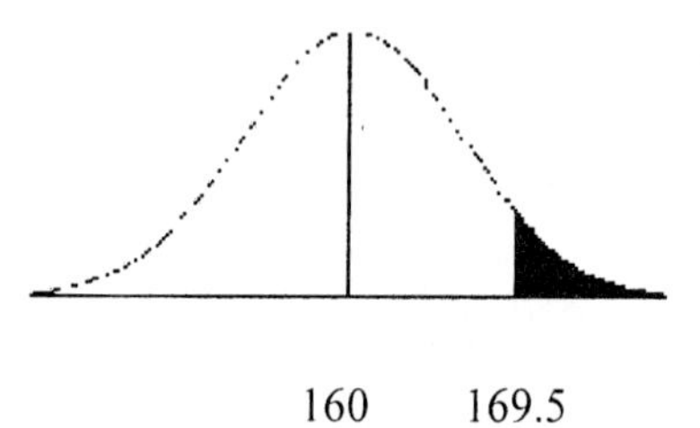

6.
$\mu = 100(0.05) = 5$
$\sigma = \sqrt{(100)(0.05)(0.95)} = 2.18$

$z = \frac{5.5-5}{2.18} = 0.23$ area $= 0.0910$

$P(X > 5.5) = 0.5 - 0.0910 = 0.4090$ or 40.9%

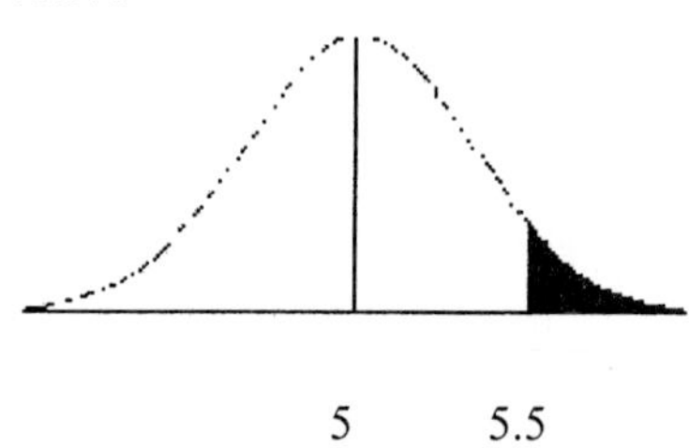

7.
$\mu = 300(0.509) = 152.7$
$\sigma = \sqrt{(300)(0.509)(0.491)} = 8.66$

$z = \frac{175.5-152.7}{8.66} = 2.63$ area $= 0.4957$

$P(X > 175.5) = 0.5 - 0.4957 = 0.0043$

7. continued

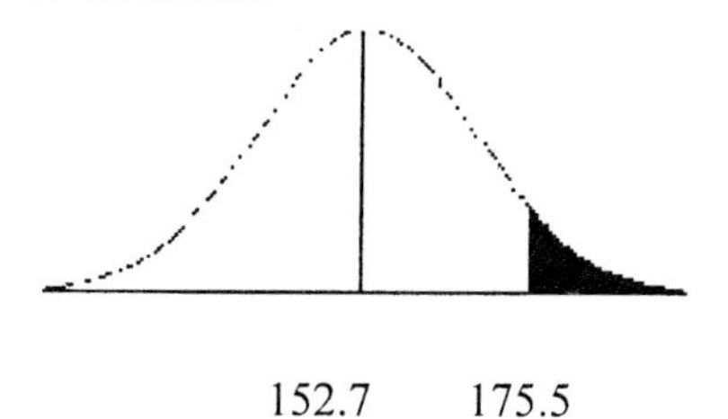

8.
$\mu = 175(0.53) = 92.75$
$\sigma = \sqrt{(175)(0.53)(0.47)} = 6.60$

$z = \frac{75.5-92.75}{6.60} = -2.61$ area $= 0.4955$

$z = \frac{109.5-92.75}{6.60} = 2.54$ area $= 0.4945$

$P(75.5 < X < 109.5) = 0.4955 + 0.4945$
$P(75.5 < X < 109.5) = 0.99$

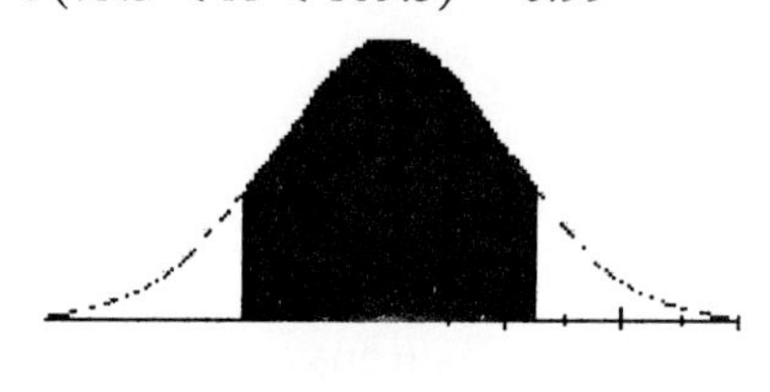

9.
$\mu = 180(0.236) = 42.48$
$\sigma = \sqrt{(180)(0.236)(0.764)} = 5.70$

$z = \frac{50.5-42.48}{5.70} = 1.41$ area $= 0.4207$

$P(X > 50.5) = 0.5 - 0.4207 = 0.0793$

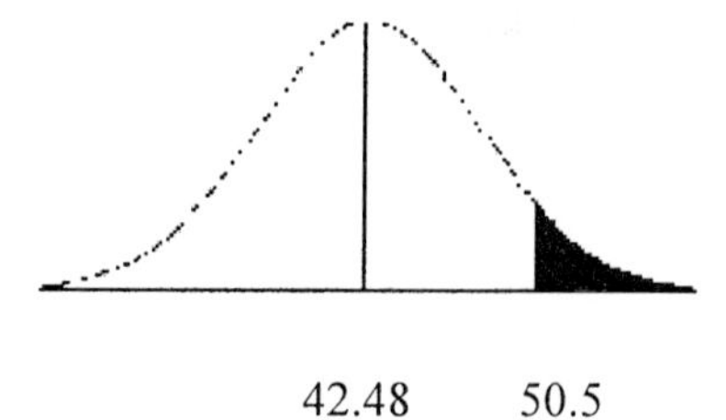

10.
$\mu = 400(0.24) = 96$
$\sigma = \sqrt{(400)(0.24)(0.76)} = 8.54$

$z = \frac{120.5-96}{8.54} = 2.87$ area $= 0.4979$

$P(X > 120.5) = 0.5 - 0.4979 = 0.0021$

10. continued

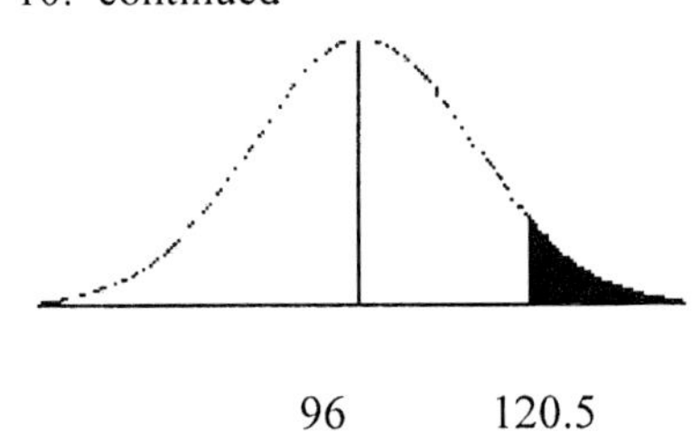

96 120.5

11.

$\mu = 300(0.167) = 50.1$

$\sigma = \sqrt{(300)(0.167)(0.833)} = 6.46$

$z = \frac{50.5-50.1}{6.46} = 0.06$ area $= 0.0239$

$P(X > 50.5) = 0.5 - 0.0239 = 0.4761$

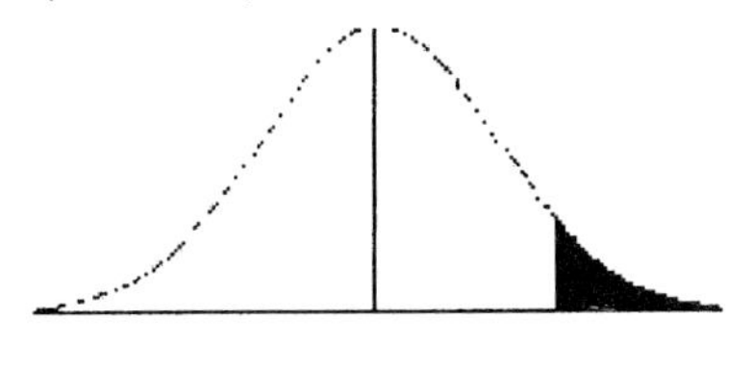

50.1 50.5

12.

$\mu = 290(0.23) = 66.7$

$\sigma = \sqrt{(290)(0.23)(0.77)} = 7.17$

$z = \frac{50.5-66.7}{7.17} = -2.26$ area $= 0.4881$

$P(X > 50.5) = 0.5 + 0.4881 = 0.9881$

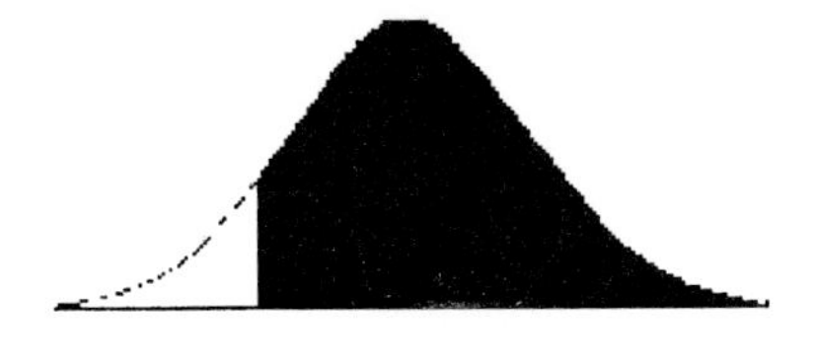

50.5 66.7

13.

$\mu = 400(0.3) = 120$

$\sigma = \sqrt{(400)(0.3)(0.7)} = 9.17$

$z = \frac{99.5-120}{9.17} = -2.24$

$P(X > 99.5) = 0.5 + 0.4875 = 0.9875$ or 98.75%

13. continued

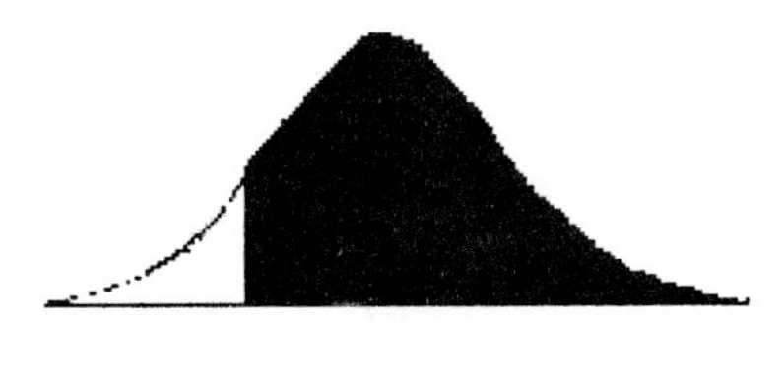

99.5 120

14.

a. $n(0.1) = 5$ $n \geq 50$

b. $n(0.3) = 5$ $n \geq 17$

c. $n(0.5) = 5$ $n \geq 10$

d. $n(0.2) = 5$ $n \geq 25$

e. $n(0.1) = 5$ $n \geq 50$

REVIEW EXERCISES - CHAPTER 6

1.

a. 0.4744

0 1.95

b. 0.1443

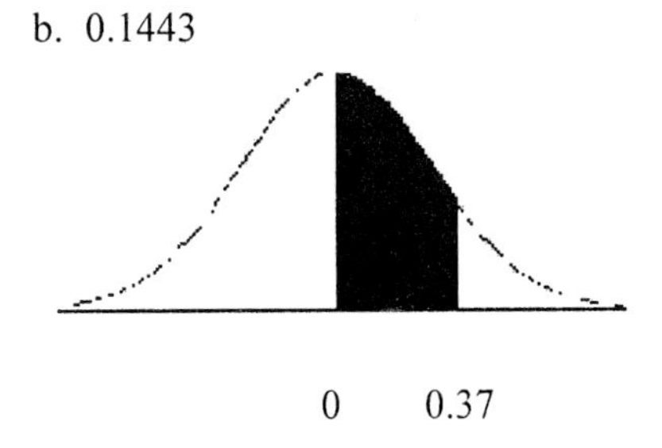

0 0.37

c. $0.4656 - 0.4066 = 0.0590$

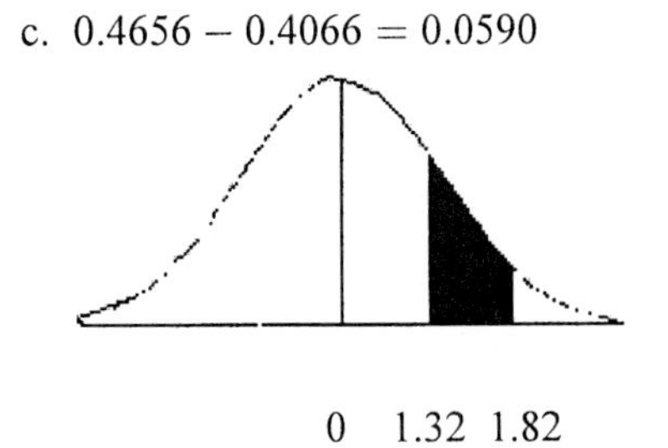

0 1.32 1.82

1. continued

d. $0.3531 + 0.4798 = 0.8329$

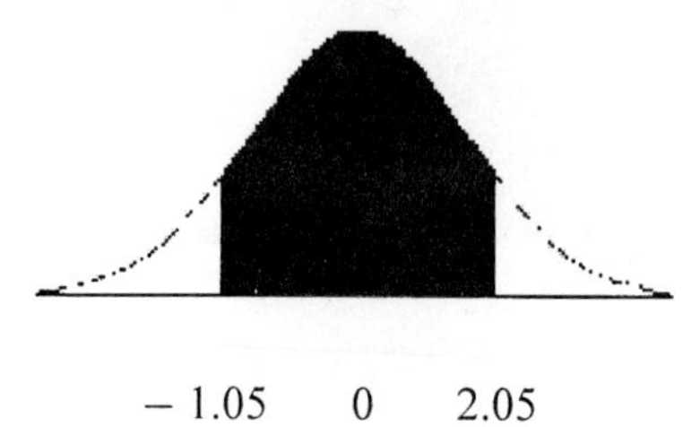

e. $0.2019 + 0.0120 = 0.2139$

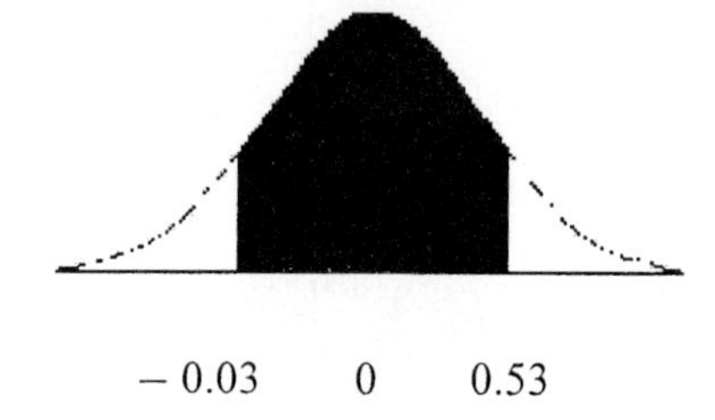

f. $0.3643 + 0.4641 = 0.8284$

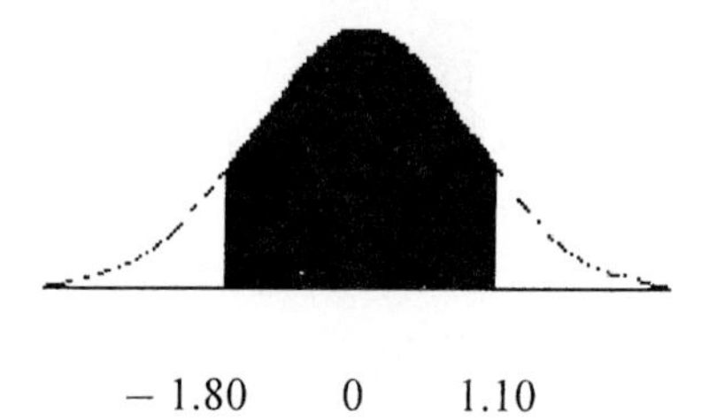

g. $0.5 - 0.4767 = 0.0233$

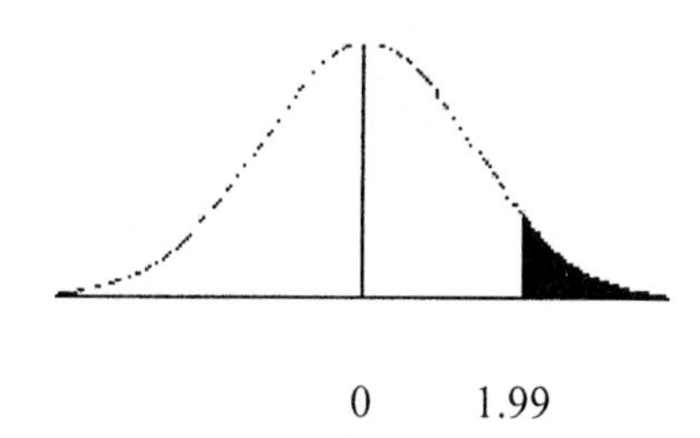

h. $0.5 + 0.4131 = 0.9131$

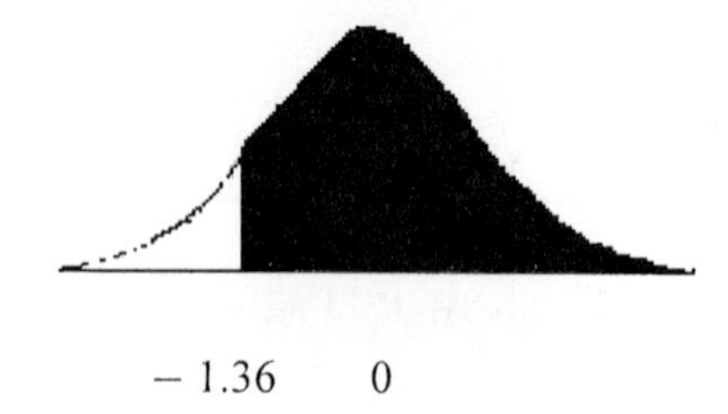

1. continued

i. $0.5 - 0.4817 = 0.0183$

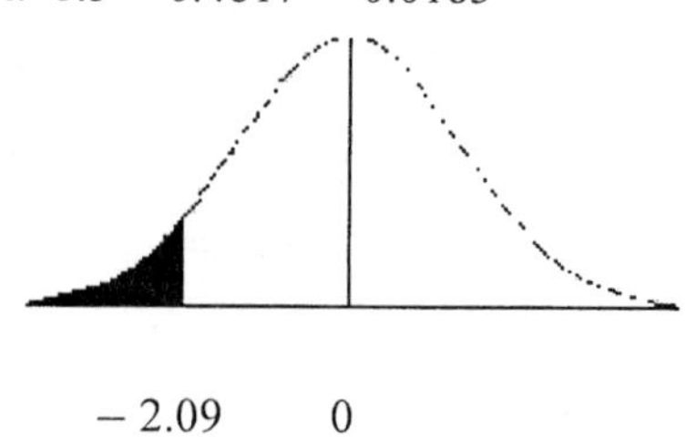

j. $0.5 + 0.4535 = 0.9535$

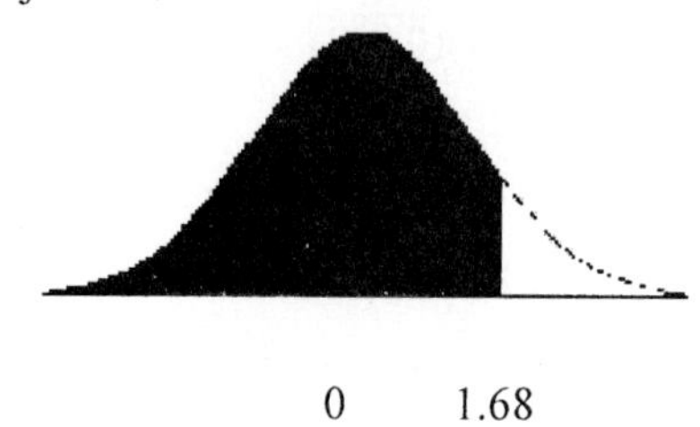

2.

a. 0.4808

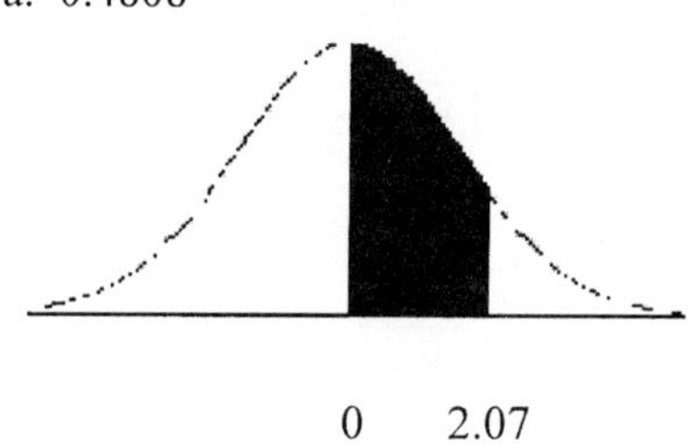

b. 0.4664

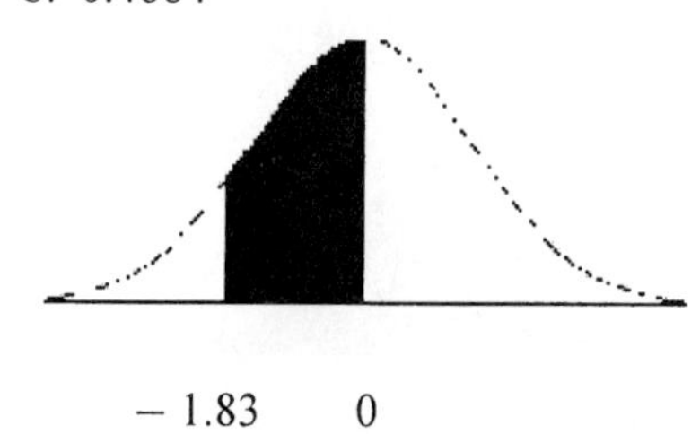

c. $0.4778 + 0.4441 = 0.9219$

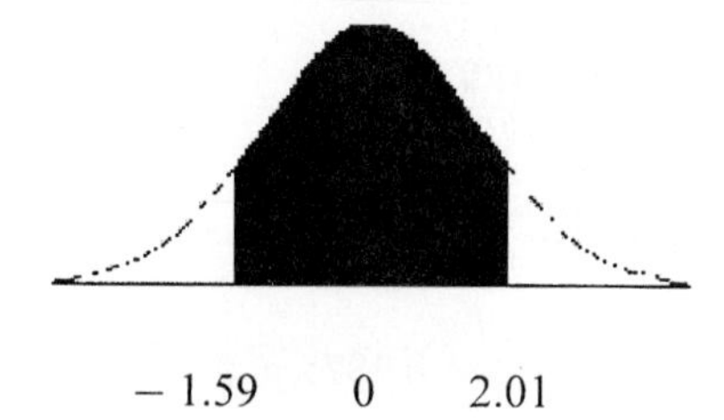

2. continued

d. $0.4699 - 0.4082 = 0.0617$

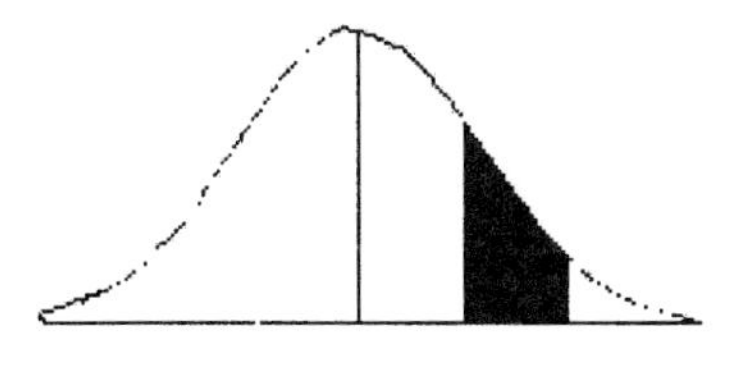

0 1.33 1.88

e. $0.1443 + 0.4948 = 0.6391$

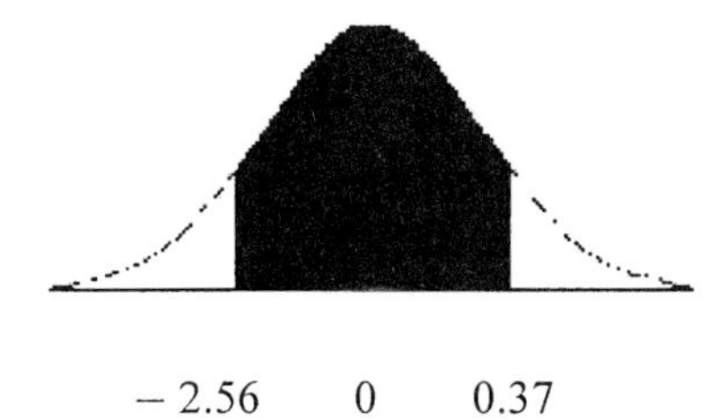

− 2.56 0 0.37

f. $0.5 - 0.4515 = 0.0485$

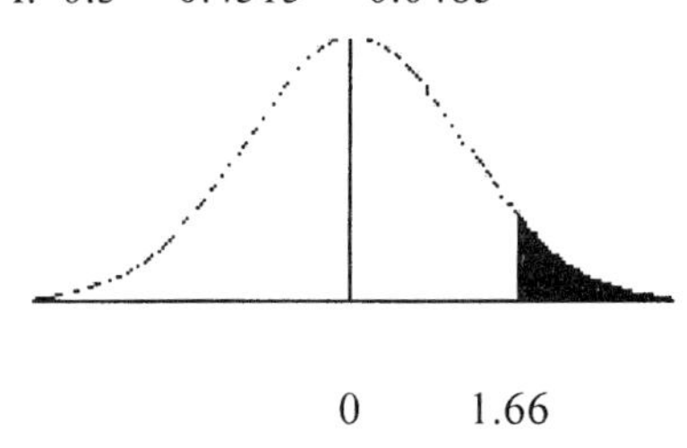

0 1.66

g. $0.5 - 0.4788 = 0.0212$

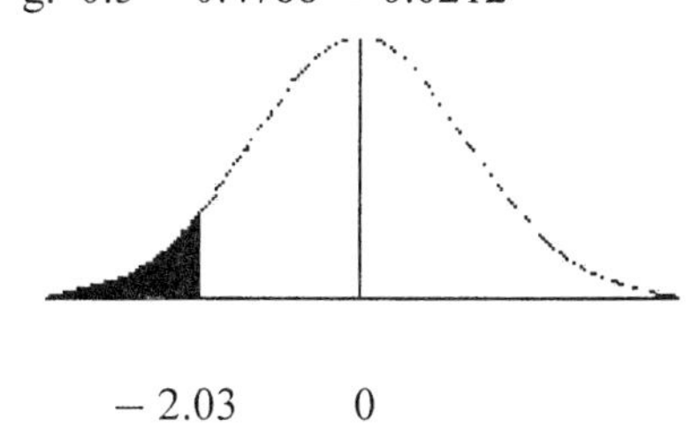

− 2.03 0

h. $0.5 + 0.3830 = 0.8830$

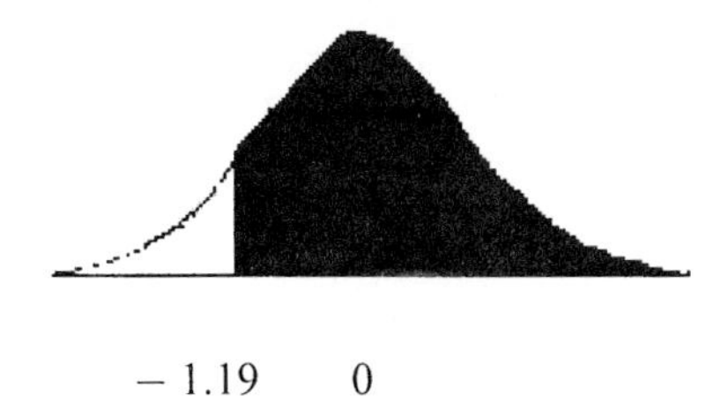

− 1.19 0

2. continued

i. $0.5 + 0.4732 = 0.9732$

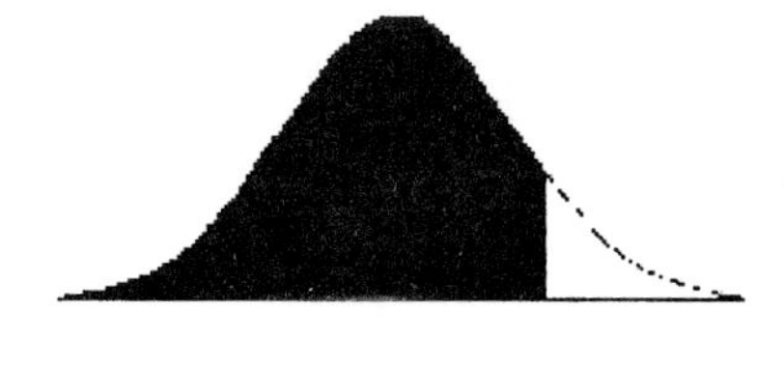

0 1.93

j. $0.5 + 0.4616 = 0.9616$

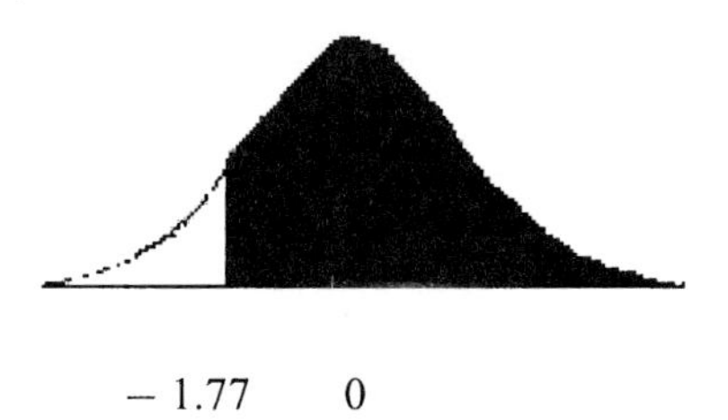

− 1.77 0

3.

a. $z = \frac{10.00-8.99}{3.00} = 0.34$ area $= 0.1331$

$P(z > 0.34) = 0.5 - 0.1331 = 0.3669$

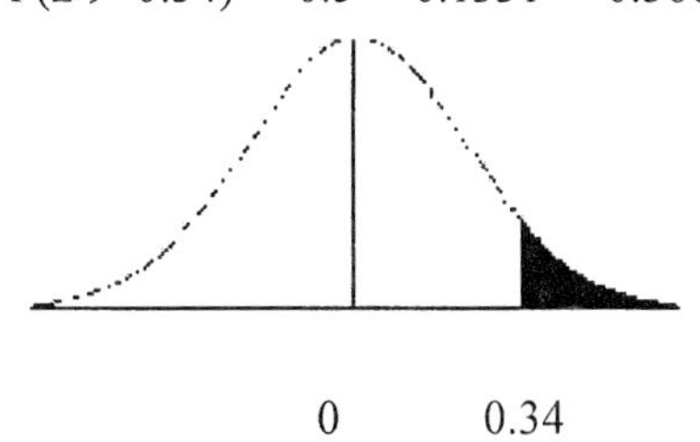

0 0.34

b. $z = \frac{11.00-8.99}{3.00} = 0.67$ area $= 0.2486$

$P(z < 0.67) = 0.5 + 0.2486 = 0.7486$

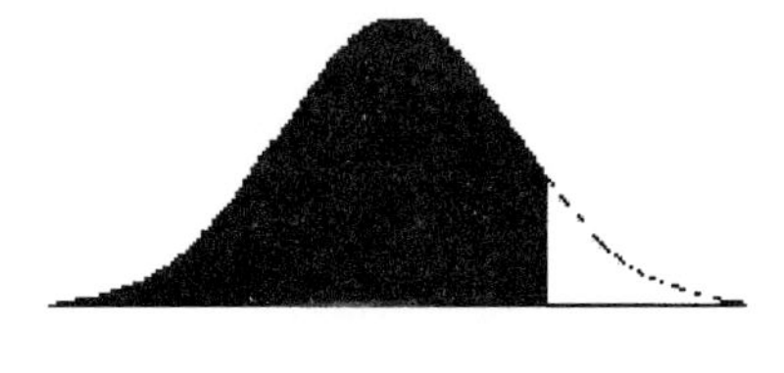

0 0.67

4.

a. $z = \frac{45{,}000-40{,}000}{5000} = 1.00$ area $= 0.3413$

$P(z > 1.00) = 0.5 - 0.3413 = 0.1587$

4a. continued

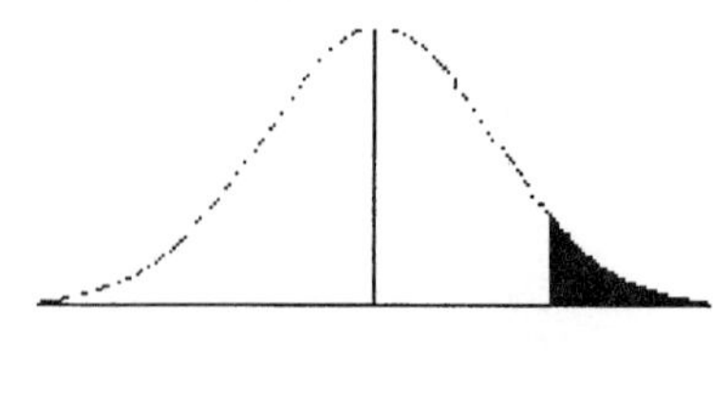

0 1.00

b. $z = \frac{45{,}000-40{,}000}{\frac{5000}{\sqrt{9}}} = 3.00$ area $= 0.4987$

$P(z > 3.00) = 0.5 - 0.4987 = 0.0013$

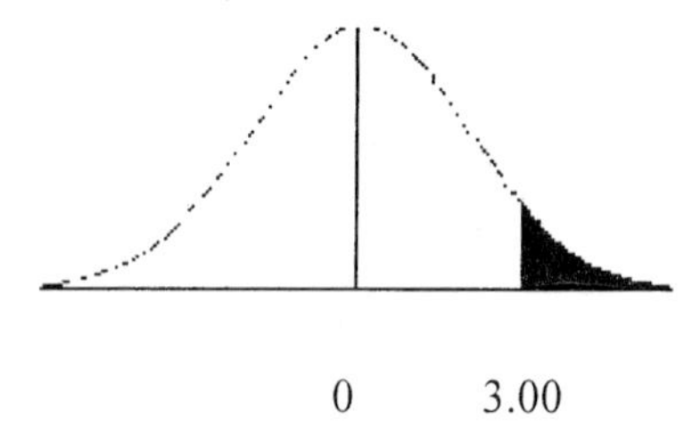

0 3.00

5.
(a) $z = \frac{65-63}{8} = 0.25$ area $= 0.0987$

$P(z > 0.25) = 0.5 - 0.0987 = 0.4013$ or 40.13%

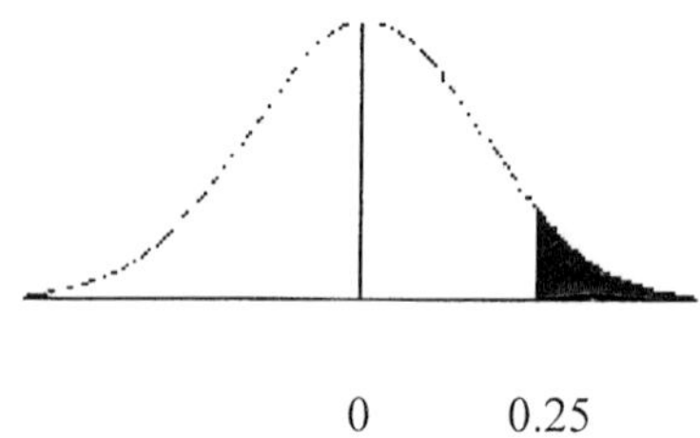

0 0.25

(b) $z = \frac{72-63}{8} = 1.13$ area $= 0.3708$

$P(z > 1.13) = 0.5 - 0.3708 = 0.1292$ or 12.92%

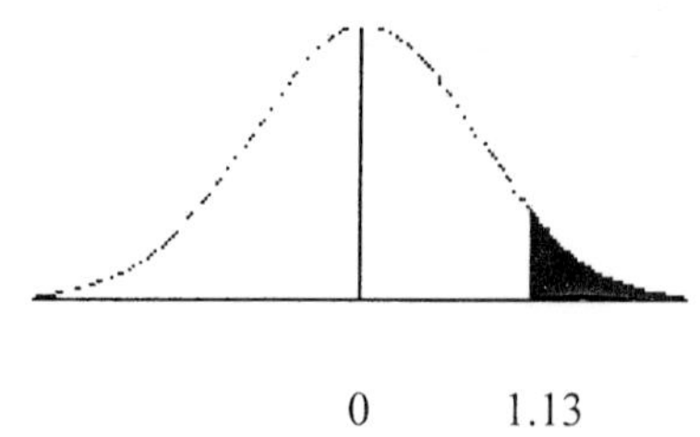

0 1.13

6.
(a) $z = \frac{9-8.28}{3.5} = 0.21$ area $= 0.0832$

$P(z < 0.21) = 0.5 + 0.0832 = 0.5832$

6a. continued

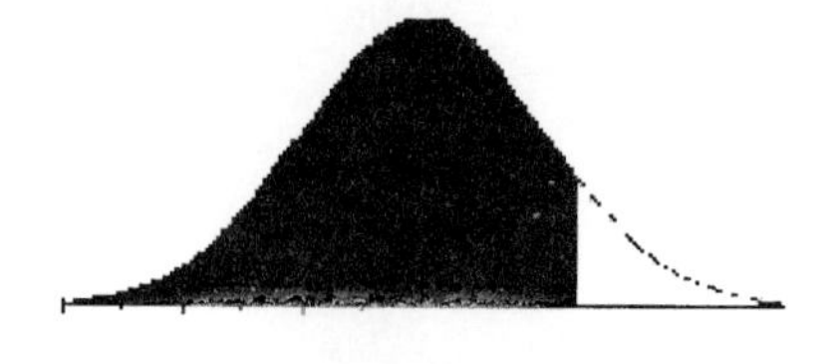

0 0.21

(b) $z = \frac{8-8.28}{3.5} = -0.08$ area $= 0.0319$

$P(z < -0.08) = 0.5 - 0.0319 = 0.4681$

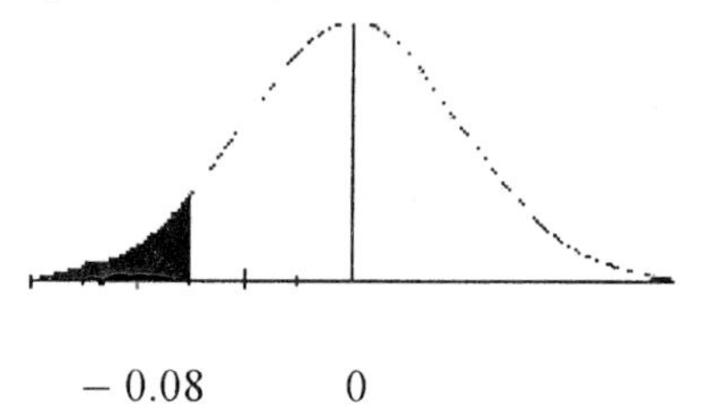

− 0.08 0

7.
(a) $z = \frac{18-19.32}{2.44} = -0.54$ area $= 0.2054$

$P(z > -0.54) = 0.5 + 0.2054 = 0.7054$

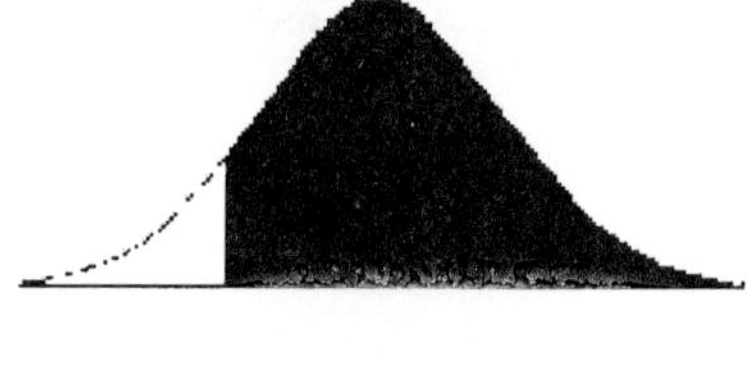

− 0.54 0

(b) $z = \frac{18-19.32}{\frac{2.44}{\sqrt{5}}} = -1.21$ area $= 0.3869$

$P(z > -1.21) = 0.5 + 0.3869 = 0.8869$

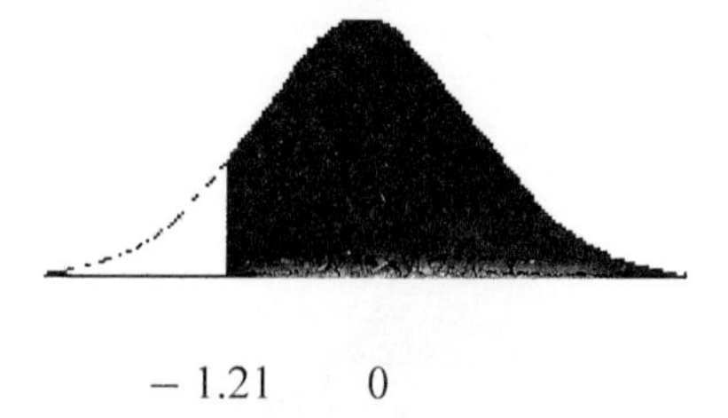

− 1.21 0

8.
The 15% overweight suitcases are in the right tail; the corresponding z score for the area is 1.04.

$X = \mu + z\sigma$

$X = 45 + (1.04)(2)$

$X = 47.08$ lbs

8. continued

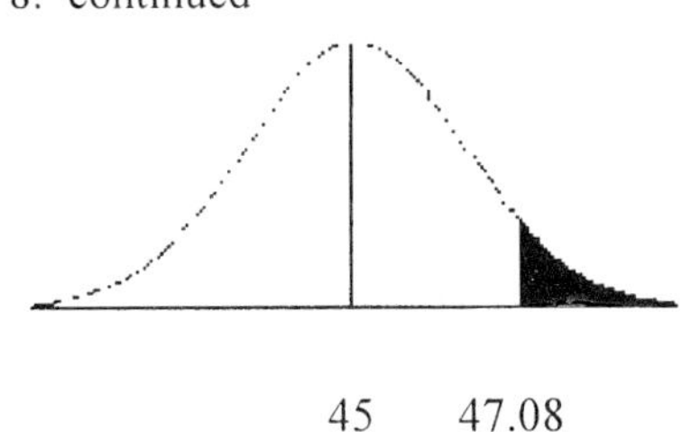

45 47.08

9.
The middle 40% means that 20% of the area is on either side of the mean. The corresponding z scores are ± 0.52.
$X_1 = 100 + (0.52)(15) = 107.8$
$X_2 = 100 + (-0.52)(15) = 92.2$
The scores should be between 92.2 and 107.8.

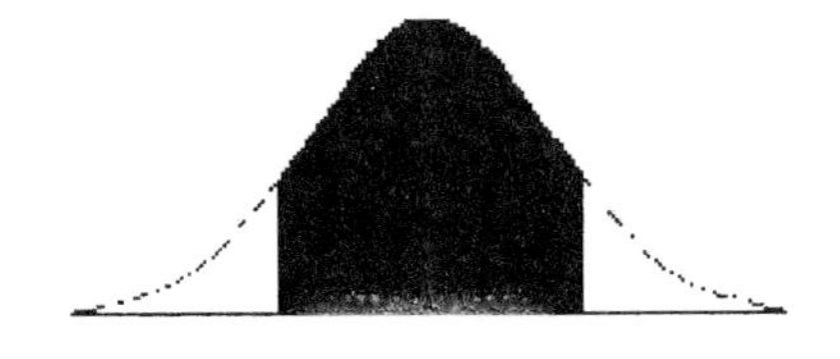

92.2 100 107.8

10.
$z = \frac{X-\mu}{\frac{\sigma}{\sqrt{n}}} = \frac{70-73}{\frac{8}{\sqrt{9}}} = -1.13$
area = 0.3708
$P(\overline{X} < 70) = 0.5 - 0.3708 = 0.1292$ or 12.92%

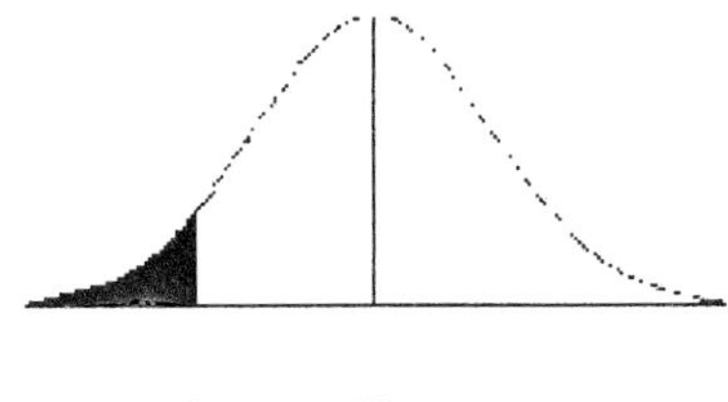

70 73

11.
$z = \frac{11-12.2}{\frac{2.3}{\sqrt{12}}} = -1.81$ area = 0.4649
$P(\overline{X} < 11) = 0.5 - 0.4649 = 0.0351$ or 3.51%

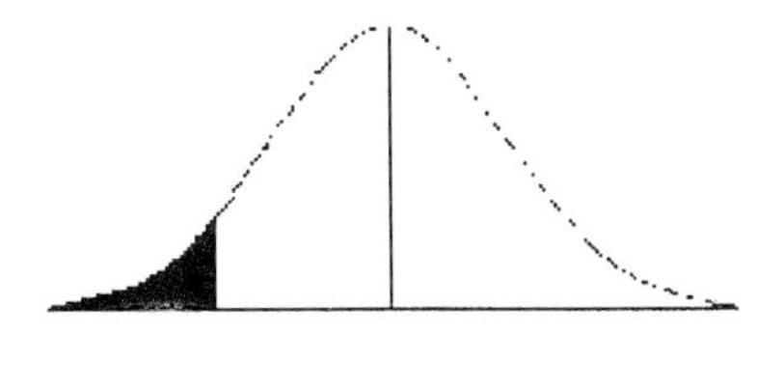

11 12.2

12.
$\mu = np = 500(0.05) = 25$
$\sigma = \sqrt{npq} = \sqrt{(500)(0.05)(0.95)} = 4.87$
$z = \frac{30.5-25}{4.87} = 1.13$ area = 0.3708

$z = \frac{29.5-25}{4.87} = 0.92$ area = 0.3212

$P(0.92 < z < 1.13) = 0.3708 - 0.3212 = 0.0496$ or 4.96%

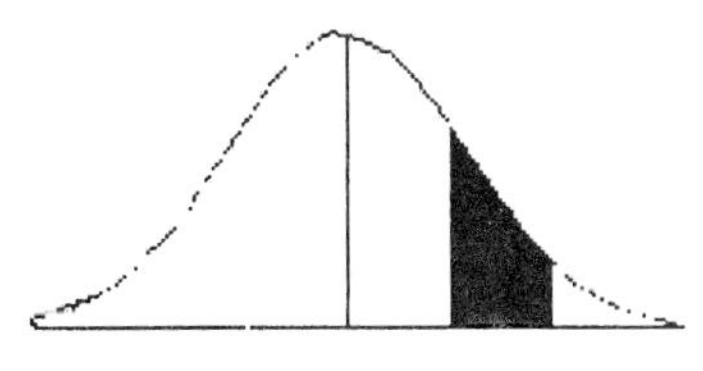

25 29.5 30.5

13.
$\mu = 200(0.18) = 36$
$\sigma = \sqrt{(200)(0.18)(0.82)} = 5.43$

$z = \frac{40.5-36}{5.43} = 0.83$ area = 0.2967

$P(X > 40.5) = 0.5 - 0.2967 = 0.2033$ or 20.33%

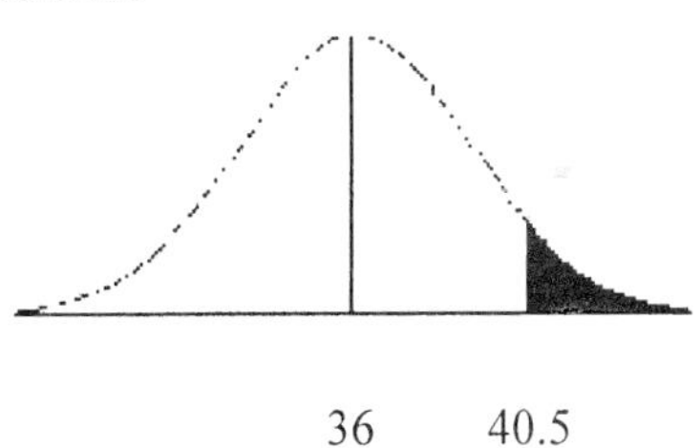

36 40.5

14.
$\mu = 800(0.30) = 240$
$\sigma = \sqrt{(800)(0.3)(0.7)} = 12.96$

$z = \frac{259.5-240}{12.96} = 1.50$ area = 0.4332

$P(X \geq 259.5) = 0.5 - 0.4332 = 0.0668$ or 6.68%

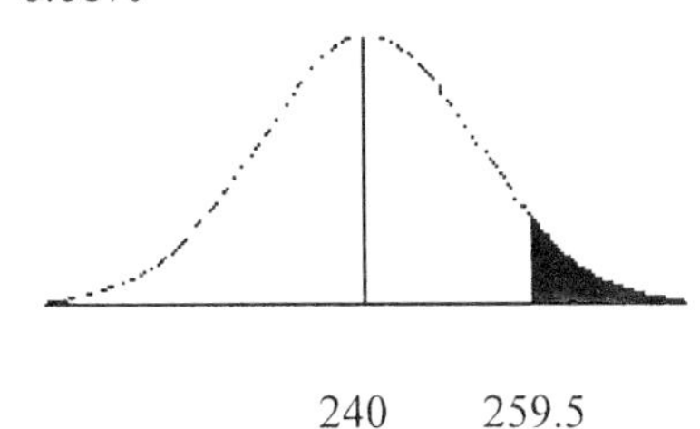

240 259.5

15.
$\mu = 200(0.2) = 40$
$\sigma = \sqrt{(200)(0.2)(0.8)} = 5.66$

15. continued

$z = \frac{49.5-40}{5.66} = 1.68$ area = 0.4535

$P(X \geq 49.5) = 0.5 - 0.4535 = 0.0465$ or 4.65%

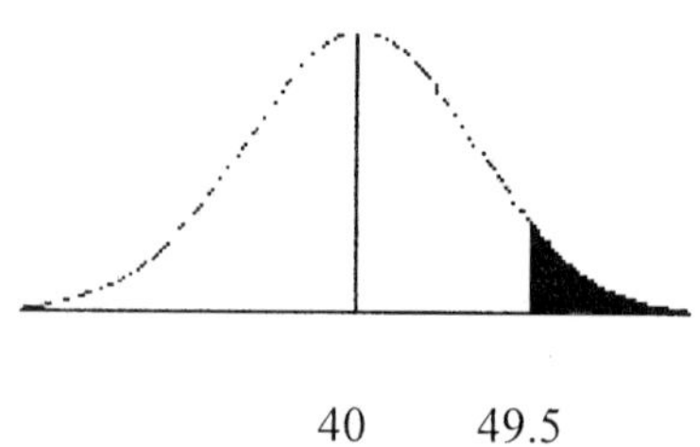

40 49.5

16.
Histogram:

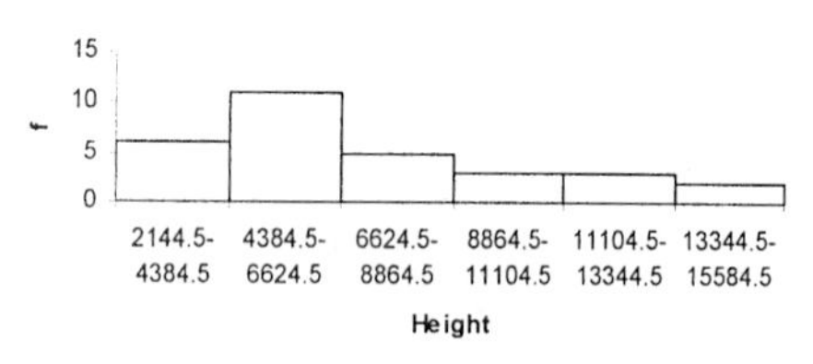

The histogram shows a positive skew.

$PI = \frac{3(6972.2-5931.5)}{3458.85} = 0.90$

$IQR = Q_3 - Q_1$
$IQR = 9348 - 5135 = 4213$
$1.5(IQR) = 1.5(4213) = 6319.5$
$Q_1 - 6319.5 = -1184.5$
$Q_3 + 6319.5 = 15{,}667.5$

There are no outliers.

Conclusion: The distribution is not normal.

17.
Histogram:

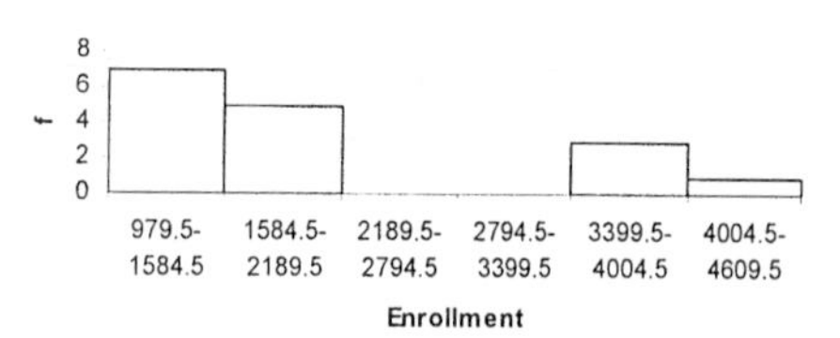

The histogram shows a positive skew.

$PI = \frac{3(2136.1-1755)}{1171.7} = 0.98$

$IQR = Q_3 - Q_1$
$IQR = 2827 - 1320 = 1507$
$1.5(IQR) = 1.5(1507) = 2260.5$
$Q_1 - 2260.5 = -940.5$
$Q_3 + 2260.5 = 5087.5$

17. continued
There are no outliers.

Conclusion: The distribution is not normal.

CHAPTER 6 QUIZ

1. False, the total area is equal to one.
2. True
3. True
4. True
5. False, the area is positive.
6. False, it applies to means taken from the same population.
7. a.
8. a.
9. b.
10. b.
11. c.
12. 0.5
13. sampling error
14. the population mean
15. the standard error of the mean
16. 5
17. 5%
18. the areas are:
a. 0.4332 f. 0.8079
b. 0.3944 g. 0.0401
c. 0.0344 h. 0..8997
d. 0.1029 i. 0.017
e. 0.2912 j. 0.9131
19. the probabilities are:
a. 0.4846 f. 0.0384
b. 0.4693 g. 0.0089
c. 0.9334 h. 0.9582
d. 0.0188 i. 0.9788
e. 0.7461 j. 0.8461
20. the probabilities are:
a. 0.0531 c. 0.1056
b. 0.1056 d. 0.0994
21. the probabilities are:
a. 0.0668 c. 0.4649
b. 0.0228 d. 0.0934
22. the probabilities are:
a. 0.4525 c. 0.3707
b. 0.3707 d. 0.019
23. the probabilities are:
a. 0.0013 c. 0.0081
b. 0.5 d. 0.5511
24. the probabilities are:
a. 0.0037 c. 0.5
b. 0.0228 d. 0.3232
25. 8.804 cm
26. The lowest acceptable score is 121.24.
27. 0.015

28. 0.9738
29. 0.2296
30. 0.0630
31. 0.8577
32. 0.0495
33. The distribution is not normal.
34. The distribution is normal.

Note: Answers may vary due to rounding.

EXERCISE SET 7-2

1.
A point estimate of a parameter specifies a specific value such as $\mu = 87$, whereas an interval estimate specifies a range of values for the parameter such as $84 < \mu < 90$. The advantage of an interval estimate is that a specific confidence level (say 95%) can be selected, and one can be 95% confident that the parameter being estimated lies in the interval.

2.
The standard deviation of the population must be known or it must be estimated or specified in terms of E. Sample size must be specified, and the degree of confidence must be selected.

3.
The maximum error of estimate is the likely range of values to the right or left of the statistic in which may contain the parameter.

4.
A 95% confidence interval means that one can be 95% confident that the parameter being estimated will be contained within the limits of the interval.

5.
A good estimator should be unbiased, consistent, and relatively efficient.

6.
$\overline{X}$

7.
To determine sample size, the maximum error of estimate and the degree of confidence must be specified and the population standard deviation must be known.

8.
No, as long as it is larger than the sample size needed.

9.
a. 2.58
b. 2.33
c. 1.96
d. 1.65
e. 1.88

10.
$\overline{X} = 46{,}970.9 \quad s = 14{,}358.2$
$46{,}970.9 - 1.96(\frac{14{,}358.2}{\sqrt{30}}) < \mu < 46{,}970.9 + 1.96(\frac{14{,}358.2}{\sqrt{30}})$
$41{,}832.9 < \mu < 52{,}108.9$

11.
a. $\overline{X} - z_{\frac{\alpha}{2}}(\frac{s}{\sqrt{n}}) < \mu < \overline{X} + z_{\frac{\alpha}{2}}(\frac{s}{\sqrt{n}})$
$82 - (1.96)(\frac{15}{\sqrt{35}}) < \mu < 82 + (1.96)(\frac{15}{\sqrt{35}})$
$82 - 4.97 < \mu < 82 + 4.97$
$77 < \mu < 87$

b. $82 - (2.58)(\frac{15}{\sqrt{35}}) < \mu < 82 + (2.58)(\frac{15}{\sqrt{35}})$
$82 - 6.54 < \mu < 82 + 6.54$
$75 < \mu < 89$

c. The 99% confidence interval is larger because the confidence level is larger.

12.
$\overline{X} = \$196.0 \quad s = \$617.3$
$196.0 - 1.65(\frac{617.3}{\sqrt{50}}) < \mu < 196.0 + 1.65(\frac{617.3}{\sqrt{50}})$
$196.0 - 144.0 < \mu < 196.0 + 144.0$
$52.0 < \mu < 340.0$

13.
a. $\overline{X} - z_{\frac{\alpha}{2}}(\frac{\sigma}{\sqrt{n}}) < \mu < \overline{X} + z_{\frac{\alpha}{2}}(\frac{\sigma}{\sqrt{n}})$
$12.6 - 1.65(\frac{2.5}{\sqrt{40}}) < \mu < 12.6 + 1.65(\frac{2.5}{\sqrt{40}})$
$12.6 - 0.652 < \mu < 12.6 + 0.652$
$11.9 < \mu < 13.3$

b. It would be highly unlikely since this is far larger than 13.3 minutes.

14.
a. $186 - 1.96(\frac{6}{\sqrt{40}}) < \mu < 186 + 1.96(\frac{6}{\sqrt{40}})$
$186 - 1.86 < \mu < 186 + 1.86$
$184 < \mu < 188$

b.
$186 - 1.96(\frac{6}{\sqrt{100}}) < \mu < 186 + 1.96(\frac{6}{\sqrt{100}})$
$186 - 1.176 < \mu < 186 + 1.176$
$185 < \mu < 187$

c. The interval found in part b is smaller because a larger sample size was used.

15.
$\overline{X} - z_{\frac{\alpha}{2}}(\frac{s}{\sqrt{n}}) < \mu < \overline{X} + z_{\frac{\alpha}{2}}(\frac{s}{\sqrt{n}})$

15. continued

$\$150,000 - 1.96(\frac{15,000}{\sqrt{35}}) < \mu <$
$\$150,000 + 1.96(\frac{15,000}{\sqrt{35}})$

$\$150,000 - 4969.51 < \mu <$
$\$150,000 + 4969.51$

$\$145,030 < \mu < \$154,970$

16.

$\overline{X} = 43.45 \quad s = 31.27$

$43.45 - 1.65(\frac{31.27}{\sqrt{31}}) < \mu <$
$43.45 + 1.65(\frac{31.27}{\sqrt{31}})$

$43.45 - 9.27 < \mu < 43.45 + 9.27$

$34.19 < \mu < 52.72$ or $34.2 < \mu < 52.7$

17.

$\overline{X} = 3222.4 \quad s = 3480.1$

$3222.43 - 1.645(\frac{3480.11}{\sqrt{40}}) < \mu < 3222.43 +$
$1.65(\frac{3480.11}{\sqrt{40}})$

$2317.3 < \mu < 4127.6$

18.

a. $38 - 2.33(\frac{4}{\sqrt{48}}) < \mu < 38 + 2.33(\frac{4}{\sqrt{48}})$

$37 < \mu < 39$

b.

a. $38 - 2.33(\frac{8}{\sqrt{48}}) < \mu < 38 + 2.33(\frac{8}{\sqrt{48}})$

$35 < \mu < 41$

c. The confidence interval for part b is larger because the standard deviation is larger.

19.

$\overline{X} - z_{\frac{\alpha}{2}}(\frac{s}{\sqrt{n}}) < \mu < \overline{X} + z_{\frac{\alpha}{2}}(\frac{s}{\sqrt{n}})$

$61.2 - 1.96(\frac{7.9}{\sqrt{84}}) < \mu < 61.2 + 1.96(\frac{7.9}{\sqrt{84}})$

$61.2 - 1.69 < \mu < 61.2 + 1.69$

$59.5 < \mu < 62.9$

20.

$190.7 - 1.96(\frac{54.2}{\sqrt{35}}) < \mu <$
$190.7 + 1.96(\frac{54.2}{\sqrt{35}})$

$190.7 - 17.96 < \mu < 190.7 + 17.96$

$172.74 < \mu < 208.66$

21.

$n = \left[\frac{z_{\frac{\alpha}{2}}\sigma}{E}\right]^2 = \left[\frac{(1.96)(54.2)}{2}\right]^2$

$= (53.116)^2 = 2821.31$ or 2822

22.

$58.0 - 1.65(\frac{4.8}{\sqrt{171}}) < \mu < 58.0 + 1.65(\frac{4.8}{\sqrt{171}})$

$58.0 - 0.61 < \mu < 58.0 + 0.61$

$57.4 < \mu < 58.6$

23.

$n = \left[\frac{z_{\frac{\alpha}{2}}\sigma}{E}\right]^2 = \left[\frac{(1.96)(2.5)}{1}\right]^2$

$= (4.9)^2 = 24.01$ or 25

24.

$n = \left[\frac{z_{\frac{\alpha}{2}}\sigma}{E}\right]^2 = \left[\frac{(2.58)(0.12)}{0.10}\right]^2$

$= (3.096)^2 = 9.5852$ or 10

25.

$n = \left[\frac{z_{\frac{\alpha}{2}}\sigma}{E}\right]^2 = \left[\frac{(1.65)(8)}{6}\right]^2$

$= (2.2)^2 = 4.84$ or 5

EXERCISE SET 7-3

1.

The characteristics of the t-distribution are: It is bell-shaped, symmetrical about the mean, and never touches the x-axis. The mean, median, and mode are equal to 0 and are located at the center of the distribution. The variance is greater than 1. The t-distribution is a family of curves based on degrees of freedom. As sample size increases the t-distribution approaches the normal distribution.

2.

The degrees of freedom are the number of values free to vary after a sample statistic has been computed.

3.

The t-distribution should be used when σ is unknown and $n < 30$.

4.

a. 2.898 where d. f. = 17
b. 2.074 where d. f. = 22
c. 2.624 where d. f. = 14
d. 1.833 where d. f. = 9
e. 2.093 where d. f. = 19

5.
$\overline{X} - t_{\frac{\alpha}{2}}(\frac{s}{\sqrt{n}}) < \mu < \overline{X} + t_{\frac{\alpha}{2}}(\frac{s}{\sqrt{n}})$
$16 - (2.861)(\frac{2}{\sqrt{20}}) < \mu < 16 + (2.861)(\frac{2}{\sqrt{20}})$
$16 - 1.28 < \mu < 16 + 1.28$
$15 < \mu < 17$

6.
$\overline{X} = 91.06 \quad s = 38.37$
$\overline{X} - t_{\frac{\alpha}{2}}(\frac{s}{\sqrt{n}}) < \mu < \overline{X} + t_{\frac{\alpha}{2}}(\frac{s}{\sqrt{n}})$
$91.06 - 2.583(\frac{38.37}{\sqrt{17}}) < \mu <$
$91.06 + 2.583(\frac{38.37}{\sqrt{17}})$
$91.06 - 24.04 < \mu < 91.06 + 24.04$
$67.0 < \mu < 115.1$

7.
$\overline{X} = 33.4 \quad s = 28.7$
$\overline{X} - t_{\frac{\alpha}{2}}(\frac{s}{\sqrt{n}}) < \mu < \overline{X} + t_{\frac{\alpha}{2}}(\frac{s}{\sqrt{n}})$
$33.4 - 1.746(\frac{28.7}{\sqrt{17}}) < \mu < 33.4 + 1.746(\frac{28.7}{\sqrt{17}})$
$33.4 - 12.2 < \mu < 33.4 + 12.2$
$21.2 < \mu < 45.6$

The point estimate is 33.4 and is close to the actual population mean of 32, which is within the 90% confidence interval. The mean may not be the best estimate since the data value 132 is large and possibly an outlier.

8.
$\overline{X} = 4.1125 \quad s = 2.2025$
$\overline{X} - t_{\frac{\alpha}{2}}(\frac{s}{\sqrt{n}}) < \mu < \overline{X} + t_{\frac{\alpha}{2}}(\frac{s}{\sqrt{n}})$
$4.1125 - 2.365\left(\frac{2.2025}{\sqrt{8}}\right) < \mu <$
$4.1125 + 2.365\left(\frac{2.2025}{\sqrt{8}}\right)$
$2.271 < \mu < 5.954$

9.
$\overline{X} - t_{\frac{\alpha}{2}}(\frac{s}{\sqrt{n}}) < \mu < \overline{X} + t_{\frac{\alpha}{2}}(\frac{s}{\sqrt{n}})$
$12{,}200 - 2.571(\frac{200}{\sqrt{6}}) < \mu <$
$12{,}200 + 2.571(\frac{200}{\sqrt{6}})$
$12{,}200 - 209.921 < \mu < 12{,}200 + 209.921$
$11{,}990 < \mu < 12{,}410$

10.
$\overline{X} = 58.9 \quad s = 5.1$
$\overline{X} - t_{\frac{\alpha}{2}}(\frac{s}{\sqrt{n}}) < \mu < \overline{X} + t_{\frac{\alpha}{2}}(\frac{s}{\sqrt{n}})$
$58.9 - 1.895(\frac{5.1}{\sqrt{8}}) < \mu <$
$58.9 + 1.895(\frac{5.1}{\sqrt{8}})$

10. continued
$58.9 - 3.4 < \mu < 58.9 + 3.4$
$55.5 < \mu < 62.3$

11.
$\overline{X} - t_{\frac{\alpha}{2}}(\frac{s}{\sqrt{n}}) < \mu < \overline{X} + t_{\frac{\alpha}{2}}(\frac{s}{\sqrt{n}})$
$9.3 - 1.703(\frac{2}{\sqrt{28}}) < \mu < 9.3 + 1.703(\frac{2}{\sqrt{28}})$
$9.3 - 0.644 < \mu < 9.3 + 0.644$
$8.7 < \mu < 9.9$

12.
$\overline{X} - t_{\frac{\alpha}{2}}(\frac{s}{\sqrt{n}}) < \mu < \overline{X} + t_{\frac{\alpha}{2}}(\frac{s}{\sqrt{n}})$
$18 - 4.032(\frac{3}{\sqrt{6}}) < \mu < 18 + 4.032(\frac{3}{\sqrt{6}})$
$18 - 4.938 < \mu < 18 + 4.938$
$13 < \mu < 23$

13.
$\overline{X} - t_{\frac{\alpha}{2}}(\frac{s}{\sqrt{n}}) < \mu < \overline{X} + t_{\frac{\alpha}{2}}(\frac{s}{\sqrt{n}})$
$18.53 - 2.064(\frac{3}{\sqrt{25}}) < \mu <$
$18.53 + 2.064(\frac{3}{\sqrt{25}})$
$18.53 - 1.238 < \mu < 18.53 + 1.238$
$\$17.29 < \mu < \$19.77$

14.
$\overline{X} - t_{\frac{\alpha}{2}}(\frac{s}{\sqrt{n}}) < \mu < \overline{X} + t_{\frac{\alpha}{2}}(\frac{s}{\sqrt{n}})$
$126 - 2.262(\frac{4}{\sqrt{10}}) < \mu < 126 + 2.262(\frac{4}{\sqrt{10}})$
$126 - 2.861 < \mu < 126 + 2.861$
$123 < \mu < 129$

15.
$\overline{X} - t_{\frac{\alpha}{2}}(\frac{s}{\sqrt{n}}) < \mu < \overline{X} + t_{\frac{\alpha}{2}}(\frac{s}{\sqrt{n}})$
$115 - 2.571(\frac{6}{\sqrt{6}}) < \mu < 115 + 2.571(\frac{6}{\sqrt{6}})$
$115 - 6.298 < \mu < 115 + 6.298$
$109 < \mu < 121$

16.
$\overline{X} - t_{\frac{\alpha}{2}}(\frac{s}{\sqrt{n}}) < \mu < \overline{X} + t_{\frac{\alpha}{2}}(\frac{s}{\sqrt{n}})$
$41.6 - 2.069(\frac{7.5}{\sqrt{24}}) < \mu <$
$41.6 + 2.069(\frac{7.5}{\sqrt{24}})$
$41.6 - 3.17 < \mu < 41.6 + 3.17$
$38.4 < \mu < 44.8$

17.
$\overline{X} = 41.6 \quad x = 5.995$
$\overline{X} - t_{\frac{\alpha}{2}}(\frac{s}{\sqrt{n}}) < \mu < \overline{X} + t_{\frac{\alpha}{2}}(\frac{s}{\sqrt{n}})$
$41.6 - 2.093(\frac{5.995}{\sqrt{20}}) < \mu <$
$41.6 + 2.093(\frac{5.995}{\sqrt{20}})$
$41.6 - 2.806 < \mu < 41.6 + 2.806$
$38.8 < \mu < 44.4$

18.
$\overline{X} - t_{\frac{\alpha}{2}}(\frac{s}{\sqrt{n}}) < \mu < \overline{X} + t_{\frac{\alpha}{2}}(\frac{s}{\sqrt{n}})$
$96 - 2.093(\frac{5}{\sqrt{20}}) < \mu < 96 + 2.093(\frac{5}{\sqrt{20}})$
$96 - 2.34 < \mu < 96 + 2.34$
$94 < \mu < 98$

19.
$\overline{X} - t_{\frac{\alpha}{2}}(\frac{s}{\sqrt{n}}) < \mu < \overline{X} + t_{\frac{\alpha}{2}}(\frac{s}{\sqrt{n}})$
$\$58{,}219 - 2.052(\frac{56}{\sqrt{28}}) < \mu <$
$\$58{,}219 + 2.052(\frac{56}{\sqrt{28}})$
$\$58{,}197 < \mu < \$58{,}241$

20.
$\overline{X} = 33.5 \quad s = 27.678$
$\overline{X} - t_{\frac{\alpha}{2}}(\frac{s}{\sqrt{n}}) < \mu < \overline{X} + t_{\frac{\alpha}{2}}(\frac{s}{\sqrt{n}})$
$33.5 - 2.821(\frac{27.678}{\sqrt{10}}) < \mu <$
$33.5 + 2.821(\frac{27.678}{\sqrt{10}})$
$33.5 - 24.691 < \mu < 33.5 + 24.691$
$8.8 < \mu < 58.2$

21.
$\overline{X} = 2.175 \quad s = 0.585$
For $\mu > \overline{X} - t_{\frac{\alpha}{2}}(\frac{s}{\sqrt{n}})$:
$\mu > 2.175 - 1.729(\frac{0.585}{\sqrt{20}})$
$\mu > 2.175 - 0.226$
Thus, $\mu > \$1.95$ means that one can be 95% confident that the mean revenue is greater than \$1.95.

For $\mu < \overline{X} + t_{\frac{\alpha}{2}}(\frac{s}{\sqrt{n}})$:
$\mu < 2.175 + 1.729(\frac{0.585}{\sqrt{20}})$
$\mu < 2.175 + 0.226$
Thus, $\mu < \$2.40$ means that one can be 95% confident that the mean revenue is less than \$2.40.

EXERCISE SET 7-4

1.
a. $\hat{p} = \frac{40}{80} = 0.5 \qquad \hat{q} = \frac{40}{80} = 0.5$

b. $\hat{p} = \frac{90}{200} = 0.45 \qquad \hat{q} = \frac{110}{200} = 0.55$

c. $\hat{p} = \frac{60}{130} = 0.46 \qquad \hat{q} = \frac{70}{130} = 0.54$

d. $\hat{p} = \frac{35}{60} = 0.58 \qquad \hat{q} = \frac{25}{60} = 0.42$

e. $\hat{p} = \frac{43}{95} = 0.45 \qquad \hat{q} = \frac{52}{95} = 0.55$

2.
For each part, change the percent to a decimal by dividing by 100, and find $\hat{q}$ using $\hat{q} = 1 - \hat{p}$.
a. $\hat{p} = 0.12 \qquad \hat{q} = 1 - 0.12 = 0.88$
b. $\hat{p} = 0.29 \qquad \hat{q} = 1 - 0.29 = 0.71$
c. $\hat{p} = 0.65 \qquad \hat{q} = 1 - 0.65 = 0.35$
d. $\hat{p} = 0.53 \qquad \hat{q} = 1 - 0.53 = 0.47$
e. $\hat{p} = 0.67 \qquad \hat{q} = 1 - 0.67 = 0.33$

3.
$\hat{p} = 0.39 \qquad \hat{q} = 0.61$
$\hat{p} - (z_{\frac{\alpha}{2}})\sqrt{\frac{\hat{p}\hat{q}}{n}} < p < \hat{p} + (z_{\frac{\alpha}{2}})\sqrt{\frac{\hat{p}\hat{q}}{n}}$
$0.39 - (1.96)\sqrt{\frac{(0.39)(0.61)}{1500}} < p <$
$0.39 + (1.96)\sqrt{\frac{(0.39)(0.61)}{1500}}$
$0.39 - 0.025 < p < 0.39 + 0.025$
$0.365 < p < 0.415$

4.
$\hat{p} = \frac{X}{n} = \frac{27}{100} = 0.27$
$\hat{q} = 1 - 0.27 = 0.73$
$\hat{p} - (z_{\frac{\alpha}{2}})\sqrt{\frac{\hat{p}\hat{q}}{n}} < p < \hat{p} + (z_{\frac{\alpha}{2}})\sqrt{\frac{\hat{p}\hat{q}}{n}}$
$0.27 - 1.65\sqrt{\frac{(0.27)(0.73)}{100}} < p <$
$0.27 + 1.65\sqrt{\frac{(0.27)(0.73)}{100}}$
$0.27 - 0.073 < p < 0.27 + 0.073$
$0.197 < p < 0.343$

5.
$\hat{p} = \frac{X}{n} = \frac{55}{450} = 0.12$
$\hat{q} = 1 - 0.12 = 0.88$
$\hat{p} - (z_{\frac{\alpha}{2}})\sqrt{\frac{\hat{p}\hat{q}}{n}} < p < \hat{p} + (z_{\frac{\alpha}{2}})\sqrt{\frac{\hat{p}\hat{q}}{n}}$
$0.12 - 1.96\sqrt{\frac{(0.12)(0.88)}{450}} < p < 0.12 + 1.96\sqrt{\frac{(0.12)(0.88)}{450}}$
$0.12 - 0.03 < p < 0.12 + 0.03$
0.09 or $9\% < p < 0.15$ or 15%
11% is contained in the confidence interval.

6.
$\hat{p} = 0.27 \quad \hat{q} = 0.73$
$n = (0.27)(0.73)\left(\frac{1.96}{0.025}\right)^2 = 1211.5$
The sample size should be 1212.

7.
$\hat{p} = 0.84 \qquad \hat{q} = 0.16$
$\hat{p} - (z_{\frac{\alpha}{2}})\sqrt{\frac{\hat{p}\hat{q}}{n}} < p < \hat{p} + (z_{\frac{\alpha}{2}})\sqrt{\frac{\hat{p}\hat{q}}{n}}$
$0.84 - 1.65\sqrt{\frac{(0.84)(0.16)}{200}} < p <$
$0.84 + 1.65\sqrt{\frac{(0.84)(0.16)}{200}}$

7. continued

$0.84 - 0.043 < p < 0.84 + 0.043$

$0.797 < p < 0.883$

8.

$\hat{p} = \frac{X}{n} = \frac{329}{763} = 0.431$

$\hat{q} = 1 - 0.431 = 0.569$

$\hat{p} - (z_{\frac{\alpha}{2}})\sqrt{\frac{\hat{p}\hat{q}}{n}} < p < \hat{p} + (z_{\frac{\alpha}{2}})\sqrt{\frac{\hat{p}\hat{q}}{n}}$

$0.431 - 1.75\sqrt{\frac{(0.431)(0.569)}{763}} < p <$

$0.431 + 1.75\sqrt{\frac{(0.431)(0.569)}{763}}$

$0.431 - 0.031 < p < 0.431 + 0.031$

$0.400 < p < 0.462$

9.

$\hat{p} = 0.23$ $\hat{q} = 0.77$

$\hat{p} - (z_{\frac{\alpha}{2}})\sqrt{\frac{\hat{p}\hat{q}}{n}} < p < \hat{p} + (z_{\frac{\alpha}{2}})\sqrt{\frac{\hat{p}\hat{q}}{n}}$

$0.23 - 2.58\sqrt{\frac{(0.23)(0.77)}{200}} < p <$

$0.23 + 2.58\sqrt{\frac{(0.23)(0.77)}{200}}$

$0.23 - 0.077 < p < 0.23 + 0.077$

$0.153 < p < 0.307$

The statement that one in five or 20% of 13 to 14 year olds is a sometime smoker is within the interval.

10.

$\hat{p} = \frac{X}{n} = \frac{46}{80} = 0.575$

$\hat{q} = 1 - 0.575 = 0.425$

$\hat{p} - (z_{\frac{\alpha}{2}})\sqrt{\frac{\hat{p}\hat{q}}{n}} < p < \hat{p} + (z_{\frac{\alpha}{2}})\sqrt{\frac{\hat{p}\hat{q}}{n}}$

$0.575 - 1.96\sqrt{\frac{(0.575)(0.425)}{80}} < p <$

$0.575 + 1.96\sqrt{\frac{(0.575)(0.425)}{80}}$

$0.575 - 0.108 < p < 0.575 + 0.108$

$0.467 < p < 0.683$

11.

$\hat{p} = \frac{40}{90} = 0.44$ $\hat{q} = \frac{50}{90} = 0.56$

$\hat{p} - (z_{\frac{\alpha}{2}})\sqrt{\frac{\hat{p}\hat{q}}{n}} < p < \hat{p} + (z_{\frac{\alpha}{2}})\sqrt{\frac{\hat{p}\hat{q}}{n}}$

$0.44 - 1.96\sqrt{\frac{(0.44)(0.56)}{90}} < p <$

$0.44 + 1.96\sqrt{\frac{(0.44)(0.56)}{90}}$

$0.44 - 0.103 < p < 0.44 + 0.103$

$0.337 < p < 0.543$

12.

$\hat{p} = \frac{X}{n} = 0.45$

$\hat{q} = 1 - 0.45 = 0.55$

$\hat{p} - (z_{\frac{\alpha}{2}})\sqrt{\frac{\hat{p}\hat{q}}{n}} < p < \hat{p} + (z_{\frac{\alpha}{2}})\sqrt{\frac{\hat{p}\hat{q}}{n}}$

$0.45 - 1.65\sqrt{\frac{(0.45)(0.55)}{500}} < p <$

$0.45 + 1.65\sqrt{\frac{(0.45)(0.55)}{500}}$

$0.45 - 0.037 < p < 0.45 + 0.037$

$0.413 < p < 0.487$

13.

$\hat{p} = 0.44975$ $\hat{q} = 0.55025$

$\hat{p} - (z_{\frac{\alpha}{2}})\sqrt{\frac{\hat{p}\hat{q}}{n}} < p < \hat{p} + (z_{\frac{\alpha}{2}})\sqrt{\frac{\hat{p}\hat{q}}{n}}$

$0.44975 - 1.96\sqrt{\frac{(0.44975)(0.55025)}{1005}} < p <$

$0.44975 + 1.96\sqrt{\frac{(0.44975)(0.55025)}{1005}}$

$0.44975 - 0.03076 < p <$

$0.44975 + 0.03076$

$0.419 < p < 0.481$

14.

$\hat{p} = \frac{560}{1000} = 0.56$ $\hat{q} = \frac{440}{1000} = 0.44$

$\hat{p} - (z_{\frac{\alpha}{2}})\sqrt{\frac{\hat{p}\hat{q}}{n}} < p < \hat{p} + (z_{\frac{\alpha}{2}})\sqrt{\frac{\hat{p}\hat{q}}{n}}$

$0.56 - 1.96\sqrt{\frac{(0.56)(0.44)}{1000}} < p <$

$0.56 + 1.96\sqrt{\frac{(0.56)(0.44)}{1000}}$

$0.56 - 0.030766 < p < 0.56 + 0.030766$

$0.529 < p < 0.591$

15.

a. $\hat{p} = 0.25$ $\hat{q} = 0.75$

$n = \hat{p}\,\hat{q}\left[\frac{z_{\frac{\alpha}{2}}}{E}\right]^2 = (0.25)(0.75)\left[\frac{2.58}{0.02}\right]^2$

$= 3120.1875$ or 3121

b. $\hat{p} = 0.5$ $\hat{q} = 0.5$

$n = \hat{p}\,\hat{q}\left[\frac{z_{\frac{\alpha}{2}}}{E}\right]^2 = (0.5)(0.5)\left[\frac{2.58}{0.02}\right]^2$

$n = 4160.25$ or 4161

16.

a. $\hat{p} = 0.29$ $\hat{q} = 0.71$

$n = \hat{p}\,\hat{q}\left[\frac{z_{\frac{\alpha}{2}}}{E}\right]^2 = (0.29)(0.71)\left[\frac{1.65}{0.05}\right]^2$

$= 224.2251$ or 225

b. $\hat{p} = 0.5$ $\hat{q} = 0.5$

$n = \hat{p}\,\hat{q}\left[\frac{z_{\frac{\alpha}{2}}}{E}\right]^2 = (0.5)(0.5)\left[\frac{1.65}{0.05}\right]^2$

$n = 272.25$ or 273

17.

a. $\hat{p} = \frac{30}{300} = 0.1$ $\hat{q} = \frac{270}{300} = 0.9$

17. continued

$n = \hat{p}\,\hat{q}\left[\frac{z_{\frac{\alpha}{2}}}{E}\right]^2 = (0.1)(0.9)\left[\frac{1.65}{0.05}\right]^2$

$= 98.01$ or 99

b. $\hat{p} = 0.5$ $\hat{q} = 0.5$

$n = \hat{p}\,\hat{q}\left[\frac{z_{\frac{\alpha}{2}}}{E}\right]^2 = (0.5)(0.5)\left[\frac{1.65}{0.05}\right]^2$

$n = 272.25$ or 273

18.

a. $\hat{p} = 0.5$ $\hat{q} = 0.5$

$n = \hat{p}\,\hat{q}\left[\frac{z_{\frac{\alpha}{2}}}{E}\right]^2 = (0.5)(0.5)\left[\frac{1.96}{0.0055}\right]^2$

$n = 31{,}748.8$ or 31,749

19.

$\hat{p} = 0.5$ $\hat{q} = 0.5$

$n = \hat{p}\,\hat{q}\left[\frac{z_{\frac{\alpha}{2}}}{E}\right]^2$

$n = (0.5)(0.5)\left[\frac{1.96}{0.03}\right]^2$

$n = 1067.11$ or 1068

20.

$n = \hat{p}\,\hat{q}\left[\frac{z_{\frac{\alpha}{2}}}{E}\right]^2$

$n = (0.27)(0.73)\left[\frac{1.96}{0.02}\right]^2$

$n = 1892.9$ or 1893

EXERCISE SET 7-5

1.

χ^2

2.

The variable must be normally distributed.

3.

	χ^2_{left}	χ^2_{right}
a.	6.262	27.488
b.	0.711	9.488
c.	8.643	42.796
d.	15.308	44.461
e.	5.892	22.362

4.

$\frac{(n-1)s^2}{\chi^2_{\text{right}}} < \sigma^2 < \frac{(n-1)s^2}{\chi^2_{\text{left}}}$

$\frac{19(1.7)^2}{32.852} < \sigma^2 < \frac{19(1.7)^2}{8.907}$

4. continued

$1.7 < \sigma^2 < 6.2$

$1.3 < \sigma < 2.5$

5.

$\frac{(n-1)s^2}{\chi^2_{\text{right}}} < \sigma^2 < \frac{(n-1)s^2}{\chi^2_{\text{left}}}$

$\frac{26(6.8)^2}{38.885} < \sigma^2 < \frac{26(6.8)^2}{15.379}$

$30.9 < \sigma^2 < 78.2$

$5.6 < \sigma < 8.8$

6.

$\frac{(n-1)s^2}{\chi^2_{\text{right}}} < \sigma^2 < \frac{(n-1)s^2}{\chi^2_{\text{left}}}$

$\frac{13(3.2)}{29.819} < \sigma^2 < \frac{13(3.2)}{3.565}$

$1.4 < \sigma^2 < 11.7$

$1.2 < \sigma < 3.4$

7.

$s^2 = 0.80997$ or 0.81

$\frac{(n-1)s^2}{\chi^2_{\text{right}}} < \sigma^2 < \frac{(n-1)s^2}{\chi^2_{\text{left}}}$

$\frac{19(0.81)}{38.582} < \sigma^2 < \frac{19(0.81)}{6.844}$

$0.40 < \sigma^2 < 2.25$

$0.63 < \sigma < 1.50$

8.

$\frac{(n-1)s^2}{\chi^2_{\text{right}}} < \sigma^2 < \frac{(n-1)s^2}{\chi^2_{\text{left}}}$

$\frac{23(2.3)^2}{35.172} < \sigma^2 < \frac{23(2.3)^2}{13.091}$

$3.5 < \sigma^2 < 9.3$

$1.9 < \sigma < 3$

9.

$\frac{(n-1)s^2}{\chi^2_{\text{right}}} < \sigma^2 < \frac{(n-1)s^2}{\chi^2_{\text{left}}}$

$\frac{19(19.1913)^2}{30.144} < \sigma^2 < \frac{19(19.1913)^2}{10.117}$

$232.1 < \sigma^2 < 691.6$

$15.2 < \sigma < 26.3$

10.

$s^2 = 411.46$

$\frac{(n-1)s^2}{\chi^2_{\text{right}}} < \sigma^2 < \frac{(n-1)s^2}{\chi^2_{\text{left}}}$

$\frac{19(411.46)}{30.144} < \sigma^2 < \frac{19(411.46)}{10.117}$

10. continued
$259.3 < \sigma^2 < 772.7$
$16.1 < \sigma < 27.8$

11.
$\frac{(n-1)s^2}{\chi^2_{right}} < \sigma^2 < \frac{(n-1)s^2}{\chi^2_{left}}$

$\frac{27(5.2)^2}{43.194} < \sigma^2 < \frac{27(5.2)^2}{14.573}$

$16.9 < \sigma^2 < 50.1$
$4.1 < \sigma < 7.1$

12.
$s^2 = \frac{\sum X^2 - \frac{(\sum X)^2}{n}}{n-1} = \frac{1155.3752 - \frac{96.14^2}{8}}{8-1}$

$s^2 = 0.0018$
$s = \sqrt{0.0018} = 0.043$
$\frac{(n-1)s^2}{\chi^2_{right}} < \sigma^2 < \frac{(n-1)s^2}{\chi^2_{left}}$

$\frac{7(0.043)^2}{16.013} < \sigma^2 < \frac{7(0.043)^2}{1.690}$

$0.001 < \sigma^2 < 0.008$
$0.032 < \sigma < 0.089$

13.
$s - z_{\frac{\alpha}{2}}\left(\frac{s}{\sqrt{2n}}\right) < \sigma < s + z_{\frac{\alpha}{2}}\left(\frac{s}{\sqrt{2n}}\right)$

$18 - 1.96\left(\frac{18}{\sqrt{400}}\right) < \sigma < 18 + 1.96\left(\frac{18}{\sqrt{400}}\right)$

$16.2 < \sigma < 19.8$

REVIEW EXERCISES - CHAPTER 7

1.
$\overline{X} - z_{\frac{\alpha}{2}}\left(\frac{s}{\sqrt{n}}\right) < \mu < \overline{X} + z_{\frac{\alpha}{2}}\left(\frac{s}{\sqrt{n}}\right)$

$2.6 - 1.96\left(\frac{0.4}{\sqrt{36}}\right) < \mu < 2.6 + 1.96\left(\frac{0.4}{\sqrt{36}}\right)$

$2.5 < \mu < 2.7$

2.
$\hat{p} - (z_{\frac{\alpha}{2}})\sqrt{\frac{\hat{p}\hat{q}}{n}} < p < \hat{p} + (z_{\frac{\alpha}{2}})\sqrt{\frac{\hat{p}\hat{q}}{n}}$

$0.44 - 1.96\sqrt{\frac{(0.44)(0.56)}{1004}} < P <$
$0.44 + 1.96\sqrt{\frac{(0.44)(0.56)}{1004}}$

$0.44 - 0.0307 < p < 0.44 + 0.0307$

$0.409 < p < 0.471$

3.
$\overline{X} - z_{\frac{\alpha}{2}}\left(\frac{s}{\sqrt{n}}\right) < \mu < \overline{X} + z_{\frac{\alpha}{2}}\left(\frac{s}{\sqrt{n}}\right)$

$7.5 - 1.96\left(\frac{0.8}{\sqrt{1500}}\right) < \mu < 7.5 + 1.96\left(\frac{0.8}{\sqrt{1500}}\right)$

$7.46 < \mu < 7.54$

4.
$\overline{X} - t_{\frac{\alpha}{2}}\left(\frac{s}{\sqrt{n}}\right) < \mu < \overline{X} + t_{\frac{\alpha}{2}}\left(\frac{s}{\sqrt{n}}\right)$

$\$1151 - 2.228\left(\frac{281.97227}{\sqrt{12}}\right) < \mu <$
$\$1151 + 2.228\left(\frac{281.97227}{\sqrt{12}}\right)$

$\$1151 - 181.36 < \mu < \$1151 + 181.36$
$\$969.64 < \mu < \$1332.36$

5.
$\overline{X} - t_{\frac{\alpha}{2}}\left(\frac{s}{\sqrt{n}}\right) < \mu < \overline{X} + t_{\frac{\alpha}{2}}\left(\frac{s}{\sqrt{n}}\right)$

$28 - 2.132\left(\frac{3}{\sqrt{5}}\right) < \mu < 28 + 2.132\left(\frac{3}{\sqrt{5}}\right)$

$25 < \mu < 31$

6.
$n = \left[\frac{z_{\frac{\alpha}{2}}\sigma}{E}\right]^2 = \left[\frac{1.96(1050)}{200}\right]^2$

$= 105.88$ or 106

7.
$n = \left[\frac{z_{\frac{\alpha}{2}}\sigma}{E}\right]^2 = \left[\frac{1.65(80)}{25}\right]^2$

$= (5.28)^2 = 27.88$ or 28

8.
$\hat{p} = 0.42 \quad \hat{q} = 0.58$

$\hat{p} - (z_{\frac{\alpha}{2}})\sqrt{\frac{\hat{p}\hat{q}}{n}} < p < \hat{p} + (z_{\frac{\alpha}{2}})\sqrt{\frac{\hat{p}\hat{q}}{n}}$

$0.42 - 1.96\sqrt{\frac{(0.42)(0.58)}{1500}} < p <$
$0.42 + 1.96\sqrt{\frac{(0.42)(0.58)}{1500}}$

$0.42 - 1.96(0.013) < p < 0.42 + 1.96(0.013)$

$0.395 < p < 0.445$

9.
$\hat{p} = 0.4 \qquad \hat{q} = 0.6$

9. continued

$\hat{p} - (z_{\frac{\alpha}{2}})\sqrt{\frac{\hat{p}\hat{q}}{n}} < p < \hat{p} + (z_{\frac{\alpha}{2}})\sqrt{\frac{\hat{p}\hat{q}}{n}}$

$0.4 - 1.65\sqrt{\frac{(0.4)(0.6)}{200}} < p <$
$0.4 + 1.65\sqrt{\frac{(0.4)(0.6)}{200}}$

$0.4 - 0.057 < p < 0.4 + 0.057$
$0.343 < p < 0.457$

10.
$\hat{p} = 0.85 \qquad \hat{q} = 0.15$

$\hat{p} - (z_{\frac{\alpha}{2}})\sqrt{\frac{\hat{p}\hat{q}}{n}} < p < \hat{p} + (z_{\frac{\alpha}{2}})\sqrt{\frac{\hat{p}\hat{q}}{n}}$

$0.85 - 1.65\sqrt{\frac{(0.85)(0.15)}{100}} < p <$
$0.85 + 1.65\sqrt{\frac{(0.85)(0.15)}{100}}$

$0.85 - 0.059 < p < 0.85 + 0.059$

$0.791 < p < 0.909$

11.
$\hat{p} = 0.88 \qquad \hat{q} = 0.12$

$n = \hat{p}\,\hat{q}\left[\frac{z_{\frac{\alpha}{2}}}{E}\right]^2 = (0.88)(0.12)\left[\frac{1.65}{0.025}\right]^2$

$n = 459.99$ or 460

12.
$\hat{p} = 0.80 \qquad \hat{q} = 0.20$

$n = \hat{p}\,\hat{q}\left[\frac{z_{\frac{\alpha}{2}}}{E}\right]^2 = (0.80)(0.20)\left[\frac{2.33}{0.03}\right]^2$

$n = 965.1$ or 966

13.
$\frac{(n-1)s^2}{\chi^2_{right}} < \sigma^2 < \frac{(n-1)s^2}{\chi^2_{left}}$

$\frac{(28-1)(0.34)^2}{49.645} < \sigma^2 < \frac{(28-1)(0.34)^2}{11.808}$

$0.06287 < \sigma^2 < 0.26433$

$0.25 < \sigma < 0.51$

14.
$\frac{(n-1)s^2}{\chi^2_{right}} < \sigma^2 < \frac{(n-1)s^2}{\chi^2_{left}}$

$\frac{(22-1)(2.6)}{35.479} < \sigma^2 < \frac{(22-1)(2.6)}{10.283}$

14. continued
$1.5 < \sigma^2 < 5.3$

15.
$\frac{(n-1)s^2}{\chi^2_{right}} < \sigma^2 < \frac{(n-1)s^2}{\chi^2_{left}}$

$\frac{(15-1)(8.6)}{23.685} < \sigma^2 < \frac{(15-1)(8.6)}{6.571}$

$5.1 < \sigma^2 < 18.3$

16.
$\frac{(n-1)s^2}{\chi^2_{right}} < \sigma^2 < \frac{(n-1)s^2}{\chi^2_{left}}$

$\frac{(28-1)(1.83)^2}{43.194} < \sigma^2 < \frac{(28-1)(1.83)^2}{14.573}$

$2.093 < \sigma^2 < 6.205$
$1.447 < \sigma < 2.491$

CHAPTER 7 QUIZ

1. True
2. True
3. False, it is consistent if, as sample size increases, the estimator approaches the parameter being estimated.
4. True
5. b.
6. a.
7. b.
8. unbiased, consistent, relatively efficient
9. maximum error of estimate
10. point
11. 90, 95, 99
12. $\overline{X} - z_{\frac{\alpha}{2}}(\frac{s}{\sqrt{n}}) < \mu < \overline{X} + z_{\frac{\alpha}{2}}(\frac{s}{\sqrt{n}})$

$\$23.45 - 1.65(\frac{2.80}{\sqrt{49}}) < \mu <$
$\$23.45 + 1.65(\frac{2.80}{\sqrt{49}})$

$\$22.79 < \mu < \$24.11$

13. $\overline{X} - t_{\frac{\alpha}{2}}(\frac{s}{\sqrt{n}}) < \mu < \overline{X} + t_{\frac{\alpha}{2}}(\frac{s}{\sqrt{n}})$

$\$44.80 - 2.093(\frac{3.53}{\sqrt{20}}) < \mu <$
$\$44.80 + 2.093(\frac{3.53}{\sqrt{20}})$

$\$43.15 < \mu < \$46.45$

14. $\overline{X} - z_{\frac{\alpha}{2}}(\frac{s}{\sqrt{n}}) < \mu < \overline{X} + z_{\frac{\alpha}{2}}(\frac{s}{\sqrt{n}})$

14. continued
$\$4150 - 2.58(\frac{480}{\sqrt{40}}) < \mu <$
$\$4150 + 2.58(\frac{480}{\sqrt{40}})$

$\$3954 < \mu < \$4346$

15. $\overline{X} - t_{\frac{\alpha}{2}}(\frac{s}{\sqrt{n}}) < \mu < \overline{X} + t_{\frac{\alpha}{2}}(\frac{s}{\sqrt{n}})$

$48.6 - 2.262(\frac{4.1}{\sqrt{10}}) < \mu < 48.6 + 2.262(\frac{4.1}{\sqrt{10}})$

$45.7 < \mu < 51.5$

16. $\overline{X} - t_{\frac{\alpha}{2}}(\frac{s}{\sqrt{n}}) < \mu < \overline{X} + t_{\frac{\alpha}{2}}(\frac{s}{\sqrt{n}})$

$438 - 3.499(\frac{16}{\sqrt{8}}) < \mu < 438 + 3.499(\frac{16}{\sqrt{8}})$

$418 < \mu < 458$

17. $\overline{X} - t_{\frac{\alpha}{2}}(\frac{s}{\sqrt{n}}) < \mu < \overline{X} + t_{\frac{\alpha}{2}}(\frac{s}{\sqrt{n}})$

$31 - 2.353(\frac{4}{\sqrt{4}}) < \mu < 31 + 2.353(\frac{4}{\sqrt{4}})$

$26 < \mu < 36$

18. $n = \left[\frac{z_{\frac{\alpha}{2}}\sigma}{E}\right]^2 = \left[\frac{2.58(2.6)}{0.5}\right]^2$

$= 179.98$ or 180

19. $n = \left[\frac{z_{\frac{\alpha}{2}}\sigma}{E}\right]^2 = \left[\frac{1.65(900)}{300}\right]^2$

$= 24.5$ or 25

20. $\hat{p} - (z_{\frac{\alpha}{2}})\sqrt{\frac{\hat{p}\hat{q}}{n}} < p < \hat{p} + (z_{\frac{\alpha}{2}})\sqrt{\frac{\hat{p}\hat{q}}{n}}$

$\hat{p} = \frac{53}{75} = 0.707 \quad \hat{q} = \frac{22}{75} = 0.293$

$0.71 - 1.96\sqrt{\frac{(0.707)(0.293)}{75}} < p <$
$0.71 + 1.96\sqrt{\frac{(0.707)(0.293)}{75}}$

$0.604 < p < 0.810$

21. $\hat{p} - (z_{\frac{\alpha}{2}})\sqrt{\frac{\hat{p}\hat{q}}{n}} < p < \hat{p} + (z_{\frac{\alpha}{2}})\sqrt{\frac{\hat{p}\hat{q}}{n}}$

$0.36 - 1.65\sqrt{\frac{(0.36)(0.64)}{150}} < p <$
$0.36 + 1.65\sqrt{\frac{(0.36)(0.64)}{150}}$

21. continued
$0.295 < p < 0.425$

22. $\hat{p} - (z_{\frac{\alpha}{2}})\sqrt{\frac{\hat{p}\hat{q}}{n}} < p < \hat{p} + (z_{\frac{\alpha}{2}})\sqrt{\frac{\hat{p}\hat{q}}{n}}$

$0.4444 - 1.96\sqrt{\frac{(0.4444)(0.5556)}{90}} < p <$
$0.4444 + 1.96\sqrt{\frac{(0.4444)(0.5556)}{90}}$

$0.342 < p < 0.547$

23. $n = \hat{p}\,\hat{q}\left[\frac{z_{\frac{\alpha}{2}}}{E}\right]^2$

$= (0.15)(0.85)\left[\frac{1.96}{0.03}\right]^2$

$= 544.22$ or 545

24. $\frac{(n-1)s^2}{\chi^2_{right}} < \sigma^2 < \frac{(n-1)s^2}{\chi^2_{left}}$

$\frac{24(9)^2}{39.364} < \sigma^2 < \frac{24(9)^2}{12.401}$

$49.4 < \sigma^2 < 156.8$

$7 < \sigma < 13$

25. $\frac{(n-1)s^2}{\chi^2_{right}} < \sigma^2 < \frac{(n-1)s^2}{\chi^2_{left}}$

$\frac{26(6.8)^2}{38.885} < \sigma^2 < \frac{26(6.8)^2}{15.379}$

$30.9 < \sigma^2 < 78.2$

$5.6 < \sigma < 8.8$

26. $\frac{(n-1)s^2}{\chi^2_{right}} < \sigma^2 < \frac{(n-1)s^2}{\chi^2_{left}}$

$\frac{19(2.3)^2}{30.144} < \sigma^2 < \frac{19(2.3)^2}{10.177}$

$3.33 < \sigma^2 < 10$

$1.8 < \sigma < 3.2$

Note: Graphs are not to scale and are intended to convey a general idea. Answers may vary due to rounding.

EXERCISE SET 8-2

1.
The null hypothesis is a statistical hypothesis that states there is no difference between a parameter and a specific value or there is no difference between two parameters. The alternative hypothesis specifies a specific difference between a parameter and a specific value, or that there is a difference between two parameters. Examples will vary.

2.
Type I error occurs by rejecting the null hypothesis when it is true. Type II error occurs when the null hypothesis is not rejected and it is false. They are related in that decreasing the probability of one type of error increases the probability of the other type of error.

3.
A statistical test uses the data obtained from a sample to make a decision as to whether or not the null hypothesis should be rejected.

4.
A one-tailed test indicates the null hypothesis should be rejected when the test statistic value is in the critical region on one side of the mean. A two-tailed test indicates the null hypothesis should be rejected when the test statistic value is in either critical region on both sides of the mean.

5.
The critical region is the region of values of the test-statistic that indicates a significant difference and the null hypothesis should be rejected. The non-critical region is the region of values of the test-statistic that indicates the difference was probably due to chance, and the null hypothesis should not be rejected.

6.
"H_0" represents the null hypothesis. "H_1" represents the alternative hypothesis.

7.
Type I is represented by α, type II is represented by β.

8.
When the difference between the sample mean and the hypothesized mean is large, then the difference is said to be significant and probably not due to chance.

9.
A one-tailed test should be used when a specific direction, such as greater than or less than, is being hypothesized, whereas when no direction is specified, a two-tailed test should be used.

10.
The steps in hypothesis testing are:

1. State the hypotheses.
2. Find the critical value(s).
3. Compute the test statistic value.
4. Make the decision.
5. Summarize the results.

11.
Hypotheses can only be proved true when the entire population is used to compute the test statistic. In most cases, this is impossible.

12.
a. $+2.58, \ -2.58$

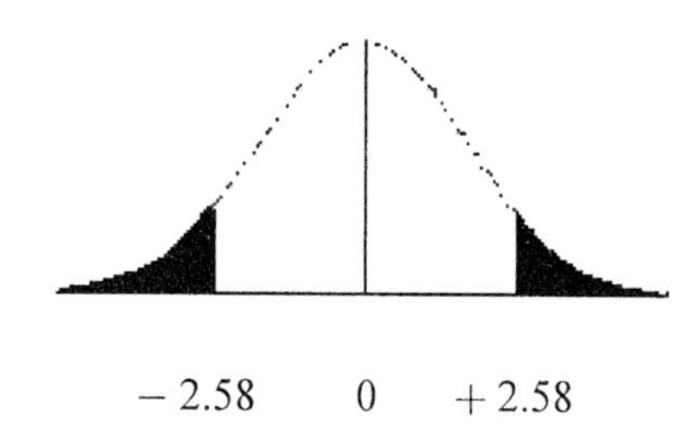

b. $+1.65$

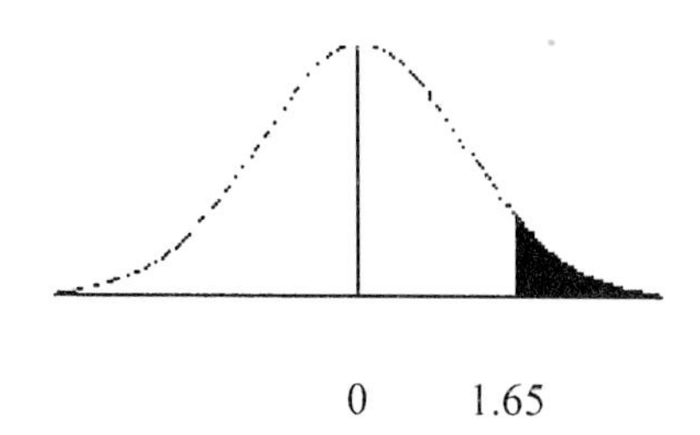

12. continued

c. -2.58

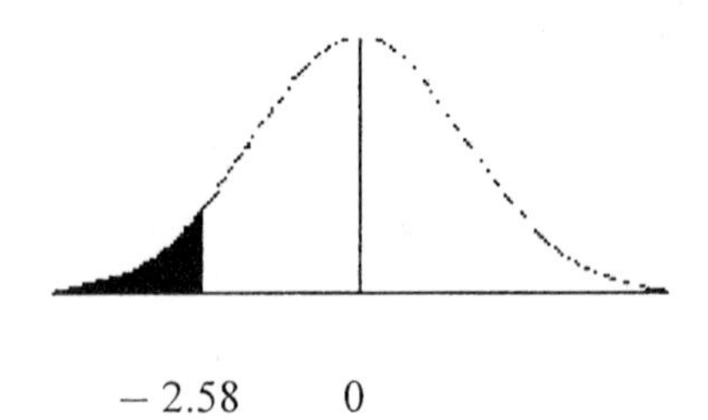

d. -1.28

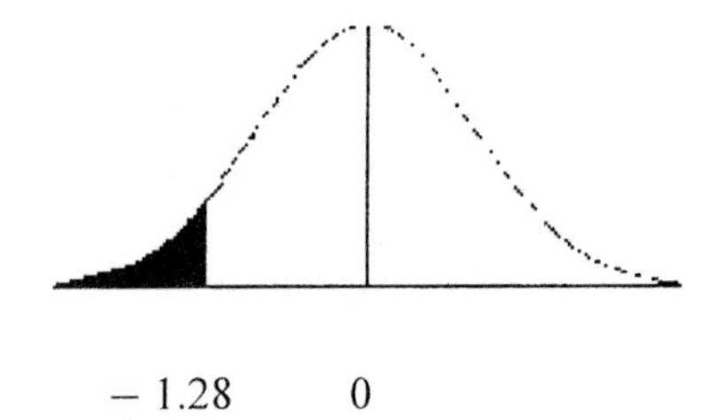

e. $+1.96$, -1.96

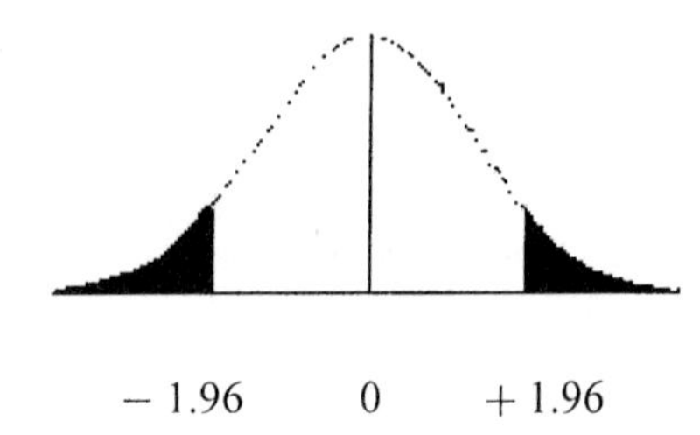

f. $+1.75$

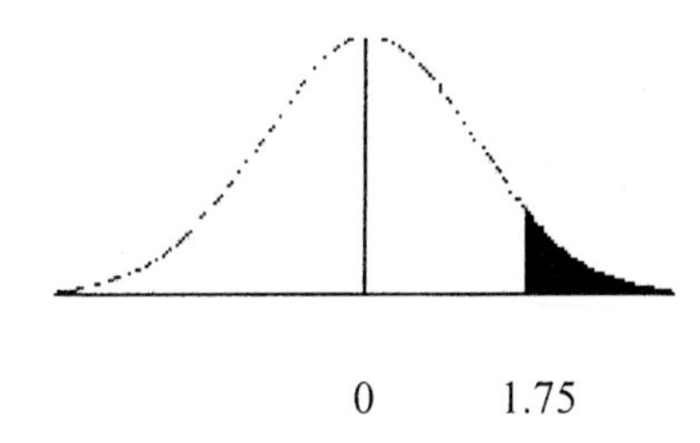

g. -2.33

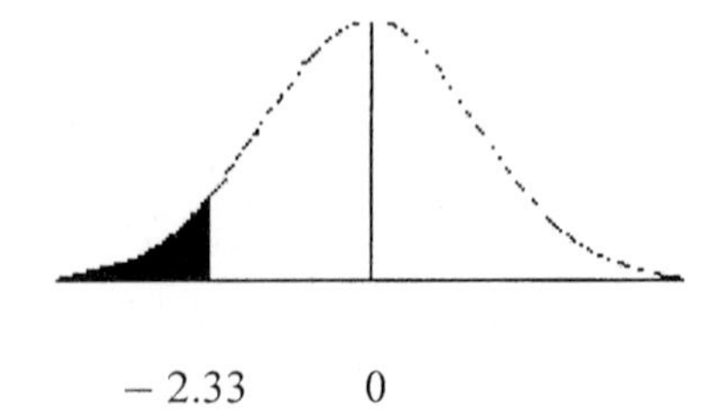

12. continued

h. $+1.65$, -1.65

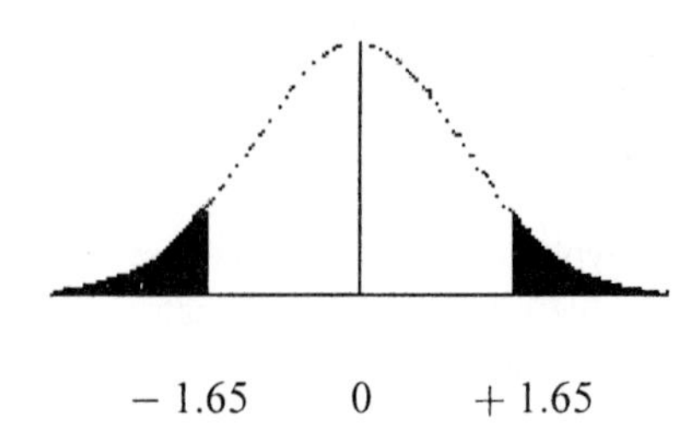

i. $+2.05$

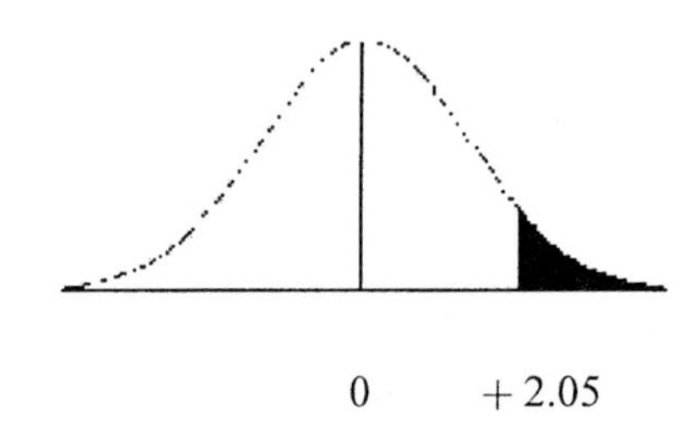

j. $+2.33$, -2.33

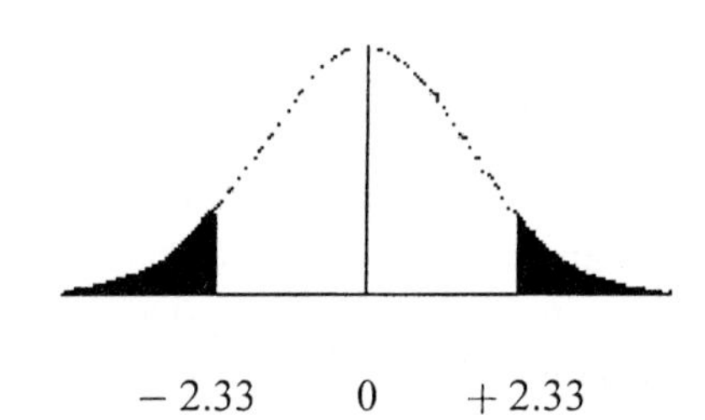

13.

a. H_0: $\mu = 36.3$ (claim)
H_1: $\mu \neq 36.3$

b. H_0: $\mu = \$36{,}250$ (claim)
H_1: $\mu \neq \$36{,}250$

c. H_0: $\mu \leq 27.6$ years
H_1: $\mu > 27.6$ years (claim)

d. H_0: $\mu \geq 72$
H_1: $\mu < 72$ (claim)

e. H_0: $\mu \geq 100$
H_1: $\mu < 100$ (claim)

f. H_0: $\mu = \$297.75$ (claim)
H_1: $\mu \neq \$297.75$

g. H_0: $\mu \leq \$52.98$
H_1: $\mu > \$52.98$ (claim)

13. continued

h. H_0: $\mu \leq 300$ (claim)
   H_1: $\mu > 300$

i. H_0: $\mu \geq 3.6$ (claim)
   H_1: $\mu < 3.6$

EXERCISE SET 8-3

1.
H_0: $\mu = \$69.21$ (claim)
H_1: $\mu \neq \$69.21$

C. V. $= \pm 1.96$

$z = \frac{\overline{X}-\mu}{\frac{\sigma}{\sqrt{n}}} = \frac{\$68.43-\$69.21}{\frac{3.72}{\sqrt{30}}} = -1.15$

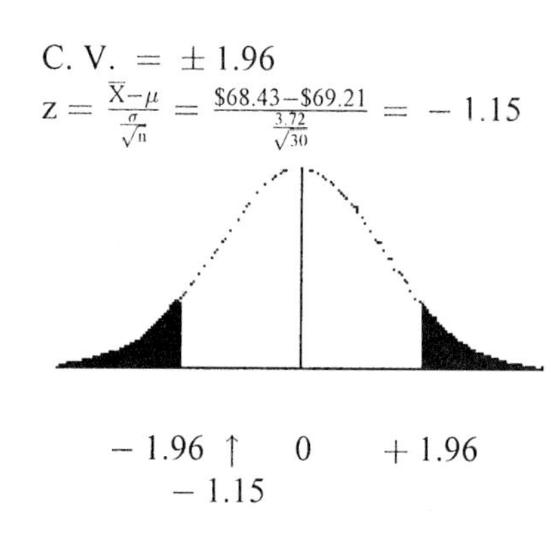

$-1.96 \uparrow$ 0 $+1.96$
-1.15

Do not reject the null hypothesis. There is not enough evidence to reject the claim that the average cost of a hotel stay in Atlanta is $69.21.

2.
H_0: $\mu \geq \$3262$
H_1: $\mu < \$3262$ (claim)

C. V. $= -1.65$

$z = \frac{\overline{X}-\mu}{\frac{\sigma}{\sqrt{n}}} = \frac{2995-3262}{\frac{1100}{\sqrt{50}}} = -1.72$

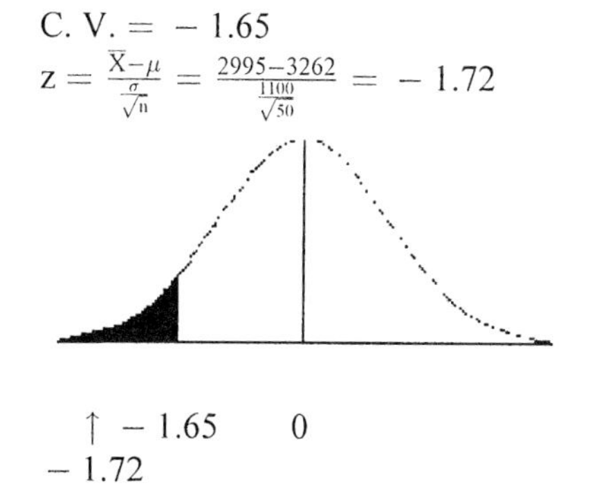

$\uparrow$ -1.65 0
-1.72

Reject the null hypothesis. There is enough evidence to support the claim that the average debt is less than $3262.

3.
H_0: $\mu \leq \$24$ billion
H1: $\mu > \$24$ billion (claim)

3. continued

C. V. $= +1.65$ $\overline{X} = \$31.5$ $s = \$28.7$

$z = \frac{\overline{X}-\mu}{\frac{\sigma}{\sqrt{n}}} = \frac{31.5-24}{\frac{28.7}{\sqrt{50}}} = 1.85$

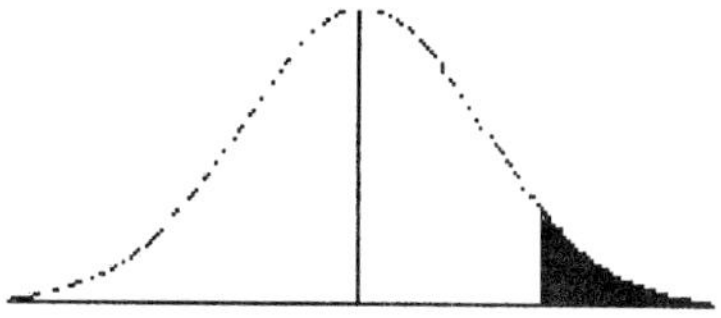

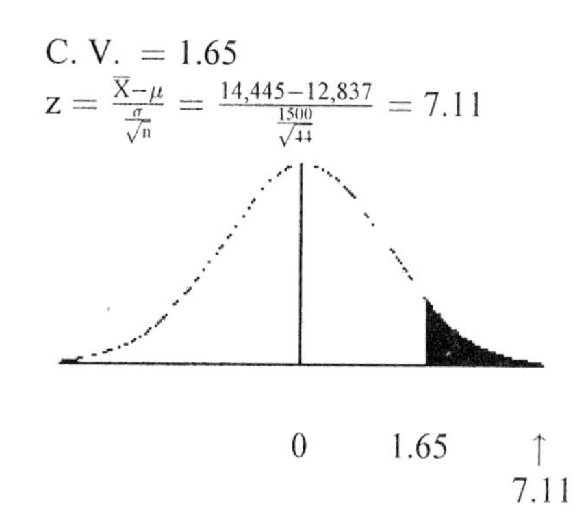

0 1.65 $\uparrow$
1.85

Reject the null hypothesis. There is enough evidence to support the claim that the average revenue exceeds $24 billion.

4.
H_0: $\mu \leq \$12,837$
H_1: $\mu > \$12,837$ (claim)

C. V. $= 1.65$

$z = \frac{\overline{X}-\mu}{\frac{\sigma}{\sqrt{n}}} = \frac{14,445-12,837}{\frac{1500}{\sqrt{44}}} = 7.11$

0 1.65 $\uparrow$
7.11

Reject the null hypothesis. There is enough evidence to support the claim that the average salary is more than $12,837.

5.
H_0: $\mu \geq 14$
H_1: $\mu < 14$ (claim)

C. V. $= -2.33$

$z = \frac{\overline{X}-\mu}{\frac{s}{\sqrt{n}}} = \frac{11.8-14}{\frac{2.7}{\sqrt{36}}} = -4.89$

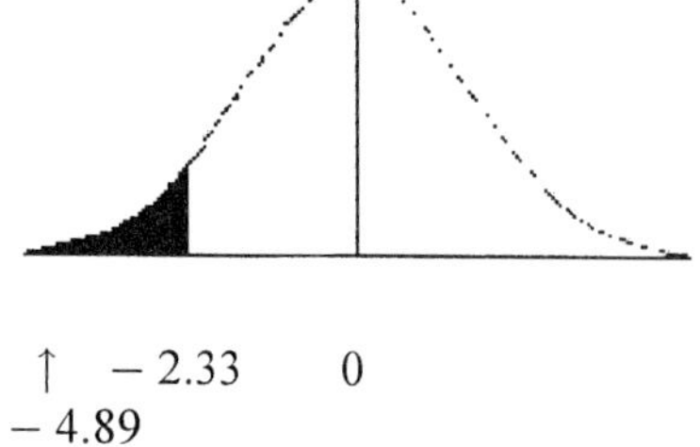

$\uparrow$ -2.33 0
-4.89

5. continued

Reject the null hypothesis. There is enough evidence to support the claim that the average age of the planes in the executive's airline is less than the national average.

6.

H_0: $\mu \leq 3000$

H_1: $\mu > 3000$ (claim)

C. V. = 1.65

$z = \frac{\overline{X}-\mu}{\frac{\sigma}{\sqrt{n}}} = \frac{3120-3000}{\frac{578}{\sqrt{60}}} = 1.61$

0 ↑ 1.65
1.61

Do not reject the null hypothesis. There is not enough evidence to support the claim that the average production has increased.

7.

H_0: $\mu = 29$

H_1: $\mu \neq 29$ (claim)

C. V. = ± 1.96 $\overline{X} = 29.45$ s = 2.61

$z = \frac{\overline{X}-\mu}{\frac{\sigma}{\sqrt{n}}} = \frac{29.45-29}{\frac{2.61}{\sqrt{30}}} = 0.944$

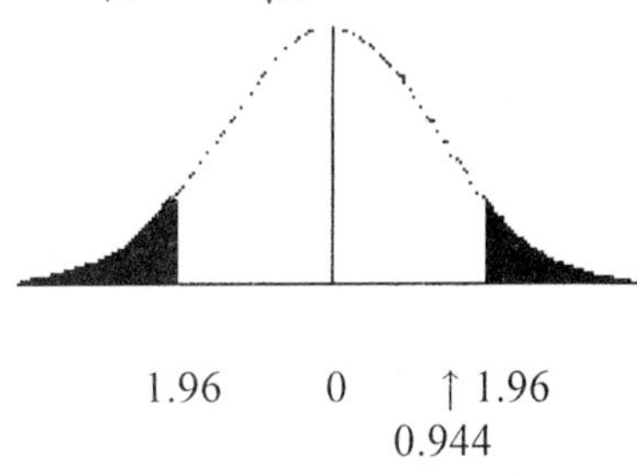

1.96 0 ↑ 1.96
0.944

Do not reject the null hypothesis. There is enough evidence to reject the claim that the average height differs from 29 inches.

8.

H_0: $\mu \leq \$91,600$

H_1: $\mu > \$91,600$ (claim)

C. V. = 1.65

8. continued

$z = \frac{\overline{X}-\mu}{\frac{s}{\sqrt{n}}} = \frac{\$96,321-\$91,600}{\frac{\$9555}{\sqrt{100}}} = 4.94$

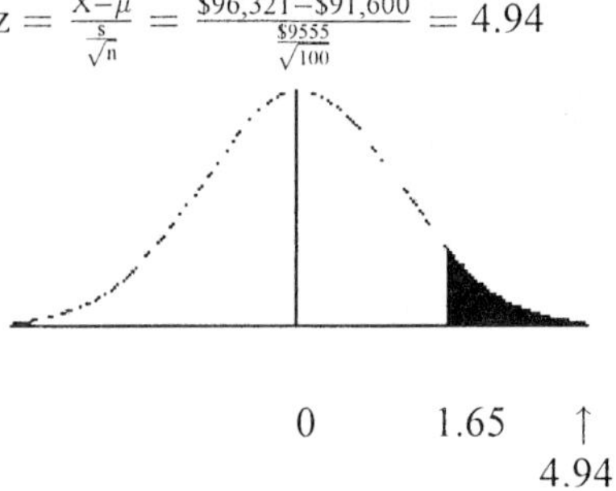

0 1.65 ↑
4.94

Reject the null hypothesis. There is enough evidence to support the claim that the average income is greater than $91,600.

9.

H_0: $\mu \leq \$19,410$

H_1: $\mu > \$19,410$ (claim)

C. V. = 2.33

$z = \frac{\overline{X}-\mu}{\frac{\sigma}{\sqrt{n}}} = \frac{\$22,098-\$19,410}{\frac{6050}{\sqrt{40}}} = 2.81$

0 2.33 ↑
2.81

Reject the null hypothesis. There is enough evidence to support the claim that the average tuition cost has increased.

10.

H_0: $\mu = \$60,000$ (claim)

H_1: $\mu \neq \$60,000$

C. V. = ± 1.96 $\overline{X} = \$82,496$

s = $76,025

$z = \frac{\overline{X}-\mu}{\frac{\sigma}{\sqrt{n}}} = \frac{82496-60,000}{\frac{76,025}{\sqrt{36}}} = 1.78$

− 1.96 0 ↑ 1.96
1.78

10. continued
Do not reject the null hypothesis. There is not enough evidence to reject the claim that the average price of a home is $60,000.

11.
H_0: $\mu = 125$
H_1: $\mu \neq 125$ (claim)

C. V. $= \pm 2.58$
$z = \frac{\bar{X}-\mu}{\frac{\sigma}{\sqrt{n}}} = \frac{110-125}{\frac{30}{\sqrt{35}}} = -2.96$

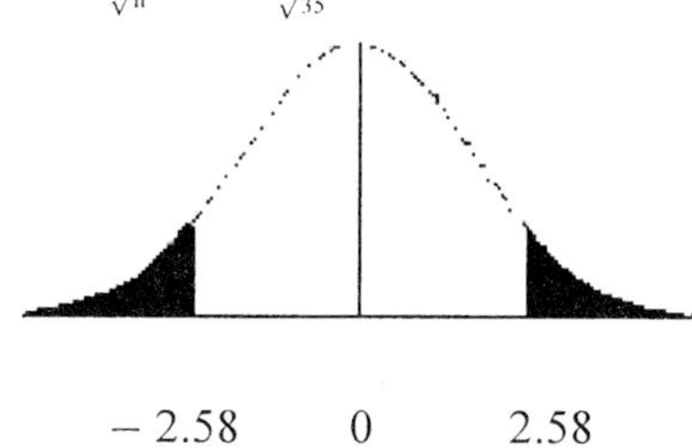

− 2.58 0 2.58
↑ − 2.96

Reject the null hypothesis. There is a significant difference in the average number of guests.

12.
H_0: $\mu = \$39,385$
H_1: $\mu \neq \$39,385$ (claim)

C. V. $= \pm 1.96$
$z = \frac{\bar{X}-\mu}{\frac{\sigma}{\sqrt{n}}} = \frac{\$41,680-\$39,385}{\frac{5975}{\sqrt{50}}} = 2.72$

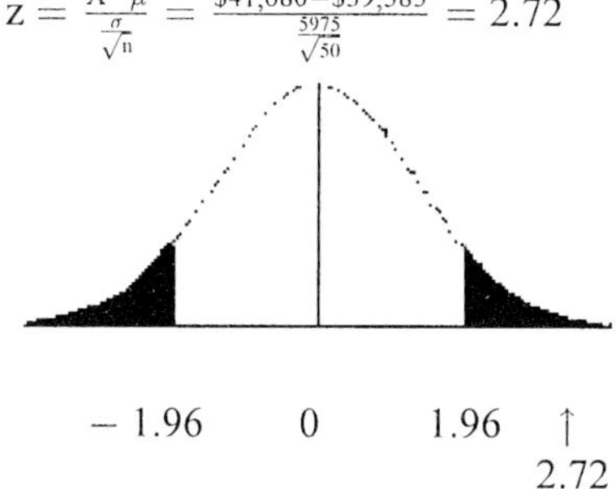

− 1.96 0 1.96 ↑
2.72

Reject the null hypothesis. There is a significant difference in the salaries.

13.
H_0: $\mu = \$24.44$
H_1: $\mu \neq \$24.44$ (claim)

C. V. $= \pm 2.33$
$z = \frac{\bar{X}-\mu}{\frac{s}{\sqrt{n}}} = \frac{22.97-24.44}{\frac{3.70}{\sqrt{33}}} = -2.28$

13. continued

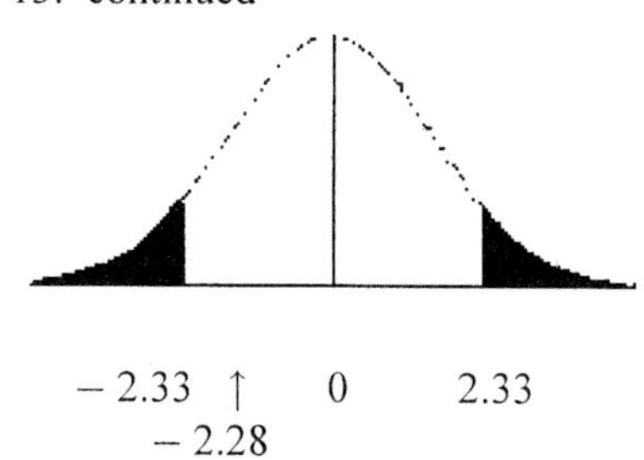

− 2.33 ↑ 0 2.33
− 2.28

Do not reject the null hypothesis. There is not enough evidence to support the claim that the amount spent at a local mall is not equal to the national average of $24.44.

14.
The *P* value is the actual probability of getting the sample mean if the null hypothesis is true.

15.
a. Do not reject.
b. Do not reject.
c. Do not reject.
d. Reject
e. Reject

16.
H_0: $\mu \leq \$27.50$
H_1: $\mu > \$27.50$ (claim)
$z = \frac{\bar{X}-\mu}{\frac{s}{\sqrt{n}}} = \frac{\$29.30-\$27.50}{\frac{5}{\sqrt{50}}} = 2.55$

The area corresponding to $z = 2.55$ is 0.4946. The P-value is $0.5 - 0.4946 = 0.0054$. Hence, the null hypothesis should be rejected at $\alpha = 0.05$ since $0.0054 < 0.05$. There is enough evidence to support the claim that the cost of the textbooks is greater than $27.50.

17.
H_0: $\mu \geq 264$
H_1: $\mu < 264$ (claim)
$z = \frac{\bar{X}-\mu}{\frac{\sigma}{\sqrt{n}}} = \frac{262.3-264}{\frac{3}{\sqrt{20}}} = -2.53$

The area corresponding to $z = 2.53$ is 0.4943. The P-value is $0.5 - 0.4943 = 0.0057$. The decision is to reject the null hypothesis since $0.0057 < 0.01$. There is enough evidence to support the claim that the average stopping distance is less than 264 feet.

18.
H_0: $\mu \geq 40$
H_1: $\mu < 40$ (claim)
$\overline{X} = 29.3$ $s = 30.9$
$z = \frac{\overline{X}-\mu}{\frac{\sigma}{\sqrt{n}}} = \frac{29.3-40}{\frac{30.9}{\sqrt{50}}} = -2.45$

The area corresponding to z = 3.14 is 0.4999. The P-value is 0.5 – 0.4929 = 0.0071. The decision is reject the null hypothesis since 0.0071 < 0.01. There is enough evidence to support the claim that the average number of copies is less than 40.

19.
H_0: $\mu \leq 84$
H_1: $\mu > 84$ (claim)
$z = \frac{\overline{X}-\mu}{\frac{\sigma}{\sqrt{n}}} = \frac{85.1-84}{\frac{10}{\sqrt{100}}} = 1.1$

The area corresponding to z = 1.1 is 0.3643. The P-value is 0.5 – 0.3643 = 0.1357. The decision is do not reject the null hypothesis since 0.1357 > 0.01. There is not enough evidence to support the claim that the average lifetime of the television sets is greater than 84 months.

20.
H_0: $\mu = 800$ (claim)
H_1: $\mu \neq 800$
$z = \frac{\overline{X}-\mu}{\frac{\sigma}{\sqrt{n}}} = \frac{793-800}{\frac{12}{\sqrt{200}}} = -2.61$

The area corresponding to z = 2.61 is 0.4955. The P-value is found by subtracting the area from 0.5 then multiplying by 2 since this is a two-tailed test. Hence, 2(0.5 – 0.4955) = 2(0.0045) = 0.009. The decision is to reject the null hypothesis since 0.009 < 0.01. There is enough evidence to reject the null hypothesis that the breaking strength is 800 pounds.

21.
H_0: $\mu = 6.32$ (claim)
H_1: $\mu \neq 6.32$
$z = \frac{\overline{X}-\mu}{\frac{\sigma}{\sqrt{n}}} = \frac{6.51-6.32}{\frac{0.54}{\sqrt{50}}} = 2.49$

The area corresponding to z = 2.49 is 0.4936. To get the P-value, subtract 0.4936 from 0.5 and then multiply by 2 since this is a two-tailed test.
2(0.5 – 0.4936) = 2(0.0064) = 0.0128

21. continued
The decision is to reject the null hypothesis since 0.0128 < 0.05. There is enough evidence to reject the claim that the average wage is $6.32.

22.
H_0: $\mu = 65$ (claim)
H_1: $\mu \neq 65$
$z = \frac{\overline{X}-\mu}{\frac{\sigma}{\sqrt{n}}} = \frac{63.2-65}{\frac{7}{\sqrt{22}}} = -1.21$

The area corresponding to z = – 1.21 is 0.3869. The P-value is 2(0.5 – 0.3869) = 2(0.1131) = 0.2262. The decision is do not reject the null hypothesis since 0.2262 > 0.10. Hence, there is not enough evidence to reject the claim that the average is 65 acres.

23.
H_0: $\mu = 30{,}000$ (claim)
H_1: $\mu \neq 30{,}000$
$z = \frac{\overline{X}-\mu}{\frac{s}{\sqrt{n}}} = \frac{30{,}456-30{,}000}{\frac{1684}{\sqrt{40}}} = 1.71$

The area corresponding to z = 1.71 is 0.4564. The P-value is 2(0.5 – 0.4564) = 2(0.0436) = 0.0872.

The decision is to reject the null hypothesis at $\alpha = 0.10$ since 0.0872 < 0.10. The conclusion is that there is enough evidence to reject the claim that customers are adhering to the recommendation. A 0.10 significance level is probably appropriate since there is little consequence of a Type I error. The dealer would be advised to increase efforts to make its customers aware of the service recommendation.

24.
H_0: $\mu = 60$ (claim)
H_1: $\mu \neq 60$

$\overline{X} = 59.93$ $s = 13.42$
$z = \frac{\overline{X}-\mu}{\frac{s}{\sqrt{n}}} = \frac{59.93-60}{\frac{13.42}{\sqrt{30}}} = -0.03$

The area corresponding to 0.03 is 0.0120. The P-value is 2(0.5 – 0.0120) = 0.976. Since 0.976 > 0.05, the decision is do not reject the null hypothesis. There is not

24. continued
enough evidence to reject the claim that the average number of speeding tickets is 60.

25.
H_0: $\mu \geq 10$
H_1: $\mu < 10$ (claim)

$\overline{X} = 5.025$ $s = 3.63$
$z = \frac{\overline{X}-\mu}{\frac{s}{\sqrt{n}}} = \frac{5.025-10}{\frac{3.63}{\sqrt{40}}} = -8.67$

The area corresponding to 8.67 is greater than 0.4999. The P-value is $0.5 - 0.4999 < 0.0001$. Since $0.0001 < 0.05$, the decision is to reject the null hypothesis. There is enough evidence to support the claim that the average number of days missed per year is less than 10.

26.
Reject the claim at $\alpha = 0.05$ but not at $\alpha = 0.01$. There is no contradiction since the value of α should be chosen before the test is conducted.

27.
The mean and standard deviation are found as follows:

	f	X_m	$f \cdot X_m$	$f \cdot X_m^2$
8.35 - 8.43	2	8.39	16.78	140.7842
8.44 - 8.52	6	8.48	50.88	431.4624
8.53 - 8.61	12	8.57	102.84	881.3388
8.62 - 8.70	18	8.66	155.88	1349.9208
8.71 - 8.79	10	8.75	87.5	765.625
8.80 - 8.88	2	8.84	17.68	156.2912
	50		431.56	3725.4224

$\overline{X} = \frac{\sum f \cdot X_m}{n} = \frac{431.56}{50} = 8.63$

$s = \sqrt{\frac{\sum f \cdot X_m^2 - \frac{\sum(f \cdot X_m)^2}{n}}{n-1}} = \sqrt{\frac{3725.4224 - \frac{(431.56)^2}{50}}{49}}$

$= 0.105$

H_0: $\mu = 8.65$ (claim)
H_1: $\mu \neq 8.65$

C. V. $= \pm 1.96$
$z = \frac{\overline{X}-\mu}{\frac{s}{\sqrt{n}}} = \frac{8.63-8.65}{\frac{0.105}{\sqrt{50}}} = -1.35$

Do not reject the null hypothesis. There is not enough evidence to reject the claim that

27. continued
the average hourly wage of the employees is \$8.65.

EXERCISE SET 8-4

1.
It is bell-shaped, symmetric about the mean, and it never touches the x axis. The mean, median, and mode are all equal to 0 and they are located at the center of the distribution. The t distribution differs from the standard normal distribution in that it is a family of curves, the variance is greater than one, and as the degrees of freedom increase the t distribution approaches the standard normal distribution.

2.
The degrees of freedom are the number of values that are free to vary after a sample statistic has been computed. They tell the researcher which specific curve to use when a distribution consists of a family of curves.

3.
a. d. f. = 9 C. V. = $+1.833$
b. d. f. = 17 C. V. = ± 1.740
c. d. f. = 5 C. V. = -3.365
d. d. f. = 8 C. V. = $+2.306$
e. d. f. = 14 C. V. = ± 2.145
f. d. f. = 22 C. V. = -2.819
g. d. f. = 27 C. V. = ± 2.771
h. d. f. = 16 C. V. = ± 2.583

4.
a. $0.01 < \text{P-value} < 0.025$ (0.018)
b. $0.05 < \text{P-value} < 0.10$ (0.062)
c. $0.10 < \text{P-value} < 0.25$ (0.123)
d. $0.10 < \text{P-value} < 0.20$ (0.138)
e. $\text{P-value} < 0.005$ (0.003)
f. $0.10 < \text{P-value} < 0.25$ (0.158)
g. $\text{P-value} = 0.05$ (0.05)
h. $\text{P-value} > 0.25$ (0.261)

5.
H_0: $\mu \geq 11.52$
H_1: $\mu < 11.52$ (claim)

C. V. $= -1.833$ d. f. $= 6$

$t = \frac{\overline{X}-\mu}{\frac{s}{\sqrt{n}}} = \frac{7.42-11.52}{\frac{1.3}{\sqrt{10}}} = -9.97$

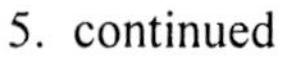

5. continued

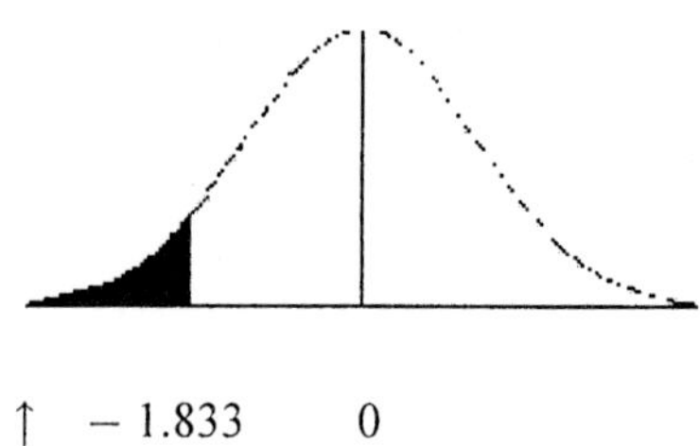

↑ -1.833 0
-9.97

Reject the null hypothesis. There is enough evidence to support the claim that the rainfall is below average.

6.
H_0: $\mu \geq 2000$
H_1: $\mu < 2000$ (claim)
C. V. $= -3.747$ d. f. $= 4$
$\overline{X} = 1885.8$ $s = 2456.3$

$$t = \frac{\overline{X}-\mu}{\frac{s}{\sqrt{n}}} = \frac{1885.8-2000}{\frac{2456.3}{\sqrt{5}}} = -0.104$$

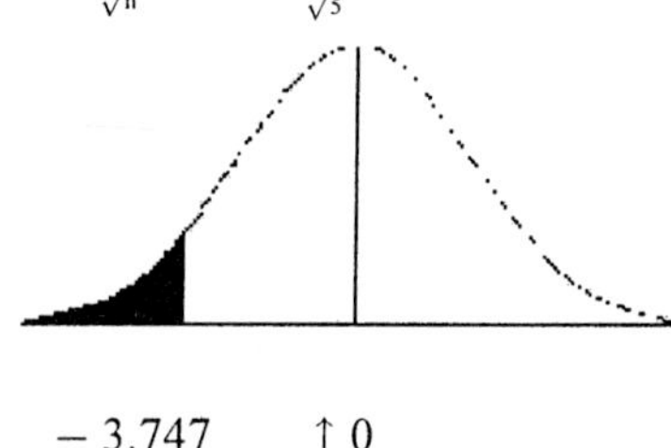

-3.747 ↑ 0
-0.104

Do not reject the null hypothesis. There is not enough evidence to support the claim that the average acreage is less than 2000.

7.
H_0: $\mu = \$40,000$
H_1: $\mu \neq \$40,000$ (claim)

C. V. $= \pm 2.093$ d. f. $= 19$

$$t = \frac{\overline{X}-\mu}{\frac{s}{\sqrt{n}}} = \frac{43{,}228-40{,}000}{\frac{4000}{\sqrt{20}}} = 3.61$$

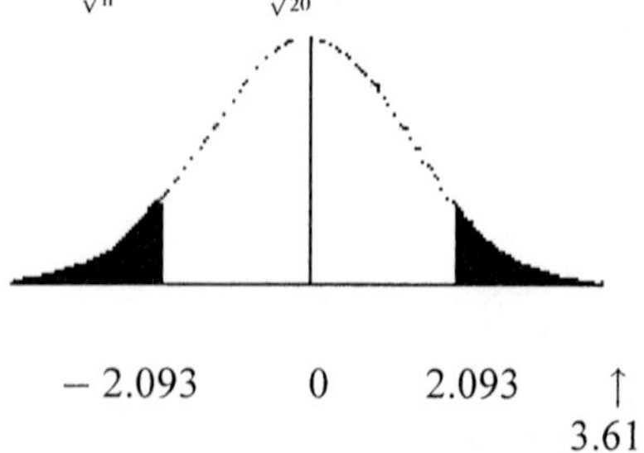

-2.093 0   2.093   ↑
3.61

Reject the null hypothesis. There is enough evidence to support the claim that the average salary is not \$40,000.

8.
H_0: $\mu \geq 25.4$
H_1: $\mu < 25.4$ (claim)

C. V. $= -1.318$ d. f. $= 24$

$$t = \frac{\overline{X}-\mu}{\frac{s}{\sqrt{n}}} = \frac{22.1-25.4}{\frac{5.3}{\sqrt{25}}} = -3.11$$

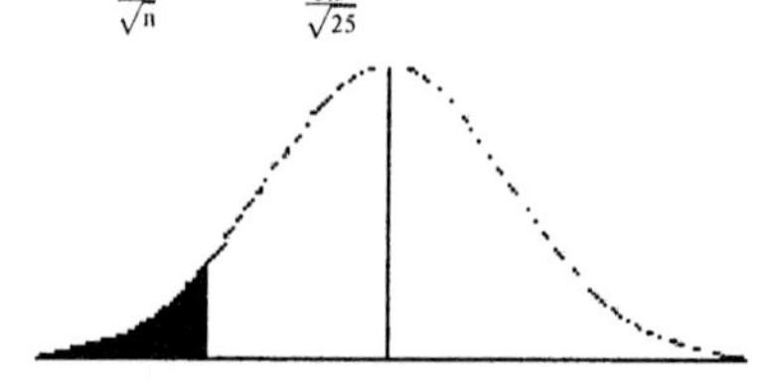

↑ -1.318 0
-3.11

Reject the null hypothesis. There is enough evidence to support the claim that the commute time is less than 25.4 minutes.

9.
H_0: $\mu \geq 700$ (claim)
H_1: $\mu < 700$
$\overline{X} = 606.5$ $s = 109.1$
C. V. $= -2.262$ d. f. $= 9$

$$t = \frac{\overline{X}-\mu}{\frac{s}{\sqrt{n}}} = \frac{606.5-700}{\frac{109.1}{\sqrt{10}}} = -2.71$$

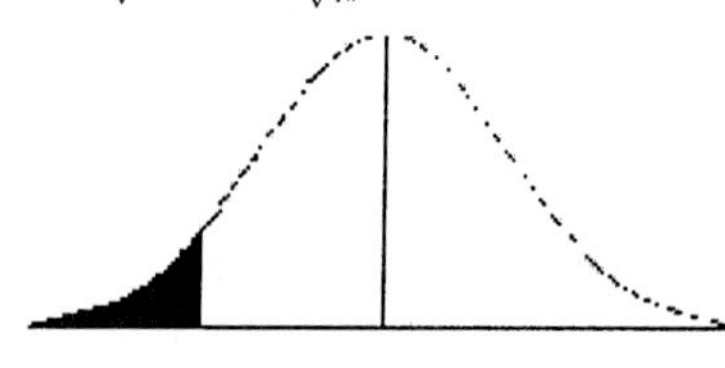

↑ -2.262 0
-2.71

Reject the null hypothesis. There is enough evidence to reject the claim that the average height of the buildings is at least 700 feet.

10.
H_0: $\mu = \$17.63$ (claim)
H_1: $\mu \neq \$17.63$

C. V. $= \pm 2.145$ d. f. $= 14$

$$t = \frac{\overline{X}-\mu}{\frac{s}{\sqrt{n}}} = \frac{18.72-17.63}{\frac{3.64}{\sqrt{15}}} = 1.16$$

10. continued

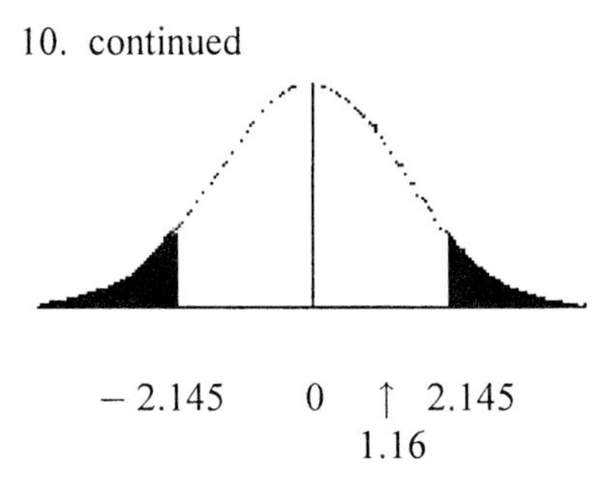

$-2.145 \quad 0 \quad \uparrow \; 2.145$
1.16

Do not reject the null hypothesis. There is not enough evidence to reject the claim that there is no difference in the rates.

11.
H_0: $\mu \leq \$13{,}252$
H_1: $\mu > \$13{,}252$ (claim)

C. V. = 2.539   d. f. = 19
$t = \frac{\overline{X}-\mu}{\frac{s}{\sqrt{n}}} = \frac{\$15{,}560-\$13{,}252}{\frac{\$3500}{\sqrt{19}}} = 2.95$

$0 \quad 2.539 \; \uparrow$
2.95

Reject the null hypothesis. There is enough evidence to support the claim that the average tuition cost has increased.

12.
H_0: $\mu = \$91{,}600$
H_1: $\mu \neq \$91{,}600$ (claim)

C. V. = $\pm$ 1.703   d. f. = 27
$t = \frac{\overline{X}-\mu}{\frac{s}{\sqrt{n}}} = \frac{\$88{,}500-\$91{,}600}{\frac{\$10{,}000}{\sqrt{28}}} = -1.64$

$-1.703 \; \uparrow \; 0 \quad 1.703$
-1.64

Do not reject the null hypothesis. There is not enough evidence to support the claim that the average income differs from \$91,600.

13.
H_0: $\mu \leq \$54.8$
H_1: $\mu > \$54.8$ (claim)

C. V. = 1.761   d. f. = 14
$t = \frac{\overline{X}-\mu}{\frac{s}{\sqrt{n}}} = \frac{\$62.3-\$54.8}{\frac{\$9.5}{\sqrt{15}}} = 3.06$

$0 \quad 1.761 \; \uparrow$
3.06

Reject the null hypothesis. There is enough evidence to support the claim that the cost to produce an action movie is more than \$54.8.

14.
H_0: $\mu \leq 110$
H_1: $\mu > 110$ (claim)

$\overline{X} = 137.33$ s = 24.1178
C. V. = 2.624   d. f. = 14

$t = \frac{\overline{X}-\mu}{\frac{s}{\sqrt{n}}} = \frac{137.3-110}{\frac{24.1178}{\sqrt{15}}} = 4.38$

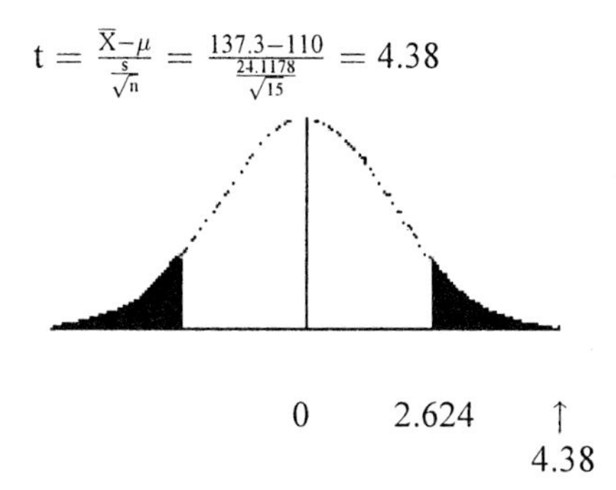

$0 \quad 2.624 \quad \uparrow$
4.38

Reject the null hypothesis. There is enough evidence to support the claim that the average calorie content is greater than 110.

15.
H_0: $\mu = 132$ min. (claim)
H_1: $\mu \neq 132$ min.
C. V. = $\pm$2.365   d. f. = 7

$t = \frac{\overline{X}-\mu}{\frac{s}{\sqrt{n}}} = \frac{125-132}{\frac{11}{\sqrt{8}}} = -1.80$

Do not reject the null hypothesis. There is enough evidence to support the claim that the average show time is 132 minutes, or 2 hours and 12 minutes.

16.
H_0: $\mu = \$30.00$
H_1: $\mu \neq \$30.00$ (claim)
d. f. = 15
$0.20 < \text{P-value} < 0.50$ (0.409)
$t = \frac{\overline{X}-\mu}{\frac{s}{\sqrt{n}}} = \frac{\$31.17-30.00}{\frac{5.51}{\sqrt{16}}} = 0.85$

Since P-value > 0.05, do not reject the null hypothesis. There is not enough evidence to support the claim that the average cost has changed.

17.
H_0: $\mu = 75$ (claim)
H_1: $\mu \neq 75$
$\overline{X} = 70.85$ $s = 6.56$
d. f. = 19
$0.01 < \text{P-value} < 0.02$ (0.011)
$t = \frac{\overline{X}-\mu}{\frac{s}{\sqrt{n}}} = \frac{70.85-75}{\frac{6.56}{\sqrt{20}}} = -2.83$

Since P-value > 0.01, do not reject the null hypothesis. There is not enough evidence to reject the claim that the average score on the real estate exam is 75.

18.
H_0: $\mu = 16$ (claim)
H_1: $\mu \neq 16$
$\overline{X} = 19.625$ $s = 2.97$
d. f. = 7
$0.01 < \text{P-value} < 0.02$ (0.011)
$t = \frac{\overline{X}-\mu}{\frac{s}{\sqrt{n}}} = \frac{19.625-16}{\frac{2.97}{\sqrt{8}}} = 3.45$

Since P-value < 0.05, reject the null hypothesis. There is enough evidence to reject the claim that the average number of students using the computer lab is 16.

19.
H_0: $\mu = \$15,000$
H_1: $\mu \neq \$15,000$ (claim)

$\overline{X} = \$14,347.17$ $s = \$2048.54$
d. f. = 11 C. V. = ± 2.201

$t = \frac{\overline{X}-\mu}{\frac{s}{\sqrt{n}}} = \frac{\$14,347.17-\$15,000}{\frac{\$2048.54}{\sqrt{12}}} = -1.10$

19. continued

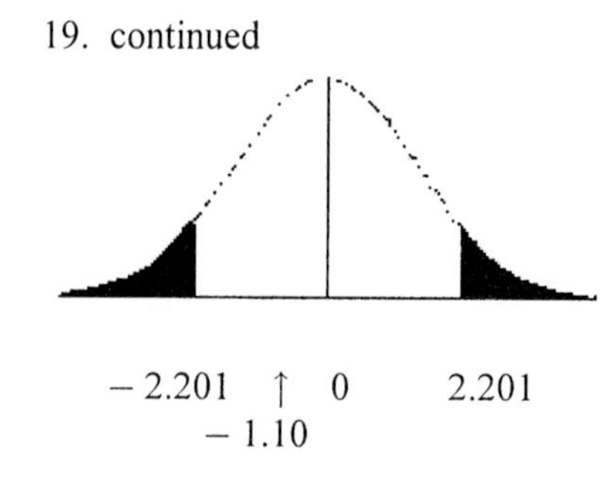

Do not reject the null hypothesis. There is not enough evidence to say that the average stipend differs from $15,000.

20.
H_0: $\mu = 3.18$
H_1: $\mu \neq 3.18$ (claim)

$\overline{X} = 3.833$ $s = 1.434563$
d. f. = 23 C. V. = ± 2.069

$t = \frac{\overline{X}-\mu}{\frac{s}{\sqrt{n}}} = \frac{3.833-3.18}{\frac{1.434563}{\sqrt{24}}} = 2.23$

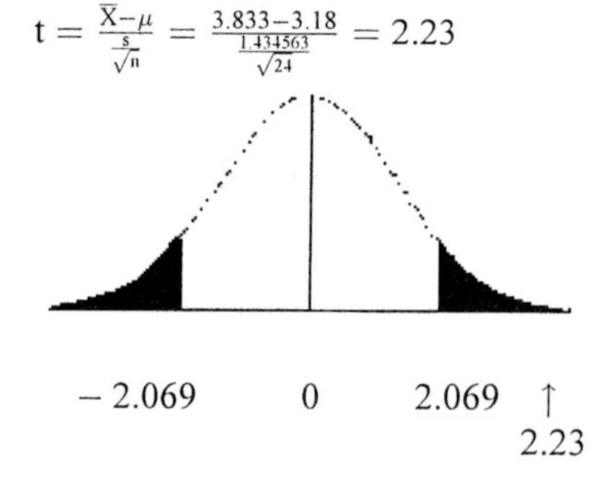

Reject the null hypothesis. There is enough evidence to support the claim that the average family size differs from the national average.

EXERCISE SET 8-5

1.
Answers will vary.

2.
The proportion of A items can be considered a success whereas the proportion of items that are not included in A can be considered a failure.

3.
$np \geq 5$ and $nq \geq 5$

4.
$\mu = np$ $\sigma = \sqrt{npq}$

5.
H_0: $p = 0.647$
H_1: $p \neq 0.647$ (claim)

$\hat{p} = \frac{92}{150} = 0.613 \quad p = 0.647 \quad q = 0.353$
C. V. = ± 2.58

$z = \frac{\hat{p} - p}{\sqrt{\frac{pq}{n}}} = \frac{0.613 - 0.647}{\sqrt{\frac{(0.647)(0.353)}{150}}} = -0.86$

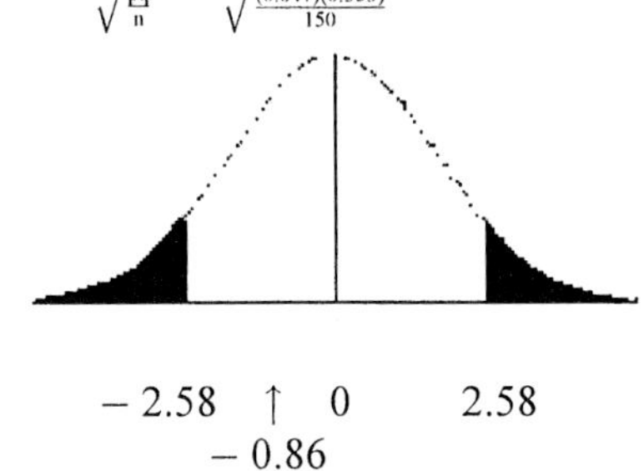

$-2.58 \uparrow 0 \quad 2.58$
-0.86

Do not reject the null hypothesis. There is not enough evidence to support the claim that the proportion of homeowners is different from 64.7%.

6.
H_0: $p = 0.488$
H_1: $p \neq 0.488$ (claim)

$\hat{p} = \frac{142}{250} = 0.568 \quad p = 0.488 \quad q = 0.512$

$z = \frac{\hat{p} - p}{\sqrt{\frac{pq}{n}}} = \frac{0.568 - 0.488}{\sqrt{\frac{(0.488)(0.512)}{250}}} = 2.53$

Since the p-value = 0.0114, it can be concluded that the null hypothesis would be rejected for any $\alpha \leq 0.0114$.

7.
H_0: $p = 0.40$
H_1: $p \neq 0.40$ (claim)

$\hat{p} = \frac{65}{180} = 0.361 \quad p = 0.40 \quad q = 0.60$
C. V. = ± 2.58
$z = \frac{\hat{p} - p}{\sqrt{\frac{pq}{n}}} = \frac{0.361 - 0.40}{\sqrt{\frac{(0.40)(0.60)}{180}}} = -1.07$

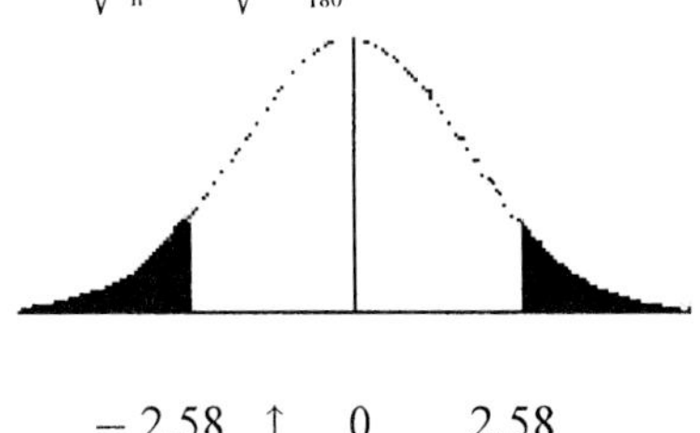

$-2.58 \uparrow 0 \quad 2.58$
-1.07

7. continued
Do not reject the null hypothesis. There is not enough evidence to conclude that the proportion differs from 40%.

8.
H_0: $p \leq 0.279$
H_1: $p > 0.279$ (claim)

$\hat{p} = \frac{45}{120} = 0.375 \quad p = 0.279 \quad q = 0.721$
C. V. = 1.65
$z = \frac{\hat{p} - p}{\sqrt{\frac{pq}{n}}} = \frac{0.375 - 0.279}{\sqrt{\frac{(0.279)(0.721)}{120}}} = 2.35$

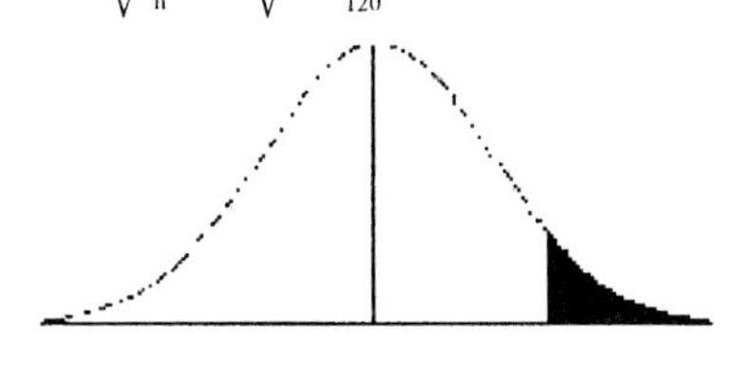

$0 \quad 1.65 \uparrow$
2.35

Reject the null hypothesis. There is enough evidence to conclude that the proportion of female physicians at the university health system is higher than 27.9%.

9.
H_0: $p = 0.63$ (claim)
H_1: $p \neq 0.63$

$\hat{p} = \frac{85}{143} = 0.5944 \quad p = 0.63 \quad q = 0.37$
C. V. = ± 1.96
$z = \frac{\hat{p} - p}{\sqrt{\frac{pq}{n}}} = \frac{0.5944 - 0.63}{\sqrt{\frac{(0.63)(0.37)}{143}}} = -0.88$

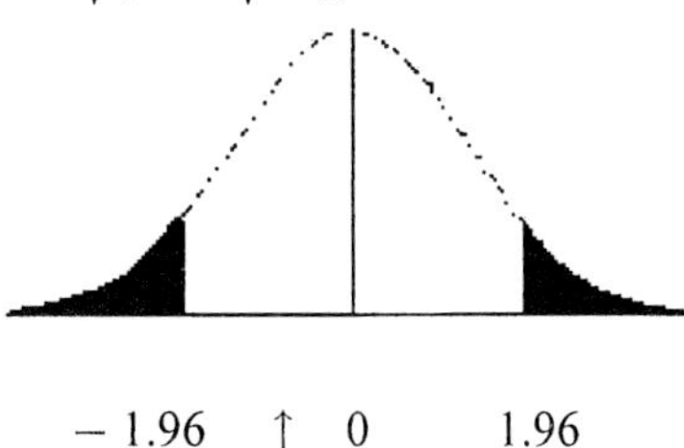

$-1.96 \uparrow 0 \quad 1.96$
-0.88

Do not reject the null hypothesis. There is not enough evidence to reject the claim that the percentage is the same.

10.
H_0: $p = 0.17$ (claim)
H_1: $p \neq 0.17$

$\hat{p} = \frac{22}{90} = 0.2444 \quad p = 0.17 \quad q = 0.83$
C. V. = ± 1.96

10. continued

$z = \frac{\hat{p}-p}{\sqrt{\frac{pq}{n}}} = \frac{0.2444-0.17}{\sqrt{\frac{(0.17)(0.83)}{90}}} = 1.88$

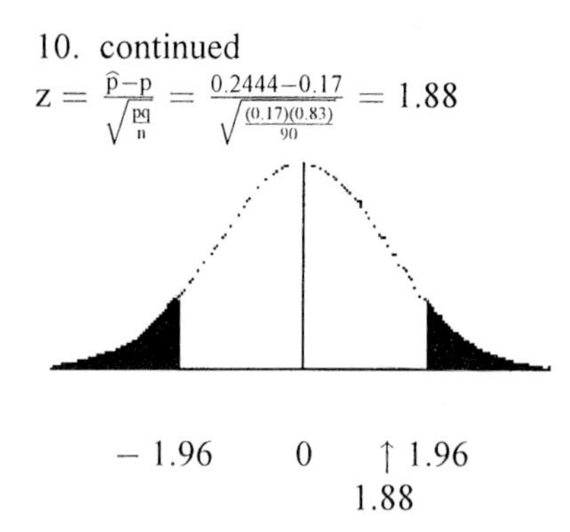

Do not reject the null hypothesis. There is not enough evidence to reject the claim that the percentage is the same.

11.
H_0: $p = 0.54$
H_1: $p \neq 0.54$ (claim)

$\hat{p} = \frac{14}{30} = 0.4667 \quad p = 0.54 \quad q = 0.46$
C. V. $= \pm 1.96$
$z = \frac{\hat{p}-p}{\sqrt{\frac{pq}{n}}} = \frac{0.4667-0.54}{\sqrt{\frac{(0.54)(0.46)}{30}}} = -0.81$

– 1.96 ↑ 0 1.96
– 0.81

Do not reject the null hypothesis. There is not enough evidence to reject the claim that 54% of fatal car/truck accidents are caused by driver error.

12.
H_0: $p \leq 0.25$ (claim)
H_1: $p > 0.25$

$\hat{p} = \frac{30}{100} = 0.30 \quad p = 0.25 \quad q = 0.75$
$z = \frac{\hat{p}-p}{\sqrt{\frac{pq}{n}}} = \frac{0.30-0.25}{\sqrt{\frac{(0.25)(0.75)}{100}}} = 1.15$
Area = 0.3749
P-value = 0.5 – 0.3749 = 0.1251
Since P-value > 0.10, do not reject the null hypothesis. There is not enough evidence to reject the claim that no more than 25% of the students who commute travel more than 14 miles to campus.

13.
H_0: $p \leq 0.30$
H_1: $p > 0.30$ (claim)

$\hat{p} = \frac{72}{200} = 0.36 \quad p = 0.30 \quad q = 0.70$
$z = \frac{\hat{p}-p}{\sqrt{\frac{pq}{n}}} = \frac{0.36-0.30}{\sqrt{\frac{(0.30)(0.70)}{200}}} = 1.85$
Area = 0.4678
P-value = 0.5 – 0.4678 = 0.0322
Since P-value < 0.05, reject the null hypothesis. There is enough evidence to support the claim that more than 30% of the customers have at least two telephones.

14.
H_0: $p = 0.10$ (claim)
H_1: $p \neq 0.10$

$\hat{p} = \frac{10}{67} = 0.149 \quad p = 0.10 \quad q = 0.90$
$z = \frac{\hat{p}-p}{\sqrt{\frac{pq}{n}}} = \frac{0.149-0.10}{\sqrt{\frac{(0.10)(0.90)}{67}}} = 1.34$
Area = 0.4099
P-value = 2(0.5 – 0.4099) = 0.1802
Since P-value > 0.01, do not reject the null hypothesis. There is not enough evidence to reject the claim that 10% of the murders are committed by women.

15.
H_0: $p = 0.18$ (claim)
H_1: $p \neq 0.18$

$\hat{p} = \frac{50}{300} = 0.1667 \quad p = 0.18 \quad q = 0.82$
$z = \frac{\hat{p}-p}{\sqrt{\frac{pq}{n}}} = \frac{0.1667-0.18}{\sqrt{\frac{(0.18)(0.82)}{300}}} = -0.60$
Area = 0.2257
P-value = 2(0.5 – 0.2257) = 0.5486
Since P-value > 0.01, do not reject the null hypothesis. There is not enough evidence to reject the claim that 18% of all high school students smoke at least a pack of cigarettes a day.

16.
H_0: $p \geq 0.83$
H_1: $p < 0.83$ (claim)

$\hat{p} = \frac{40}{50} = 0.8 \quad p = 0.83 \quad q = 0.17$
C. V. $= -1.75$
$z = \frac{\hat{p}-p}{\sqrt{\frac{pq}{n}}} = \frac{0.8-0.83}{\sqrt{\frac{(0.83)(0.17)}{50}}} = -0.56$

16. continued

$-1.75 \quad \uparrow \quad 0$

-0.56

Do not reject the null hypothesis. There is not enough evidence to support the claim that the percentage is less than 83%.

17.

H_0: $p = 0.67$

H_1: $p \neq 0.67$ (claim)

$\hat{p} = \frac{82}{100} = 0.82 \quad p = 0.67 \quad q = 0.33$

C. V. $= \pm 1.96$

$$z = \frac{\hat{p}-p}{\sqrt{\frac{pq}{n}}} = \frac{0.82-0.67}{\sqrt{\frac{(0.67)(0.33)}{100}}} = 3.19$$

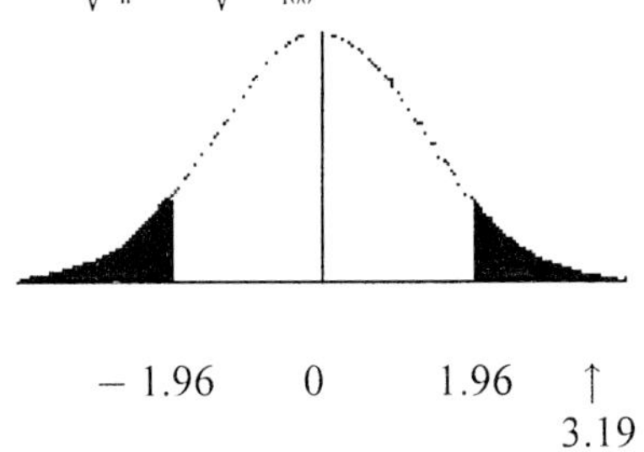

$-1.96 \quad 0 \quad 1.96 \quad \uparrow$

3.19

Reject the null hypothesis. There is enough evidence to support the claim that the percentage is not 67%.

18.

H_0: $p \geq 0.6$

H_1: $p < 0.6$ (claim)

$\hat{p} = \frac{26}{50} = 0.52 \quad p = 0.6 \quad q = 0.4$

C. V. $= -1.65$

$$z = \frac{\hat{p}-p}{\sqrt{\frac{pq}{n}}} = \frac{0.52-0.6}{\sqrt{\frac{(0.6)(0.4}{50}}} = -1.15$$

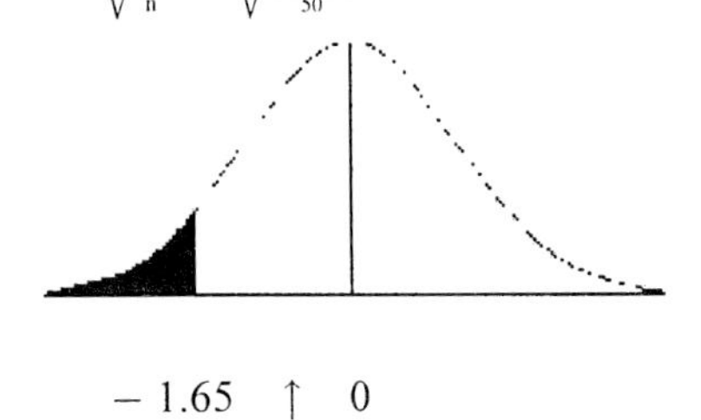

$-1.65 \quad \uparrow \quad 0$

-1.15

18. continued

Do not reject the null hypothesis. There is not enough evidence to support the claim that the percentage of paid assistantships is below 60%.

19.

H_0: $p \geq 0.576$

H_1: $p < 0.576$ (claim)

$\hat{p} = \frac{17}{36} = 0.472 \quad p = 0.576 \quad q = 0.424$

C. V. $= -1.65$

$$z = \frac{\hat{p}-p}{\sqrt{\frac{pq}{n}}} = \frac{0.472-0.576}{\sqrt{\frac{(0.576)(0.424)}{36}}} = -1.26$$

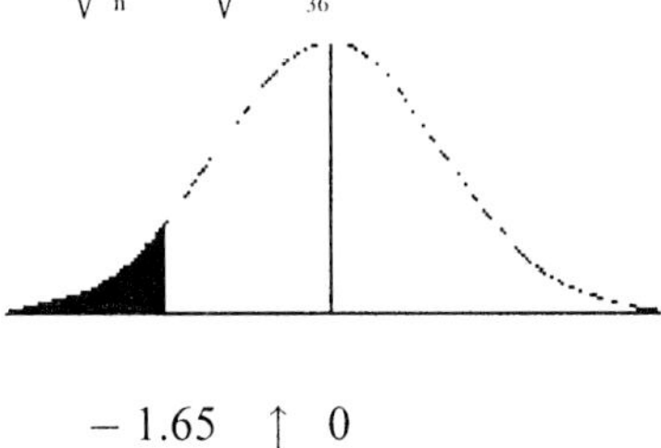

$-1.65 \quad \uparrow \quad 0$

-1.26

Do not reject the null hypothesis. There is not enough evidence to support the claim that the percentage of injuries during practice is below 57.6%.

20.

H_0: $p \geq 0.45$

H_1: $p < 0.45$ (claim)

$\hat{p} = \frac{58}{150} = 0.387 \quad p = 0.45 \quad q = 0.55$

C. V. $= -1.65$

$$z = \frac{\hat{p}-p}{\sqrt{\frac{pq}{n}}} = \frac{0.387-0.45}{\sqrt{\frac{(0.45)(0.55)}{150}}} = -1.55$$

$-1.65 \quad \uparrow \quad 0$

-1.55

Do not reject the null hypothesis. There is not enough evidence to support the claim that the proportion is below 45%.

21.

$P(X = 3, p = 0.5, n = 9) = 0.164$

Since $0.164 > \frac{0.10}{2}$ or 0.05, the conclusion that the coin is not balanced is probably

21. continued
false. Note that α must be split since this is a two-tailed test.

22.
Using the binomial distribution, P(X = 5, p = 0.2, n = 15) = 0.103. Since $0.103 > \frac{0.10}{2}$ or 0.05, there is not enough evidence to indicate that the 20% estimate has changed.

23.
$z = \frac{X-\mu}{\sigma}$

$z = \frac{X-np}{\sqrt{npq}}$

$z = \frac{\frac{X}{n}-\frac{np}{n}}{\frac{1}{n}\sqrt{npq}}$

$z = \frac{\frac{X}{n}-\frac{np}{n}}{\sqrt{\frac{npq}{n^2}}}$

$z = \frac{\hat{p}-p}{\sqrt{\frac{pq}{n}}}$

EXERCISE SET 8-6

1.
a. H_0: $\sigma^2 \leq 225$
H_1: $\sigma^2 > 225$

C. V. = 27.587 d. f. = 17

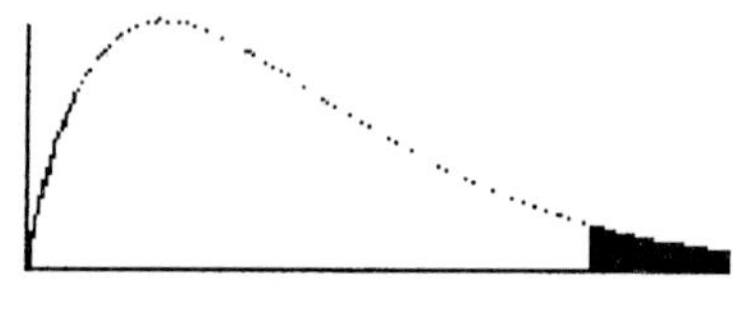

0 27.587

b. H_0: $\sigma^2 \geq 225$
H_1: $\sigma^2 < 225$

C. V. = 14.042 d. f. = 22

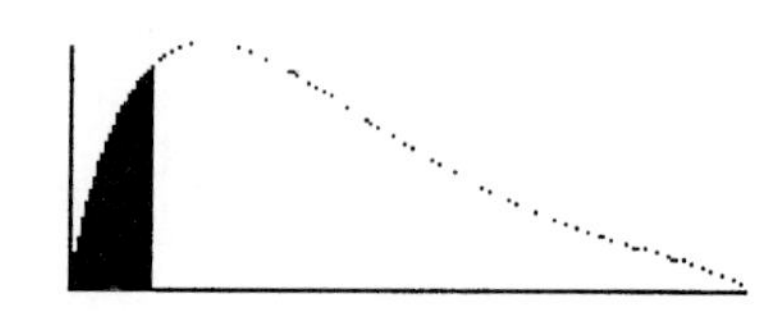

0 14.042

1. continued
c. H_0: $\sigma^2 = 225$
H_1: $\sigma^2 \neq 225$

C. V. = 5.629, 26.119 d. f. = 14

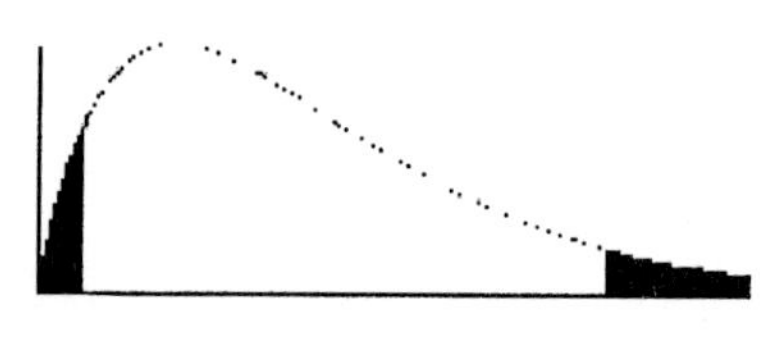

0 5.629 26.119

d. H_0: $\sigma^2 = 225$
H_1: $\sigma^2 \neq 225$

C. V. = 2.167, 14.067 d. f. = 7

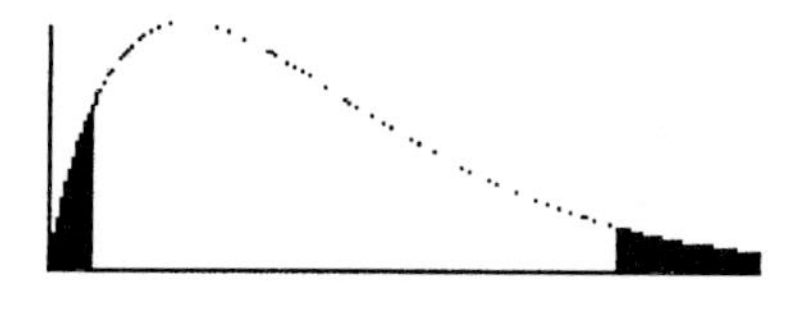

0 2.167 14.067

e. H_0: $\sigma^2 \leq 225$
H_1: $\sigma^2 > 225$

C. V. = 32.000 d. f. = 16

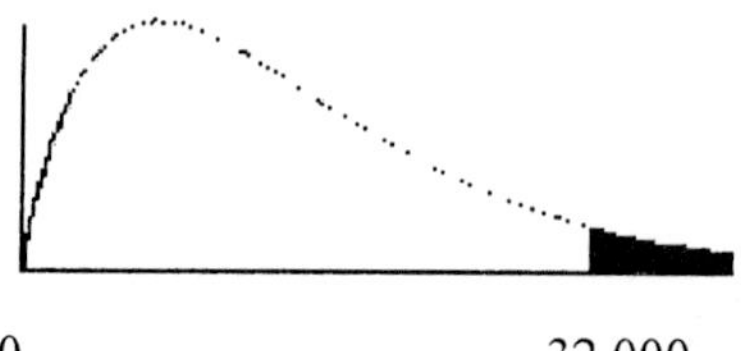

0 32.000

f. H_0: $\sigma^2 \geq 225$
H_1: $\sigma^2 < 225$

C. V. = 8.907 d. f. = 19

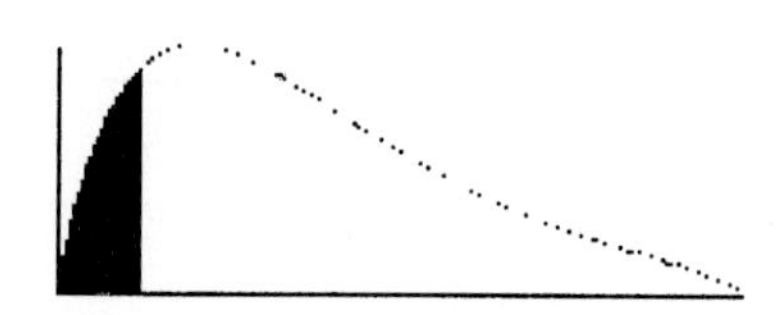

0 8.907

g. H_0: $\sigma^2 = 225$
H_1: $\sigma^2 \neq 225$

C. V. = 3.074, 28.299 d. f. = 12

1g. continued

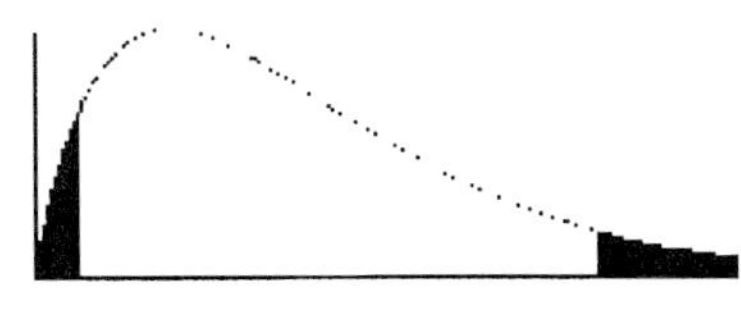

h. H_0: $\sigma^2 \geq 225$
H_1: $\sigma^2 < 225$

C. V. = 15.308 d. f. = 28

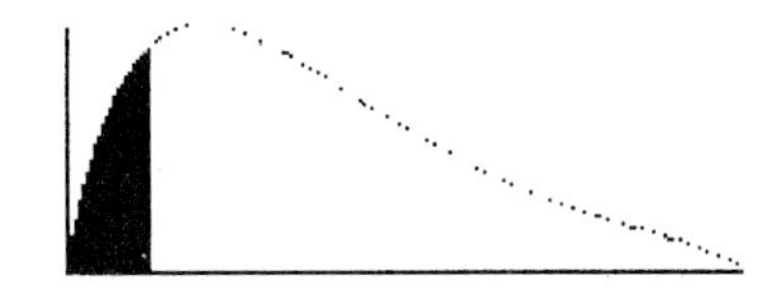

2.
a. 0.01 < P-value < 0.025 (0.015)
b. 0.005 < P-value < 0.01 (0.006)
c. 0.01 < P-value < 0.025 (0.012)
d. P-value < 0.005 (0.003)
e. 0.025 < P-value < 0.05 (0.037)
f. 0.05 < P-value < 0.10 (0.088)
g. 0.05 < P-value < 0.10 (0.066)
h. P-value < 0.01 (0.007)

3.
H_0: $\sigma = 60$ (claim)
H_1: $\sigma \neq 60$

C. V. = 8.672, 27.587 $\alpha = 0.10$
d. f. = 17
s = 64.6
$\chi^2 = \frac{(n-1)s^2}{\sigma^2} = \frac{(18-1)(64.6)^2}{(60)^2} = 19.707$

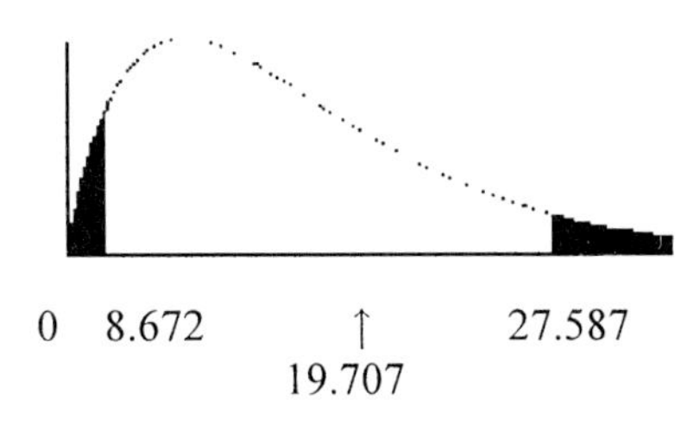

Do not reject the null hypothesis. There is not enough evidence to reject the claim that the standard deviation is 60.

4.
H_0: $\sigma^2 \leq 6.2$
H_1: $\sigma^2 > 6.2$ (claim)

C. V. = 33.409 $\alpha = 0.01$ d. f. = 17

$\chi^2 = \frac{(n-1)s^2}{\sigma^2} = \frac{(18-1)(6.5)}{6.2} = 17.823$

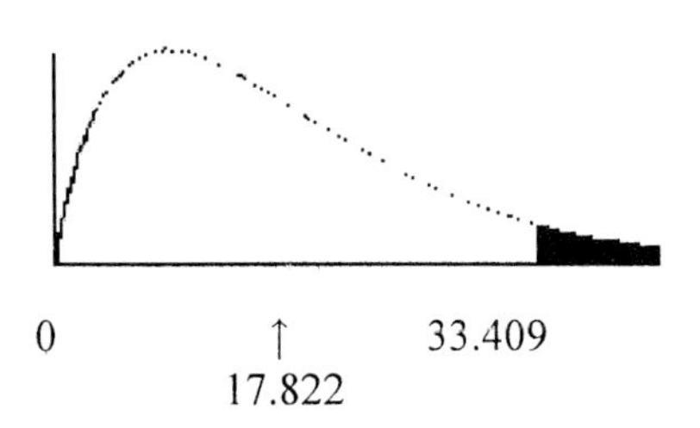

Do not reject the null hypothesis. There is not enough evidence to support the claim that the variance is greater than 6.2.

5.
H_0: $\sigma^2 \leq 25$ (claim)
H_1: $\sigma^2 > 25$

C. V. = 27.204 $\alpha = 0.10$ d. f. = 19

$\chi^2 = \frac{(n-1)s^2}{\sigma^2} = \frac{(20-1)(36)}{25} = 27.36$

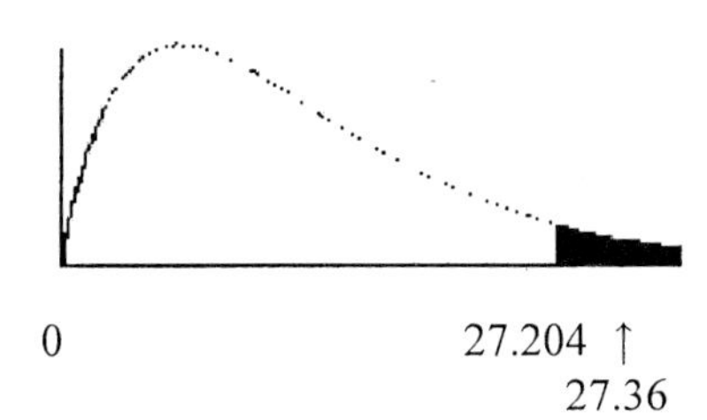

Reject the null hypothesis. There is enough evidence to reject the claim that the variance is less than or equal to 25.

6.
H_0: $\sigma \geq 10$
H_1: $\sigma < 10$ (claim)

C. V. = 8.672 $\alpha = 0.05$ d. f. = 17

$\chi^2 = \frac{(n-1)s^2}{\sigma^2} = \frac{(18-1)(72.222)}{10} = 12.278$

6. continued

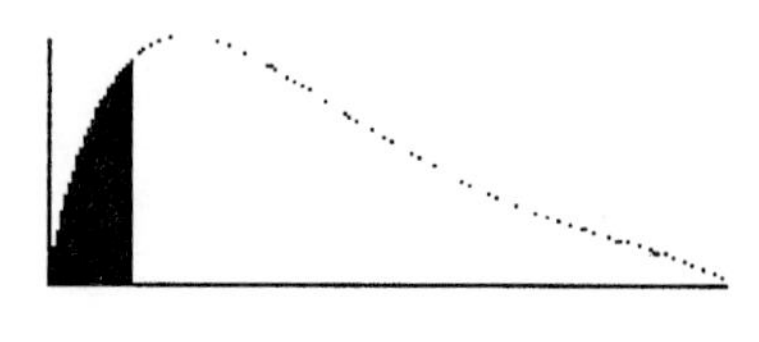

0 8.672 ↑
12.278

Reject the null hypothesis. There is enough evidence to support the claim that the standard deviation of the weights is less than 10 pounds.

7.
H_0: $\sigma \leq 1.2$ (claim)
H_1: $\sigma > 1.2$

$\alpha = 0.01$ d. f. = 14
$\chi^2 = \frac{(n-1)s^2}{\sigma^2} = \frac{(15-1)(1.8)^2}{(1.2)^2} = 31.5$
P-value < 0.005 (0.0047)

Since P-value < 0.01, reject the null hypothesis. There is enough evidence to reject the claim that the standard deviation is less than or equal to 1.2 minutes.

8.
H_0: $\sigma \leq 0.03$ (claim)
H_1: $\sigma > 0.03$

s = 0.043
$\alpha = 0.05$ d. f. = 7
$\chi^2 = \frac{(n-1)s^2}{\sigma^2} = \frac{(7(0.043)^2}{0.03^2} = 14.381$
0.025 < P-value < 0.05 (0.045)

Since P-value < 0.05, reject the null hypothesis. There is enough evidence to reject the claim that the standard deviation is less than or equal to 0.03 ounce.

9.
H_0: $\sigma \leq 20$
H_1: $\sigma > 20$ (claim)

s = 35.11
C. V. = 36.191 $\alpha = 0.01$ d. f. = 19
$\chi^2 = \frac{(n-1)s^2}{\sigma^2} = \frac{(20-1)(35.11)^2}{20^2} = 58.55$

9. continued

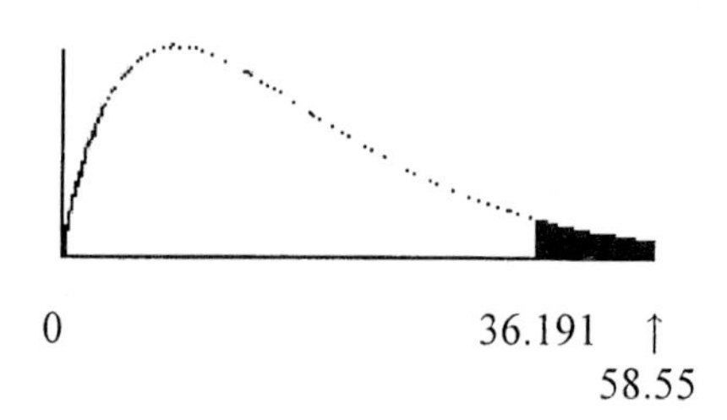

0 36.191 ↑
58.55

Reject the null hypothesis. There is enough evidence to support the claim that the standard deviation is more than 20 calories.

10.
H_0: $\sigma \geq 10$
H_1: $\sigma < 10$ (claim)

s = 5.407
C. V. = 2.204 $\alpha = 0.10$ d. f. = 6
$\chi^2 = \frac{(n-1)s^2}{\sigma^2} = \frac{(7-1)(5.407)^2}{20^2} = 1.754$

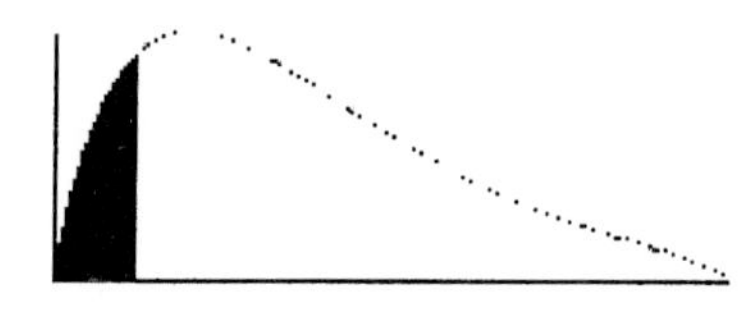

0 ↑ 2.204
1.754

Do not reject the null hypothesis. There is not enough evidence to support the claim that the standard deviation is less than 10°.

11.
H_0: $\sigma \leq 100$
H_1: $\sigma > 100$ (claim)

C. V. = 124.342 $\alpha = 0.05$ d. f. = 299
$\chi^2 = \frac{(n-1)s^2}{\sigma^2} = \frac{(300-1)(110)^2}{100^2} = 361.79$

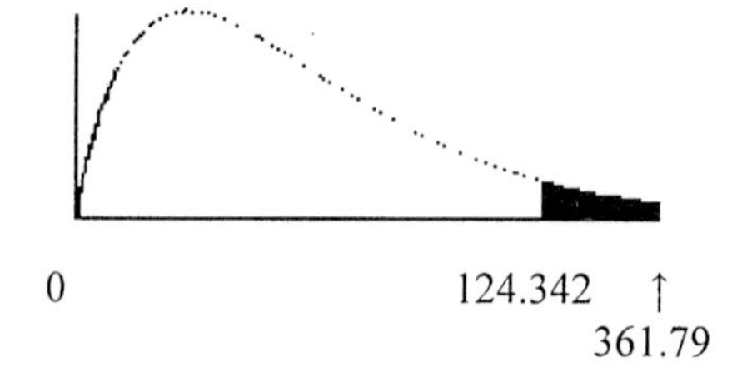

0 124.342 ↑
361.79

Reject the null hypothesis. There is enough evidence to support the claim that the standard deviation is more than 100.

12.
H_0: $\sigma \leq 8$
H_1: $\sigma > 8$ (claim)

C. V. = 55.758 $\alpha = 0.05$ d. f. = 49
$\chi^2 = \frac{(n-1)s^2}{\sigma^2} = \frac{(50-1)(10.5)^2}{8^2} = 84.41$

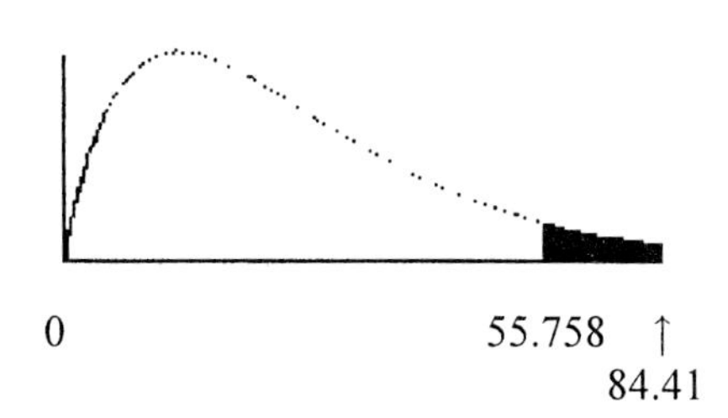

0 55.758 ↑
84.41

Reject the null hypothesis. There is enough evidence to support the claim that the standard deviation is more than 8.

13.
H_0: $\sigma \leq 25$
H_1: $\sigma > 25$ (claim)

C. V. = 22.362 $\alpha = 0.05$ d. f. = 13
$\chi^2 = \frac{(n-1)s^2}{\sigma^2} = \frac{(14-1)(6.74)^2}{25} = 23.622$

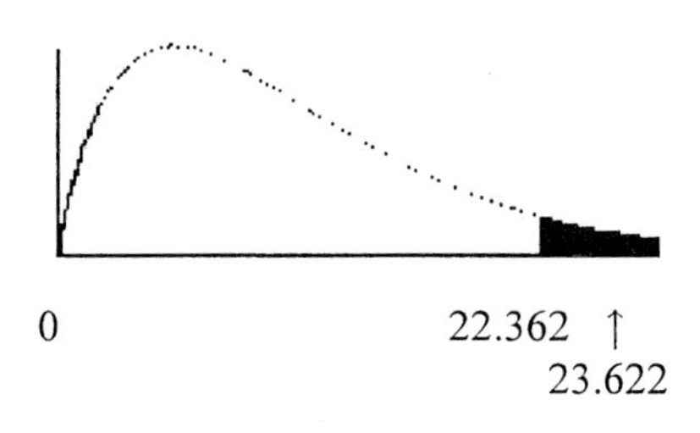

0 22.362 ↑
23.622

Reject the null hypothesis. There is enough evidence to support the claim that the variance is greater than 25.

14.
H_0: $\sigma \leq 4$
H_1: $\sigma > 4$ (claim)

C. V. = 24.996 $\alpha = 0.05$ d. f. = 15
$\chi^2 = \frac{(n-1)s^2}{\sigma^2} = \frac{(16-1)(4.2303)^2}{4^2} = 16.777$

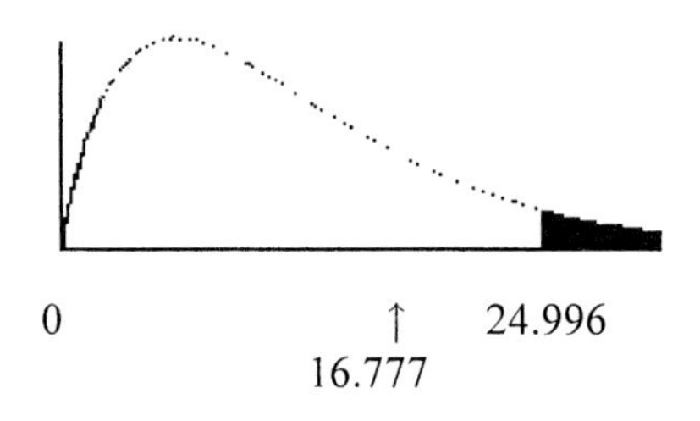

0 ↑ 24.996
16.777

14. continued
Do not reject the null hypothesis. There is not enough evidence to support the claim that the standard deviation exceeds 4.

EXERCISE SET 8-7

1.
H_0: $\mu = 1800$ (claim)
H_1: $\mu \neq 1800$

C. V. = ± 1.96
$z = \frac{\overline{X}-\mu}{\frac{\sigma}{\sqrt{n}}} = \frac{1830-1800}{\frac{200}{\sqrt{10}}} = 0.47$

− 1.96 0 ↑ 1.96
0.47

The 95% confidence interval of the mean is:
$\overline{X} - z_{\frac{\alpha}{2}} \frac{\sigma}{\sqrt{n}} < \mu < \overline{X} + z_{\frac{\alpha}{2}} \frac{\sigma}{\sqrt{n}}$

$1830 - 1.96\left(\frac{200}{\sqrt{10}}\right) < \mu <$
$1830 + 1.96\left(\frac{200}{\sqrt{10}}\right)$
$1706.04 < \mu < 1953.96$

The hypothesized mean is within the interval, thus we can be 95% confident that the average sales will be between $1706.94 and $1953.96.

2.
H_0: $\mu = 42$ (claim)
H_1: $\mu \neq 42$

C. V. = ± 1.65
$z = \frac{\overline{X}-\mu}{\frac{\sigma}{\sqrt{n}}} = \frac{48-42}{\frac{8}{\sqrt{10}}} = 2.37$

− 1.65 0 1.65 ↑
2.37

The 90% confidence interval of the mean is:

2. continued

$\overline{X} - z_{\frac{\alpha}{2}} \frac{\sigma}{\sqrt{n}} < \mu < \overline{X} + z_{\frac{\alpha}{2}} \frac{\sigma}{\sqrt{n}}$

$48 - 1.65 \cdot \frac{8}{\sqrt{10}} < \mu < 48 + 1.65 \cdot \frac{8}{\sqrt{10}}$

$43.83 < \mu < 52.17$

The decision is to reject the null hypothesis at $\alpha = 0.10$ since $2.37 > 1.65$ and the 90% confidence interval of the mean does not contain the hypothesized mean of 42.

There is agreement between the z-test and the confidence interval. The conclusion then is that there is enough evidence to support the claim that the mean time has changed.

3.

H_0: $\mu = 86$ (claim)
H_1: $\mu \neq 86$

C. V. $= \pm 2.58$

$z = \frac{\overline{X} - \mu}{\frac{\sigma}{\sqrt{n}}} = \frac{84 - 86}{\frac{6}{\sqrt{15}}} = -1.29$

-2.58 ↑ 0 2.58
-1.29

$\overline{X} - z_{\frac{\alpha}{2}} \frac{\sigma}{\sqrt{n}} < \mu < \overline{X} + z_{\frac{\alpha}{2}} \frac{\sigma}{\sqrt{n}}$

$84 - 2.58 \cdot \frac{6}{\sqrt{15}} < \mu < 84 + 1.58 \cdot \frac{6}{\sqrt{15}}$

$80.00 < \mu < 88.00$

The decision is do not reject the null hypothesis since $-1.29 > -2.58$ and the 99% confidence interval contains the hypothesized mean. There is not enough evidence to reject the claim that the monthly maintenance is \$86.

4.

H_0: $\mu = 47$
H_1: $\mu \neq 47$ (claim)

C. V. $= \pm 1.65$

$z = \frac{\overline{X} - \mu}{\frac{\sigma}{\sqrt{n}}} = \frac{42 - 47}{\frac{7}{\sqrt{10}}} = -2.26$

4. continued

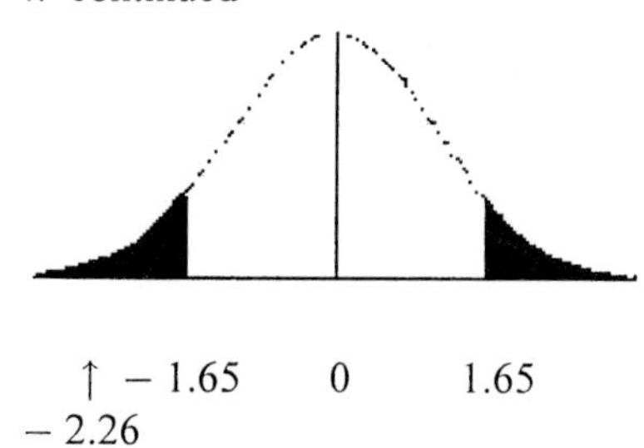

↑ -1.65 0 1.65
-2.26

The 90% confidence interval of the mean is:

$\overline{X} - z_{\frac{\alpha}{2}} \frac{\sigma}{\sqrt{n}} < \mu < \overline{X} + z_{\frac{\alpha}{2}} \frac{\sigma}{\sqrt{n}}$

$42 - 1.65 \cdot \frac{7}{\sqrt{10}} < \mu < 42 + 1.65 \cdot \frac{7}{\sqrt{10}}$

$38.35 < \mu < 45.65$

The decision is to reject the null hypothesis since $-2.26 < -1.65$ and the confidence interval does not contain the hypothesized mean of 47. There is enough evidence to support the claim that the mean has changed.

5.

H_0: $\mu = 22$
H_1: $\mu \neq 22$ (claim)

C. V. $= \pm 2.58$

$z = \frac{\overline{X} - \mu}{\frac{\sigma}{\sqrt{n}}} = \frac{20.8 - 22}{\frac{4}{\sqrt{60}}} = -2.32$

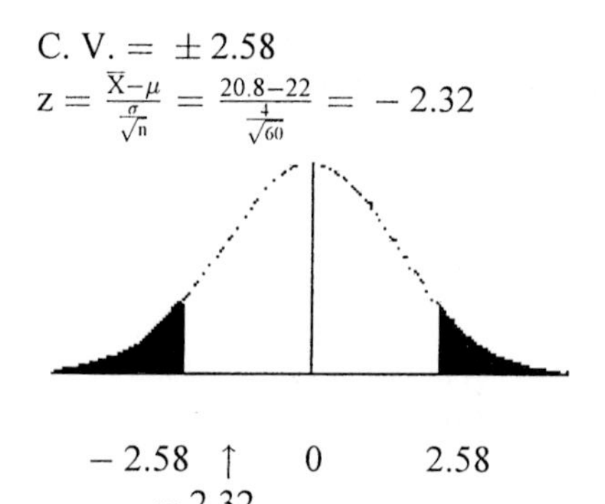

-2.58 ↑ 0 2.58
-2.32

The 99% confidence interval of the mean is:

$\overline{X} - z_{\frac{\alpha}{2}} \frac{\sigma}{\sqrt{n}} < \mu < \overline{X} + z_{\frac{\alpha}{2}} \frac{\sigma}{\sqrt{n}}$

$20.8 - 2.58 \cdot \frac{4}{\sqrt{60}} < \mu < 20.8 + 2.58 \cdot \frac{4}{\sqrt{60}}$

$19.47 < \mu < 22.13$

The decision is do not reject the null hypothesis since $-2.32 > -2.58$ and the 99% confidence interval does contain the hypothesized mean of 22. The conclusion is

5. continued
that there is not enough evidence to support the claim that the average studying time has changed.

6.
H_0: $\mu = 10.8$ (claim)
H_1: $\mu \neq 10.8$

C. V. $= \pm 2.33$

$z = \frac{\overline{X}-\mu}{\frac{\sigma}{\sqrt{n}}} = \frac{12.2-10.8}{\frac{3}{\sqrt{36}}} = 2.80$

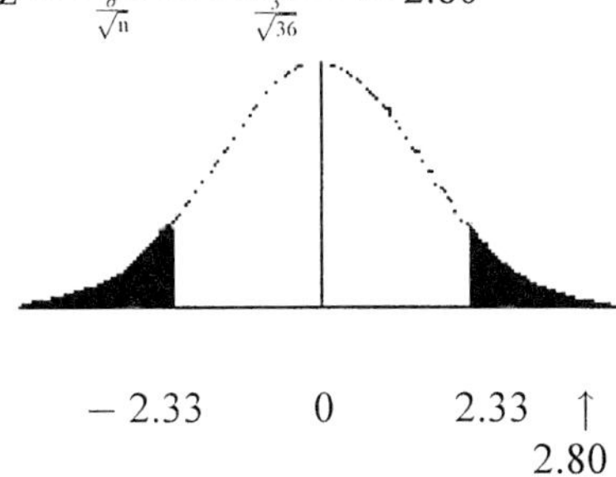

-2.33 0 2.33 ↑
2.80

$\overline{X} - z_{\frac{\alpha}{2}} \frac{\sigma}{\sqrt{n}} < \mu < \overline{X} + z_{\frac{\alpha}{2}} \frac{\sigma}{\sqrt{n}}$

$12.2 - 2.33 \cdot \frac{3}{\sqrt{36}} < \mu < 12.2 + 2.33 \cdot \frac{3}{\sqrt{36}}$

$11.035 < \mu < 13.365$

The decision is to reject the null hypothesis since $2.80 > 2.33$ and the confidence interval does not contain the hypothesized mean 10.8. The conclusion is that there is enough evidence to reject the claim that the average time a person spends reading a newspaper is 10.8 minutes.

7.
The power of a statistical test is the probability of rejecting the null hypothesis when it is false.

8.
The power of a test is equal to $1 - \beta$ where β is the probability of a type II error.

9.
The power of a test can be increased by increasing α or selecting a larger sample size.

REVIEW EXERCISES - CHAPTER 8

1.
H_0: $\mu = 98°$ (claim)
H_1: $\mu \neq 98°$

1. continued
C. V. $= \pm 1.96$

$z = \frac{\overline{X}-\mu}{\frac{s}{\sqrt{n}}} = \frac{95.8-98}{\frac{7.71}{\sqrt{50}}} = -2.02$

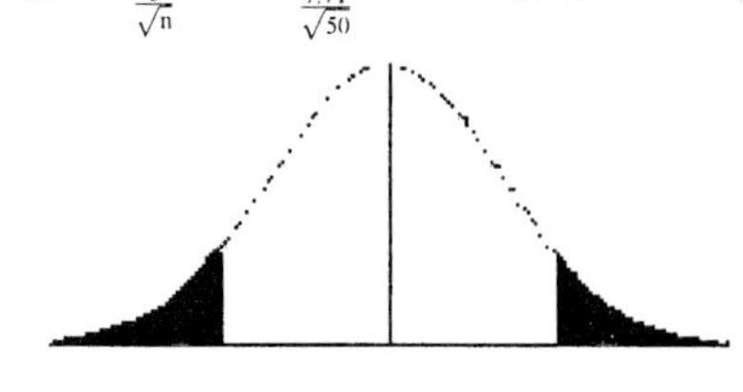

↑ -1.96 0 1.96
-2.02

Reject the null hypothesis. There is enough evidence to reject the claim that the average high temperature is 98°.

2.
H_0: $\mu = 14$ (claim)
H_1: $\mu \neq 14$

C. V. $= \pm 1.96$

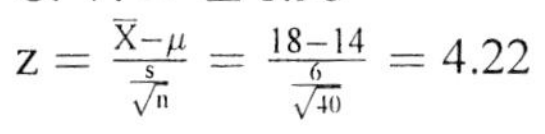

$z = \frac{\overline{X}-\mu}{\frac{s}{\sqrt{n}}} = \frac{18-14}{\frac{6}{\sqrt{40}}} = 4.22$

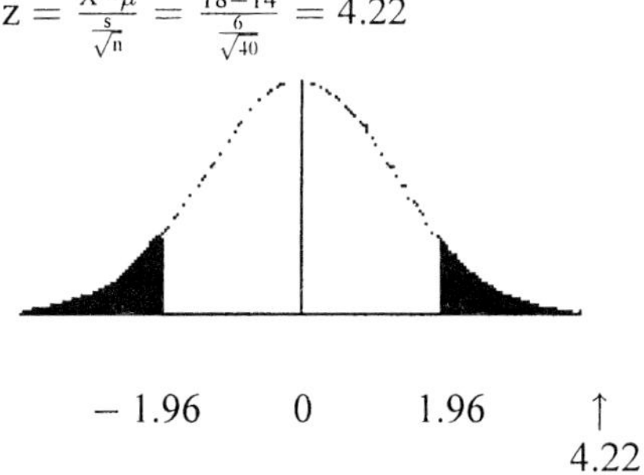

-1.96 0 1.96 ↑
4.22

Reject the null hypothesis. There is enough evidence to reject the claim that the average number of cigarettes a person smokes a day is 14.

3.
H_0: $\mu \leq \$40,000$
H_1: $\mu > \$40,000$ (claim)

C. V. $= 1.65$

$z = \frac{\overline{X}-\mu}{\frac{\sigma}{\sqrt{n}}} = \frac{\$41,000-\$40,000}{\frac{\$3000}{\sqrt{36}}} = 2.00$

0 1.65 ↑
2.00

3. continued

Reject the null hypothesis. There is enough evidence to support the claim that the average salary is more than \$40,000.

4.

H_0: $\mu \leq \$150{,}000$

H_1: $\mu > \$150{,}000$ (claim)

C. V. = 1.895 d. f. = 7

$t = \frac{\overline{X}-\mu}{\frac{s}{\sqrt{n}}} = \frac{155{,}500-150{,}000}{\frac{15{,}000}{\sqrt{8}}} = 1.04$

0 ↑ 1.895

1.04

Do not reject the null hypothesis. There is not enough evidence to support the claim that the average salary is more than \$150,000.

5.

H_0: $\mu \leq 67$

H_1: $\mu > 67$ (claim)

C. V. = 1.383 d. f. = 9

$t = \frac{\overline{X}-\mu}{\frac{s}{\sqrt{n}}} = \frac{69.6-67.0}{\frac{1.1}{\sqrt{10}}} = 7.47$

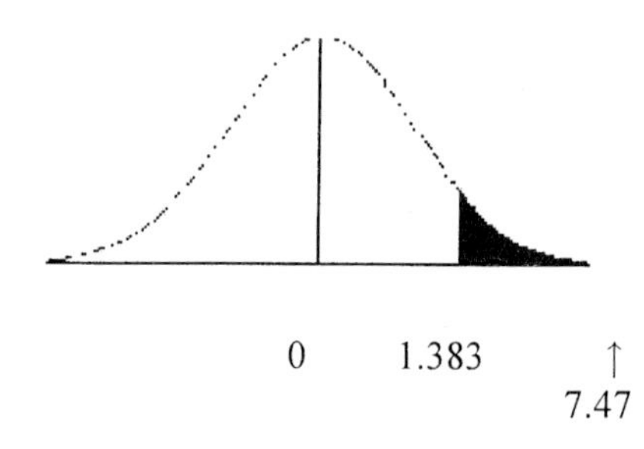

0 1.383 ↑

7.47

Reject the null hypothesis. There is enough evidence to support the claim that 1995 was warmer than average.

6.

H_0: $\mu \leq 23.2$ (claim)

H_1: $\mu > 23.2$

d. f. = 17

$t = \frac{\overline{X}-\mu}{\frac{s}{\sqrt{n}}} = \frac{22.6-23.2}{\frac{2}{\sqrt{18}}} = -1.27$

6. continued

0.10 < P-value < 0.25 (0.111)

Do not reject the null hypothesis. There is not enough evidence to reject the claim that the average age is less than or equal to 23.2 years.

7.

H_0: $\mu = 6$ (claim)

H_1: $\mu \neq 6$

C. V. = ± 2.821 $\overline{X} = 8.42$ $s = 4.17$

$t = \frac{\overline{X}-\mu}{\frac{s}{\sqrt{n}}} = \frac{8.42-6}{\frac{4.17}{\sqrt{10}}} = 1.835$

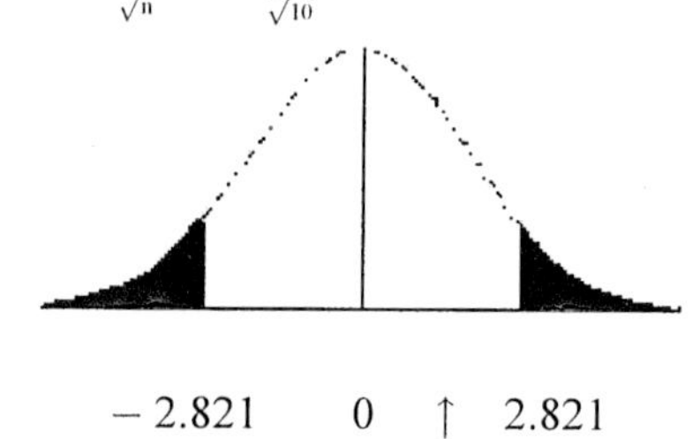

− 2.821 0 ↑ 2.821

1.835

Do not reject the null hypothesis. There is not enough evidence to support the claim that the average attendance has changed.

8.

H_0: $p \leq 0.585$

H_1: $p > 0.585$ (claim)

C. V. = 1.65

$\hat{p} = \frac{622}{1000} = 0.622$ $p = 0.585$ $q = 0.415$

$z = \frac{\hat{p}-p}{\sqrt{\frac{pq}{n}}} = \frac{0.622-0.585}{\sqrt{\frac{(0.585)(0.415)}{1000}}} = 2.37$

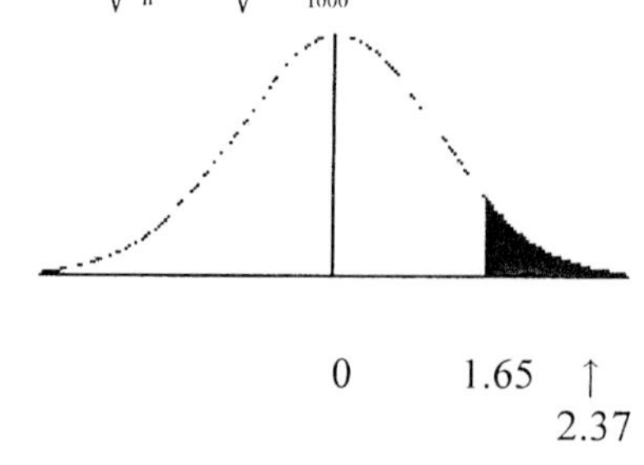

0 1.65 ↑

2.37

Reject the null hypothesis. There is enough evidence to support the claim that the percentage of women working is more than 58.5%.

9.

H_0: $p \leq 0.602$

H_1: $p > 0.602$ (claim)

C. V. = 1.65

9. continued

$\widehat{p} = 0.65 \quad p = 0.602 \quad q = 0.398$

$z = \frac{\widehat{p}-p}{\sqrt{\frac{pq}{n}}} = \frac{0.65-0.602}{\sqrt{\frac{(0.602)(0.398)}{400}}} = 1.96$

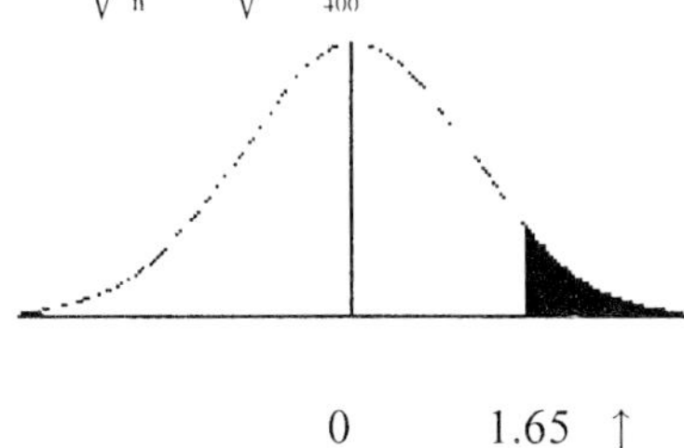

0 1.65 ↑ 1.96

Reject the null hypothesis. There is enough evidence to support the claim that the percentage of drug offenders is higher than 60.2%.

10.

H_0: $p = 0.80$ (claim)

H_1: $p \neq 0.80$

C. V. $= \pm 2.33$

$\widehat{p} = \frac{20}{30} = 0.6667 \quad p = 0.80 \quad q = 0.20$

$z = \frac{\widehat{p}-p}{\sqrt{\frac{pq}{n}}} = \frac{0.6667-0.80}{\sqrt{\frac{(0.8)(0.2)}{30}}} = -1.83$

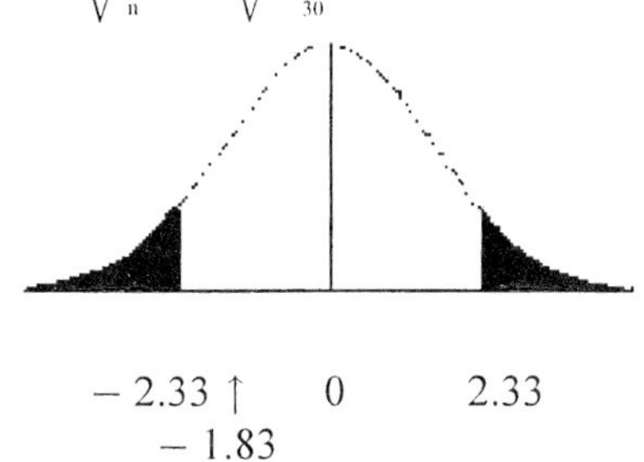

-2.33 ↑ 0 2.33
-1.83

Do not reject the null hypothesis. There is not enough evidence to reject the claim that 80% of the new home buyers wanted a fireplace.

11.

H_0: $p = 0.65$ (claim)

H_1: $p \neq 0.65$

$\widehat{p} = \frac{57}{80} = 0.7125 \quad p = 0.65 \quad q = 0.35$

$z = \frac{\widehat{p}-p}{\sqrt{\frac{pq}{n}}} = \frac{0.7125-0.65}{\sqrt{\frac{(0.65)(0.35)}{80}}} = 1.17$

Area = 0.3790

P-value = 2(0.5 – 0.3790) = 0.242

Since P-value > 0.05, do not reject the null hypothesis. There is not enough evidence to reject the claim that 65% of the teenagers own their own radios.

12.

H_0: $\mu = 225$ (claim)

H_1: $\mu \neq 225$

$z = \frac{\overline{X}-\mu}{\frac{s}{\sqrt{50}}} = \frac{230-225}{\frac{15}{\sqrt{50}}} = 2.36$

Area = 0.4909

P-value = 2(0.5 – 0.4909) = 0.0182

Since 0.0182 > 0.01 the decision is do not reject the null hypothesis. The conclusion is that there is not evidence to reject the claim that the mean is 225 pounds.

13.

H_0: $\mu \geq 10$

H_1: $\mu < 10$ (claim)

$z = \frac{\overline{X}-\mu}{\frac{\sigma}{\sqrt{n}}} = \frac{9.25-10}{\frac{2}{\sqrt{35}}} = -2.22$

Area = 0.4868

P-value = 0.5 – 0.4699 = 0.0132

Since 0.0132 < 0.05, reject the null hypothesis. The conclusion is that there is enough evidence to support the claim that the average time is less than 10 minutes.

14.

H_0: $\sigma = 3.4$ (claim)

H_1: $\sigma \neq 3.4$

C. V. = 11.689, 38.076 d. f. = 23

$\chi^2 = \frac{(n-1)s^2}{\sigma^2} = \frac{(24-1)(4.2)^2}{(3.4)^2} = 35.1$

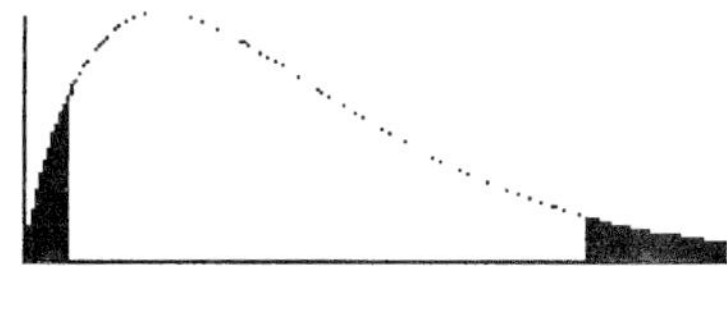

11.689 35.1 ↑ 38.076

Do not reject the null hypothesis. There is not enough evidence to reject the claim that the standard deviation is 3.4 minutes.

15.

H_0: $\sigma \geq 4.3$ (claim)

H_1: $\sigma < 4.3$

d. f. = 19

$\chi^2 = \frac{(n-1)s^2}{\sigma^2} = \frac{(20-1)(2.6)^2}{(4.3^2)} = 6.95$

0.005 < P-value < 0.01 (0.006)

15. continued
Since P-value < 0.05, reject the null hypothesis. There is enough evidence to reject the claim that the standard deviation is greater than or equal to 4.3 miles per gallon.

16.
H_0: $\sigma = \$95$ (claim)
H_1: $\sigma \neq \$95$

s = 89.3
C. V. = 6.408, 33.409 d. f. = 17
$\chi^2 = \frac{(n-1)s^2}{\sigma^2} = \frac{(18-1)(89.3)^2}{95^2} = 15.0212$

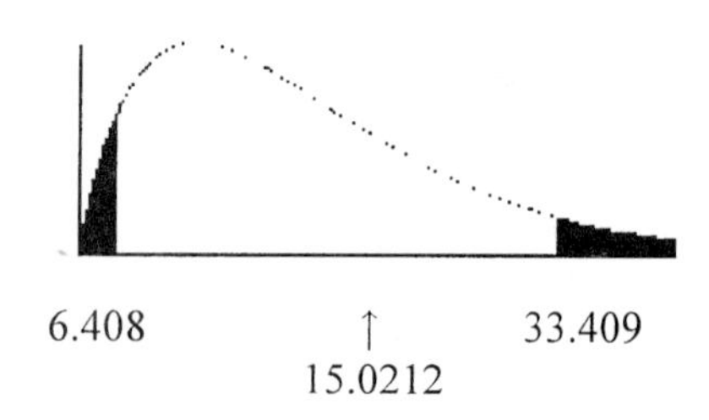

6.408 ↑ 33.409
15.0212

Do not reject the null hypothesis. There is not enough evidence to reject the claim that the standard deviation of rental rates is $95.

17.
H_0: $\sigma = 18$ (claim)
H_1: $\sigma \neq 18$

C. V. = 11.143 and 0.484 d. f. = 4

$\chi^2 = \frac{(n-1)s^2}{\sigma^2} = \frac{(5-1)(21)^2}{18^2} = 5.44$

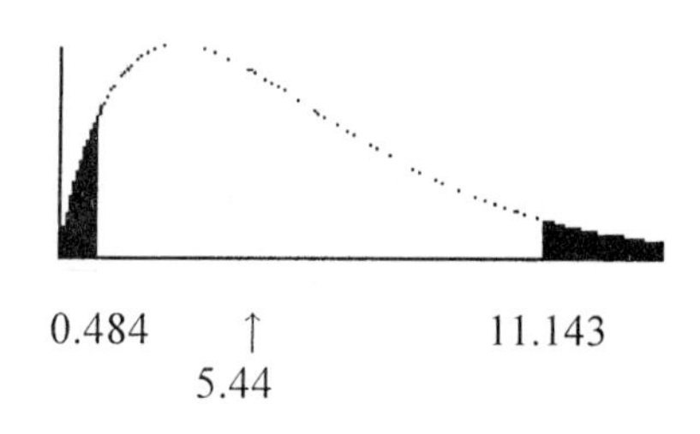

0.484 ↑ 11.143
5.44

Do not reject the null hypothesis. There is not enough evidence to reject the claim that the standard deviation is 21 minutes.

18.
H_0: $\mu = 35$ (claim)
H_1: $\mu \neq 35$

C. V. = ± 1.65
$z = \frac{\overline{X}-\mu}{\frac{s}{\sqrt{n}}} = \frac{33.5-35}{\frac{3}{\sqrt{36}}} = -3.00$

18. continued
The 90% confidence interval of the mean is:

$\overline{X} - z_{\frac{\alpha}{2}} \frac{\sigma}{\sqrt{n}} < \mu < \overline{X} + z_{\frac{\alpha}{2}} \frac{\sigma}{\sqrt{n}}$

$33.5 - 1.65 \cdot \frac{3}{\sqrt{36}} < \mu < 33.5 + 1.65 \cdot \frac{3}{\sqrt{36}}$

$32.675 < \mu < 34.325$

The decision is to reject the null hypothesis since $-3.00 < -1.65$ and the 90% confidence interval does not contain the hypothesized mean of 35. The conclusion is that there is enough evidence to reject the claim that the mean is 35 pounds.

19.
H_0: $\mu = 4$
H_1: $\mu \neq 4$ (claim)

C. V. = ± 2.58
$z = \frac{\overline{X}-\mu}{\frac{s}{\sqrt{n}}} = \frac{4.2-4}{\frac{0.6}{\sqrt{20}}} = 1.49$

The 99% confidence interval of the mean is:

$\overline{X} - z_{\frac{\alpha}{2}} \frac{\sigma}{\sqrt{n}} < \mu < \overline{X} + z_{\frac{\alpha}{2}} \frac{\sigma}{\sqrt{n}}$

$4.2 - 2.58 \cdot \frac{0.6}{\sqrt{20}} < \mu < 4.2 + 2.58 \cdot \frac{0.6}{\sqrt{20}}$
$3.85 < \mu < 4.55$

The decision is do not reject the null hypothesis since $1.49 < 2.58$ and the confidence interval does contain the hypothesized mean of 4. There is not enough evidence to support the claim that the growth has changed.

CHAPTER 8 QUIZ

1. True
2. True
3. False, the critical value separates the critical region from the noncritical region.
4. True
5. False, it can be one-tailed or two-tailed.
6. b.
7. d.
8. c.
9. b.
10. type I
11. β
12. statistical hypothesis

13. right
14. $n - 1$

15. H_0: $\mu = 28.6$ (claim)
H_1: $\mu \neq 28.6$
C. V. $= \pm 1.96$
$z = 2.14$
Reject the null hypothesis. There is enough evidence to reject the claim that the average age is 28.6.

16. H_0: $\mu = \$6{,}500$ (claim)
H_1: $\mu \neq \$6{,}500$
C. V. $= \pm 1.96$
$z = 5.27$
Reject the null hypothesis. There is enough evidence to reject the agent's claim.

17. H_0: $\mu \leq 8$
H_1: $\mu > 8$ (claim)
C. V. $= 1.65$
$z = 6.00$
Reject the null hypothesis. There is enough evidence to support the claim that the average number of sticks is greater than 8.

18. H_0: $\mu = 21$ (claim)
H_1: $\mu \neq 21$
C. V. $= \pm 2.921$
$t = -2.06$
Do not reject the null hypothesis. There is not enough evidence to reject the claim that the average number of dropouts is 21.

19. H_0: $\mu \geq 67$
H_1: $\mu < 67$ (claim)
$t = -3.1568$
P-value < 0.005 (0.003)
Since P-value < 0.05, reject the null hypothesis. There is enough evidence to support the claim that the average height is less than 67 inches.

20. H_0: $\mu \geq 12.4$
H_1: $\mu < 12.4$ (claim)
C. V. $= -1.345$
$t = -0.328$
Reject the null hypothesis. There is enough evidence to support the claim that the average is less than what the company claimed.

21. H_0: $\mu \leq 63.5$
H_1: $\mu > 63.5$ (claim)
$t = 0.47075$

21. continued
P-value > 0.25 (0.322)
Since P-value > 0.05, do not reject the null hypothesis. There is not enough evidence to support the claim that the average is greater than 63.5.

22. H_0: $\mu = 26$ (claim)
H_1: $\mu \neq 26$
C. V. $= \pm 2.492$
$t = -1.5$
Do not reject the null hypothesis. There is not enough evidence to reject the claim that the average age is 26.

23. H_0: $p \geq 0.25$ (claim)
H_1: $p < 0.25$
$\mu = 25$ $\sigma = 4.33$
C. V. $= -1.65$
$z = -0.6928$
Do not reject the null hypothesis. There is not enough evidence to reject the claim that the proportion is at least 0.25.

24. H_0: $p \geq 0.55$ (claim)
H_1: $p < 0.55$
$\mu = 44$ $\sigma = 4.45$
C. V. $= -1.28$
$z = -0.899$
Do not reject the null hypothesis. There is not enough evidence to reject the dietitian's claim.

25. H_0: $p = 0.7$ (claim)
H_1: $p \neq 0.7$
$\mu = 21$ $\sigma = 2.51$
C. V. $= \pm 2.33$
$z = 0.7968$
Do not reject the null hypothesis. There is not enough evidence to reject the claim that the proportion is 0.7.

26. H_0: $p = 0.75$ (claim)
H_1: $p \neq 0.75$
$\mu = 45$ $\sigma = 3.35$
C. V. $= \pm 2.58$
$z = 2.6833$
Reject the null hypothesis. there is enough evidence to reject the claim.

27. The area corresponding to $z = 2.14$ is 0.4838.
P-value $= 2(0.5 - 0.4838) = 0.0324$

28. The area corresponding to z = 5.27 is greater than 0.4999.
Thus, P-value $\leq 2(0.5 - 0.4999) \leq 0.0002$.
(Note: Calculators give 0.0001)

29. H_0: $\sigma \leq 6$
H_1: $\sigma > 6$ (claim)
C. V. = 36.415
$\chi^2 = 54$
Reject the null hypothesis. There is enough evidence to support the claim that the standard deviation is more than 6 pages.

30. H_0: $\sigma = 8$ (claim)
H_1: $\sigma \neq 8$
C. V. = 27.991, 79.490
$\chi^2 = 33.2$
Do not reject the null hypothesis. There is not enough evidence to reject the claim that $\sigma = 8$.

31. H_0: $\sigma \geq 2.3$
H_1: $\sigma < 2.3$ (claim)
C. V. = 10.117
$\chi^2 = 13$
Reject the null hypothesis. There is enough evidence to support the claim that the standard deviation is less than 2.3.

32. H_0: $\sigma = 9$ (claim)
H_1: $\sigma \neq 9$
$\chi^2 = 13.4$
P-value > 0.20 (0.290)
Since P-value > 0.05, do not reject the null hypothesis. There is not enough evidence to reject the claim that $\sigma = 9$.

33. $28.3 < \mu < 30.1$

34. $\$6562.81 < \mu < \$6,637.19$

Note: Graphs are not to scale and are intended to convey a general idea.
Answers may vary due to rounding, TI-83's, or computer programs.

EXERCISE SET 9-2

1.
Testing a single mean involves comparing a sample mean to a specific value such as $\mu = 100$; whereas testing the difference between means means comparing the means of two samples such as $\mu_1 = \mu_2$.

2.
When both samples are greater than or equal to 30 the distribution will be approximately normal. The mean of the differences will be equal to zero. The standard deviation of the differences will be $\sqrt{\frac{\sigma_1^2}{n_1} + \frac{\sigma_2^2}{n_2}}$.

3.
The populations must be independent of each other and they must be normally distributed. s_1 and s_2 can be used in place of σ_1 and σ_2 when σ_1 and σ_2 are unknown and both samples are each greater than or equal to 30.

4.
H_0: $\mu_1 = \mu_2$ or H_0: $\mu_1 - \mu_2 = 0$

5.
H_0: $\mu_1 = \mu_2$ (claim)
H_1: $\mu_1 \neq \mu_2$

C. V. = ± 2.58

$\overline{X}_1 = 662.6111$ $\overline{X}_2 = 758.875$
$s_1 = 449.8703$ $s_2 = 474.1258$

$$z = \frac{(\overline{X}_1 - \overline{X}_2) - (\mu_1 - \mu_2)}{\sqrt{\frac{\sigma_1^2}{n_1} + \frac{\sigma_2^2}{n_2}}} = \frac{(662.6111 - 758.875) - 0}{\sqrt{\frac{449.8703^2}{36} + \frac{474.1258^2}{36}}} =$$

$z = -0.88$

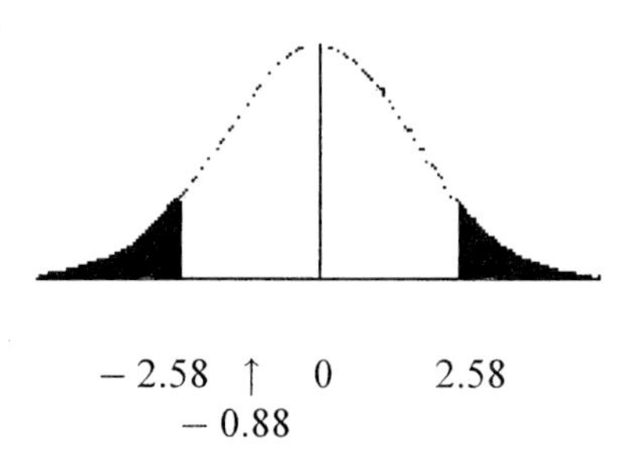

-2.58 ↑ 0 2.58
-0.88

5. continued
Do not reject the null hypothesis. There is not enough evidence to reject the claim that the average lengths of the rivers is the same.

6.
H_0: $\mu_1 = \mu_2$
H_1: $\mu_1 \neq \mu_2$ (claim)

C. V. = ± 1.65

$$z = \frac{(2 - 1.7) - 0}{\sqrt{\frac{0.6^2}{120} + \frac{0.7^2}{34}}} = 2.274$$

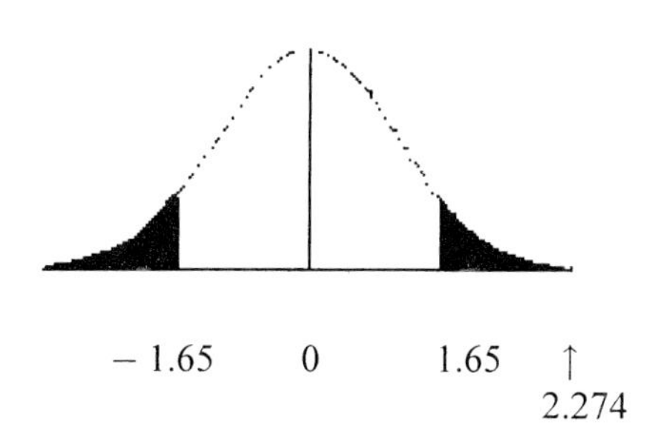

-1.65 0 1.65 ↑
2.274

Reject the null hypothesis. There is enough evidence to support the claim that there is a difference in coping skills.

7.
H_0: $\mu_1 \leq \mu_2$
H_1: $\mu_1 > \mu_2$ (claim)

C. V. = 1.65

$$z = \frac{(\overline{X}_1 - \overline{X}_2) - (\mu_1 - \mu_2)}{\sqrt{\frac{s_1^2}{n_1} + \frac{s_2^2}{n_2}}} = \frac{(90 - 88) - 0}{\sqrt{\frac{5^2}{100} + \frac{6^2}{100}}} = 2.56$$

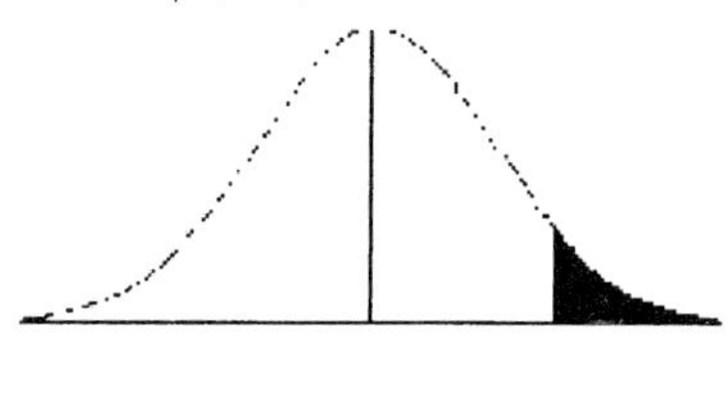

0 1.65 ↑
2.56

Reject the null hypothesis. There is enough evidence to support the claim that pulse rates of smokers are higher than the pulse rates of non-smokers.

8.
$\overline{D} = 126.2 - 123.5 = 2.7$

8. continued

$$(\overline{X}_1 - \overline{X}_2) - z_{\frac{\alpha}{2}}\sqrt{\frac{\sigma_1^2}{n_1} + \frac{\sigma_2^2}{n_2}} < \mu_1 - \mu_2 < (\overline{X}_1 - \overline{X}_2) + z_{\frac{\alpha}{2}}\sqrt{\frac{\sigma_1^2}{n_1} + \frac{\sigma_2^2}{n_2}}$$

$$2.7 - 1.96\sqrt{\frac{98}{60} + \frac{120}{50}} < \mu_1 - \mu_2 < 2.7 + 1.96\sqrt{\frac{98}{60} + \frac{120}{50}}$$

$$2.7 - 3.9363 < \mu_1 - \mu_2 < 2.7 + 3.9363$$
$$-1.2363 < \mu_1 - \mu_2 < 6.6363$$

Yes, the interval supports the claim since 0 is contained in the interval.

9.
H_0: $\mu_1 \leq \mu_2$
H_1: $\mu_1 > \mu_2$ (claim)

C. V. = 2.05

$$z = \frac{(\overline{X}_1 - \overline{X}_2) - (\mu_1 - \mu_2)}{\sqrt{\frac{s_1^2}{n_1} + \frac{s_2^2}{n_2}}} = \frac{(61.2 - 59.4) - 0}{\sqrt{\frac{7.9^2}{84} + \frac{7.9^2}{34}}} = 1.12$$

0 ↑ 2.05
1.12

Do not reject the null hypothesis. There is not enough evidence to support the claim that noise levels in the corridors is higher than in the clinics.

10.
H_0: $\mu_1 = \mu_2$ (claim)
H_1: $\mu_1 \neq \mu_2$

C. V. = ± 2.58

$$z = \frac{(63{,}255 - 59{,}102) - 0}{\sqrt{\frac{5602^2}{35} + \frac{4731^2}{40}}} = 3.44$$

−2.58 0 2.58 ↑
3.44

10. continued
Reject the null hypothesis. There is enough evidence to reject the claim that the average costs are the same.

11.
H_0: $\mu_1 \geq \mu_2$
H_1: $\mu_1 < \mu_2$ (claim)

C. V. = −1.65

$$z = \frac{(\overline{X}_1 - \overline{X}_2) - (\mu_1 - \mu_2)}{\sqrt{\frac{s_1^2}{n_1} + \frac{s_2^2}{n_2}}} = \frac{(3.16 - 3.28) - 0}{\sqrt{\frac{0.52^2}{103} + \frac{0.46^2}{225}}} = -2.01$$

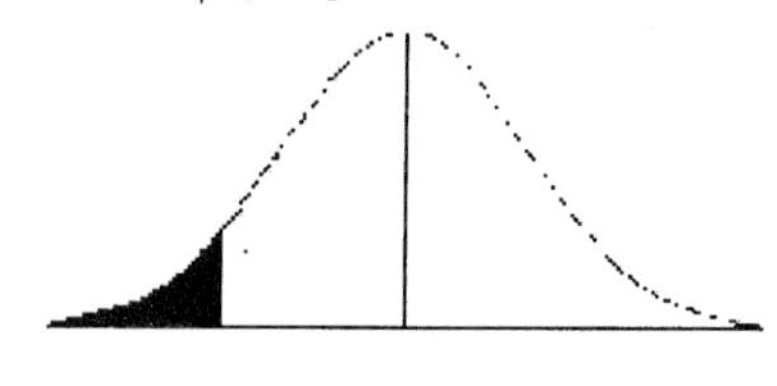

↑ −1.65 0
−2.01

Reject the null hypothesis. There is enough evidence to support the claim that leavers have a lower GPA than stayers.

12.
H_0: $\mu_1 \leq \mu_2$
H_1: $\mu_1 > \mu_2$ (claim)

C. V. = 1.65
$\overline{X}_1 = 21.4$ $\overline{X}_2 = 20.8$
$s_1 = 3$ $s_2 = 3$

$$z = \frac{(21.4 - 20.8) - 0}{\sqrt{\frac{3^2}{1000} + \frac{3^2}{500}}} = 3.65$$

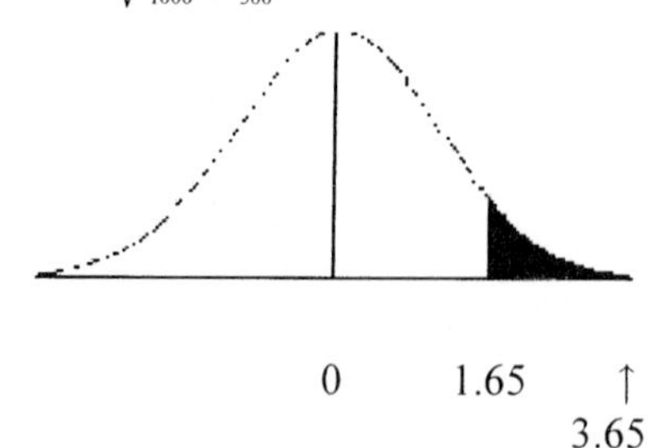

0 1.65 ↑
3.65

Reject the null hypothesis. There is enough evidence to support the claim that ACT scores for Ohio students are below the national average.

13.
H_0: $\mu_1 \leq \mu_2$
H_1: $\mu_1 > \mu_2$ (claim)

13. continued
C. V. = 2.33
$\overline{X}_1 = \$9224$ $\overline{X}_2 = \$8497.5$
$s_1 = 3829.826$ $s_2 = 2745.293$

$$z = \frac{(\overline{X}_1 - \overline{X}_2) - (\mu_1 - \mu_2)}{\sqrt{\frac{s_1^2}{n_1} + \frac{s_2^2}{n_2}}}$$

$$z = \frac{(9224 - 8497.5) - 0}{\sqrt{\frac{3829.826^2}{50} + \frac{2745.293^2}{50}}} = 1.09$$

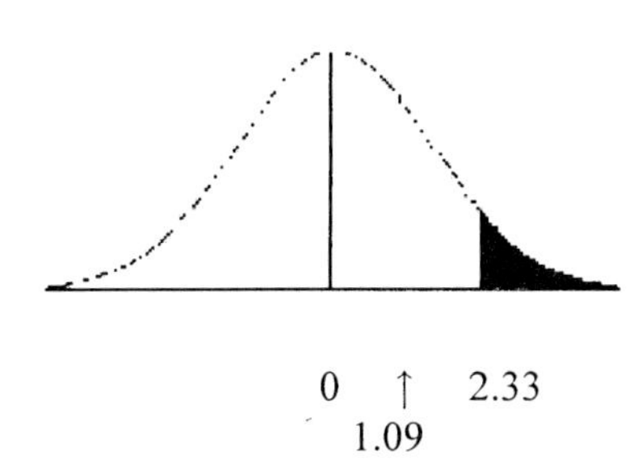

Do not reject the null hypothesis. There is not enough evidence to support the claim that colleges spent more money on men's sports than women's.

14.
H_0: $\mu_1 = \mu_2$
H_1: $\mu_1 \neq \mu_2$ (claim)

$$z = \frac{(837 - 753) - 0}{\sqrt{\frac{30^2}{35} + \frac{40^2}{40}}} = 10.36$$

Area = 0.4999
P-value = 2(0.5 – 0.4999) = 0.0002
Since P-value < 0.05, reject the null hypothesis. There is enough evidence to support the claim that there is a difference in the average miles traveled for each of the two taxi companies.

15.
H_0: $\mu_1 = \mu_2$
H_1: $\mu_1 \neq \mu_2$ (claim)

$$z = \frac{(\overline{X}_1 - \overline{X}_2) - (\mu_1 - \mu_2)}{\sqrt{\frac{s_1^2}{n_1} + \frac{s_2^2}{n_2}}} = \frac{(3.05 - 2.96) - 0}{\sqrt{\frac{0.75^2}{103} + \frac{0.75^2}{225}}}$$

z = 1.01
Area = 0.3438
P-value = 2(0.5 – 0.3438) = 0.3124
Since P-value > 0.05, do not reject the null hypothesis. There is not enough evidence to support the claim that there is a difference in scores.

16.
Residents: $\overline{X}_1 = 22.12$, $s_1 = 3.68$, $n_1 = 50$
Commuters: $\overline{X}_2 = 22.76$, $s_2 = 4.70$, $n_2 = 50$

H_0: $\mu_1 = \mu_2$ (claim)
H_1: $\mu_1 \neq \mu_2$

$$z = \frac{(22.12 - 22.76) - 0}{\sqrt{\frac{3.68^2}{50} + \frac{4.70^2}{50}}} = -0.76$$

Area = 0.2764
P-value = 2(0.5 – 0.2764) = 0.4472
Since P-value > 0.05, do not reject the null hypothesis. There is not enough evidence to reject the claim that there is no difference in the ages.

17.
$\overline{D} = 83.6 - 79.2 = 4.4$

$$(\overline{X}_1 - \overline{X}_2) - z_{\frac{\alpha}{2}}\sqrt{\frac{\sigma_1^2}{n_1} + \frac{\sigma_2^2}{n_2}} < \mu_1 - \mu_2 < (\overline{X}_1 - \overline{X}_2) + z_{\frac{\alpha}{2}}\sqrt{\frac{\sigma_1^2}{n_1} + \frac{\sigma_2^2}{n_2}}$$

$$4.4 - (1.65)\sqrt{\frac{4.3^2}{36} + \frac{3.8^2}{36}} < \mu_1 - \mu_2 < 4.4 + (1.65)\sqrt{\frac{4.3^2}{36} + \frac{3.8^2}{36}}$$

$2.8 < \mu_1 - \mu_2 < 6.0$

18.
H_0: $\mu_1 \leq \mu_2$
H_1: $\mu_1 > \mu_2$ (claim)

C. V. = 1.65
$\overline{X}_1 = 67$ $\overline{X}_2 = 64$
$s_1 = 8$ $s_2 = 7$
$n_1 = 50$ $n_2 = 50$

$$z = \frac{(67 - 64) - 0}{\sqrt{\frac{8^2}{50} + \frac{7^2}{50}}} = 2.00$$

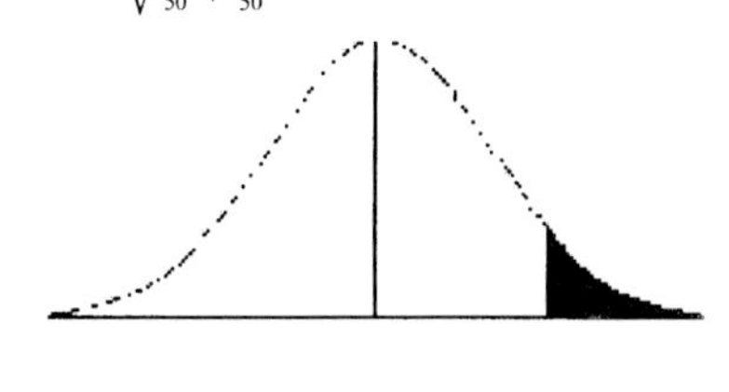

Reject the null hypothesis. There is enough evidence to support the claim that Michigan drivers drive faster than Ohio drivers.

19.
$\overline{D} = 28.6 - 32.9 = -4.3$

$$(\overline{X}_1 - \overline{X}_2) - z_{\frac{\alpha}{2}}\sqrt{\frac{\sigma_1^2}{n_1} + \frac{\sigma_2^2}{n_2}} < \mu_1 - \mu_2 < (\overline{X}_1 - \overline{X}_2) + z_{\frac{\alpha}{2}}\sqrt{\frac{\sigma_1^2}{n_1} + \frac{\sigma_2^2}{n_2}}$$

$$-4.3 - (2.58)\sqrt{\frac{5.1^2}{30} + \frac{4.4^2}{40}} < \mu_1 - \mu_2 < -4.3 + (2.58)\sqrt{\frac{5.2^2}{30} + \frac{4.4^2}{40}}$$

$-7.3 < \mu_1 - \mu_2 < -1.3$

20.
$\overline{D} = 9.2 - 8.8 = 0.4$

$$0.4 - (1.96)\sqrt{\frac{0.3^2}{27} + \frac{0.1^2}{30}} < \mu_1 - \mu_2 < 0.4 + (1.96)\sqrt{\frac{0.3^2}{27} + \frac{0.1^2}{30}}$$

$0.281 < \mu_1 - \mu_2 < 0.519$
or $0.3 < \mu_1 - \mu_2 < 0.5$

21.
H_0: $\mu_1 - \mu_2 \leq 8$ (claim)
H_1: $\mu_1 - \mu_2 > 8$

C. V. = 1.65

$$z = \frac{(\overline{X}_1 - \overline{X}_2) - K}{\sqrt{\frac{s_1^2}{n_1} + \frac{s_2^2}{n_2}}} = \frac{(110 - 104) - 8}{\sqrt{\frac{15^2}{60} + \frac{15^2}{60}}} = -0.73$$

↑ 0 1.65
−0.73

Do not reject the null hypothesis. There is not enough evidence to reject the claim that private school students have exam scores that are at most 8 points higher than public school students.

EXERCISE SET 9-3

1.
It should be the larger of the two variances.

2.
The critical region is on the right side because the F test value is always greater than or equal to 1, since the larger variance is always placed in the numerator.

3.
One d.f. is used for the variance associated with the numerator and one is used for the variance associated with the denominator.

4.
The characteristics of the F-distribution are:
1. The values of the F cannot be negative.
2. The distribution is positively skewed.
3. The mean value of the F is approximately equal to one.
d. The F is a family of curves based upon the degrees of freedom.

5.
a. d. f. N = 15, d. f. D = 22; C. V. = 3.36
b. d. f. N = 24, d. f. D = 13; C. V. = 3.59
c. d. f. N = 45, d. f. D = 29; C. V. = 2.03
d. d. f. N = 20, d. f. D = 16; C. V. = 2.28
e. d. f. N = 10, d. f. D = 10; C. V. = 2.98

6.
Note: Specific P-values are in parentheses.
a. 0.025 < P-value < 0.05 (0.033)
b. 0.05 < P-value < 0.10 (0.072)
c. P-value = 0.05
d. 0.005 < P-value < 0.01 (0.006)
e. P-value = 0.05
f. P > 0.10 (0.112)
g. 0.05 < P-value < 0.10 (0.068)
h. 0.01 < P-value < 0.02 (0.015)

7.
H_0: $\sigma_1^2 = \sigma_2^2$
H_1: $\sigma_1^2 \neq \sigma_2^2$ (claim)

C. V. = 2.53 $\alpha = \frac{0.10}{2}$
d. f. N = 14 d. f. D = 14
$F = \frac{s_1^2}{s_2^2} = \frac{13.12^2}{6.17^2} = 4.52$

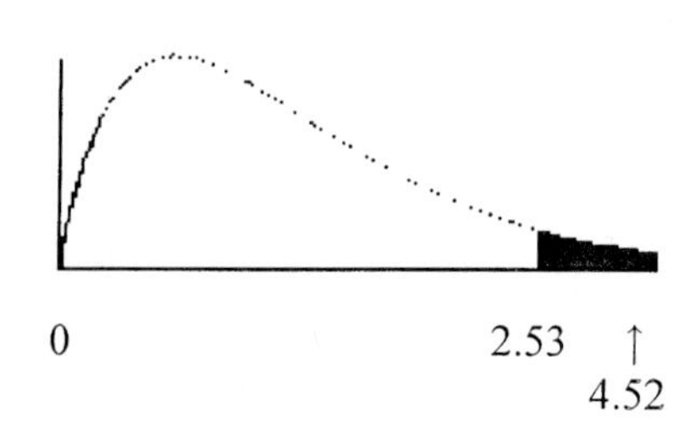

0 2.53 ↑
4.52

7. continued
Reject the null hypothesis. There is enough evidence to support the claim that there is a difference in the variances of the best seller lists for fiction and non-fiction.

8.
H_0: $\sigma_1^2 \geq \sigma_2^2$

H_1: $\sigma_1^2 < \sigma_2^2$ (claim)

C. V. = 1.74 $\alpha = 0.05$
d. f. N = 39 d. f. D = 49

$F = \frac{37}{32} = 1.16$

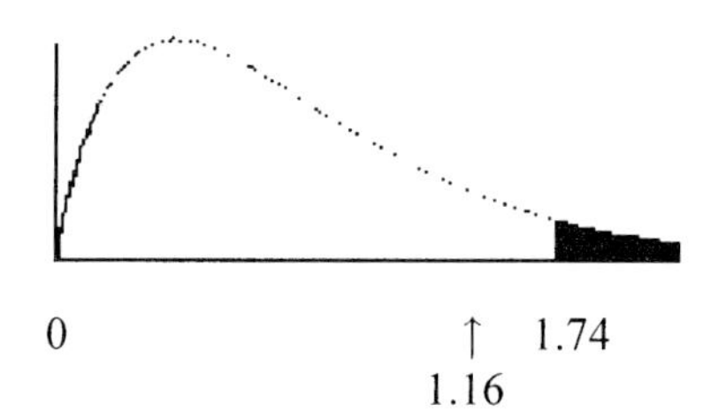

0 ↑ 1.74
1.16

Do not reject the null hypothesis. There is not enough evidence to support the claim that the variance of the scores for students using graphing calculators is smaller than the variance of the scores for students using scientific calculators.

9.
H_0: $\sigma_1^2 = \sigma_2^2$

H_1: $\sigma_1^2 \neq \sigma_2^2$ (claim)

$s_1 = 25.97$ $s_2 = 72.74$
C. V. = 2.86 $\alpha = \frac{0.05}{2}$
d. f. N = 15 d. f. D = 15

$F = \frac{s_1^2}{s_2^2} = \frac{72.74^2}{25.97^2} = 7.85$

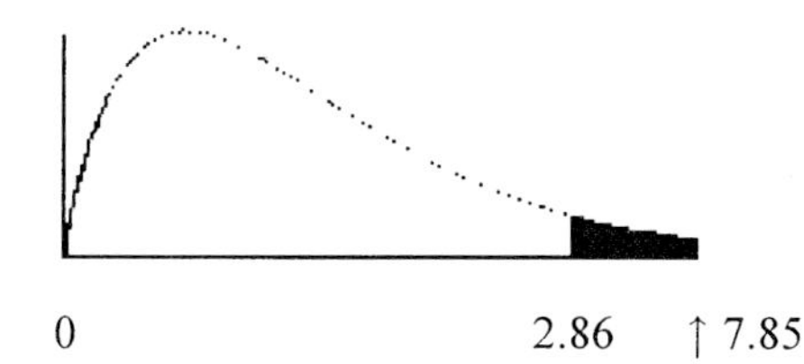

0 2.86 ↑ 7.85

Reject the null hypothesis. There is enough evidence to support the claim that the

9. continued
variances of the values of tax exempt properties are different.

10.
H_0: $\sigma_1^2 = \sigma_2^2$ (claim)

H_1: $\sigma_1^2 \neq \sigma_2$

C. V. = 2.51 $\alpha = \frac{0.05}{2}$
d. f. N = 23 d. f. D = 19

$F = \frac{(7.5)^2}{(4.1)^2} = 3.35$

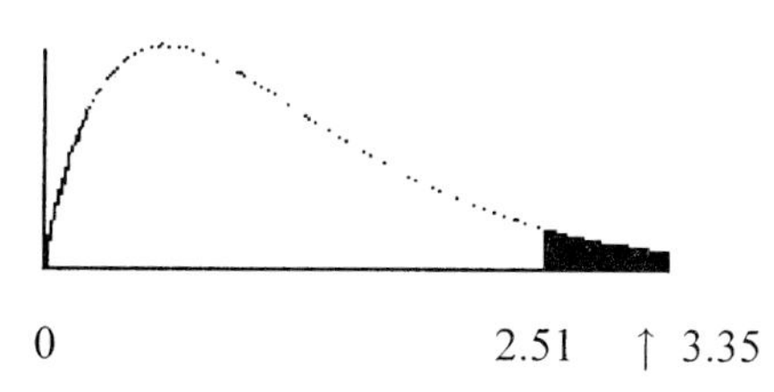

0 2.51 ↑ 3.35

Reject the null hypothesis. There is enough evidence to reject the claim that there is no difference in the standard deviations.

11.
H_0: $\sigma_1^2 = \sigma_2^2$
H_1: $\sigma_1^2 \neq \sigma_2^2$ (claim)

$s_1 = 33.99$ $s_2 = 33.99$
C. V. = 4.99 $\alpha = \frac{0.05}{2}$
d. f. N = 7 d. f. D = 7
$F = \frac{s_1^2}{s_2^2} = \frac{(33.99)^2}{(33.99)^2} = 1$

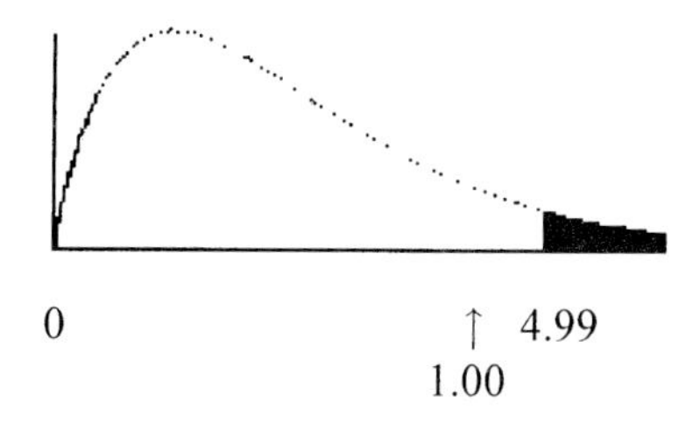

0 ↑ 4.99
1.00

Do not reject the null hypothesis. There is not enough evidence to support the claim that the variance in the number of calories differs between the two brands.

12.
H_0: $\sigma_1^2 = \sigma_2^2$
H_1: $\sigma_1^2 \neq \sigma_2$ (claim)

12. continued

$s_1 = 1.79$ $s_2 = 1.305$
C. V. = 3.53 $\alpha = \frac{0.05}{2}$
d. f. N = 11 d. f. D = 11

$F = \frac{(1.79)^2}{(1.305)^2} = 1.88$

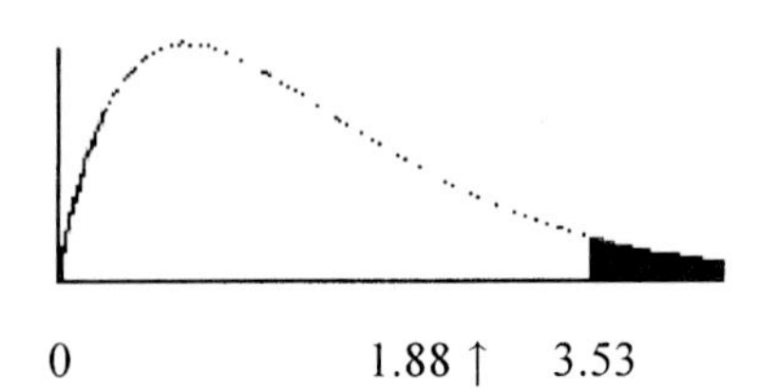

Do not reject the null hypothesis. There is not enough evidence to support the claim that the variances in the number of vehicles are different.

13.

H_0: $\sigma_1^2 \leq \sigma_2^2$

H_1: $\sigma_1^2 > \sigma_2^2$ (claim)

$s_1 = 32$ $s_2 = 28$
$n_1 = 100$ $n_2 = 100$
C. V. = 1.53 $\alpha = 0.05$
d. f. N = 99 d. f. D = 99

$F = \frac{s_1^2}{s_2^2} = \frac{(32)^2}{(28)^2} = 1.306$

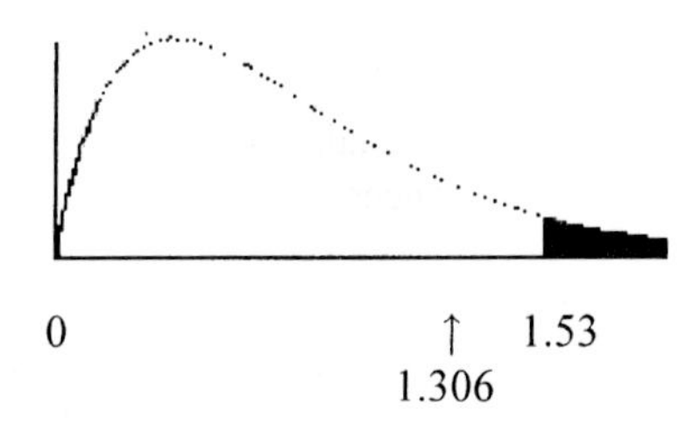

Reject the null hypothesis. There is enough evidence to support the claim that the variation of blood pressure of overweight individuals is greater than the variation of blood pressure of normal weight individuals.

14.

H_0: $\sigma_1^2 = \sigma_2^2$

H_1: $\sigma_1^2 \neq \sigma_2^2$ (claim)

Chocolate: s = 6.4985
Non-chocolate: s = 11.2006

14. continued

C. V. = 2.75 $\alpha = \frac{0.10}{2}$
d. f. N = 10 d. f. D = 12

$F = \frac{11.2006^2}{6.4985^2} = 2.97$

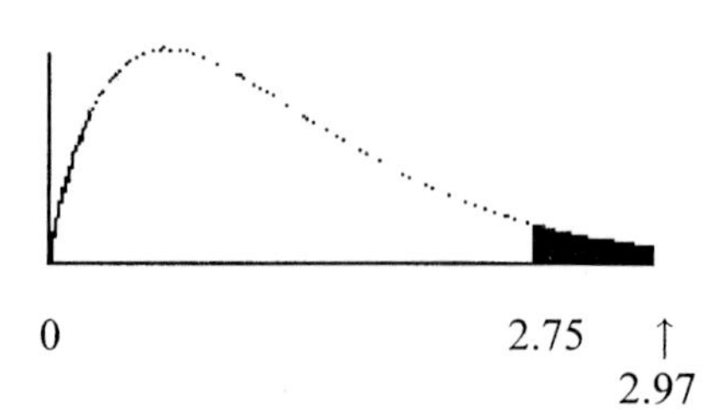

Reject the null hypothesis. There is enough evidence to support the claim that the variances in carbohydrate grams of chocolate candy and non-chocolate candy are not the same.

15.

H_0: $\sigma_1^2 = \sigma_2^2$

H_1: $\sigma_1^2 \neq \sigma_2^2$ (claim)

Research: $s_1 = 5501.118$
Primary Care: $s_2 = 5238.809$

C. V. = 4.03 $\alpha = \frac{0.05}{2}$
d. f. N = 9 d. f. D = 9

$F = \frac{s_1^2}{s_2^2} = \frac{(5501.118)^2}{(5238.809)^2} = 1.10$

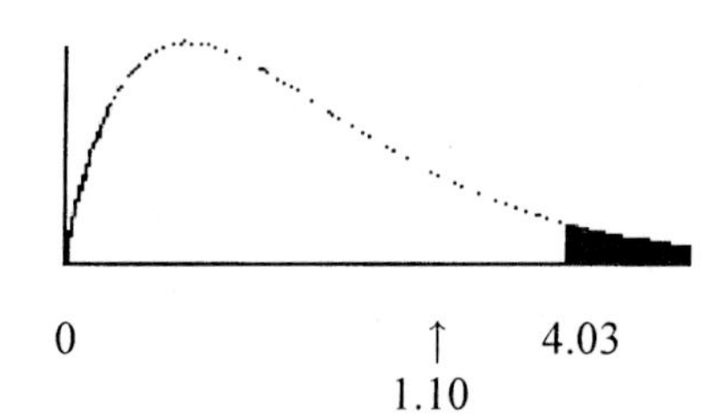

Do not reject the null hypothesis. There is not enough evidence to support the claim that there is a difference between the variances in tuition costs.

16.

H_0: $\sigma_1^2 \geq \sigma_2^2$
H_1: $\sigma_1^2 < \sigma_2^2$ (claim)

$s_1 = 98.2$ $s_2 = 118.4$
C. V. = 3.15 $\alpha = 0.01$
d. f. N = 19 d. f. D = 19

$F = \frac{(118.4)^2}{(98.2)^2} = 1.45$

16. continued

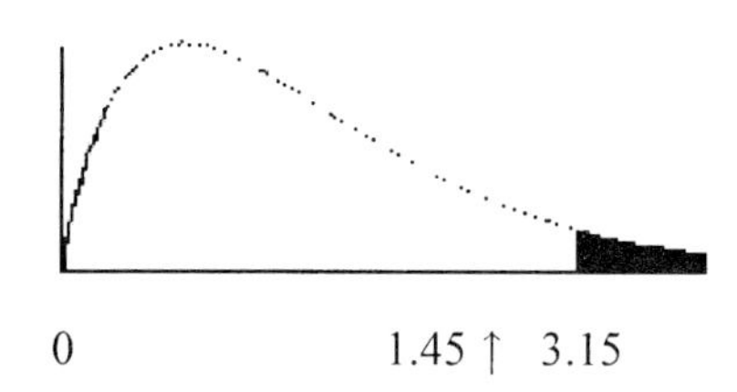

Do not reject the null hypothesis. There is not enough evidence to support the claim that the variance of the areas in Indiana is less than the variance of the areas in Iowa.

17.
H_0: $\sigma_1^2 = \sigma_2^2$ (claim)
H_1: $\sigma_1^2 \neq \sigma_2^2$

$s_1 = 130.496$ $s_2 = 73.215$
C. V. = 3.87 $\alpha = \frac{0.10}{2}$
d. f. N = 6 d. f. D = 7
$F = \frac{s_1^2}{s_2^2} = \frac{(130.496)^2}{(73.215)^2} = 3.18$

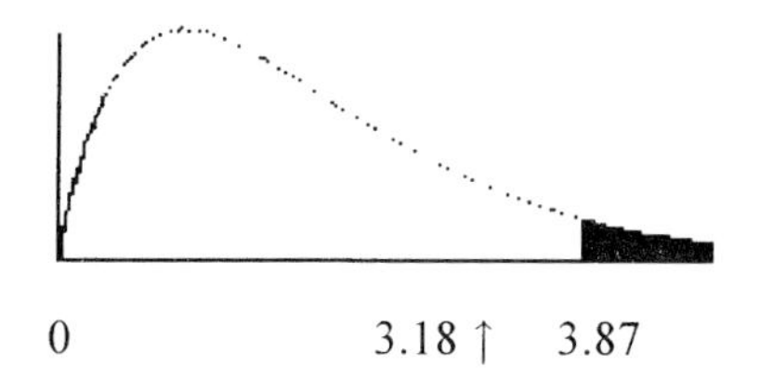

Do not reject the null hypothesis. There is not enough evidence to reject the claim that the variances of the heights are equal.

18.
H_0: $\sigma_1^2 \leq \sigma_2^2$
H_1: $\sigma_1^2 > \sigma_2^2$ (claim)

$\alpha = 0.05$ P-value = 0.003
d. f. N = 29 d. f. D = 29
$F = \frac{8324}{2862} = 2.91$
Since P-value < 0.05, reject the null hypothesis. There is enough evidence to support the claim that the variation in the salaries of the elementary school teachers is greater than the variation in salaries of the secondary teachers.

19.

Men	Women
$s_1^2 = 2.363$	$s_2^2 = 0.444$
$n_1 = 15$	$n_2 = 15$

19. continued
H_0: $\sigma_1^2 = \sigma_2^2$ (claim)
H_1: $\sigma_1^2 \neq \sigma_2$

$\alpha = 0.05$ P-value = 0.004
d. f. N = 14 d. f. D = 14
$F = \frac{s_1^2}{s_2^2} = \frac{2.363}{0.444} = 5.32$

Since P-value < 0.05, reject the null hypothesis. There is enough evidence to reject the claim that the variances in weights are equal.

20.

Hard Body	Soft Body
$s_1^2 = 6.007$	$s_2^2 = 22.667$
$n_1 = 17$	$n_2 = 6$

H_0: $\sigma_1^2 = \sigma_2^2$
H_1: $\sigma_1^2 \neq \sigma_2$ (claim)

C. V. = 3.50 $\alpha = \frac{0.05}{2}$
d. f. N = 5 d. f. D = 16
$F = \frac{22.667}{6.007} = 3.77$

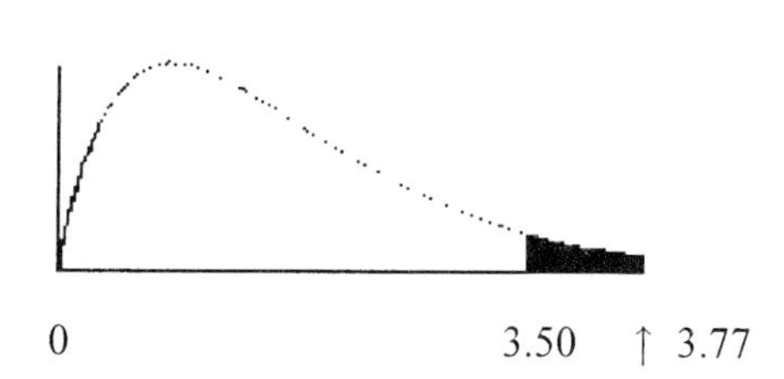

Reject the null hypothesis. There is enough evidence to support the claim that there is a difference in the variances.

EXERCISE SET 9-4

1.
H_0: $\sigma_1^2 = \sigma_2^2$
H_1: $\sigma_1^2 \neq \sigma_2^2$
d. f. N = 9 d. f. D = 9 $\alpha = \frac{0.05}{2}$
$F = \frac{3256^2}{2341^2} = 1.93$ C. V. = 4.03
Do not reject. The variances are equal.

H_0: $\mu_1 = \mu_2$
H_1: $\mu_1 \neq \mu_2$ (claim)
C. V. = ± 2.101 d. f. = 18

$$t = \frac{(\bar{X}_1 - \bar{X}_2) - (\mu_1 - \mu_2)}{\sqrt{\frac{(n_1 - 1)s_1^2 + (n_2 - 1)s_2^2}{n_1 + n_2 - 2}}\sqrt{\frac{1}{n_1} + \frac{1}{n_2}}}$$

1. continued

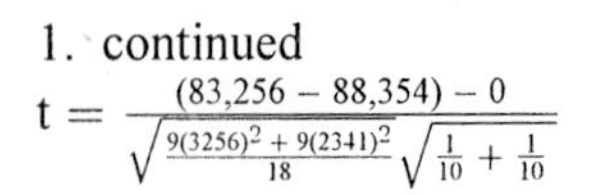

$$t = \frac{(83{,}256 - 88{,}354) - 0}{\sqrt{\frac{9(3256)^2 + 9(2341)^2}{18}}\sqrt{\frac{1}{10} + \frac{1}{10}}}$$

$t = -4.02$

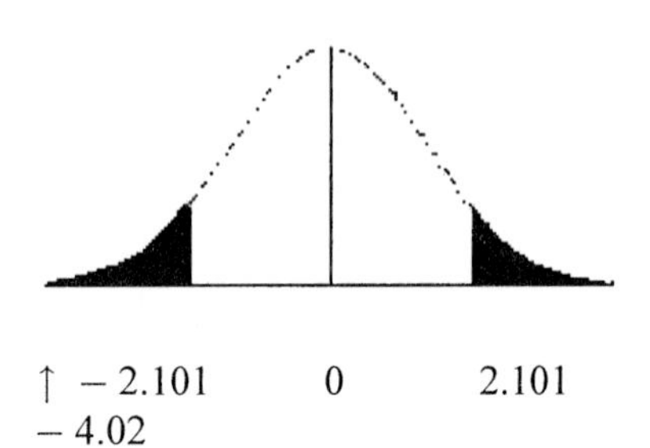

$\uparrow$ -2.101 0   2.101
-4.02

Reject the null hypothesis. There is enough evidence to support the claim that there is a significant difference in the values of the homes based upon the appraisers' values.

Confidence Interval:

$$-5098 - 2.101\left(\sqrt{\frac{9(3256)^2 + 9(2341)^2}{18}}\sqrt{\frac{1}{10} + \frac{1}{10}}\right) < \mu_1 - \mu_2 <$$

$$-5098 + 2.101\left(\sqrt{\frac{9(3256)^2 + 9(2341)^2}{18}}\sqrt{\frac{1}{10} + \frac{1}{10}}\right) =$$

$$-5098 - 2.101(1268.14) < \mu_1 - \mu_2 < -5098 + 2.101(1268.14)$$

$$-\$7762 < \mu_1 - \mu_2 < -\$2434$$

2.
H_0: $\sigma_1^2 = \sigma_2^2$
H_1: $\sigma_1^2 \neq \sigma_2^2$
d. f. N = 15   d. f. D = 19 $\alpha = \frac{0.05}{2}$
$F = \frac{300^2}{250^2} = 1.44$ C. V. = 2.62
Do not reject. The variances are equal.

H_0: $\mu_1 \leq \mu_2$
H_1: $\mu_1 > \mu_2$ (claim)

C. V. = 1.65   d. f. = 34

$$t = \frac{(23{,}800 - 23{,}750) - 0}{\sqrt{\frac{(16-1)300^2 + (20-1)250^2}{16 + 20 - 2}}\sqrt{\frac{1}{16} + \frac{1}{20}}}$$

$$t = \frac{50}{\sqrt{74632.35}\sqrt{0.1125}} = 0.55$$

2. continued

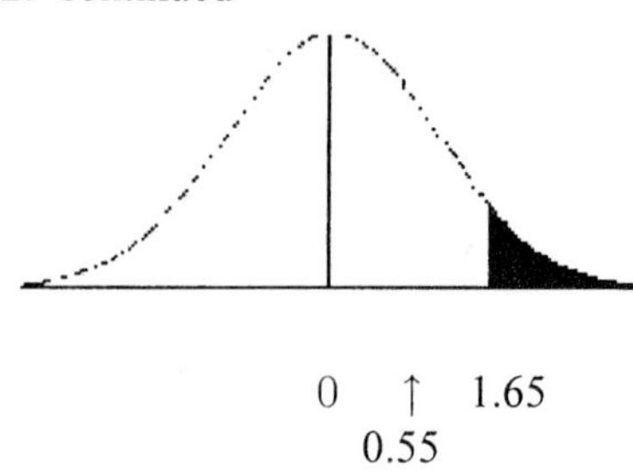

0   $\uparrow$ 1.65
0.55

Do not reject the null hypothesis. There is not enough evidence to support the claim that male nurses earn more than female nurses.

3.
H_0: $\sigma_1^2 = \sigma_2^2$
H_1: $\sigma_1^2 \neq \sigma_2^2$

d. f. N = 14   d. f. D = 14 $\alpha = \frac{0.05}{2}$
$F = \frac{20{,}000^2}{20{,}000^2} = 1$ C. V. = 3.05
Do not reject. The variances are equal.

H_0: $\mu_1 = \mu_2$
H_1: $\mu_1 \neq \mu_2$ (claim)

C. V. $= \pm 2.048$
d. f. $= 14 + 14 - 2 = 28$

$$t = \frac{(\overline{X}_1 - \overline{X}_2) - (\mu_1 - \mu_2)}{\sqrt{\frac{(n_1 - 1)s_1^2 + (n_2 - 1)s_2^2}{n_1 + n_2 - 2}}\sqrt{\frac{1}{n_1} + \frac{1}{n_2}}}$$

$$t = \frac{(501{,}580 - 513{,}360) - 0}{\sqrt{\frac{14(20{,}000^2) + 14(20{,}000^2)}{15 + 15 - 2}}\sqrt{\frac{1}{15} + \frac{1}{15}}}$$

$t = -1.61$

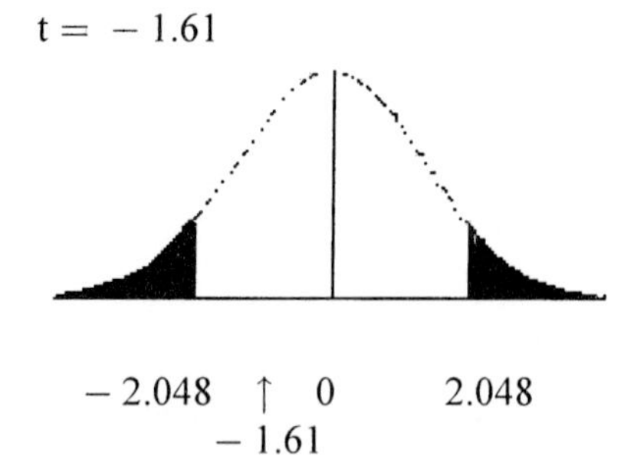

-2.048 $\uparrow$ 0   2.048
-1.61

Do not reject the null hypothesis. There is enough evidence to support the claim that there is no difference between the salaries.

4.
H_0: $\sigma_1^2 = \sigma_2^2$
H_1: $\sigma_1^2 \neq \sigma_2^2$

4. continued
d. f. N = 21 d. f. D = 15 $\alpha = 0.01$
$F = \frac{0.8^2}{0.6^2} = 1.78$ C. V. = 3.88

Do not reject. The variances are equal.

$H_0: \mu_1 \le \mu_2$
$H_1: \mu_1 > \mu_2$ (claim)

C. V. = 2.33 d. f. = 36

$$t = \frac{(3.9 - 3.6) - 0}{\sqrt{\frac{(16-1)(0.6)^2 + (22-1)(0.8)^2}{16+22-2}}\sqrt{\frac{1}{16} + \frac{1}{22}}}$$

$t = 1.26$

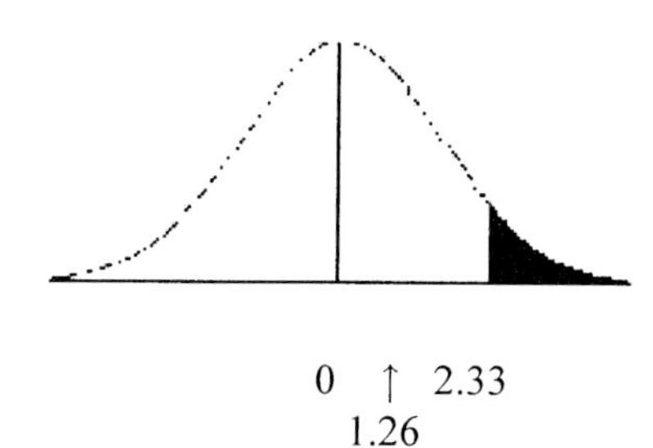

0 ↑ 2.33
1.26

Do not reject the null hypothesis. There is not enough evidence to support the claim that high school girls miss more days of school than high school boys.

5.
$H_0: \sigma_1^2 = \sigma_2^2$
$H_1: \sigma_1^2 \ne \sigma_2^2$
$\overline{X}_1 = 37.167$ $\overline{X}_2 = 25$
$s_1 = 13.2878$ $s_2 = 15.7734$
d. f. N = 5 d. f. D = 5 $\alpha = 0.01$
$F = \frac{15.7734^2}{13.2878^2} = 1.41$ C. V. = 14.94
Do not reject. The variances are equal.

$H_0: \mu_1 \le \mu_2$
$H_1: \mu_1 > \mu_2$ (claim)

C. V. = 2.764 d. f. = 10

$$t = \frac{(\overline{X}_1 - \overline{X}_2) - (\mu_1 - \mu_2)}{\sqrt{\frac{(n_1-1)s_1^2 + (n_2-1)s_2^2}{n_1+n_2-2}}\sqrt{\frac{1}{n_1} + \frac{1}{n_2}}}$$

$$t = \frac{(37.167 - 25) - 0}{\sqrt{\frac{5(13.2878)^2 + 5(15.7734)^2}{6+6-2}}\sqrt{\frac{1}{6} + \frac{1}{6}}} = 1.45$$

5. continued

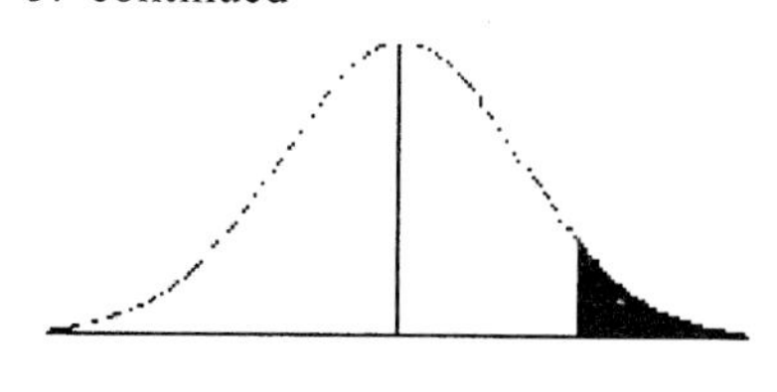

0 1.45 ↑ 2.764

Do not reject the null hypothesis. There is not enough evidence to support the claim that the average number of family day care centers is greater than the average number of day care centers.

6.
$H_0: \sigma_1^2 = \sigma_2^2$
$H_1: \sigma_1^2 \ne \sigma_2^2$
$\overline{X}_1 = 63{,}356.2$ $\overline{X}_2 = 35{,}386.8$
$s_1 = 2808.31385$ $s_2 = 2631.03947$
d. f. N = 4 d. f. D = 4 $\alpha = 0.10$
$F = \frac{2808.31385^2}{2631.03947^2} = 1.14$ C. V. = 6.39
Do not reject. The variances are equal.

$H_0: \mu_1 \le \mu_2$
$H_1: \mu_1 > \mu_2$ (claim)

C. V. = 1.397 d. f. = 8

$$t = \frac{(63{,}356.2 - 35{,}386.8) - 0}{\sqrt{\frac{4(2808.31385)^2 + 4(2631.03947)^2}{5+5-2}}\sqrt{\frac{1}{5} + \frac{1}{5}}}$$

$= 16.252$

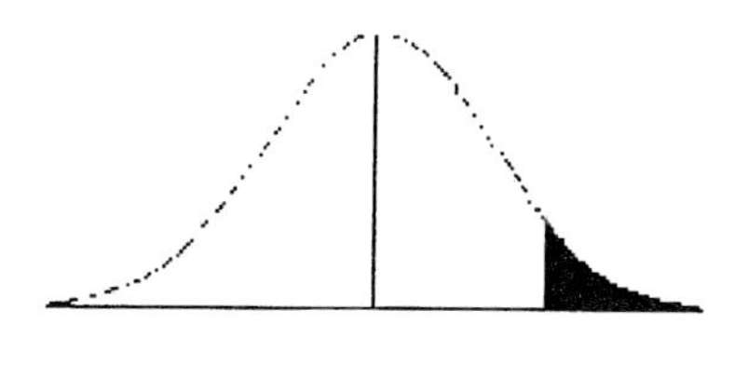

0 1.397 ↑ 16.252

Reject the null hypothesis. There is enough evidence to support the claim that more juveniles are reported missing than adults.

7.
$H_0: \sigma_1^2 = \sigma_2^2$
$H_1: \sigma_1^2 \ne \sigma_2^2$
d. f. N = 9 d. f. D = 13 $\alpha = 0.02$
$F = \frac{5.6^2}{4.3^2} = 1.7$ C. V. = 4.19
Do not reject. The variances are equal.

7. continued
H_0: $\mu_1 = \mu_2$
H_1: $\mu_1 \neq \mu_2$ (claim)

C. V. = ± 2.508 d. f. = 22

$$t = \frac{(\overline{X}_1 - \overline{X}_2) - (\mu_1 - \mu_2)}{\sqrt{\frac{(n_1 - 1)s_1^2 + (n_2 - 1)s_2^2}{n_1 + n_2 - 2}}\sqrt{\frac{1}{n_1} + \frac{1}{n_2}}}$$

$$t = \frac{(21 - 27) - 0}{\sqrt{\frac{9(5.6)^2 + 13(4.3)^2}{10 + 14 - 2}}\sqrt{\frac{1}{10} + \frac{1}{14}}} = -2.97$$

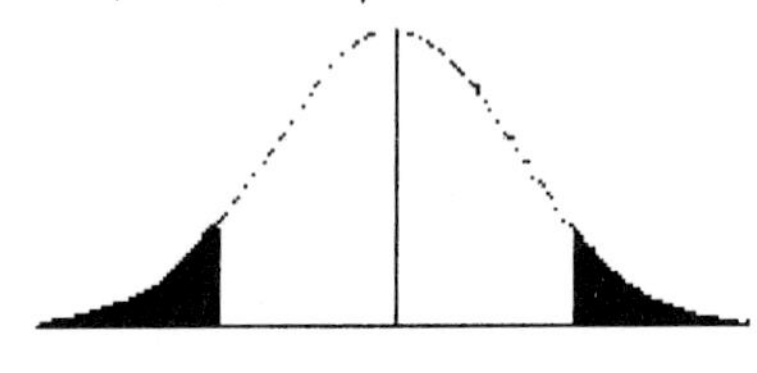

↑ − 2.508 0 2.508
− 2.97

Reject the null hypothesis. There is enough evidence to support the claim that there is a difference in the average times of the two groups.

Confidence Interval:
$-6 - 2.508(2.02) < \mu_1 - \mu_2 < -6 + 2.508(2.02)$
$-11.1 < \mu_1 - \mu_2 < -0.93$

8.
H_0: $\sigma_1^2 = \sigma_2^2$
H_1: $\sigma_1^2 \neq \sigma_2^2$
d. f. N = 19 d. f. D = 17 $\alpha = \frac{0.01}{2}$
$F = \frac{3.5}{2.2} = 1.59$ C. V. = 3.79
Do not reject. The variances are equal.

H_0: $\mu_1 = \mu_2$
H_1: $\mu_1 \neq \mu_2$ (claim)

C. V. = ± 2.58
d. f. = 18 + 20 − 2 = 36

$$t = \frac{(2.5 - 3.8) - 0}{\sqrt{\frac{17(2.2) + 19(3.5)}{18 + 20 - 2}}\sqrt{\frac{1}{18} + \frac{1}{20}}} = -2.36$$

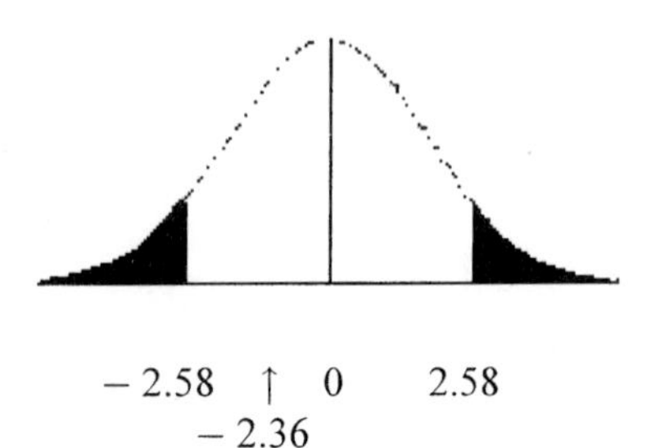

− 2.58 ↑ 0 2.58
− 2.36

8. continued
Do not reject the null hypothesis. There is not enough evidence to support the claim that there is a difference between the number of volunteer hours.

9.
H_0: $\sigma_1^2 = \sigma_2^2$
H_1: $\sigma_1^2 \neq \sigma_2^2$
d. f. N = 26 d. f. D = 11 $\alpha = 0.05$
$F = \frac{5.75^2}{3^2} = 3.67$ P-value = 0.021

0.02 < P-value < 0.05
Reject. The variances are unequal.

H_0: $\mu_1 \geq \mu_2$
H_1: $\mu_1 < \mu_2$ (claim)

P-value < 0.005 d. f. = 11

$$t = \frac{(\overline{X}_1 - \overline{X}_2) - (\mu_1 - \mu_2)}{\sqrt{\frac{s_1^2}{n_1} + \frac{s_2^2}{n_2}}} = \frac{(56 - 63) - 0}{\sqrt{\frac{3^2}{12} + \frac{5.75^2}{27}}}$$

t = − 4.98

Since P-value < 0.05, reject the null hypothesis. There is enough evidence to support the claim that the nurses pay more for insurance than the administrators.

10.
H_0: $\sigma_1^2 = \sigma_2^2$
H_1: $\sigma_1^2 \neq \sigma_2^2$
d. f. N = 15 d. f. D = 15 $\alpha = 0.01$
$F = \frac{0.6^2}{0.6^2} = 1$ P-value > 0.20
Do not reject since P-value > 0.01. The variances are equal.

H_0: $\mu_1 = \mu_2$ (claim)
H_1: $\mu_1 \neq \mu_2$

P-value = 0.069 d. f. = 30

$$t = \frac{(2.3 - 1.9) - 0}{\sqrt{\frac{15(0.6)^2 + 15(0.6)^2}{30}}\sqrt{\frac{1}{16} + \frac{1}{16}}} = 1.89$$

0.05 < P-value < 0.10
Since P-value > 0.01, do not reject the null hypothesis. There is not enough evidence to reject the claim that the mean hospital stay is the same.

10. continued
Confidence Interval:
$0.4 - 2.58(0.2121) < \mu_1 - \mu_2 < 0.4 + 2.58(0.2121)$
$-0.15 < \mu_1 - \mu_2 < 0.95$

11.

White Mice	Brown Mice
$\overline{X}_1 = 17$	$\overline{X}_2 = 16.67$
$s_1 = 4.56$	$s_2 = 5.05$
$n_1 = 6$	$n_2 = 6$

H_0: $\sigma_1^2 = \sigma_2^2$
H_1: $\sigma_1^2 \neq \sigma_2^2$
d. f. N = 5 d. f. D = 5 $\alpha = \frac{0.05}{2}$
$F = \frac{5.05^2}{4.56^2} = 1.23$ C. V. = 7.15
Do not reject. The variances are equal.

H_0: $\mu_1 = \mu_2$
H_1: $\mu_1 \neq \mu_2$ (claim)

C. V. = ± 2.228 d. f. = 10

$$t = \frac{(17 - 16.67) - 0}{\sqrt{\frac{5(4.56)^2 + 5(5.05)^2}{6+6-2}}\sqrt{\frac{1}{6} + \frac{1}{6}}} = 0.119$$

−2.228 0 ↑ 2.228
0.119

Do not reject the null hypothesis. There is not enough evidence to support the claim that the color of the mice made a difference.

Confidence Interval:
$0.33 - 2.228(2.78) < \mu_1 - \mu_2 < 0.33 + 2.228(2.78)$
$-5.9 < \mu_1 - \mu_2 < 6.5$

12.
Research: $\overline{X}_1 = 596.2353$ $s_1 = 163.2362$
Primary Care: $\overline{X}_2 = 481.5$ $s_2 = 179.3957$

F test:
d. f. N = 16 − 1 = 15
d. f. D = 17 − 1 = 16
C. V. = 2.35

$F = \frac{179.3957^2}{163.2362^2} = 1.21$

12. continued
Do not reject. The variances are equal.

Confidence Interval:
$$t_{\frac{\alpha}{2}}\sqrt{\frac{(n_1-1)s_1^2 + (n_2-1)s_2^2}{n_1+n_2-2}}\sqrt{\frac{1}{n_1} + \frac{1}{n_2}} =$$

$$1.645\sqrt{\frac{(17\text{-}1)(163.2362)^2 + (16\text{-}1)(179.3957)^2}{17+16-2}}\sqrt{\frac{1}{17} + \frac{1}{16}}$$

$= 98.1202$

$(596.2353 - 481.5) - 98.1202 < \mu_1 - \mu_2 < (596.2353 - 481.5) + 98.1202$

$114.7353 - 98.1202 < \mu_1 - \mu_2 < 114.7353 + 98.1202$
$16.62 < \mu_1 - \mu_2 < 212.86$

13.
Private: $\overline{X} = \$16{,}147.5$ s = 4023.7
Public: $\overline{X} = \$9039.9$ s = 3325.5

F test:
d. f. N = 6 − 1 = 5
d. f. D = 7 − 1 = 6
C. V. = 5.99
$F = \frac{4023.7^2}{3325.5^2} = 1.46$
Do not reject. The variances are equal.

Confidence Interval:

$$t_{\frac{\alpha}{2}}\sqrt{\frac{(n_1-1)s_1^2 + (n_2-1)s_2^2}{n_1+n_2-2}}\sqrt{\frac{1}{n_1} + \frac{1}{n_2}} =$$

$$2.201\sqrt{\frac{5(4023.7)^2 + 6(3325.5)^2}{6+7-2}}\sqrt{\frac{1}{6} + \frac{1}{7}}$$

$= 4481.04$

$16{,}147.5 - 9039.9 = 7107.6$

$7107.6 - 4481.04 < \mu_1 - \mu_2 < 7107.6 + 4481.04$

$\$2626.60 < \mu_1 - \mu_2 < \$11{,}588.64$

EXERCISE SET 9-5

1.
a. dependent
b. dependent
c. independent
d. dependent
e. independent

2.

Before	After	D	D^2
2	1	1	1
3	4	-1	1
6	3	3	9
7	8	-1	1
4	3	1	1
5	3	2	4
3	1	2	4
1	0	1	1
0	1	-1	1
0	0	0	0
		$\sum D = 7$	$\sum D^2 = 23$

H_0: $\mu_D \leq 0$
H_1: $\mu_D > 0$ (claim)

C. V. = 1.833 d. f. = 9

$\overline{D} = \frac{\sum D}{n} = \frac{7}{10} = 0.7$

$s_D = \sqrt{\frac{\sum D^2 - \frac{(\sum D)^2}{n}}{n-1}} = \sqrt{\frac{23 - \frac{7^2}{10}}{9}} = 1.42$

$t = \frac{\overline{D} - \mu_D}{\frac{s_D}{\sqrt{n}}} = \frac{0.7 - 0}{\frac{1.42}{\sqrt{10}}} = 1.56$

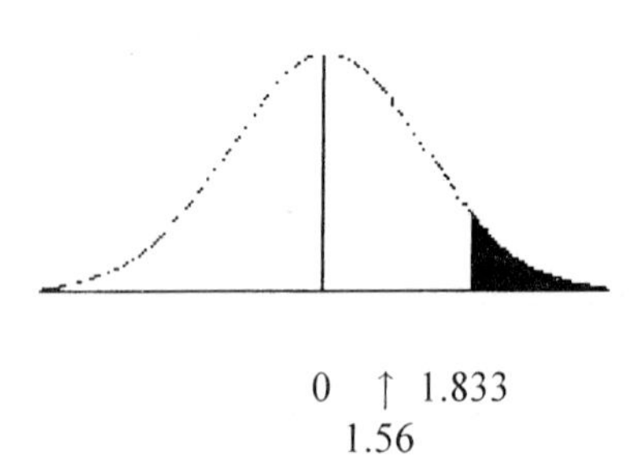

0 ↑ 1.833
1.56

Do not reject the null hypothesis. There is not enough evidence to support the claim that the workers missed fewer days after completing the program.

3.

Before	After	D	D^2
9	9	0	0
12	17	-5	25
6	9	-3	9
15	20	-5	25
3	2	1	1
18	21	-3	9
10	15	-5	25
13	22	-9	81
7	6	1	1
		$\sum D = -28$	$\sum D^2 = 176$

H_0: $\mu_D \geq 0$
H_1: $\mu_D < 0$ (claim)

C. V. = − 1.397 d. f. = 8

3. continued

$\overline{D} = \frac{\sum D}{n} = -3.11$

$s_D = \sqrt{\frac{\sum D^2 - \frac{(\sum D)^2}{n}}{n-1}} = \sqrt{\frac{176 - \frac{(-28)^2}{9}}{8}} = 3.33$

$t = \frac{-3.11 - 0}{\frac{3.33}{\sqrt{9}}} = -2.8$

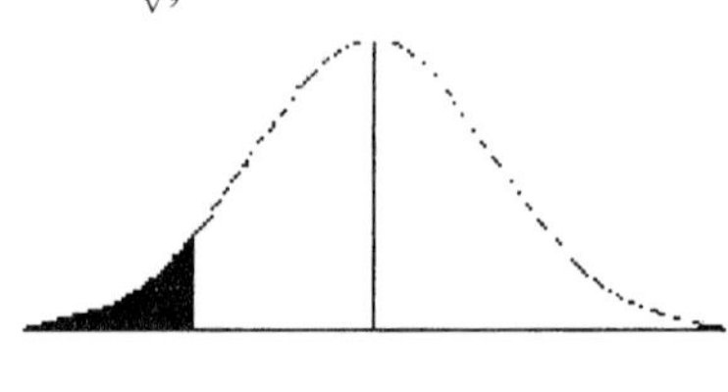

↑ − 1.397 0
− 2.8

Reject the null hypothesis. There is enough evidence to support the claim that the seminar increased the number of hours students studied.

4.

Before	After	D	D^2
12	13	-1	1
11	12	-1	1
14	10	4	16
9	9	0	0
8	8	0	0
6	8	-2	4
8	7	1	1
5	6	-1	1
4	5	-1	1
7	5	2	4
		$\sum D = 1$	$\sum D^2 = 29$

H_0: $\mu_D = 0$
H_1: $\mu_D \neq 0$ (claim)

C. V. = ± 2.262 d. f. = 9

$\overline{D} = \frac{1}{10} = 0.1$

$s_D = \sqrt{\frac{29 - \frac{(1)^2}{10}}{9}} = 1.79$

$t = \frac{0.1 - 0}{\frac{1.79}{\sqrt{10}}} = 0.176$

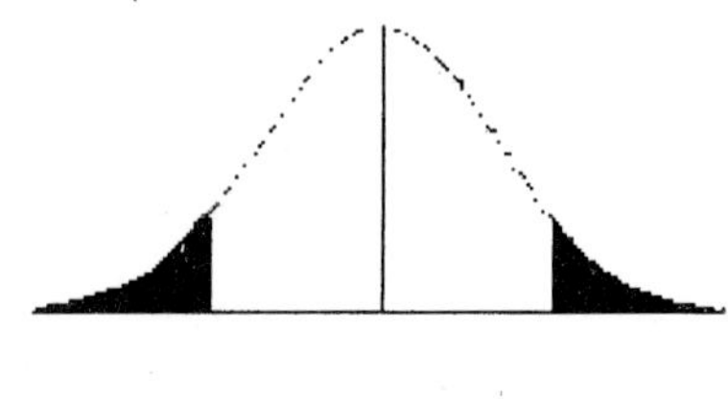

− 2.262 0 ↑ 2.262
0.176

4. continued

Do not reject the null hypothesis. There is not enough evidence to support the claim that there was a change in attitude.

Confidence Interval:

$$0.1 - 2.262\left(\frac{1.79}{\sqrt{10}}\right) < \mu_D < 0.1 + 2.262\left(\frac{1.79}{\sqrt{10}}\right)$$

$$-1.18 < \mu_D < 1.38$$

5.

F - S	S - Th	D	D^2
4	8	-4	16
7	5.5	1.5	2.25
10.5	7.5	3	9
12	8	4	16
11	7	4	16
9	6	3	9
6	6	0	0
9	8	1	1
		$\sum D = 12.5$	$\sum D^2 = 69.25$

H_0: $\mu_D = 0$
H_1: $\mu_D \neq 0$ (claim)

C. V. $= \pm 2.365$ d. f. $= 7$

$$\overline{D} = \frac{\sum D}{n} = \frac{12.5}{8} = 1.5625$$

$$s_D = \sqrt{\frac{\sum D^2 - \frac{(\sum D)^2}{n}}{n-1}}$$

$$= \sqrt{\frac{69.25 - \frac{(12.5)^2}{8}}{7}} = 2.665$$

$$t = \frac{1.5625 - 0}{\frac{2.665}{\sqrt{8}}} = 1.6583$$

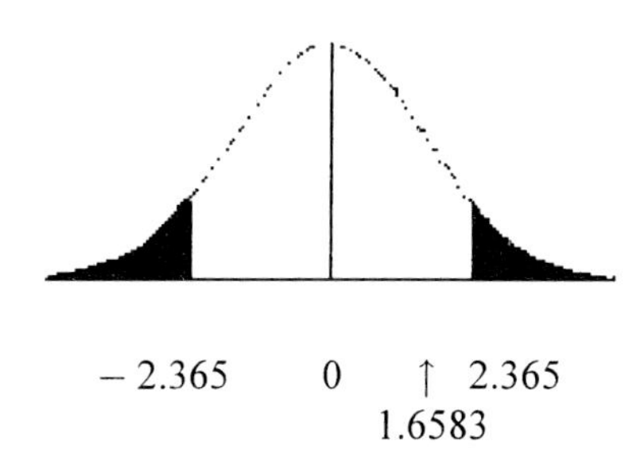

-2.365 0 ↑ 2.365
1.6583

Do not reject the null hypothesis. There is not enough evidence to support the claim that there is a difference in the mean number of hours slept.

6.

First Day	Last Day	D	D^2
80	78	2	4
72	75	-3	9
78	72	6	36
68	70	-2	4
75	72	3	9
69	68	1	1
72	73	-1	1
		$\sum D = 6$	$\sum D^2 = 64$

H_0: $\mu_D \leq 0$
H_1: $\mu_D > 0$ (claim)

C. V. $= 1.943$ d. f. $= 6$

$\overline{D} = 0.8571$
$s_D = 3.13$

$$t = \frac{0.8571 - 0}{\frac{3.13}{\sqrt{7}}} = 0.724$$

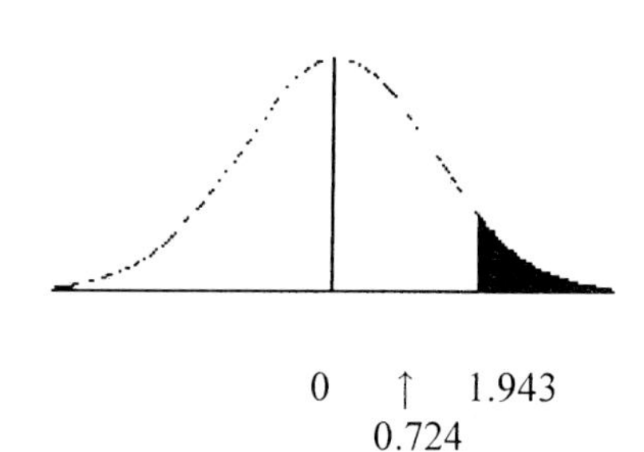

0 ↑ 1.943
0.724

Do not reject the null hypothesis. There is not enough evidence to support the claim that the golf scores improved.

7.

Before	After	D	D^2
12	9	3	9
9	6	3	9
0	1	-1	1
5	3	2	4
4	2	2	4
3	3	0	0
		$\sum D = 9$	$\sum D^2 = 27$

H_0: $\mu_D \leq 0$
H_1: $\mu_D > 0$ (claim)

C. V. $= 2.571$ d. f. $= 5$

$$\overline{D} = \frac{\sum D}{n} = \frac{9}{6} = 1.5$$

$$s_D = \sqrt{\frac{\sum D^2 - \frac{(\sum D)^2}{n}}{n-1}} = \sqrt{\frac{27 - \frac{9^2}{6}}{5}} = 1.64$$

$$t = \frac{1.5 - 0}{\frac{1.64}{\sqrt{6}}} = 2.24$$

7. continued

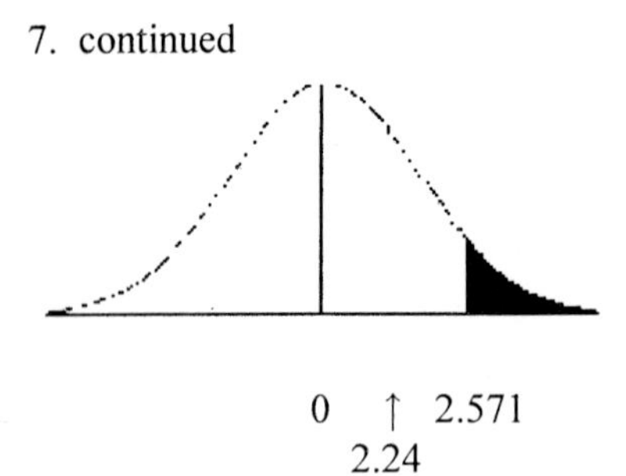

Do not reject the null hypothesis. There is not enough evidence to support the claim that the errors have been reduced.

8.

Before	After	D	D^2
10	12	-2	4
16	15	1	1
12	15	-3	9
12	12	0	0
18	17	1	1
20	20	0	0
		$\sum D = -3$	$\sum D^2 = 15$

H_0: $\mu_D \geq 0$
H_1: $\mu_D < 0$ (claim)

C. V. $= -2.015$ d. f. $= 5$

$\overline{D} = -0.5$
$s_D = 1.643$

$t = \frac{-0.5 - 0}{\frac{1.643}{\sqrt{6}}} = -0.745$

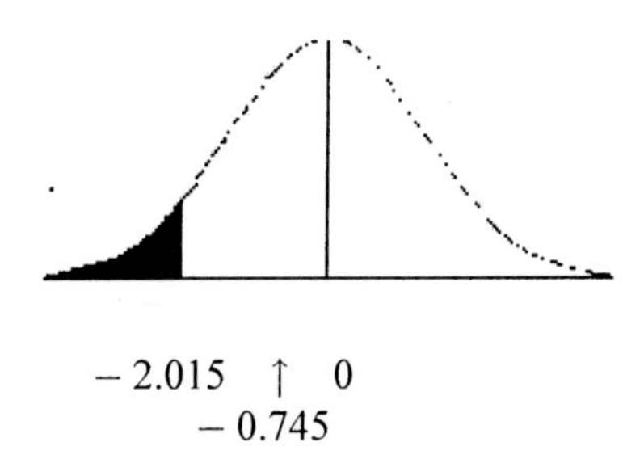

Do not reject the null hypothesis. There is not enough evidence to support the claim that the refresher course helped.

9.

A	B	D	D^2
87	83	4	16
92	95	-3	9
78	79	-1	1
83	83	0	0
88	86	2	4
90	93	-3	9
84	80	4	16
93	86	7	49
		$\sum D = 10$	$\sum D^2 = 104$

H_0: $\mu_D = 0$
H_1: $\mu_D \neq 0$ (claim)

P-value $= 0.361$ d. f. $= 7$

$\overline{D} = \frac{\sum D}{n} = \frac{10}{8} = 1.25$

$s_D = \sqrt{\frac{\sum D^2 - \frac{(\sum D)^2}{n}}{n-1}} = \sqrt{\frac{104 - \frac{10^2}{8}}{7}} = 3.62$

$t = \frac{1.25 - 0}{\frac{3.62}{\sqrt{8}}} = 0.978$

$0.20 < \text{P-value} < 0.50$ Do not reject the null hypothesis since P-value > 0.01. There is not enough evidence to support the claim that there is a difference in the pulse rates.

Confidence Interval:
$1.25 - 3.499\left(\frac{3.62}{\sqrt{8}}\right) < \mu_D <$
$1.25 + 3.499\left(\frac{3.62}{\sqrt{8}}\right)$
$-3.23 < \mu_D < 5.73$

10.

1994	1999	D	D^2
184	161	23	529
414	382	32	1024
22	22	0	0
99	109	-10	100
116	120	-4	16
49	52	-3	9
24	28	-4	16
50	50	0	0
282	297	-15	225
25	40	-15	225
141	148	-7	49
45	56	-11	121
12	20	-8	64
37	38	-1	1
9	9	0	0
17	19	-2	4
		$\sum D = -25$	$\sum D^2 = 2383$

H_0: $\mu_D = 0$
H_1: $\mu_D \neq 0$ (claim)

P-value $= 0.624$ d. f. $= 15$

10. continued

$\overline{D} = -1.5625$

$s_D = 12.5$

$t = \frac{-1.5625 - 0}{\frac{12.5}{\sqrt{16}}} = -0.5$

Since P-value > 0.50, do not reject the null hypothesis. There is not enough evidence to support the claim that the average of the assessed values has changed.

11.

Using the previous problem, $\overline{D} = -1.5625$ whereas the mean of the 1994 values is 95.375 and the mean of the 1999 values is 96.9375; hence,

$\overline{D} = 95.375 - 96.9375 = -1.5625$

EXERCISE SET 9-6

1A.

Use $\hat{p} = \frac{X}{n}$ and $\hat{q} = 1 - \hat{p}$

a. $\hat{p} = \frac{34}{48}$ $\hat{q} = \frac{14}{48}$

b. $\hat{p} = \frac{28}{75}$ $\hat{q} = \frac{47}{75}$

c. $\hat{p} = \frac{50}{100}$ $\hat{q} = \frac{50}{100}$

d. $\hat{p} = \frac{6}{24}$ $\hat{q} = \frac{18}{24}$

e. $\hat{p} = \frac{12}{144}$ $\hat{q} = \frac{132}{144}$

1B.

a. $x = 0.16(100) = 16$

b. $x = 0.08(50) = 4$

c. $x = 0.06(80) = 4.8$

d. $x = 0.52(200) = 104$

e. $x = 0.20(150) = 30$

2.

For each part, use the formulas $\overline{p} = \frac{X_1 + X_2}{n_1 + n_2}$ and $\overline{q} = 1 - \overline{p}$

a. $\overline{p} = \frac{60 + 40}{100 + 100} = 0.5$

$\overline{q} = 1 - 0.5 = 0.5$

b. $\overline{p} = \frac{22 + 18}{50 + 30} = 0.5$

$\overline{q} = 1 - 0.5 = 0.5$

c. $\overline{p} = \frac{18 + 20}{60 + 80} = 0.27$

$\overline{q} = 1 - 0.27 = 0.73$

2. continued

d. $\overline{p} = \frac{5 + 12}{32 + 48} = 0.2125$

$\overline{q} = 1 - 0.2125 = 0.7875$

e. $\overline{p} = \frac{12 + 15}{75 + 50} = 0.216$

$\overline{q} = 1 - 0.216 = 0.784$

3.

$\hat{p}_1 = \frac{X_1}{n_1} = \frac{80}{150} = 0.533$ $\hat{p}_2 = \frac{30}{100} = 0.3$

$\overline{p} = \frac{X_1 + X_2}{n_1 + n_2} = \frac{80 + 30}{150 + 100} = \frac{110}{250} = 0.44$

$\overline{q} = 1 - \overline{p} = 1 - 0.44 - 0.56$

H_0: $p_1 = p_2$

H_1: $p_1 \neq p_2$ (claim)

C. V. $= \pm 1.96$

$z = \frac{(\hat{p}_1 - \hat{p}_2) - (p_1 - p_2)}{\sqrt{(\overline{p})(\overline{q})(\frac{1}{n_1} + \frac{1}{n_2})}} = \frac{(0.533 - 0.3) - 0}{\sqrt{(0.44)(0.56)(\frac{1}{150} + \frac{1}{100})}}$

$z = 3.64$

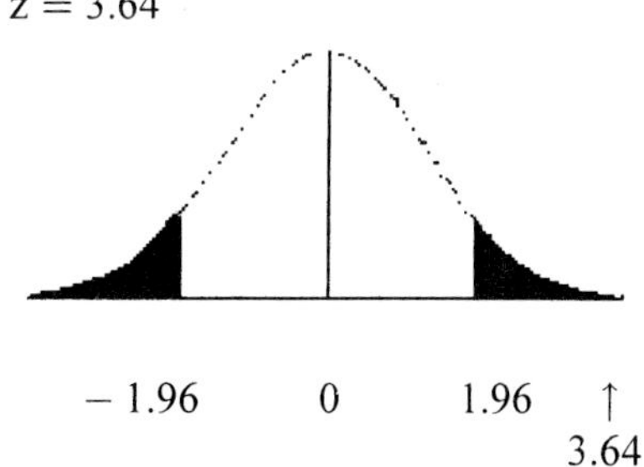

Reject the null hypothesis. There is enough evidence to support the claim that there is a significant difference in the proportions.

4.

$\hat{p}_1 = \frac{X_1}{n_1} = \frac{180}{300} = 0.6$ $\hat{p}_2 = \frac{200}{250} = 0.8$

$\overline{p} = \frac{X_1 + X_2}{n_1 + n_2} = \frac{180 + 200}{300 + 250} = 0.691$

$\overline{q} = 1 - \overline{p} = 1 - 0.691 = 0.309$

H_0: $p_1 = p_2$

H_1: $p_1 \neq p_2$ (claim)

C. V. $= \pm 2.58$

$z = \frac{(\hat{p}_1 - \hat{p}_2) - (p_1 - p_2)}{\sqrt{(\overline{p})(\overline{q})(\frac{1}{n_1} + \frac{1}{n_2})}} = \frac{(0.6 - 0.8) - 0}{\sqrt{(0.691)(0.309)(\frac{1}{300} + \frac{1}{250})}}$

$z = -5.054$

4. continued

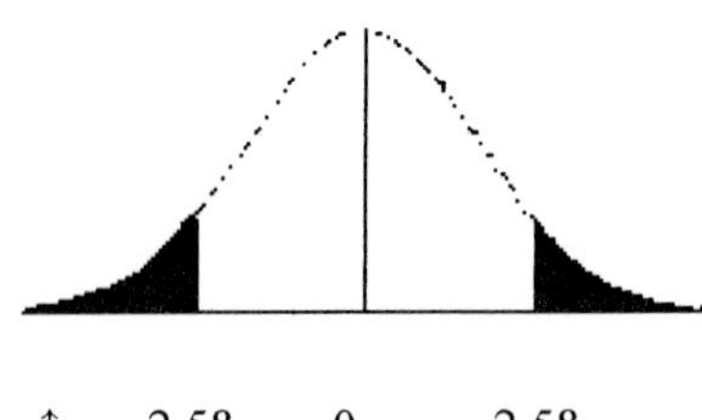

↑ -2.58 0 2.58
-5.05

Reject the null hypothesis. There is enough evidence to support the claim that the proportion of students receiving aid has changed.

5.
$\hat{p}_1 = \frac{X_1}{n_1} = \frac{112}{150} = 0.7467 \quad \hat{p}_2 = \frac{150}{200} = 0.75$

$\bar{p} = \frac{X_1 + X_2}{n_1 + n_2} = \frac{112 + 150}{150 + 200} = 0.749$

$\bar{q} = 1 - \bar{p} = 1 - 0.749 = 0.251$

H_0: $p_1 = p_2$
H_1: $p_1 \neq p_2$ (claim)

C. V. = ± 1.96

$z = \frac{(\hat{p}_1 - \hat{p}_2) - (p_1 - p_2)}{\sqrt{(\bar{p})(\bar{q})(\frac{1}{n_1} + \frac{1}{n_2})}} = \frac{(0.7467 - 0.75) - 0}{\sqrt{(0.749)(0.251)(\frac{1}{150} + \frac{1}{200})}}$

$z = -0.07$

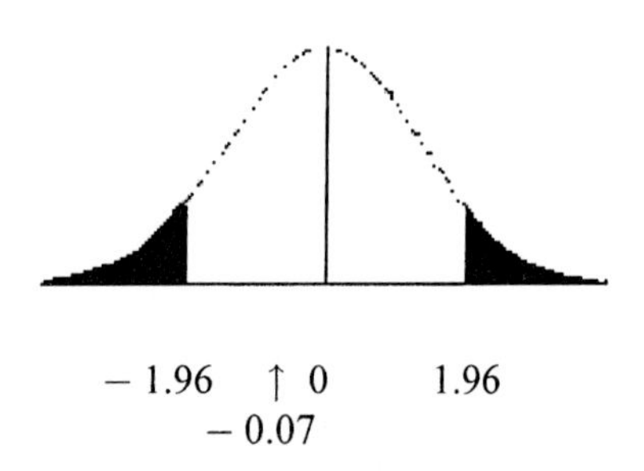

-1.96 ↑ 0 1.96
-0.07

Do not reject the null hypothesis. There is not enough evidence to support the claim that the proportions are different.

6.
$\hat{p}_1 = \frac{10}{73} = 0.14 \quad \hat{p}_2 = \frac{16}{80} = 0.20$

$\bar{p} = \frac{10 + 16}{73 + 80} = 0.17$

$\bar{q} = 1 - 0.17 = 0.83$

6. continued
H_0: $p_1 = p_2$
H_1: $p_1 \neq p_2$ (claim)

C. V. = ± 1.96

$z = \frac{(0.14 - 0.20) - 0}{\sqrt{(0.17)(0.83)(\frac{1}{73} + \frac{1}{80})}} = -0.99$

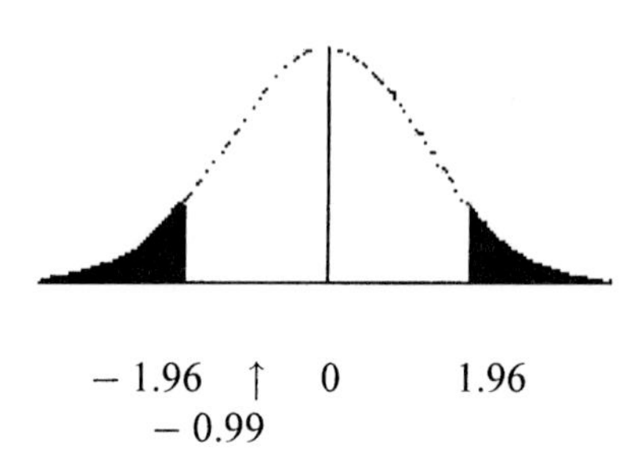

-1.96 ↑ 0 1.96
-0.99

Do not reject the null hypothesis. There is not enough evidence to support the claim that there is a difference in the proportions.

$(\hat{p}_1 - \hat{p}_2) - z_{\frac{\alpha}{2}}\sqrt{\frac{\hat{p}_1\hat{q}_1}{n_1} + \frac{\hat{p}_2\hat{q}_2}{n_2}} < p_1 - p_2 <$
$(\hat{p}_1 - \hat{p}_2) + z_{\frac{\alpha}{2}}\sqrt{\frac{\hat{p}_1\hat{q}_1}{n_1} + \frac{\hat{p}_2\hat{q}_2}{n_2}}$

$(0.14 - 0.2) - 1.96\sqrt{\frac{0.14(0.86)}{73} + \frac{0.2(0.8)}{80}} <$

$p_1 - p_2 < (0.14 - 0.2) + 1.96\sqrt{\frac{0.14(0.86)}{73} + \frac{0.2(0.8)}{80}}$

$-0.06 - 0.12 < p_1 - p_2 < -0.06 + 0.12$
$-0.18 < p_1 - p_2 < 0.06$

7.
$\hat{p}_1 = 0.83 \quad \hat{p}_2 = 0.75$
$X_1 = 0.83(100) = 83$
$X_2 = 0.75(100) = 75$

$\bar{p} = \frac{83 + 75}{100 + 100} = 0.79 \quad \bar{q} = 1 - 0.79 = 0.21$

H_0: $p_1 = p_2$ (claim)
H_1: $p_1 \neq p_2$
C. V. = ± 1.96 $\alpha = 0.05$

$z = \frac{(\hat{p}_1 - \hat{p}_2) - (p_1 - p_2)}{\sqrt{(\bar{p})(\bar{q})(\frac{1}{n_1} + \frac{1}{n_2})}} = \frac{(0.83 - 0.75) - 0}{\sqrt{(0.79)(0.21)(\frac{1}{100} + \frac{1}{100})}}$

$z = 1.39$

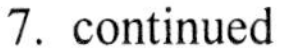

7. continued

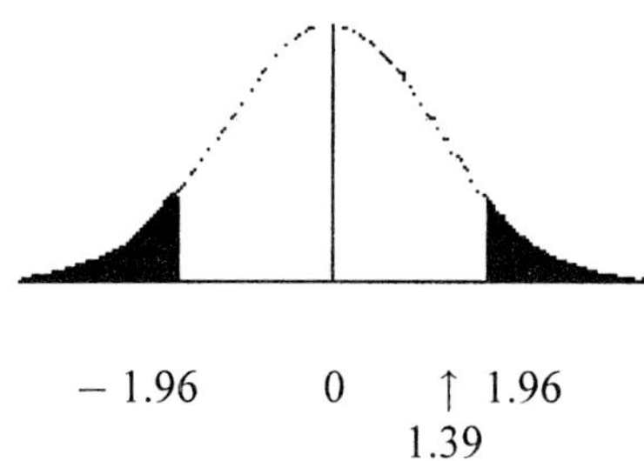

Do not reject the null hypothesis. There is not enough evidence to reject the claim that the proportions are equal.

$$(\hat{p}_1 - \hat{p}_2) - z_{\frac{\alpha}{2}}\sqrt{\frac{\hat{p}_1\hat{q}_1}{n_1} + \frac{\hat{p}_2\hat{q}_2}{n_2}} < p_1 - p_2 <$$

$$(\hat{p}_1 - \hat{p}_2) + z_{\frac{\alpha}{2}}\sqrt{\frac{\hat{p}_1\hat{q}_1}{n_1} + \frac{\hat{p}_2\hat{q}_2}{n_2}}$$

$$0.08 - 1.96\sqrt{\frac{0.83(0.17)}{100} + \frac{0.75(0.25)}{100}} < p_1 - p_2$$

$$< 0.08 + 1.96\sqrt{\frac{0.83(0.17)}{100} + \frac{0.75(0.25)}{100}}$$

$$-0.032 < p_1 - p_2 < 0.192$$

8.
$\hat{p}_1 = 0.15 \qquad \hat{p}_2 = 0.21$
$X_1 = 0.15(200) = 30$
$X_2 = 0.21(200) = 42$

$$\bar{p} = \frac{X_1 + X_2}{n_1 + n_2} = \frac{30 + 42}{200 + 200} = 0.18$$

$$\bar{q} = 1 - 0.18 = 0.82$$

H_0: $p_1 = p_2$
H_1: $p_1 \neq p_2$ (claim)

C. V. = ± 1.65

$$z = \frac{(0.15 - 0.21) - 0}{\sqrt{(0.18)(0.82)(\frac{1}{200} + \frac{1}{200})}} = -1.56$$

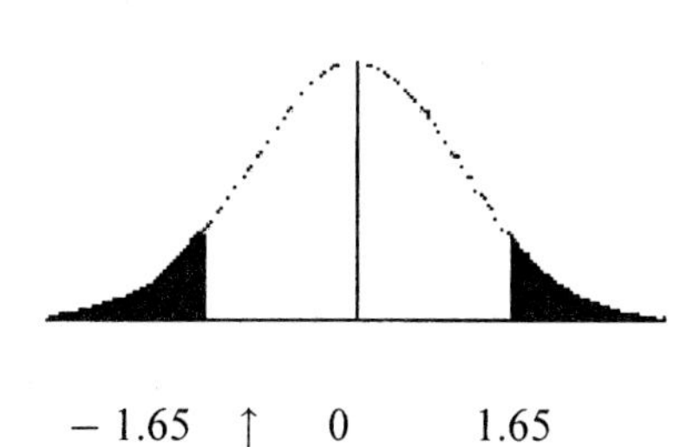

8. continued
Do not reject the null hypothesis. There is not enough evidence to support the claim that the proportions are different.

$$(\hat{p}_1 - \hat{p}_2) - z_{\frac{\alpha}{2}}\sqrt{\frac{\hat{p}_1\hat{q}_1}{n_1} + \frac{\hat{p}_2\hat{q}_2}{n_2}} < p_1 - p_2 <$$

$$(\hat{p}_1 - \hat{p}_2) + z_{\frac{\alpha}{2}}\sqrt{\frac{\hat{p}_1\hat{q}_1}{n_1} + \frac{\hat{p}_2\hat{q}_2}{n_2}}$$

$$(0.15 - 0.21) - 1.65\sqrt{\frac{0.15(0.85)}{200} + \frac{0.21(0.79)}{200}}$$

$$< p_1 - p_2 < (0.15 - 0.21) + 1.65\sqrt{\frac{0.15(0.85)}{200} + \frac{0.21(0.79)}{200}}$$

$$-0.06 - 1.65(0.038) < p_1 - p_2 <$$

$$-0.06 + 1.65(0.038)$$

$$-0.123 < p_1 - p_2 < 0.003$$

9.
$\hat{p}_1 = 0.55 \qquad \hat{p}_2 = 0.45$

$X_1 = 0.55(80) = 44 \quad X_2 = 0.45(90) = 40.5$

$$\bar{p} = \frac{X_1 + X_2}{n_1 + n_2} = \frac{44 + 40.5}{80 + 90} = 0.497$$

$$\bar{q} = 1 - \bar{p} = 1 - 0.497 = 0.503$$

H_0: $p_1 = p_2$
H_1: $p_1 \neq p_2$ (claim)

C. V. = $\pm 2.58 \qquad \alpha = 0.01$

$$z = \frac{(\hat{p}_1 - \hat{p}_2) - (p_1 - p_2)}{\sqrt{(\bar{p})(\bar{q})(\frac{1}{n_1} + \frac{1}{n_2})}} = \frac{(0.55 - 0.45) - 0}{\sqrt{(0.497)(0.503)(\frac{1}{80} + \frac{1}{90})}}$$

$z = 1.302$

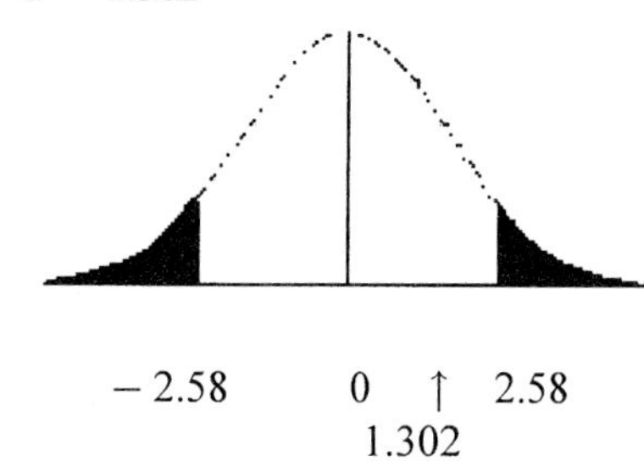

Do not reject the null hypothesis. There is not enough evidence to support the claim that the proportions are different.

9. continued

$(\hat{p}_1 - \hat{p}_2) - z_{\frac{\alpha}{2}}\sqrt{\frac{\hat{p}_1\hat{q}_1}{n_1} + \frac{\hat{p}_2\hat{q}_2}{n_2}} < p_1 - p_2 <$

$(\hat{p}_1 - \hat{p}_2) + z_{\frac{\alpha}{2}}\sqrt{\frac{\hat{p}_1\hat{q}_1}{n_1} + \frac{\hat{p}_2\hat{q}_2}{n_2}}$

$0.1 - 2.58\sqrt{\frac{0.55(0.45)}{80} + \frac{0.45(0.55)}{90}} < p_1 - p_2$

$< 0.1 + 2.58\sqrt{\frac{0.55(0.45)}{80} + \frac{0.45(0.55)}{90}}$

$-0.097 < p_1 - p_2 < 0.297$

10.

$\hat{p}_1 = \frac{X_1}{n_1} = \frac{130}{200} = 0.65$

$\hat{p}_2 = \frac{X_2}{n_2} = \frac{63}{300} = 0.21$

$\bar{p} = \frac{X_1 + X_2}{n_1 + n_2} = \frac{130 + 63}{200 + 300} = 0.386$

$\bar{q} = 1 - \bar{p} = 1 - 0.386 = 0.614$

H_0: $p_1 \leq p_2$

H_1: $p_1 > p_2$ (claim)

$z = \frac{(0.65 - 0.21) - 0}{\sqrt{(0.386)(0.614)(\frac{1}{200} + \frac{1}{300})}} = 9.90$

P-value < 0001

Since P-value < 0.01, reject the null hypothesis. There is enough evidence to support the claim that men are more safety-conscious than women.

11.

$\hat{p}_1 = \frac{45}{80} = 0.5625$ $\hat{p}_2 = \frac{63}{120} = 0.525$

$\bar{p} = \frac{X_1 + X_2}{n_1 + n_2} = \frac{45 + 63}{80 + 120} = 0.54$

$\bar{q} = 1 - \bar{p} = 1 - 0.54 = 0.46$

H_0: $p_1 = p_2$

H_1: $p_1 \neq p_2$ (claim)

C. V. = ± 1.96 $\alpha = 0.05$

$z = \frac{(\hat{p}_1 - \hat{p}_2) - (p_1 - p_2)}{\sqrt{(\bar{p})(\bar{q})(\frac{1}{n_1} + \frac{1}{n_2})}} = \frac{(0.5625 - 0.525) - 0}{\sqrt{(0.54)(0.46)(\frac{1}{80} + \frac{1}{120})}}$

$z = 0.521$

11. continued

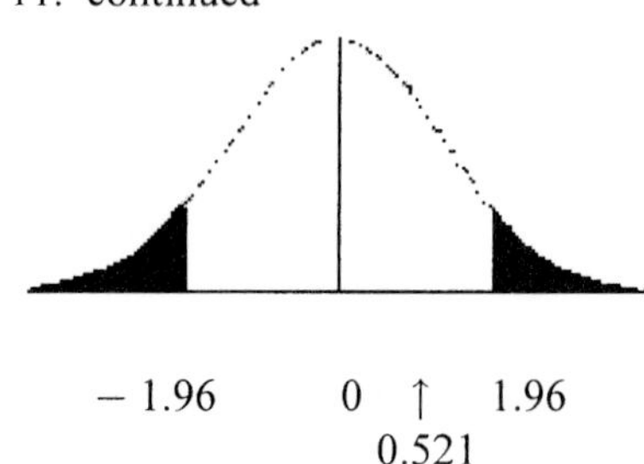

− 1.96 0 ↑ 1.96
0.521

Do not reject the null hypothesis. There is not enough evidence to support the claim that there is a difference in the proportions.

$(\hat{p}_1 - \hat{p}_2) - z_{\frac{\alpha}{2}}\sqrt{\frac{\hat{p}_1\hat{q}_1}{n_1} + \frac{\hat{p}_2\hat{q}_2}{n_2}} < p_1 - p_2 <$

$(\hat{p}_1 - \hat{p}_2) + z_{\frac{\alpha}{2}}\sqrt{\frac{\hat{p}_1\hat{q}_1}{n_1} + \frac{\hat{p}_2\hat{q}_2}{n_2}}$

$0.0375 - 1.96\sqrt{\frac{0.5625(0.4375)}{80} + \frac{0.525(0.475)}{120}} <$

$p_1 - p_2 < 0.0375 + 1.96\sqrt{\frac{0.5625(0.4375)}{80} + \frac{0.525(0.475)}{120}}$

$-0.103 < p_1 - p_2 < 0.178$

12.

$\hat{p}_1 = 0.30$ $\hat{p}_2 = 0.24$

$X_1 = 0.30(100) = 30$

$X_2 = 0.24(100) = 24$

$\bar{p} = \frac{30 + 24}{100 + 100} = 0.27$

$\bar{q} = 1 - \bar{p} = 1 - 0.27 = 0.73$

H_0: $p_1 = p_2$

H_1: $p_1 \neq p_2$ (claim)

$z = \frac{(0.30 - 0.24) - 0}{\sqrt{(0.27)(0.73)(\frac{1}{100} + \frac{1}{100})}} = 0.96$

P-value = 0.337

Since P-value > 0.02, do not reject the null hypothesis. There is not enough evidence to support the claim that the proportions are different.

13.

$\hat{p}_1 = \frac{X_1}{n_1} = \frac{50}{200} = 0.25$

$\hat{p}_2 = \frac{X_2}{n_2} = \frac{93}{300} = 0.31$

13. continued

$\bar{p} = \frac{X_1 + X_2}{n_1 + n_2} = \frac{50 + 93}{200 + 300} = 0.286$

$\bar{q} = 1 - \bar{p} = 1 - 0.286 = 0.714$

H_0: $p_1 = p_2$
H_1: $p_1 \neq p_2$ (claim)

C. V. = ± 2.58 $\alpha = 0.01$

$z = \frac{(\hat{p}_1 - \hat{p}_2) - (p_1 - p_2)}{\sqrt{(\bar{p})(\bar{q})(\frac{1}{n_1} + \frac{1}{n_2})}} = \frac{(0.25-0.31)-0}{\sqrt{(0.286)(0.714)(\frac{1}{200}+\frac{1}{300})}}$

$z = -1.45$

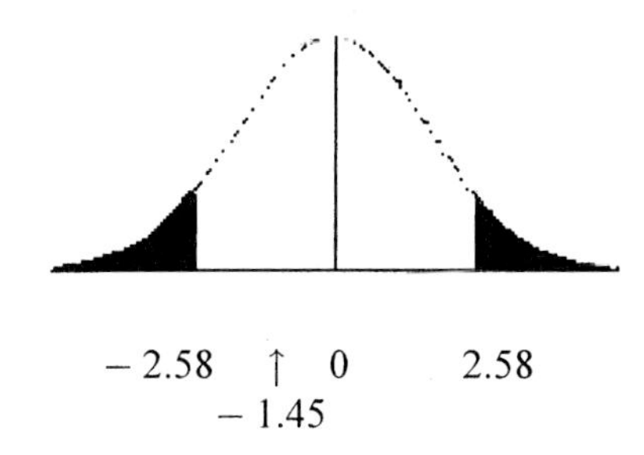

-2.58 ↑ 0 2.58
-1.45

Do not reject the null hypothesis. There is not enough evidence to support the claim that the proportions are different.

$(\hat{p}_1 - \hat{p}_2) - z_{\frac{\alpha}{2}}\sqrt{\frac{\hat{p}_1\hat{q}_1}{n_1} + \frac{\hat{p}_2\hat{q}_2}{n_2}} < p_1 - p_2 <$

$(\hat{p}_1 - \hat{p}_2) + z_{\frac{\alpha}{2}}\sqrt{\frac{\hat{p}_1\hat{q}_1}{n_1} + \frac{\hat{p}_2\hat{q}_2}{n_2}}$

$-0.06 - 2.58\sqrt{\frac{0.25(0.75)}{200} + \frac{0.31(0.69)}{300}} <$

$p_1 - p_2 < -0.06 + 2.58\sqrt{\frac{0.25(0.75)}{200} + \frac{0.31(0.69)}{300}}$

$-0.165 < p_1 - p_2 < 0.045$

14.

$\hat{p}_1 = \frac{8}{50} = 0.16$ $\hat{p}_2 = \frac{70}{75} = 0.267$

$\bar{p} = \frac{8 + 20}{50 + 75} = 0.224$

$\bar{q} = 1 - \bar{p} = 0.776$

H_0: $p_1 \geq p_2$
H_1: $p_1 < p_2$ (claim)

$z = \frac{(0.16 - 0.267) - 0}{\sqrt{(0.224)(0.776)(\frac{1}{50} + \frac{1}{75})}} = -1.41$

14. continued

P-value = 0.0793

Since P-value > 0.05, do not reject the null hypothesis. There is not enough evidence to support the claim that the proportion of college freshmen who have their own cars is higher than the proportion of high school seniors who have their own cars.

15.

$\alpha = 0.01$

$\hat{p}_1 = 0.8$ $\hat{q}_1 = 0.2$
$\hat{p}_2 = 0.6$ $\hat{q}_2 = 0.4$

$\hat{p}_1 - \hat{p}_2 = 0.8 - 0.6 = 0.2$

$(\hat{p}_1 - \hat{p}_2) - z_{\frac{\alpha}{2}}\sqrt{\frac{\hat{p}_1\hat{q}_1}{n_1} + \frac{\hat{p}_2\hat{q}_2}{n_2}} < p_1 - p_2 <$

$(\hat{p}_1 - \hat{p}_2) + z_{\frac{\alpha}{2}}\sqrt{\frac{\hat{p}_1\hat{q}_1}{n_1} + \frac{\hat{p}_2\hat{q}_2}{n_2}}$

$0.2 - 2.58\sqrt{\frac{(0.8)(0.2)}{150} + \frac{(0.6)(0.4)}{200}} < p_1 - p_2 <$

$0.2 + 2.58\sqrt{\frac{(0.8)(0.2)}{150} + \frac{(0.6)(0.4)}{200}}$

$0.077 < p_1 - p_2 < 0.323$

16.

$\hat{p}_1 = \frac{X_1}{n_1} = \frac{622}{1000} = 0.622$

$\hat{p}_2 = \frac{594}{1000} = 0.594$

$\bar{p} = \frac{X_1 + X_2}{n_1 + n_2} = \frac{622 + 594}{1000 + 1000} = 0.608$

$\bar{q} = 1 - \bar{p} = 1 - 0.608 = 0.392$

H_0: $p_1 \leq p_2$
H_1: $p_1 > p_2$ (claim)

C. V. = 1.65

$z = \frac{(\hat{p}_1 - \hat{p}_2) - (p_1 - p_2)}{\sqrt{(\bar{p})(\bar{q})(\frac{1}{n_1} + \frac{1}{n_2})}} = \frac{(0.622 - 0.594) - 0}{\sqrt{(0.608)(0.392)(\frac{1}{1000} + \frac{1}{1000})}}$

$z = 1.28$

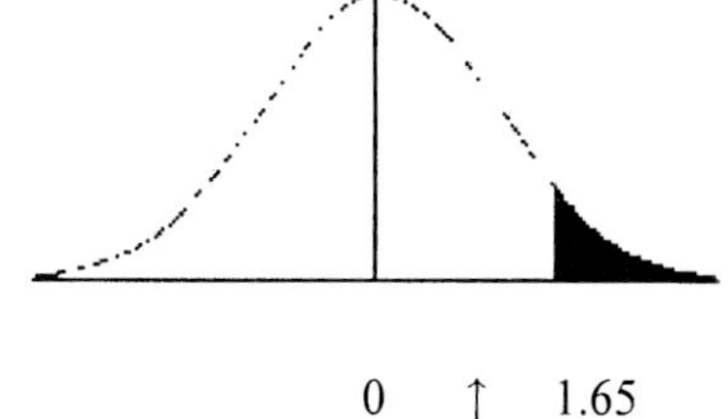

0 ↑ 1.65
1.28

16. continued
Do not reject the null hypothesis. There is not enough evidence to support the claim that the proportion of women working in Miami County is higher than in Greene County.

17.
$\hat{p}_1 = \frac{X_1}{n_1} = \frac{43}{100} = 0.43 \quad \hat{p}_2 = \frac{58}{100} = 0.58$

$\bar{p} = \frac{X_1 + X_2}{n_1 + n_2} = \frac{43 + 58}{100 + 100} = 0.505$

$\bar{q} = 1 - \bar{p} = 1 - 0.505 = 0.495$

H_0: $p_1 = p_2$
H_1: $p_1 \neq p_2$ (claim)

C. V. = ± 1.96

$$z = \frac{(\hat{p}_1 - \hat{p}_2) - (p_1 - p_2)}{\sqrt{(\bar{p})(\bar{q})(\frac{1}{n_1} + \frac{1}{n_2})}} = \frac{(0.43 - 0.58) - 0}{\sqrt{(0.505)(0.495)(\frac{1}{100} + \frac{1}{100})}}$$
$z = -2.12$

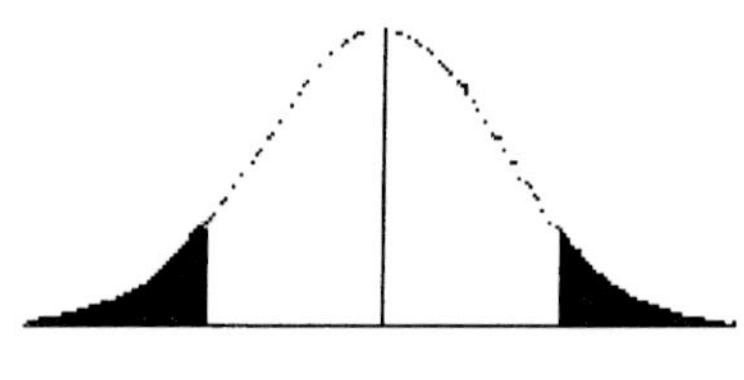

↑ -1.96 0 1.96
-2.12

Reject the null hypothesis. There is enough evidence to support the claim that the proportions are different.

18.
$\hat{p}_1 = 0.4 \qquad \hat{q}_1 = 0.6$

$\hat{p}_2 = 0.56 \qquad \hat{q}_2 = 0.44$

$\hat{p}_1 - \hat{p}_2 = 0.4 - 0.56 = -0.16$

$$(\hat{p}_1 - \hat{p}_2) - z_{\frac{\alpha}{2}}\sqrt{\frac{\hat{p}_1\hat{q}_1}{n_1} + \frac{\hat{p}_2\hat{q}_2}{n_2}} < p_1 - p_2 < (\hat{p}_1 - \hat{p}_2) + z_{\frac{\alpha}{2}}\sqrt{\frac{\hat{p}_1\hat{q}_1}{n_1} + \frac{\hat{p}_2\hat{q}_2}{n_2}}$$

$$-0.16 - 1.96\sqrt{\frac{(0.4)(0.6)}{200} + \frac{(0.56)(0.44)}{100}} < p_1 - p_2 < -0.16 + 1.96\sqrt{\frac{(0.4)(0.6)}{200} + \frac{(0.56)(0.44)}{100}}$$

$-0.279 < p_1 - p_2 < -0.041$

19.
$\hat{p}_1 = 0.2875 \qquad \hat{q}_1 = 0.7125$
$\hat{p}_2 = 0.2857 \qquad \hat{q}_2 = 0.7143$

$\hat{p}_1 - \hat{p}_2 = 0.0018$

$$(\hat{p}_1 - \hat{p}_2) - z_{\frac{\alpha}{2}}\sqrt{\frac{\hat{p}_1\hat{q}_1}{n_1} + \frac{\hat{p}_2\hat{q}_2}{n_2}} < p_1 - p_2 < (\hat{p}_1 - \hat{p}_2) + z_{\frac{\alpha}{2}}\sqrt{\frac{\hat{p}_1\hat{q}_1}{n_1} + \frac{\hat{p}_2\hat{q}_2}{n_2}}$$

$$0.0018 - 1.96\sqrt{\frac{(0.2875)(0.7125)}{400} + \frac{(0.2857)(0.7143)}{350}} < p_1 - p_2 < 0.0018 + 1.96\sqrt{\frac{(0.2875)(0.7125)}{400} + \frac{(0.2857)(0.7143)}{350}}$$

$-0.0631 < p_1 - p_2 < 0.0667$

20.
No, because p_1 could equal p_2.

REVIEW EXERCISES - CHAPTER 9

1.
H_0: $\mu_1 \leq \mu_2$
H_1: $\mu_1 > \mu_2$ (claim)

CV = 2.33 $\quad \alpha = 0.01$
$\bar{X}_1 = 120.1 \qquad \bar{X}_2 = 117.8$
$s_1 = 16.722 \qquad s_2 = 16.053$

$$z = \frac{(\bar{X}_1 - \bar{X}_2) - (\mu_1 - \mu_2)}{\sqrt{\frac{s_1^2}{n_1} + \frac{s_2^2}{n_2}}} = \frac{(120.1 - 117.8) - 0}{\sqrt{\frac{16.722^2}{36} + \frac{16.053^2}{35}}}$$

$z = 0.587$

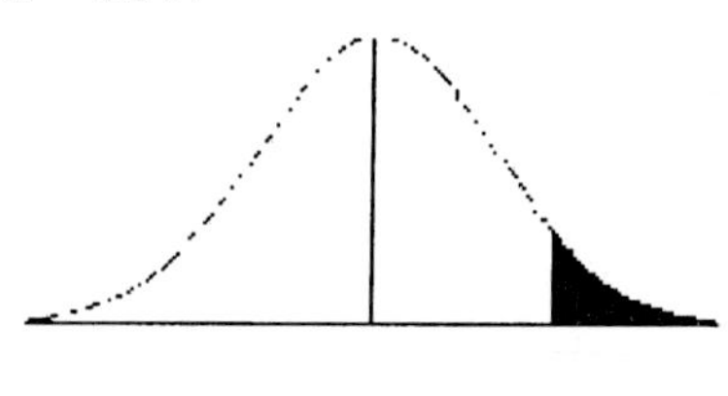

0 ↑ 2.33
0.59

Do not reject the null hypothesis. There is not enough evidence to support the claim that single people do more pleasure driving than married people.

2.
H_0: $\sigma_1^2 = \sigma_2^2$
H_1: $\sigma_1^2 \neq \sigma_2^2$ (claim)

2. continued
C. V. = 2.86 $\alpha = 0.05$
d. f. N = 17 d. f. D = 15

$F = \frac{s_1^2}{s_2^2} = \frac{585}{261} = 2.24$

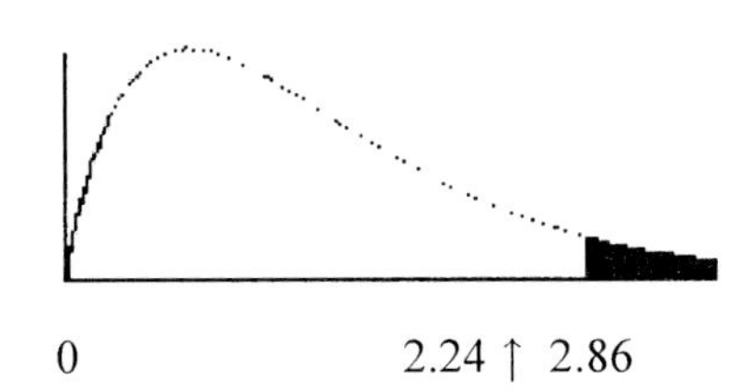

0 2.24 ↑ 2.86

Do not reject the null hypothesis. There is not enough evidence to support the claim that the variances are different.

3.
H_0: $\sigma_1 = \sigma_2$
H_1: $\sigma_1 \neq \sigma_2$ (claim)

C. V. = 2.77 $\alpha = 0.10$
d. f. N = 23 d. f. D = 10

$F = \frac{13.2^2}{4.1^2} = 10.365$

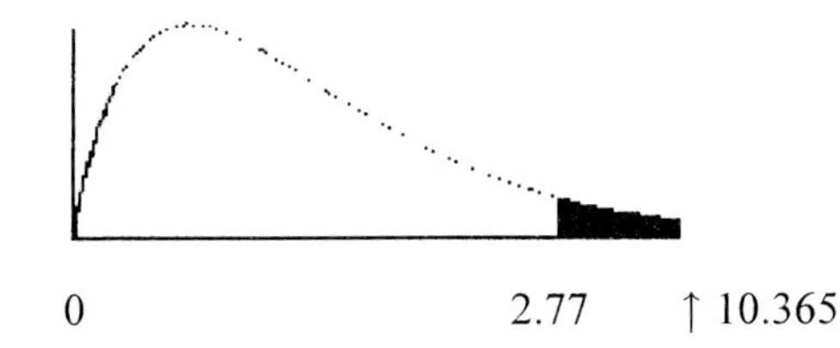

0 2.77 ↑ 10.365

Reject the null hypothesis. There is enough evidence to support the claim that there is a difference in standard deviations.

4.
H_0: $\sigma_1^2 = \sigma_2^2$
H_1: $\sigma_1^2 \neq \sigma_2^2$ (claim)

C. V. = 1.98 $\alpha = 0.10$
d. f. N = 24 d. f. D = 24

$F = \frac{4.85^2}{2.25^2} = 2.16$

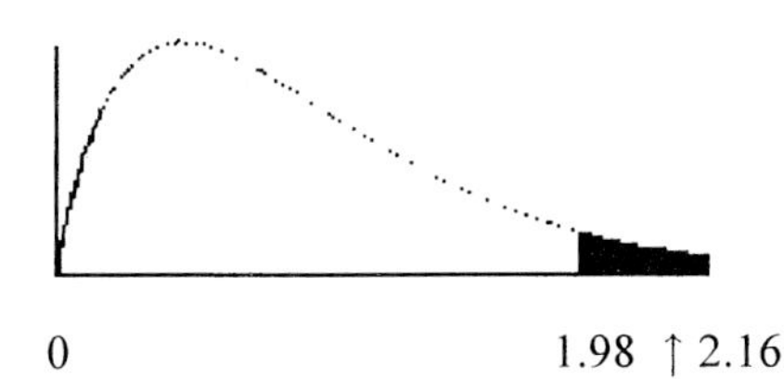

0 1.98 ↑ 2.16

4. continued
Reject the null hypothesis. There is enough evidence to support the claim that there is a significant difference between the variances of the heights for the two leagues.

5.
H_0: $\sigma_1^2 \leq \sigma_2^2$
H_1: $\sigma_1^2 > \sigma_2^2$ (claim)

$\alpha = 0.05$
d. f. N = 9 d. f. D = 9
$F = \frac{s_1^2}{s_2^2} = \frac{6.3^2}{2.8^2} = 5.06$

The P-value for the F test is $0.01 < \text{P-value} < 0.025$ (0.012). Since P-value < 0.05, reject the null hypothesis. There is enough evidence to support the claim that the variance of the number of speeding tickets on Route 19 is greater than the variance of the number of speeding tickets issued on Route 22.

6.
H_0: $\sigma_1 = \sigma_2$
H_1: $\sigma_1 \neq \sigma_2$ (claim)

C. V. = 2.76 $\alpha = 0.01$
d. f. N = 29 d. f. D = 29

$F = \frac{4.9^2}{2.5^2} = 3.84$

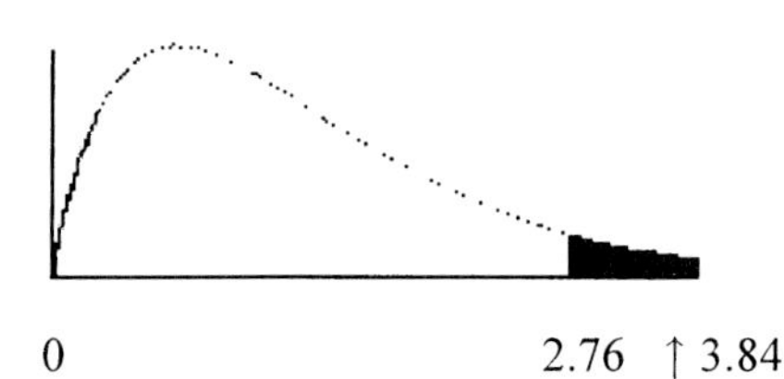

0 2.76 ↑ 3.84

Reject the null hypothesis. There is enough evidence to support the claim that there is a difference in the standard deviations in the number of absentees of the two schools.

7.
H_0: $\sigma_1^2 \leq \sigma_2^2$
H_1: $\sigma_1^2 > \sigma_2^2$ (claim)

C. V. = 1.47 $\alpha = 0.10$
d. f. N = 64 d. f. D = 41

$F = \frac{s_1^2}{s_2^2} = \frac{3.2^2}{2.1^2} = 2.32$

7. continued

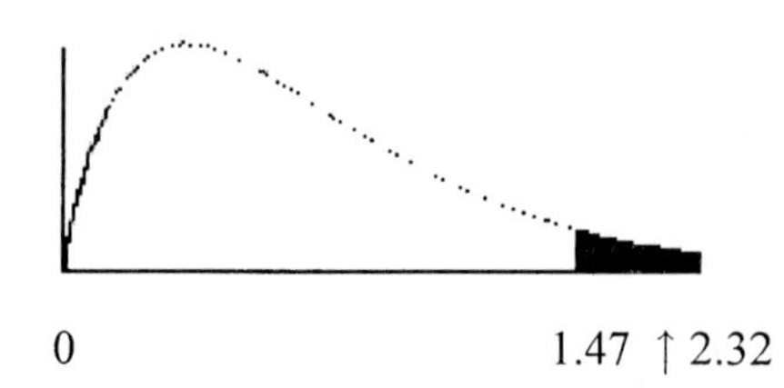

0 1.47 ↑ 2.32

Reject the null hypothesis. There is enough evidence to support the claim that the variation in the number of days factory workers miss per year due to illness is greater than the variation in the number of days hospital workers miss per year.

8.
H_0: $\sigma_1^2 = \sigma_2^2$ (claim)
H_1: $\sigma_1^2 \neq \sigma_2^2$
C. V. = 3.47 $F = \frac{(0.05)^2}{(0.03)^2} = 2.78$
Do not reject. The variances are equal.

H_0: $\mu_1 = \mu_2$
H_1: $\mu_1 \neq \mu_2$ (claim)

C. V. = ± 2.58 d. f. = 37

$$t = \frac{(0.73-0.91)-0}{\sqrt{\frac{14(0.05)^2 + 23(0.03)^2}{15+24-2}}\sqrt{\frac{1}{15}+\frac{1}{24}}} = -14.09$$

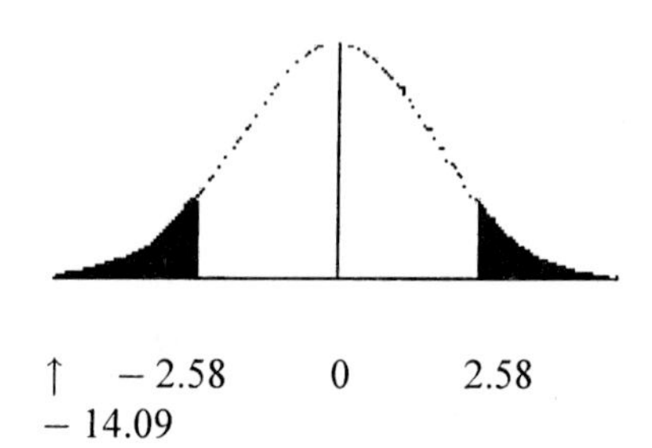

↑ − 2.58 0 2.58
− 14.09

Reject the null hypothesis. There is enough evidence to support the claim that there is a difference in the prices of soups.

9.
H_0: $\sigma_1^2 = \sigma_2^2$
H_1: $\sigma_1^2 \neq \sigma_2^2$

$\overline{X}_1 = 72.9$ $\overline{X}_2 = 70.8$
$s_1 = 5.5$ $s_2 = 5.8$
CV = 1.98 $\alpha = 0.01$
dfN = 24 dfD = 24
$F = \frac{5.8^2}{5.5^2} = 1.11$

9. continued
Do not reject H_0. The variances are equal.

H_0: $\mu_1 \leq \mu_2$
H_1: $\mu_1 > \mu_2$ (claim)

C. V. = 1.28 d. f. = 48 $\alpha = 0.10$

$$t = \frac{(\overline{X}_1-\overline{X}_2)-(\mu_1-\mu_2)}{\sqrt{\frac{(n_1-1)s_1^2+(n_2-1)s_2^2}{n_1+n_2-2}}\sqrt{\frac{1}{n_1}+\frac{1}{n_2}}}$$

$$t = \frac{(72.9-70.8)-0}{\sqrt{\frac{24(5.5)^2+24(5.8)^2}{25+25-2}}\sqrt{\frac{1}{25}+\frac{1}{25}}} = 1.31$$

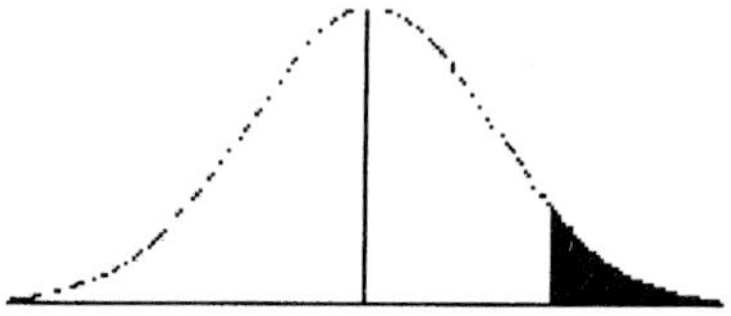

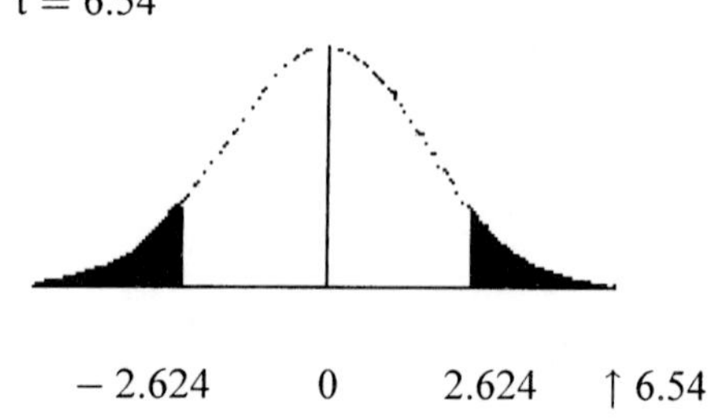

0 1.28 ↑ 1.31

Reject the null hypothesis. There is enough evidence to support the claim that it is warmer in Birmingham.

10.
H_0: $\sigma_1^2 = \sigma_2^2$ (claim)
H_1: $\sigma_1^2 \neq \sigma_2^2$
C. V. = 2.87 $F = \frac{3256^2}{1432^2} = 5.17$
Reject. The variances are unequal.

H_0: $\mu_1 = \mu_2$
H_1: $\mu_1 \neq \mu_2$ (claim)

C. V. = ± 2.624 d. f. = 14

$$t = \frac{(\overline{X}_1-\overline{X}_2)-(\mu_1-\mu_2)}{\sqrt{\frac{s_1^2}{n_1}+\frac{s_2^2}{n_2}}} = \frac{(35{,}270-29{,}512)-0}{\sqrt{\frac{3256^2}{15}+\frac{1432^2}{15}}}$$

t = 6.54

− 2.624 0 2.624 ↑ 6.54

Reject the null hypothesis. There is enough evidence to support the claim that there is a difference in the teachers' salaries.

Confidence Interval:
$\$3{,}447.80 < \mu_1 - \mu_2 < \$8{,}068.20$

11.
H_0: $\sigma_1^2 = \sigma_2^2$
H_1: $\sigma_1^2 \neq \sigma_2^2$
$\alpha = 0.05$
dfN = 15   dfD = 11
$F = \frac{8256^2}{1311^2} = 39.66$
Reject H_0 since P-value < 0.05. The variances are unequal.

H_0: $\mu_1 \leq \mu_2$
H_1: $\mu_1 > \mu_2$ (claim)

$\alpha = 0.05$ P-value < 0.005

$$t = \frac{(54{,}356 - 46{,}512) - 0}{\sqrt{\frac{8256^2}{16} + \frac{1311^2}{12}}} = 3.74$$

Reject the null hypothesis since P-value < 0.05. There is enough evidence to support the claim that incomes of the city residents are greater than the incomes of the suburban residents.

12.

Pre-Test	Post-Test	D	D^2
83	88	-5	25
76	82	-6	36
92	100	-8	64
64	72	-8	64
82	81	1	1
68	75	-7	49
70	79	-9	81
71	68	3	9
72	81	-9	81
63	70	-7	49
		$\sum D = -55$	$\sum D^2 = 459$

H_0: $\mu_D \geq 0$
H_1: $\mu_D < 0$ (claim)

C. V. $= -2.821$ d. f. = 9

$\overline{D} = \frac{\sum D}{n} = -5.5$

$$s_D = \sqrt{\frac{\sum D^2 - \frac{(\sum D)^2}{n}}{n-1}} = \sqrt{\frac{459 - \frac{(-55)^2}{10}}{9}}$$

$s_D = 4.17$

$$t = \frac{\overline{D} - \mu_D}{\frac{s_D}{\sqrt{n}}} = \frac{-5.5 - 0}{\frac{4.17}{\sqrt{10}}} = -4.17$$

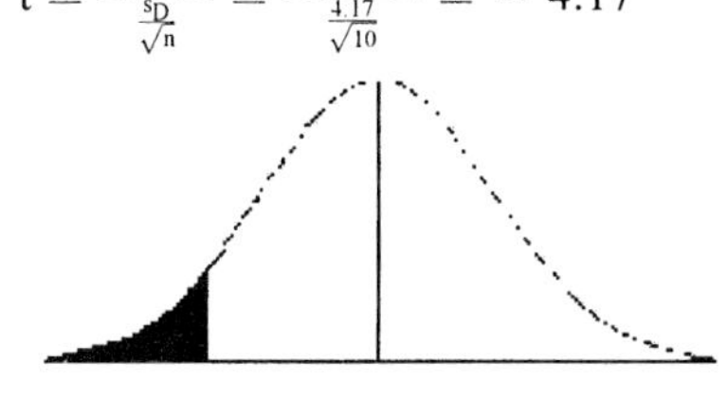

↑ -2.821 0
-4.17

12. continued
Reject the null hypothesis. There is enough evidence to support the claim that the tutoring sessions helped to improve the students' vocabulary.

13.

Before	After	D	D^2
6	10	-4	16
8	12	-4	16
10	9	1	1
9	12	-3	9
5	8	-3	9
12	13	-1	1
9	8	1	1
7	10	-3	9
		$\sum D = -16$	$\sum D^2 = 62$

H_0: $\mu_D \geq 0$
H_1: $\mu_D < 0$ (claim)

C. V. $= -1.895$ d. f. = 7 $\alpha = 0.05$

$\overline{D} = \frac{\sum D}{n} = \frac{-16}{8} = -2$

$$s_D = \sqrt{\frac{62 - \frac{(-16)^2}{8}}{7}} = 2.07$$

$$t = \frac{-2 - 0}{\frac{2.07}{\sqrt{8}}} = -2.73$$

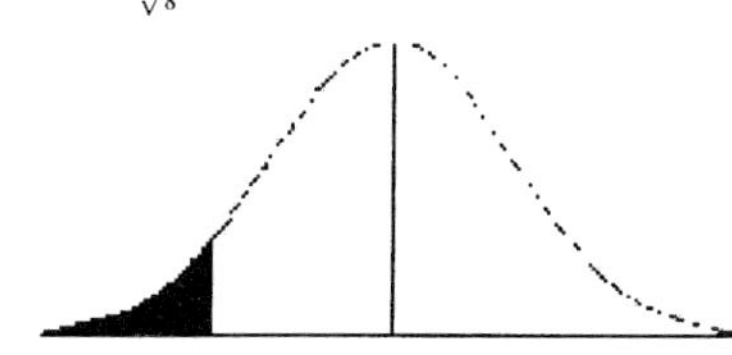

↑ -1.895 0
-2.73

Reject the null hypothesis. There is enough evidence to support the claim that the music has increased production.

14.
$\hat{p}_1 = \frac{207}{365} = 0.567$ $\hat{p}_2 = \frac{166}{365} = 0.455$

$\overline{p} = \frac{207 + 166}{365 + 365} = 0.511$

$\overline{q} = 1 - 0.51 = 0.489$

H_0: $p_1 = p_2$
H_1: $p_1 \neq p_2$ (claim)

C. V. $= \pm 2.33$

$$z = \frac{(0.567 - 0.455) - 0}{\sqrt{(0.511)(0.489)(\frac{1}{365} + \frac{1}{365})}} = 3.03$$

14. continued

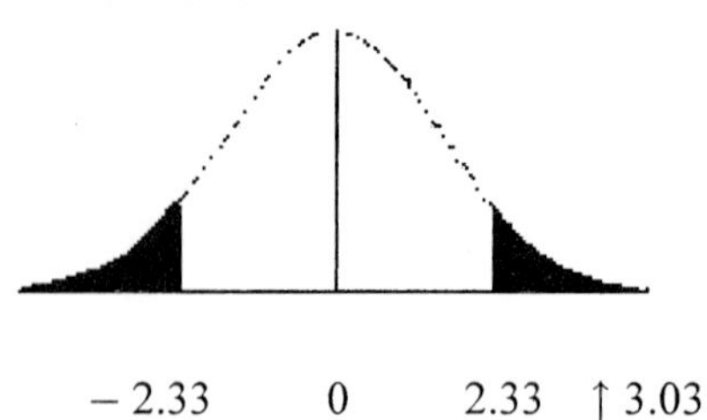

Reject the null hypothesis. There is enough evidence to reject the claim that the proportions are equal.

For the 98% confidence interval:

$$(\hat{p}_1 - \hat{p}_2) - z_{\frac{\alpha}{2}}\sqrt{\frac{\hat{p}_1\hat{q}_1}{n_1} + \frac{\hat{p}_2\hat{q}_2}{n_2}} < p_1 - p_2 < (\hat{p}_1 - \hat{p}_2) + z_{\frac{\alpha}{2}}\sqrt{\frac{\hat{p}_1\hat{q}_1}{n_1} + \frac{\hat{p}_2\hat{q}_2}{n_2}}$$

$\hat{p}_1 = 0.567 \qquad \hat{q}_1 = 0.433$

$\hat{p}_2 = 0.455 \qquad \hat{q}_2 = 0.545$

$$(0.567 - 0.455) - 2.33\sqrt{\frac{(0.567)(0.433)}{365} + \frac{(0.455)(0.545)}{365}}$$

$$< p_1 - p_2 < (0.567 - 0.455) + 2.33\sqrt{\frac{(0.567)(0.433)}{365} + \frac{(0.455)(0.545)}{365}}$$

$0.112 - 0.086 < p_1 - p_2 < 0.112 + 0.086$

$0.026 < p_1 - p_2 < 0.198$

15.
$\hat{p}_1 = \frac{32}{50} = 0.64 \qquad \hat{p}_2 = \frac{24}{60} = 0.40$

$\bar{p} = \frac{32 + 24}{50 + 60} = 0.509$

$\bar{q} = 1 - 0.509 = 0.491$

H_0: $p_1 = p_2$ (claim)
H_1: $p_1 \neq p_2$

C. V. = $\pm 1.96 \qquad \alpha = 0.05$

$$z = \frac{(0.64 - 0.40) - 0}{\sqrt{(0.509)(0.491)(\frac{1}{50} + \frac{1}{60})}} = 2.51$$

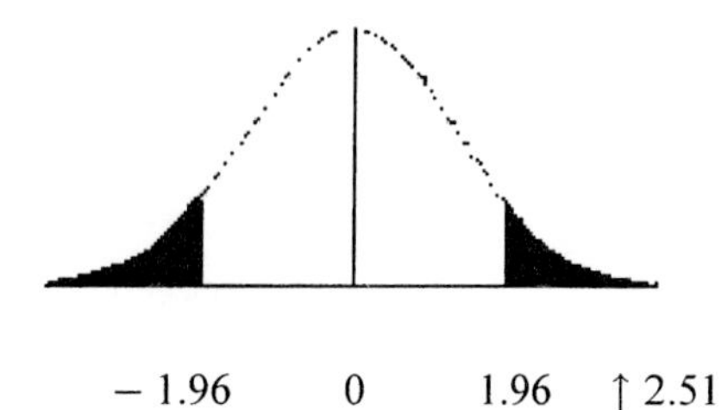

15. continued
Reject the null hypothesis. There is enough evidence to reject the claim that the proportions are equal.

For the 95% confidence interval:
$\hat{p}_1 = 0.64 \qquad \hat{q}_1 = 0.36$
$\hat{p}_2 = 0.4 \qquad \hat{q}_2 = 0.6$

$$(0.64 - 0.4) - 1.96\sqrt{\frac{(0.64)(0.36)}{50} + \frac{(0.4)(0.6)}{60}}$$

$$< p_1 - p_2 < (0.64 - 0.4) + 1.96\sqrt{\frac{(0.64)(0.36)}{50} + \frac{(0.4)(0.6)}{60}}$$

$0.24 - 1.96(0.0928) < p_1 - p_2 <$

$0.24 + 1.96(0.0928)$

$0.058 < p_1 - p_2 < 0.422$

CHAPTER 9 QUIZ

1. False, there are different formulas for independent and dependent samples.
2. False, the samples are independent.
3. True
4. False, they can be right, left, or two tailed.
5. d.
6. a.
7. c.
8. b.
9. $\mu_1 = \mu_2$
10. pooled
11. normal
12. negative
13. $\frac{s_1^2}{s_2^2}$

14. H_0: $\mu_1 = \mu_2$
H_1: $\mu_1 \neq \mu_2$ (claim)
C. V. = $\pm 2.58 \quad z = -3.69$
Reject the null hypothesis. There is enough evidence to support the claim that there is a difference in the cholesterol levels of the two groups.
99% Confidence Interval:
$-10.2 < \mu_1 - \mu_2 < -1.8$

15. H_0: $\mu_1 \leq \mu_2$
H_1: $\mu_1 > \mu_2$ (claim)
C. V. = $1.28 \qquad z = 1.60$
Reject the null hypothesis. There is enough evidence to support the claim that average rental fees for the east apartments is greater

15. continued
than the average rental fees for the west apartments.

16. H_0: $\sigma_1^2 = \sigma_2^2$
H_1: $\sigma_1^2 \neq \sigma_1^2$ (claim)
F = 1.637 P-value > 0.20 (0.357)
Do not reject the null hypothesis since P-value > 0.05. There is not enough evidence to support the claim that the variances are different.

17. H_0: $\sigma_1^2 = \sigma_2^2$
H_1: $\sigma_1^2 \neq \sigma_2^2$ (claim)
C. V. = 1.90 F = 1.296
Reject the null hypothesis. There is enough evidence to support the claim that the variances are different.

18. H_0: $\sigma_1^2 = \sigma_2^2$ (claim)
H_1 $\sigma_1^2 \neq \sigma_2^2$
C. V. = 3.53 F = 1.13
Do not reject the null hypothesis. There is not enough evidence to reject the claim that the standard deviations or the number of hours of television viewing are the same.

19. H_0: $\sigma_1^2 = \sigma_2^2$
H_1 $\sigma_1^2 \neq \sigma_2^2$ (claim)
C. V. = 3.01 F = 1.94
Do not reject the null hypothesis. There is enough evidence to support the claim that the variances are different.

20. H_0: $\sigma_1^2 \leq \sigma_2^2$
H_1 $\sigma_1^2 > \sigma_2^2$ (claim)
C. V. = 1.44 F = 1.474
Reject the null hypothesis. There is enough evidence to support the claim that the variance of days missed by teachers is greater than the variance of days missed by nurses.

21. H_0: $\sigma_1 = \sigma_2$
H_1 $\sigma_1 \neq \sigma_2$ (claim)
C. V. = 2.46 F = 1.65
Do not reject the null hypothesis. There is not enough evidence to support the claim that the standard deviations are different.

22. H_0: $\sigma_1^2 = \sigma_2^2$
H_1 $\sigma_1^2 \neq \sigma_2^2$
C. V. = 5.05 F = 1.23
Do not reject. The variances are equal.

22. continued
H_0: $\mu_1 = \mu_2$
H_1: $\mu_1 \neq \mu_2$ (claim)
C. V. = ± 2.779 t = 10.92

Reject the null hypothesis. There is enough evidence to support the claim that the average prices are different.

99% Confidence Interval:
$0.298 < \mu_1 - \mu_2 < 0.502$

23. H_0: $\sigma_1^2 = \sigma_2^2$
H_1 $\sigma_1^2 \neq \sigma_2^2$
C. V. = 9.6 F = 12.41
Reject. The variances are not equal.

H_0: $\mu_1 \geq \mu_2$
H_1: $\mu_1 < \mu_2$ (claim)
C. V. = -2.131 t = -2.07
Do not reject the null hypothesis. There is not enough evidence to support the claim that accidents have increased.

24. H_0: $\sigma_1^2 = \sigma_2^2$
H_1 $\sigma_1^2 \neq \sigma_2^2$
C. V. = 4.02 F = 6.155
Reject. The variances are unequal.

H_0: $\mu_1 = \mu_2$
H_1: $\mu_1 \neq \mu_2$ (claim)
C. V. = ± 2.718 t = 9.807
Reject the null hypothesis. There is enough evidence to support the claim that the salaries are different.

98% Confidence Interval:
$\$6652 < \mu_1 - \mu_2 < \$11,757$

25. H_0: $\sigma_1^2 = \sigma_2^2$
H_1 $\sigma_1^2 \neq \sigma_2^2$
F = 23.08 P-value < 0.05
Reject. The variances are unequal.

H_0: $\mu_1 \leq \mu_2$
H_1: $\mu_1 > \mu_2$ (claim)
t = 0.874 0.10 < P-value < 0.25 (0.198)
Do not reject the null hypothesis since P-value > 0.05. There is not enough evidence to support the claim that incomes of city residents is greater than incomes of rural residents.

26. H_0: $\mu_1 \geq \mu_2$
H_1: $\mu_1 < \mu_2$ (claim)
$\overline{D} = -6.5$ $s_D = 4.93$
C. V. = -2.821 $t = -4.17$
Reject the null hypothesis. There is enough evidence to support the claim that the sessions improved math skills.

27. H_0: $\mu_1 \geq \mu_2$
H_1: $\mu_1 < \mu_2$ (claim)
$\overline{D} = -0.8$ $s_D = 1.48$
C. V. = -1.833 $t = -1.71$
Do not reject the null hypothesis. There is not enough evidence to support the claim that egg production increased.

28. H_0: $p_1 = p_2$
H_1: $p_1 \neq p_2$ (claim)
C. V. = ± 1.65 $z = -0.69$
Do not reject the null hypothesis. There is not enough evidence to support the claim that the proportions are different.

90% Confidence Interval:
$-0.101 < p_1 - p_2 < 0.041$

29. H_0: $p_1 = p_2$ (claim)
H_1: $p_1 \neq p_2$
C. V. = ± 1.96 $z = 2.58$
Reject the null hypothesis. There is enough evidence to reject the claim that the proportions are equal.

95% Confidence Interval:
$0.067 < p_1 - p_2 < 0.445$

Note: Graphs are not to scale and are intended to convey a general idea.

Answers may vary due to rounding.

EXERCISE SET 10-3

1.
Two variables are related when there exists a discernible pattern between them.

2.
Relationships are measured by the correlation coefficient, r. When r is near + 1, there is a strong positive linear relationship between the variables. When r is near − 1, there is a strong negative linear relationship. When r is near zero, there is no linear relationship between the variables.

3.
r, ρ (rho)

4.
The range of r is from − 1 to + 1.

5.
A positive relationship means that as x increases, y also increases.
A negative relationship means that as x increases, y decreases.

6.
Answers will vary.

7.
Answers will vary.

8.
The diagram is called a scatter plot. It shows the nature of the relationship.

9.
Pearson's Product Moment Correlation Coefficient.

10.
t test

11.
There are many other possibilities, such as chance, relationship to a third variable, etc.

12.

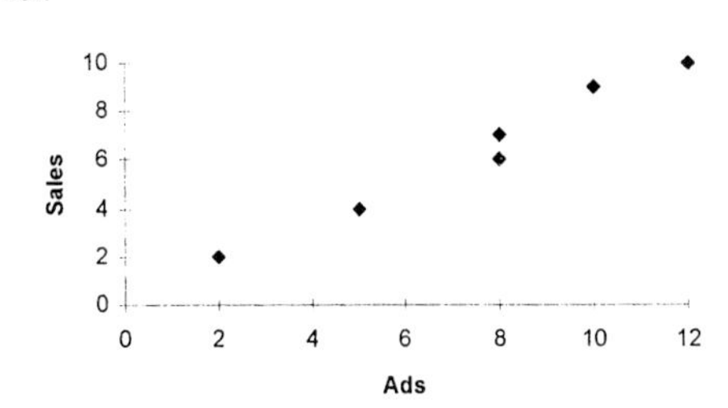

$\sum x = 45$ $\sum y^2 = 286$
$\sum y = 38$ $\sum xy = 338$
$\sum x^2 = 401$ $n = 6$

$$r = \frac{6(338)-(45)(38)}{\sqrt{[6(401)-45^2][6(286)-38^2]}} = 0.988$$

H_0: $\rho = 0$ and H_1: $\rho \neq 0$; d. f. = 4
C. V. = ± 0.811; Decision: Reject. There is a significant relationship between the number of ads and the amount of sales.

13.

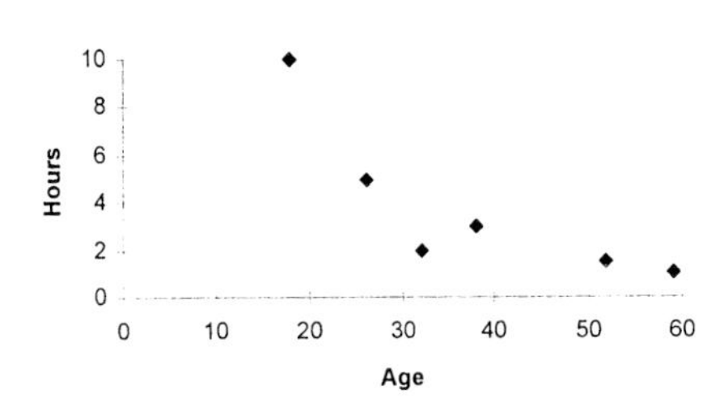

$\sum x = 225$
$\sum y = 22.5$
$\sum x^2 = 9653$
$\sum y^2 = 141.25$
$\sum xy = 625$
$n = 6$

$$r = \frac{n(\sum xy)-(\sum x)(\sum y)}{\sqrt{[n(\sum x^2)-(\sum x)^2]\,[n(\sum y^2)-(\sum y)^2]}}$$

$$r = \frac{6(625)-(225)(22.5)}{\sqrt{[6(9653)-(225)^2]\,[6(141.25)-(22.5)^2]}}$$

$r = -0.832$

H_0: $\rho = 0$ and H_1: $\rho \neq 0$; C. V. = ± 0.811; d. f. = 4; Decision: reject; there is a significant relationship between a person's age and the number of hours a person watches television.

14.

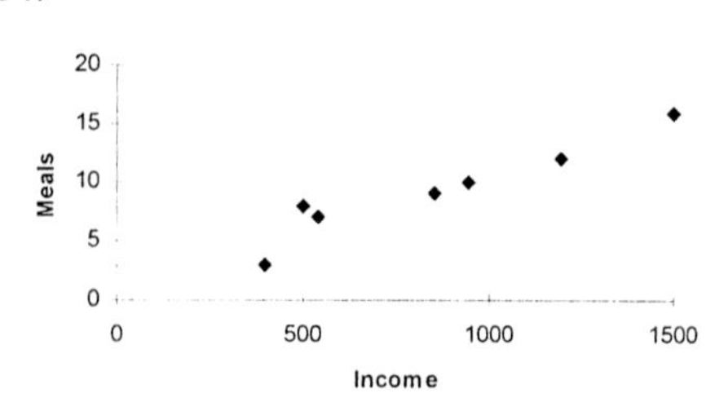

$\sum x = 5935$
$\sum y = 65$
$\sum x^2 = 6,007,125$
$\sum y^2 = 703$
$\sum xy = 64,480$
$n = 7$

$$r = \frac{7(64,480)-(5935)(65)}{\sqrt{[7(6,007,125)-(5935)^2][7(703)-(65)^2]}}$$

$r = 0.952$

H_0: $\rho = 0$
H_1: $\rho \neq 0$
C. V. = ± 0.754 d. f. = 5

Decision: Reject. There is a significant relationship between a person's monthly income and the number of meals that a person eats away from home each month.

15.

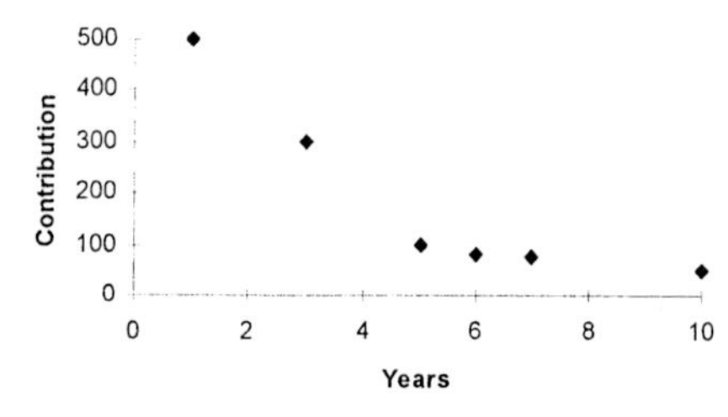

$\sum x = 32$
$\sum y = 1105$
$\sum x^2 = 220$
$\sum y^2 = 364,525$
$\sum xy = 3405$
$n = 6$

$$r = \frac{n(\sum xy)-(\sum x)(\sum y)}{\sqrt{[n(\sum x^2)-(\sum x)^2]\,[n(\sum y^2)-(\sum y)^2]}}$$

$$r = \frac{6(3405)-(32)(1105)}{\sqrt{[6(220)-(32)^2][6(364525)-(1105)^2]}}$$

$r = -0.883$

H_0: $\rho = 0$
H_1: $\rho \neq 0$
C. V. = ± 0.811 d. f. = 4

15. continued
Decision: Reject. There is a significant relationship between a person's age and his or her contribution.

16.

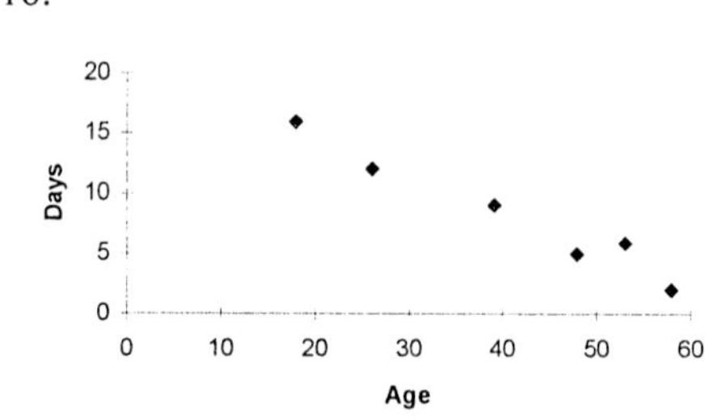

$\sum x = 242$
$\sum y = 50$
$\sum x^2 = 10,998$
$\sum y^2 = 546$
$\sum xy = 1625$
$n = 6$

$$r = \frac{6(1625)-(242)(50)}{\sqrt{[6(10,998)-(242)^2][6(546)-(50)^2]}}$$

$r = -0.979$

H_0: $\rho = 0$
H_1: $\rho \neq 0$
C. V. = ± 0.811 d. f. = 4

Decision: Reject. There is a significant relationship between a person's age and the number of sick days that a person takes each year.

17.

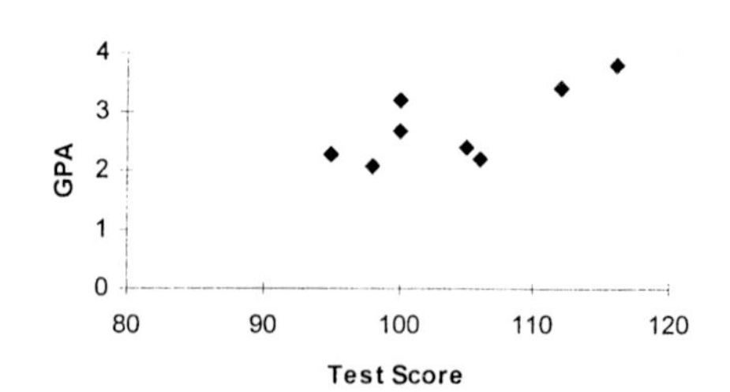

$\sum x = 832$
$\sum y = 22.1$
$\sum x^2 = 86,890$
$\sum y^2 = 63.83$
$\sum xy = 2321.1$
$n = 8$

$$r = \frac{n(\sum xy)-(\sum x)(\sum y)}{\sqrt{[n(\sum x^2)-(\sum x)^2]\,[n(\sum y^2)-(\sum y)^2]}}$$

$$r = \frac{8(2321.1)-(832)(22.1)}{\sqrt{[8(86,890)-(832)^2][8(63.83)-(22.1)^2]}}$$

17. continued
r = 0.716

H_0: $\rho = 0$
H_1: $\rho \neq 0$
C. V. = ± 0.707 d. f. = 6

Decision: Reject. There is a significant relationship between test scores and G.P. A.

18.

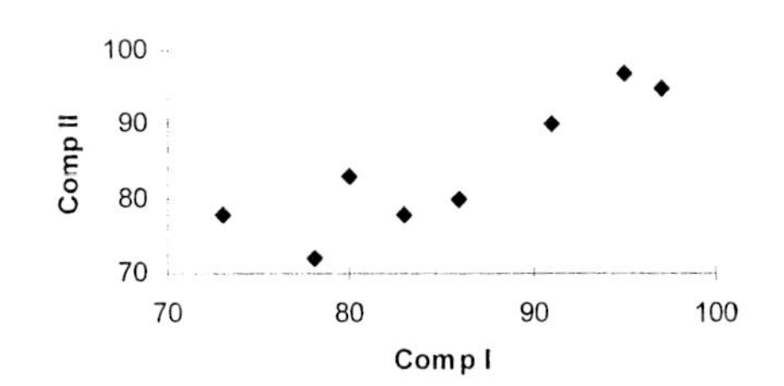

$\sum x = 683$
$\sum y = 673$
$\sum x^2 = 58{,}813$
$\sum y^2 = 57{,}175$
$\sum xy = 57{,}924$
n = 8

$$r = \frac{8(57924) - (683)(673)}{\sqrt{[8(58813) - (683)^2][8(57175) - (673)^2]}}$$

r = 0.881

H_0: $\rho = 0$
H_1: $\rho \neq 0$
C. V. = ± 0.707 d. f. = 6

Decision: Reject. There is a significant relationship between the final exam grades of students in Composition I and Composition II classes.

19.

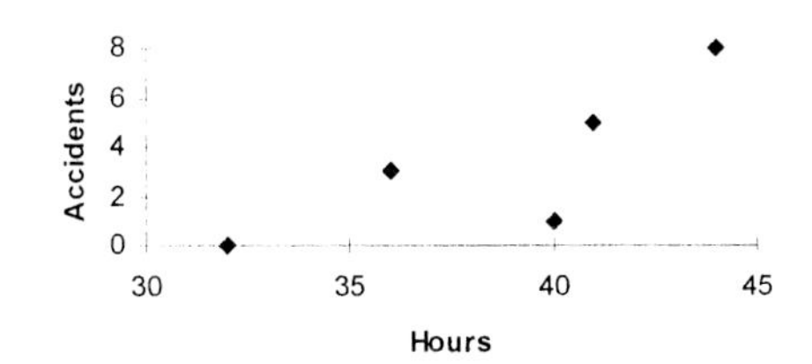

$\sum x = 193$
$\sum y = 17$
$\sum x^2 = 7537$
$\sum y^2 = 99$
$\sum xy = 705$
n = 5

$$r = \frac{n(\sum xy) - (\sum x)(\sum y)}{\sqrt{[n(\sum x^2) - (\sum x)^2]\,[n(\sum y^2) - (\sum y)^2]}}$$

19. continued
r = 0.814

H_0: $\rho = 0$
H_1: $\rho \neq 0$
C. V. = ± 0.878 d. f. = 3

Decision: Do not reject. There is not a significant relationship between the variables.

20.

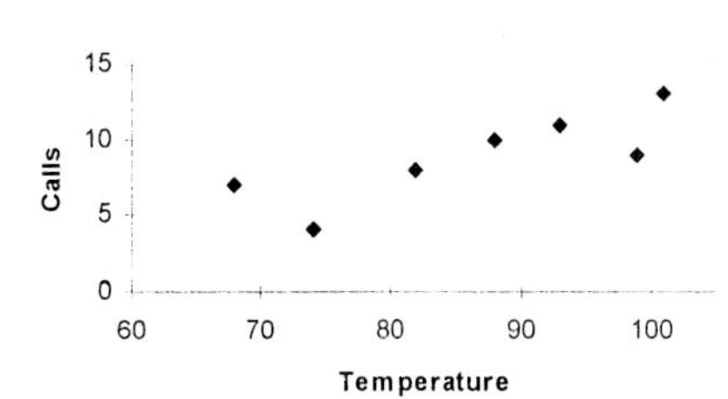

$\sum x = 605$
$\sum y = 62$
$\sum x^2 = 53219$
$\sum y^2 = 600$
$\sum xy = 5535$
n = 7

$$r = \frac{7(5535) - (605)(62)}{\sqrt{[7(53219) - (605)^2][7(600) - (62)^2]}}$$

r = 0.811

H_0: $\rho = 0$
H_1: $\rho \neq 0$
C. V. = ± 0.754 d. f. = 5

Decision: Reject. There is a significant relationship between the temperature and the number of emergency calls received.

21.

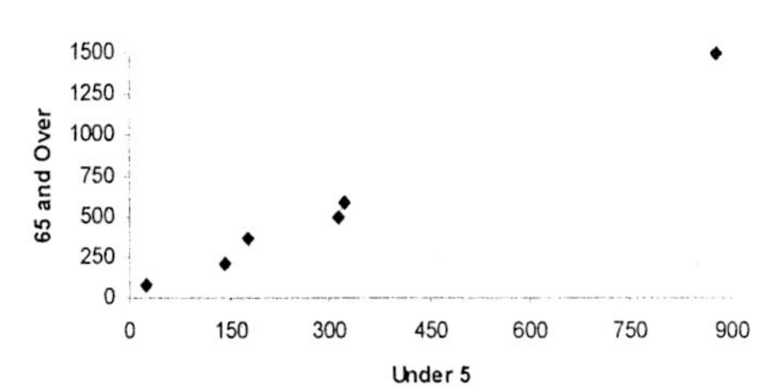

$\sum x = 1862$
$\sum y = 3222$
$\sum x^2 = 1{,}026{,}026$
$\sum y^2 = 3{,}009{,}596$
$\sum xy = 1{,}754{,}975$

21. continued

$n = 6$

$$r = \frac{n(\sum xy) - (\sum x)(\sum y)}{\sqrt{[n(\sum x^2) - (\sum x)^2][n(\sum y^2) - (\sum y)^2]}}$$

$$r = \frac{6(1,754,975) - (1862)(3222)}{\sqrt{[6(1,026,026) - 1862^2][6(3,009,596) - 3222^2]}}$$

$r = 0.997$

H_0: $\rho = 0$
H_1: $\rho \neq 0$
C. V. $= \pm 0.811$ d. f. $= 4$

Decision: Reject. There is a significant linear relationship between the under 5 age group and the 65 and over age group.

22.

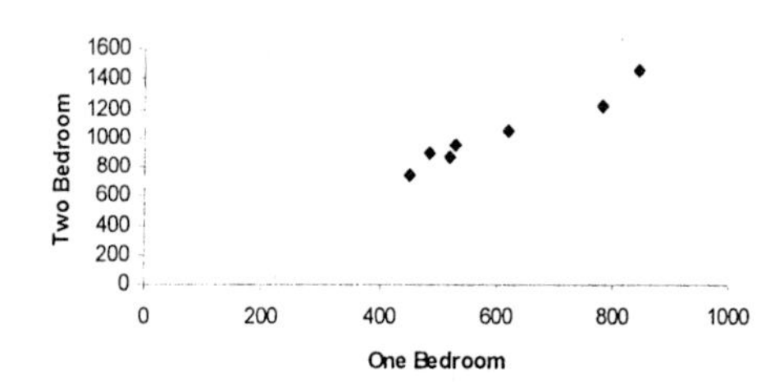

$\sum x = 4231$
$\sum y = 7203$
$\sum x^2 = 2,697,311$
$\sum y^2 = 7,761,245$
$\sum xy = 4,569,178$
$n = 7$

$$r = \frac{7(4,569,178) - (4231)(7203)}{\sqrt{[7(2,697,311) - (4231)^2][7(7,761,245) - (7203)^2]}}$$

$r = 0.974$

H_0: $\rho = 0$
H_1: $\rho \neq 0$
C. V. $= \pm 0.754$ d. f. $= 5$

Decision: Reject. There is a significant linear relationship between monthly rents for one bedroom and two bedroom apartments.

23.

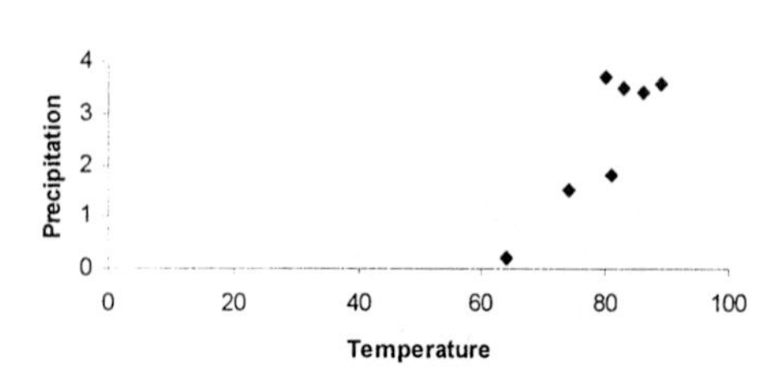

23. continued

$\sum x = 557$
$\sum y = 17.7$
$\sum x^2 = 44,739$
$\sum y^2 = 55.99$
$\sum xy = 1468.9$
$n = 7$

$$r = \frac{n(\sum xy) - (\sum x)(\sum y)}{\sqrt{[n(\sum x^2) - (\sum x)^2][n(\sum y^2) - (\sum y)^2]}}$$

$$r = \frac{7(1468.9) - (557)(17.7)}{\sqrt{[7(44,739) - 557^2][7(55.99) - 17.7^2]}}$$

$r = 0.883$

H_0: $\rho = 0$
H_1: $\rho \neq 0$
C. V. $= \pm 0.754$ d. f. $= 5$

Decision: Reject. There is a significant linear relationship between temperature and precipitation.

24.

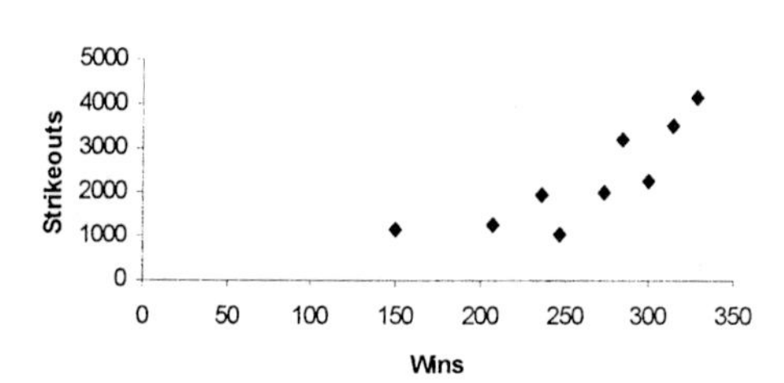

$\sum x = 2664$
$\sum y = 24,145$
$\sum x^2 = 739,052$
$\sum y^2 = 69,572,231$
$\sum xy = 6,920,176$
$n = 10$

$$r = \frac{10(6,920,176) - (2664)(24,145)}{\sqrt{[10(739,052) - 2664^2][10(69,572,231) - 24,145^2]}}$$

$r = 0.848$

H_0: $\rho = 0$
H_1: $\rho \neq 0$
C. V. $= \pm 0.632$ d. f. $= 8$

Decision: Reject. There is a significant linear relationship between wins and strikeouts.

25.

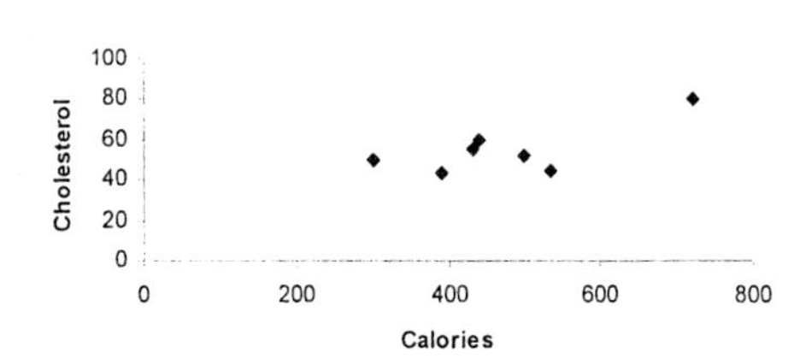

$\sum x = 3315$
$\sum y = 385$
$\sum x^2 = 1,675,225$
$\sum y^2 = 22,103$
$\sum xy = 189,495$
$n = 7$

$$r = \frac{n(\sum xy) - (\sum x)(\sum y)}{\sqrt{[n(\sum x^2) - (\sum x)^2][n(\sum y^2) - (\sum y)^2]}}$$

$$r = \frac{7(189,495) - (3315)(385)}{\sqrt{[7(1,675,225) - (3315)^2][7(22,103) - (385)^2]}}$$

$r = 0.725$

H_0: $\rho = 0$
H_1: $\rho \neq 0$
C. V. = ± 0.754 d. f. = 5

Decision: Do not reject. There is a not a significant linear relationship between calories and cholesterol.

26.

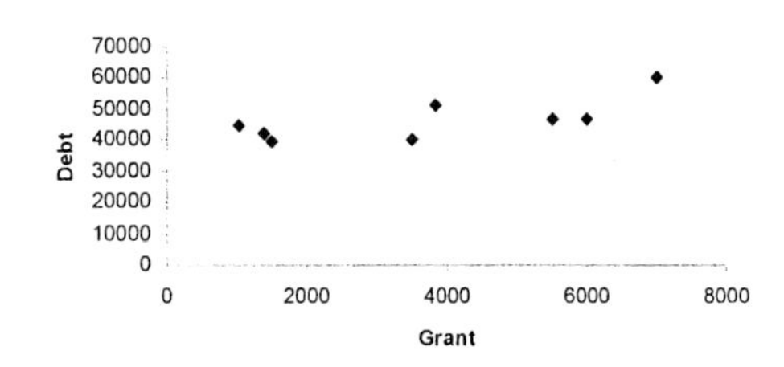

$\sum x = 29,771$
$\sum y = 371,280$
$\sum x^2 = 147,462,993$
$\sum y^2 = 1.75444 \times 10^{10}$
$\sum xy = 1,457,929,227$
$n = 8$

$$r = \frac{8(1457929227) - (29771)(371280)}{\sqrt{[8(147462993) - (29771)^2][8(1.75 \times 10^{10}) - (371280)^2]}}$$

$r = 0.711$

H_0: $\rho = 0$
H_1: $\rho \neq 0$
C. V. = ± 0.707 d. f. = 6

26. continued
Decision: Reject. There is a significant linear relationship between the median grand and average indebtedness at graduation.

27.

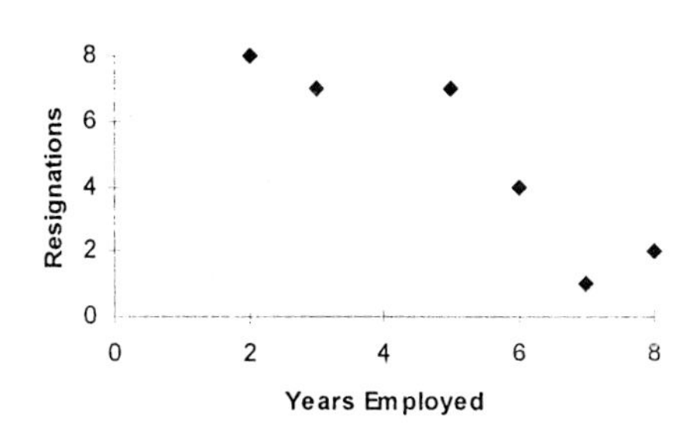

$\sum x = 31$
$\sum y = 29$
$\sum x^2 = 187$
$\sum y^2 = 183$
$\sum xy = 119$
$n = 6$

$$r = \frac{n(\sum xy) - (\sum x)(\sum y)}{\sqrt{[n(\sum x^2) - (\sum x)^2][n(\sum y^2) - (\sum y)^2]}}$$

$r = -0.909$

H_0: $\rho = 0$
H_1: $\rho \neq 0$
C. V. = ± 0.811 d. f. = 4

Decision: Reject.
There is a significant relationship between the years of service and the number of resignations.

28.
$\bar{x} = 5.17$
$\bar{y} = 4.83$
$(x - \bar{x})(y - \bar{y}) = -30.8334$
$s_x = 2.32$
$s_y = 2.93$

$$r = \frac{(-30.8334)}{(6-1)(2.32)(2.93)} = -0.907$$

(differs from -0.909 due to rounding)

29.

$$r = \frac{n(\sum xy) - (\sum x)(\sum y)}{\sqrt{[n(\sum x^2) - (\sum x)^2][n(\sum y^2) - (\sum y)^2]}}$$

$$r = \frac{5(125) - (15)(35)}{\sqrt{[5(55) - (15)^2][5(285) - (35)^2]}} = 1$$

$$r = \frac{5(125) - (35)(15)}{\sqrt{[5(285) - (35)^2][5(55) - (15)^2]}} = 1$$

29. continued
The value of r does not change when the values for x and y are interchanged.

30.

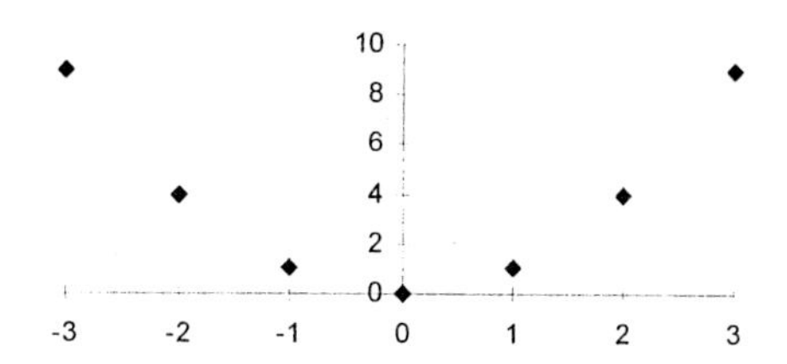

$\sum x = 0$
$\sum y = 28$
$\sum x^2 = 28$
$\sum y^2 = 196$
$\sum xy = 0$
$n = 7$

$$r = \frac{7(0)-(0)(28)}{\sqrt{[7(28)-0^2][7(196)-28^2]}} = 0$$

The relationship is non-linear, as shown in the scatter plot.

EXERCISE 10-4

1.
Draw the scatter plot and test the significance of the correlation coefficient.

2. The assumptions are:

1. For any specific value of the independent variable x, the value of the dependent variable y must be normally distributed about the regression line.

2. The standard deviation of each of the dependent variables must be the same for each value of the independent variable.

3.
$y' = a + bx$

4.
b, a

5.
It is the line that is drawn through the points on the scatter plot such that the sum of the squares of the vertical distances each point is from the line is at a minimum.

6.
r would equal $+1$ or -1

7.
When r is positive, b will be positive. When r is negative, b will be negative.

8.
They would be clustered closer to the line.

9.
The closer r is to $+1$ or -1, the more accurate the predicted value will be.

10.
If the value of r is not significant, no regression should be done. Any regression line is meaningless.

11.
When r is not significant, the mean of the y' values should be used to predict y.

12.
$$a = \frac{(38)(401)-(45)(338)}{6(401)-(45)^2} = 0.073$$

$$b = \frac{6(338)-(45)(38)}{6(401)-(45)^2} = 0.835$$

$y' = 0.073 + 0.835x$

$y' = 0.073 + 0.835(7) = 5.918$ or \$5,918

13.
$$a = \frac{(\sum y)(\sum x^2)-(\sum x)(\sum xy)}{n(\sum x^2)-(\sum x)^2}$$

$$a = \frac{(22.5)(9653) - (225)(625)}{6(9653) - (225)^2} = 10.499$$

$$b = \frac{n(\sum xy)-(\sum x)(\sum y)}{n(\sum x^2)-(\sum x)^2}$$

$$b = \frac{6(625) - (225)(22.5)}{6(9653) - (225)^2} = -0.18$$

$y' = a + bx$
$y' = 10.499 - 0.18x$
$y' = 10.499 - 0.18(35) = 4.199$ hours

14.
$$a = \frac{(65)(6007125)-(5935)(64480)}{7(6007125)-(5935)^2}$$

$$a = \frac{7774325}{6825650} = 1.139$$

$$b = \frac{7(64480)-(5935)(65)}{6825650} = \frac{65585}{6825650}$$

14. continued
$b = 0.00961$

$y' = 1.139 + 0.00961x$
$y' = 1.139 + 0.00961(1100) = 11.71 \approx 12$

15.
$a = \frac{(\sum y)(\sum x^2)-(\sum x)(\sum xy)}{n(\sum x^2)-(\sum x)^2}$

$a = \frac{(1105)(220)-(32)(3405)}{6(220)-(32)^2}$

$a = \frac{243100-108960}{1320-1024} = \frac{134140}{296} = 453.176$

$b = \frac{n(\sum xy)-(\sum x)(\sum y)}{n(\sum x^2)-(\sum x)^2}$

$b = \frac{6(3405)-(32)(1105)}{6(220)-(32)^2} = \frac{20430-35360}{296}$

$b = \frac{-14930}{296} = -50.439$

$y' = a + bx$
$y' = 453.176 - 50.439x$
$y' = 453.176 - 50.439(4) = \$251.42$

16.
$a = \frac{(50)(10998)-(242)(1625)}{6(10998)-(242)^2}$

$a = \frac{549900-393250}{7424} = \frac{156650}{7424} = 21.1$

$b = \frac{6(1625)-(242)(50)}{7424} = \frac{9750-12100}{7424}$

$b = \frac{-2350}{7424} = -0.317$

$y' = 21.1 - 0.317x$
$y' = 21.1 - 0.317(47) = 6.201 \approx 6$ days

17.
$a = \frac{(\sum y)(\sum x^2)-(\sum x)(\sum xy)}{n(\sum x^2)-(\sum x)^2}$

$a = \frac{(22.1)(86890)-(832)(2321.1)}{8(86890)-(832)^2}$

$a = \frac{1920269-1931155.2}{695120-692224} = \frac{10886.2}{2896} = -3.759$

$b = \frac{n(\sum xy)-(\sum x)(\sum y)}{n(\sum x^2)-(\sum x)^2}$

$b = \frac{8(2321.1)-(832)(22.1)}{8(86890)-(832)^2}$

$b = \frac{18568.8-18387.2}{2986} = \frac{181.6}{2896} = 0.063$

$y' = a + bx$
$y' = -3.759 + 0.063x$

17. continued
$y' = -3.759 + 0.063(104) = 2.793$

18.
$a = \frac{673(58813)-683(57924)}{8(58813)-(683)^2} = 4.746$

$b = \frac{8(57924)-(683)(673)}{4015} = 0.930$

$y' = 4.746 + 0.930x$
$y' = 4.746 + 0.930(88) = 86.586 \approx 87\%$

19.
Since r is not significant, the regression line should not be computed.

20.
$a = \frac{(62)(53219)-(605)(5535)}{7(53219)-(605)^2} = -7.544$

$b = \frac{7(5535)-(605)(62)}{7(53219)-(605)^2} = 0.190$

$y' = -7.554 + 0.190x$
$y' = -7.554 + 0.190(80) = 7.656 \approx 8$ calls

21.
$a = \frac{(\sum y)(\sum x^2)-(\sum x)(\sum xy)}{n(\sum x^2)-(\sum x)^2}$

$a = \frac{(3222)(1026026)-(1862)(1754975)}{6(1026026)-(1862)^2}$

$a = 14.165$

$b = \frac{n(\sum xy)-(\sum x)(\sum y)}{n(\sum x^2)-(\sum x)^2}$

$b = \frac{6(1754975)-(1862)(3222)}{6(1026026)-(1862)^2} = 1.685$

$y' = a + bx$
$y' = 14.165 + 1.685x$
$y' = 14.165 + 1.685(200) = 351$ under 5.

22.
$a = \frac{(7203)(2697311)-(4231)(4569178)}{7(2697311)-(4231)^2} = 98.528$

$b = \frac{7(4569178)-(4231)(7203)}{7(2697311)-(4231)^2} = 1.539$

$y' = a + bx$
$y' = 98.528 + 1.539x$
$y' = 98.528 + 1.539(700) = \$1175.83$ rent for a two-bedroom apartment.

23.
$a = \frac{(\sum y)(\sum x^2)-(\sum x)(\sum xy)}{n(\sum x^2)-(\sum x)^2}$

23. continued

$a = \frac{(17.7)(44739) - (557)(1468.9)}{7(44739) - (557)^2} = -8.994$

$b = \frac{n(\sum xy) - (\sum x)(\sum y)}{n(\sum x^2) - (\sum x)^2}$

$b = \frac{7(1468.9) - (557)(17.7)}{7(44739) - (557)^2} = 0.1448$

$y' = a + bx$
$y' = -8.994 + 0.1448x$
$y' = -8.994 + 0.1448(70) = 1.1$ inches

24.

$a = \frac{(24145)(739052) - (2664)(6920176)}{10(739052) - (2664)^2}$
$a = -2012.568$

$b = \frac{10(6920176) - (2664)(24145)}{10(739052) - (2664)^2} = 16.618$

$y' = a + bx$
$y' = -2012.568 + 16.618x$
$y' = -2012.568 + 16.618(260)$
$y' = 2308$ strikeouts

25.
Since r is not significant, no regression should be done.

26.

$a = \frac{(371280)(147462993) - (29771)(1457929227)}{8(147462993) - (29771)^2}$
$a = 38{,}672.044$

$b = \frac{8(1457929227) - (29771)(371280)}{8(147462993) - (29771)^2} = 2.079$

$y' = 38{,}672.044 + 2.079x$
$y' = 38{,}672.044 + 2.079(2500)$
$y' = \$43{,}869.54$ or $\$43{,}870$ average debt

27.

$a = \frac{(\sum y)(\sum x^2) - (\sum x)(\sum xy)}{n(\sum x^2) - (\sum x)^2}$

$a = \frac{(29)(187) - (31)(119)}{6(187) - (31)^2} = 10.770$

$b = \frac{n(\sum xy) - (\sum x)(\sum y)}{n(\sum x^2) - (\sum x)^2}$

$b = \frac{6(119) - (31)(29)}{6(187) - (31)^2} = -1.149$

$y' = a + bx$
$y' = 10.770 - 1.149x$
$y' = 10.770 - 1.149(4) = 6.174$

28.

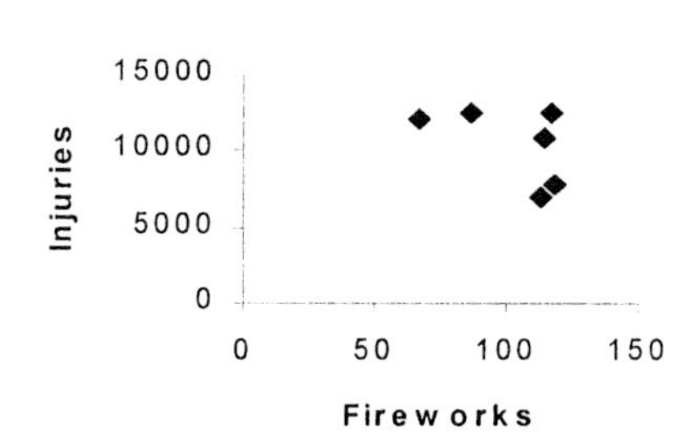

$\sum x = 617.7$
$\sum y = 62900$
$\sum x^2 = 65763.17$
$\sum y^2 = 690{,}070{,}000$
$\sum xy = 6{,}342{,}820$
$n = 6$

$r = \frac{6(6342820) - (617.7)(62900)}{\sqrt{[6(65763.17) - (617.7)^2][6(690070000) - (62900)^2]}}$

$r = -0.514$

H_0: $\rho = 0$
H_1: $\rho \neq 0$

C. V. = ± 0.811 d. f. = 4
Decision: Do not reject.
There is no significant relationship between the number of fireworks in use and the number of related injuries. No regression should be done.

29.

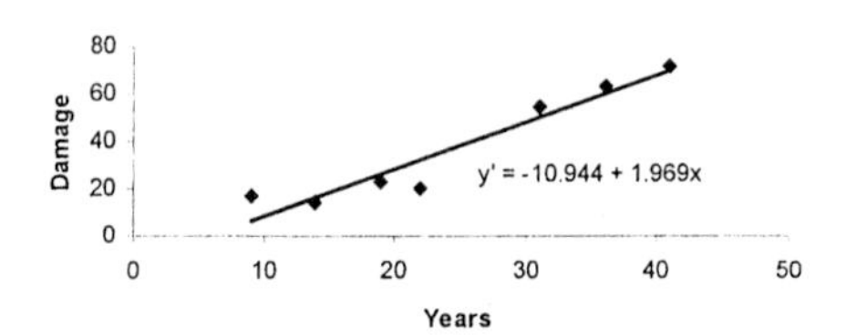

$\sum x = 172$
$\sum y = 262$
$\sum x^2 = 5060$
$\sum y^2 = 13340$
$\sum xy = 8079$
$n = 7$

$r = \frac{n(\sum xy) - (\sum x)(\sum y)}{\sqrt{[n(\sum x^2) - (\sum x)^2][n(\sum y^2) - (\sum y)^2]}}$

$r = \frac{7(8079) - (172)(262)}{\sqrt{[7(5060) - (172)^2][7(13340) - (262)^2]}} = 0.956$

29. continued

H_0: $\rho = 0$
H_1: $\rho \neq 0$
C. V. = ± 0.754 d. f. = 5

Decision: Reject
There is a significant relationship between the number of years a person smokes and the amount of lung damage.

$$a = \frac{(\sum y)(\sum x^2) - (\sum x)(\sum xy)}{n(\sum x^2) - (\sum x)^2}$$

$$a = \frac{(262)(5060) - (172)(8079)}{7(5060) - (172)^2} = -10.944$$

$$b = \frac{n(\sum xy) - (\sum x)(\sum y)}{n(\sum x^2) - (\sum x)^2}$$

$$b = \frac{7(8079) - (172)(262)}{7(5060) - (172)^2} = 1.969$$

$y' = a + bx$
$y' = -10.944 + 1.969x$
$y' = -10.944 + 1.969(30) = 48.126$

30.

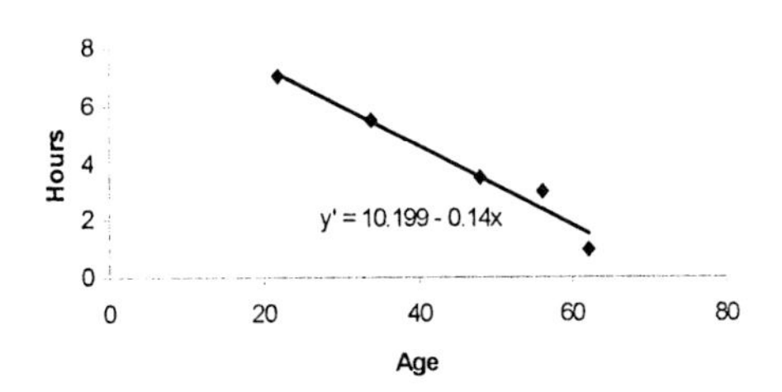

$\sum x = 222$
$\sum y = 20$
$\sum x^2 = 10924$
$\sum y^2 = 101.5$
$\sum xy = 739$
$n = 5$

$$r = \frac{n(\sum xy) - (\sum x)(\sum y)}{\sqrt{[n(\sum x^2) - (\sum x)^2][n(\sum y^2) - (\sum y)^2]}}$$

$$r = \frac{5(739) - (222)(20)}{\sqrt{[5(10924) - (222)^2][5(101.5) - (20)^2]}}$$

$r = -0.984$

H_0: $\rho = 0$
H_1: $\rho \neq 0$

C. V. = ± 0.878 d. f. = 3
Decision: Reject.

There is a significant relationship between age and the number of hours spent jogging.

30. continued

$$a = \frac{(\sum y)(\sum x^2) - (\sum x)(\sum xy)}{n(\sum x^2) - (\sum x)^2}$$

$$a = \frac{(20)(10924) - (222)(739)}{5(10924) - (222)^2} = 10.199$$

$$b = \frac{n(\sum xy) - (\sum x)(\sum y)}{n(\sum x^2) - (\sum x)^2}$$

$$b = \frac{5(739) - (222)(20)}{5(10924) - (222)^2} = -0.14$$

$y' = a + bx$
$y' = 10.199 - 0.14x$

31.

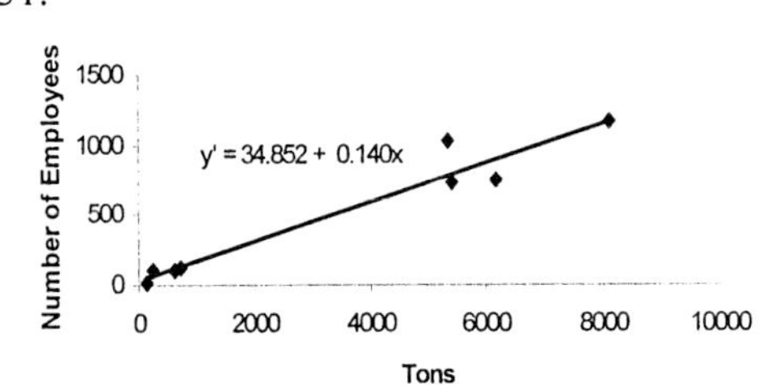

$\sum x = 26{,}728$
$\sum y = 4027$
$\sum x^2 = 162{,}101{,}162$
$\sum y^2 = 3{,}550{,}103$
$\sum xy = 23{,}663{,}669$
$n = 8$

$$r = \frac{n(\sum xy) - (\sum x)(\sum y)}{\sqrt{[n(\sum x^2) - (\sum x)^2][n(\sum y^2) - (\sum y)^2]}}$$

$$r = \frac{8(23662669) - (26728)(4027)}{\sqrt{[8(162101162) - 26728^2][8(3550103) - (4027)^2]}}$$

$r = 0.970$

H_0: $\rho = 0$
H_1: $\rho \neq 0$
C. V. = ± 0.707 d. f. = 6

Decision: Reject.
There is a significant relationship between the number of tons of coal produced and the number of employees.

$$a = \frac{(\sum y)(\sum x^2) - (\sum x)(\sum xy)}{n(\sum x^2) - (\sum x)^2}$$

$$a = \frac{(4027)(162101162) - (26728)(23663669)}{8(162101162) - (26728)^2}$$

$a = 34.852$

$$b = \frac{n(\sum xy) - (\sum x)(\sum y)}{n(\sum x^2) - (\sum x)^2}$$

$$b = \frac{8(23663669) - (26728)(4027)}{8(162101162) - (26728)^2} = 0.140$$

31. continued
$y' = a + bx$
$y' = 34.852 + 0.140x$
$y' = 34.852 + 0.140(500) = 104.8$

32.

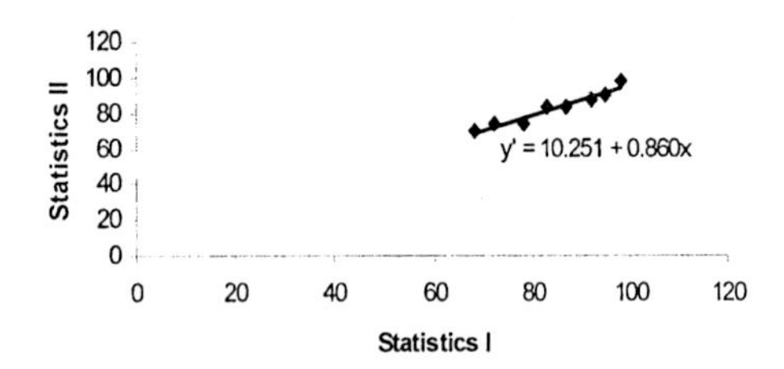

$\sum x = 673$
$\sum y = 661$
$\sum x^2 = 57443$
$\sum y^2 = 55275$
$\sum xy = 56318$
$n = 8$

$$r = \frac{8(56318)-(673)(661)}{\sqrt{[8(57443)-(673)^2][8(55275)-(661)^2]}}$$

$r = 0.963$

H_0: $\rho = 0$
H_1: $\rho \neq 0$
C. V. = ± 0.707 d. f. = 6

Decision: Reject.
There is a significant relationship between the final exam scores for the courses.

$$a = \frac{(661)(57443)-(673)(56318)}{8(57443)-(673)^2} = 10.251$$

$$b = \frac{8(56318)-(673)(661)}{8(57443)-(673)^2} = 0.860$$

$y' = 10.251 + 0.860x$

33.

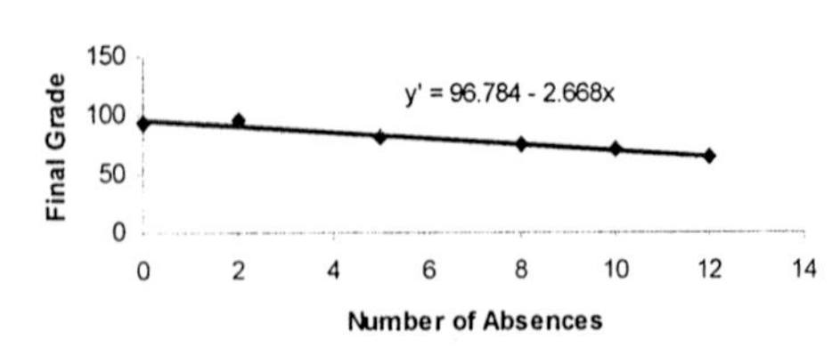

$\sum x = 37$
$\sum y = 482$
$\sum x^2 = 337$
$\sum y^2 = 39526$
$\sum xy = 2682$
$n = 6$

33. continued

$$r = \frac{n(\sum xy)-(\sum x)(\sum y)}{\sqrt{[n(\sum x^2)-(\sum x)^2]\,[n(\sum y^2)-(\sum y)^2]}}$$

$$r = \frac{6(2682)-(37)(482)}{\sqrt{[6(337)-(37)^2][6(39526)-(482)^2]}}$$

$r = -0.981$

H_0: $\rho = 0$
H_1: $\rho \neq 0$
C. V. = ± 0.811 d. f. = 4

Decision: Reject.
There is a significant negative relationship between the number of absences and the final grade.

$$a = \frac{(\sum y)(\sum x^2)-(\sum x)(\sum xy)}{n(\sum x^2)-(\sum x)^2}$$

$$a = \frac{(482)(337)-(37)(2682)}{6(337)-(37)^2} = 96.784$$

$$b = \frac{n(\sum xy)-(\sum x)(\sum y)}{n(\sum x^2)-(\sum x)^2}$$

$$b = \frac{6(2682)-(37)(482)}{6(337)-(37)^2} = -2.668$$

$y' = a + bx$
$y' = 96.784 - 2.668x$

34.

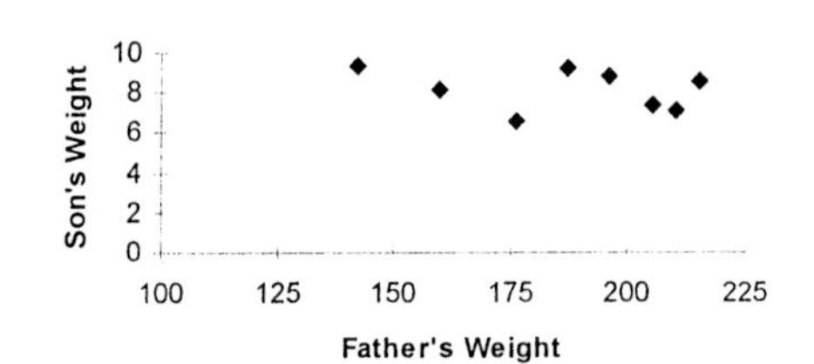

$\sum x = 1491$
$\sum y = 65.2$
$\sum x^2 = 282,475$
$\sum y^2 = 538.5$
$\sum xy = 12096.4$
$n = 8$

$$r = \frac{n(\sum xy)-(\sum x)(\sum y)}{\sqrt{[n(\sum x^2)-(\sum x)^2]\,[n(\sum y^2)-(\sum y)^2]}}$$

$$r = \frac{8(12096.4)-(1491)(65.2)}{\sqrt{[8(282475)-(1491)^2][8(538.5)-(65.2)^2]}}$$

$r = -0.306$

H_0: $\rho = 0$
H_1: $\rho \neq 0$

34. continued
$t = -0.787$; $0.20 < \text{P-value} < 0.50$ (0.462)

Decision: Do not reject since P-value > 0.05. There is no significant relationship between the weights of the fathers and sons. Since r is not significant, no regression analysis should be done.

35.

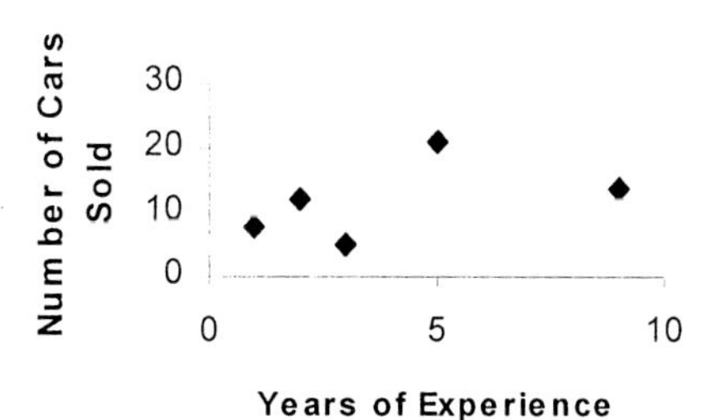

$\sum x = 20$
$\sum y = 60$
$\sum x^2 = 120$
$\sum y^2 = 870$
$\sum xy = 278$
$n = 5$

$$r = \frac{n(\sum xy) - (\sum x)(\sum y)}{\sqrt{[n(\sum x^2) - (\sum x)^2][n(\sum y^2) - (\sum y)^2]}}$$

$$r = \frac{5(278) - (20)(60)}{\sqrt{[5(120) - 20^2][5(870) - 60^2]}}$$

$r = 0.491$

H_0: $\rho = 0$
H_1: $\rho \neq 0$
$t = 0.976$; $0.20 < \text{P-value} < 0.50$ (0.401)

Decision: Do not reject since P-value > 0.05. There is no significant relationship between the number of years of experience and the number of cars sold per month. Since r is not significant, no regression analysis should be done.

36.
For 11 – 12
$\bar{x} = 7.5$
$\bar{y} = 6.33$
$y' = 0.073 + 0.835x$
$y' = 0.073 + 0.835(7.5)$
$y' = 0.073 + 6.26$
$y' = 6.333$
$\bar{y} = y'$

36. continued
For 11 – 13
$\bar{x} = 37.5$
$\bar{y} = 3.75$
$y' = 10.499 + -0.18x$
$y' = 10.499 + -0.18(37.5)$
$y' = 10.499 - 6.75$
$y' = 3.749$ or 3.75
$\bar{y} = y'$

For 11 – 14
$\bar{x} = 847.86$
$\bar{y} = 9.29$
$y' = 1.139 + 0.00961x$
$y' = 1.139 + 0.00961(847.86)$
$y' = 1.139 + 8.1479$
$y' = 9.287$ or 9.29
$\bar{y} = y'$

In all cases $\bar{y} = y'$, hence the regression line will always pass through the point $(\bar{x}, \bar{y})$.

37.
For 11 – 15
$\bar{x} = 5.3333$
$\bar{y} = 184.1667$
$b = -50.439$
$a = \bar{y} - b\bar{x}$
$a = 184.1667 - (-50.439)(5.3333)$
$a = 184.1667 + 269.0063$
$a = 453.173$ (differs due to rounding)

For 11 – 16
$\bar{x} = 40.33$
$\bar{y} = 8.33$
$b = -0.317$
$a = \bar{y} - b\bar{x}$
$a = 8.33 - (-0.317)(40.33)$
$a = 8.33 + 12.78$
$a = 21.11$ or 21.1

For 11 – 17
$\bar{x} = 104$
$\bar{y} = 2.7625$
$b = 0.063$
$a = \bar{y} - b\bar{x}$
$a = 2.7625 - (0.063)(104)$
$a = 2.7625 - 6.552$
$a = -3.7895$ (differs due to rounding)

38.
For 11 – 18
$b = 0.930$
$s_x = 8.467$
$s_y = 8.935$

38. continued

$r = \frac{bs_x}{s_y} = \frac{0.930(8.467)}{8.935} = 0.881$

For 11 – 19

$b = \frac{5(705)-(193)(17)}{5(7537)-(193)^2} = 0.5596$

$s_x = 4.6690$

$s_y = 3.2094$

$r = \frac{0.5596(4.6690)}{3.2094} = 0.814$

For 11 – 20

$b = 0.190$

$s_x = 12.448$

$s_y = 2.911$

$r = \frac{0.190(12.448)}{2.911} = 0.812$

(differs due to rounding)

EXERCISE SET 10-5

1.
Explained variation is the variation obtained from the predicted y' values, and is computed by $\sum(y' - \bar{y})^2$.

2.
Unexplained variation is the variation of the observed values and the predicted values and is computed by $\sum(y - y')^2$.

3.
Total variation is the sum of the explained and unexplained variation and is computed by $\sum(y - \bar{y})^2 = \sum(y' - \bar{y})^2 + \sum(y - y')^2$.

4.
The coefficient of determination is a measure of variation of the dependent variable that is explained by the regression line and the independent variable.

5.
It is found by squaring r.

6.
It is the percent of the variation in y that is not due to the variation in x.

7.
The coefficient of non-determination is $1 - r^2$.

8.
For $r = 0.81$, $r^2 = 0.656$, $1 - r^2 = 0.344$

8. continued
65.6% of the variation of y is due to the variation of x, and 34.4% of the variation of y is due to chance.

9.
For $r = 0.70$, $r^2 = 0.49$, $1 - r^2 = 0.51$
49% of the variation of y is due to the variation of x, and 51% of the variation of y is due to chance.

10.
For $r = 0.45$, $r^2 = 0.2025$, $1 - r^2 = 0.7975$
20.25% of the variation of y is due to the variation of x, and 79.75% of the variation of y is due to chance.

11.
For $r = 0.37$, $r^2 = 0.1369$, $1 - r^2 = 0.8631$
13.69% of the variation of y is due to the variation of x, and 86.31% of the variation of y is due to chance.

12.
For $r = 0.15$, $r^2 = 0.0225$, $1 - r^2 = 0.9775$
2.25% of the variation of y is due to the variation of x, and 97.75% of the variation of y is due to chance.

13.
For $r = 0.05$, $r^2 = 0.0025$, $1 - r^2 = 0.9975$
0.25% of the variation of y is due to the variation of x, and 99.75% of the variation of y is due to chance.

14.
The standard error of estimate is the standard deviation of the observed y values about the predicted y' values.

15.

$S_{est} = \sqrt{\frac{\sum y^2 - a\sum y - b\sum xy}{n-2}}$

$S_{est} = \sqrt{\frac{141.25 - 10.499(22.5) - (-0.18)(625)}{6-2}}$

$S_{est} = \sqrt{4.380625}$

$S_{est} = 2.09$

16.

$S_{est} = \sqrt{\frac{703 - 1.139(65) - (0.00961)(64480)}{7-2}}$

$S_{est} = \sqrt{1.86244} = 1.365$

17.

$S_{est} = \sqrt{\frac{\sum y^2 - a\sum y - b\sum xy}{n-2}} =$

$S_{est} = \sqrt{\frac{364525-(453.176)(1105)-(-50.439)(3405)}{6-2}}$

$S_{est} = 94.22$

18.

$S_{est} = \sqrt{\frac{546-21.1(50)-(-0.317)(1625)}{6-2}}$

$S_{est} = \sqrt{1.531} = 1.237$

19.

$y' = 10.499 - 0.18x$

$y' = 10.499 - 0.18(20)$

$y' = 6.899$

$y' - t_{\frac{\alpha}{2}} \cdot s_{est}\sqrt{1 + \frac{1}{n} + \frac{n(x-\bar{X})}{n\sum x^2 - (\sum x)^2}} < y < y' +$

$t_{\frac{\alpha}{2}} \cdot s_{est}\sqrt{1 + \frac{1}{n} + \frac{n(x-\bar{X})^2}{n\sum x^2 - (\sum x)^2}}$

$6.899 - (2.132)(2.09)\sqrt{1 + \frac{1}{6} + \frac{6(20-37.5)^2}{6(9653)-225^2}}$

$< y < 6.899 +$

$(2.132)(2.09)\sqrt{1 + \frac{1}{6} + \frac{6(20-37.5)^2}{6(9653)-225^2}}$

$6.899 - (2.132)(2.09)(1.191) < y < 6.899 +$
$(2.132)(2.09)(1.91)$

$1.59 < y < 12.21$

20.

$y' = 1.139 + 0.00961x$

$y' = 1.139 + 0.00961(\$1100)$

$y' = 11.710$

$11.710 - (2.571)(1.371)\sqrt{1 + \frac{1}{7} + \frac{7(1100-847.86)^2}{7(6007125)-(5935)^2}}$

$< y < 11.710 +$

$(2.571)(1.371)\sqrt{1 + \frac{1}{7} + \frac{7(1100-847.86)^2}{7(6007125)-(5935)^2}}$

$11.710 - (2.571)(1.371)(1.0991) < y <$

$11.710 + (2.571)(1.371)(1.0991)$

$7.83435 < y < 15.58265$ or $8 < y < 16$

21.

$y' = 453.176 - 50.439x$

21. continued

$y' = 453.176 - 50.439(4)$

$y' = 251.42$

$y' - t_{\frac{\alpha}{2}} \cdot s_{est}\sqrt{1 + \frac{1}{n} + \frac{n(x-\bar{X})^2}{n\sum x^2 - (\sum x)^2}} < y <$

$y' + t_{\frac{\alpha}{2}} \cdot s_{est}\sqrt{1 + \frac{1}{n} + \frac{n(x-\bar{X})^2}{n\sum x^2 - (\sum x)^2}}$

$251.42 - 2.132(94.22)\sqrt{1 + \frac{1}{6} + \frac{6(4-5.33)^2}{6(220)-32^2}}$

$< y < 251.42 + 2.132(94.22)\sqrt{1 + \frac{1}{6} + \frac{6(4-5.33)^2}{6(220)-32^2}}$

$251.42 - (2.132)(94.22)(1.1) < y <$
$251.42 + (2.132(94.22)(1.1)$

$\$30.46 < y < \$472.38$

22.

$y' = 21.1 - 0.317x$

$y' = 21.1 - 0.317(47)$

$y' = 6.201$

$6.201 - (3.747)(1.157)\sqrt{1 + \frac{1}{6} + \frac{6(47-40.333)^2}{6(10998)-(242)^2}}$

$< y < 6.201 + (3.747)(1.157)\sqrt{1 + \frac{1}{6} + \frac{6(47-40.333)^2}{6(10998)-(242)^2}}$

$6.201 - (3.747)(1.157)(1.0966) < y <$

$6.201 - (3.747)(1.157)(1.0966)$

$1.44675 < y < 10.95525$ or $1 < y < 11$

EXERCISE SET 10-6

1.

Simple linear regression has one independent variable and one dependent variable. Multiple regression has one dependent variable and two or more independent variables.

2.

$y' = a + b_1x_1 + b_2x_2 + \cdots + b_nx_n$

a represents the slope and the b's represent the partial regression coefficients.

3.

The relationship would include all variables.

4.
The variables must be normally distributed, variances must be equal, independent, not correlated, and linear.

5.
The multiple correlation coefficient R is always higher than the individual correlation coefficients. Also, the value of R can range from 0 to $+1$.

6.
$y' = -34{,}127 + 132(32) + 20{,}805(3.4)$
$y' = \$40{,}834$

7.
$y' = 9.6 + 2.2x_1 - 1.08x_2$
$y' = 9.6 + 2.2(9) - 1.08(24) = 3.48$

8.
$y' = 44.9 - 0.0266(371) + 7.56(6)$
$y' = \$80.3914$ thousand

9.
$y' = 5000 + 97x_1 + 35x_2$
$y' = 5000 + 97(120) + 35(650)$
$y' = \$39{,}390$

10.
$y' = 97.7 + 0.691(35) + 219(194) - 299(142)$
$y' = 149.885 \approx 149$

11.
R is a measure of the strength of the relationship between the dependent variables and all the independent variables.

12.
0 to $+1$

13.
R^2 is the coefficient of multiple determination. R^2_{adj} is adjusted for sample size.

14.
H_0: $\rho = 0$ and H_1: $\rho \neq 0$

15.
The F test is used to est the significance of R.

16.
The adjusted R^2 is the adjusted coefficient of multiple determination. It takes into account the fact that when the sample sizes are approximately equal, the value of R may be artificially high due to sampling error rather than a true relationship between variables.

REVIEW EXERCISES - CHAPTER 10

1.

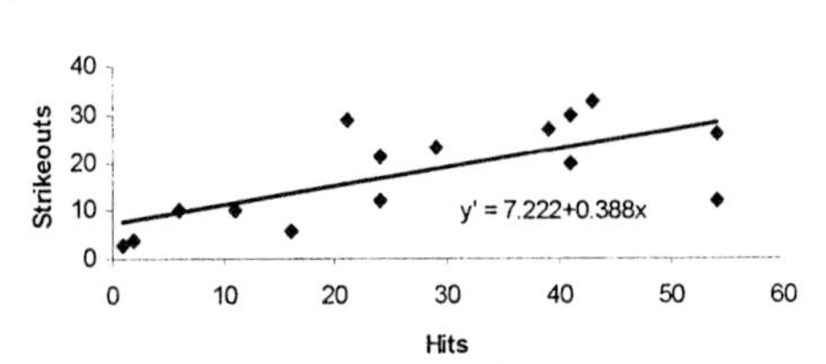

$\sum x = 406$
$\sum y = 266$
$\sum x^2 = 15{,}416$
$\sum y^2 = 6154$
$\sum xy = 8919$
$n = 15$

$$r = \frac{n(\sum xy) - (\sum x)(\sum y)}{\sqrt{[n(\sum x^2) - (\sum x)^2][n(\sum y^2) - (\sum y)^2]}}$$

$$r = \frac{15(8919) - (406)(266)}{\sqrt{[15(15416) - (406)^2][15(6154) - (266)^2]}}$$

$r = 0.682$

H_0: $\rho = 0$
H_1: $\rho \neq 0$
C. V. $= \pm 0.641$ d. f. $= 13$

Decision: Reject. There is a significant relationship between hits and strikeouts.

$$a = \frac{(\sum y)(\sum x^2) - (\sum x)(\sum xy)}{n(\sum x^2) - (\sum x)^2}$$

$$a = \frac{(266)(15416) - (406)(8919)}{15(15416) - (406)^2} = 7.222$$

$$b = \frac{n(\sum xy) - (\sum x)(\sum y)}{n(\sum x^2) - (\sum x)^2}$$

$$b = \frac{15(8919) - (406)(266)}{15(15416) - (406)^2} = 0.388$$

$y' = a + bx$
$y' = 7.222 + 0.388x$
$y' = 7.222 + 0.388(30) = 18.86$ or 19 strikeouts

2.

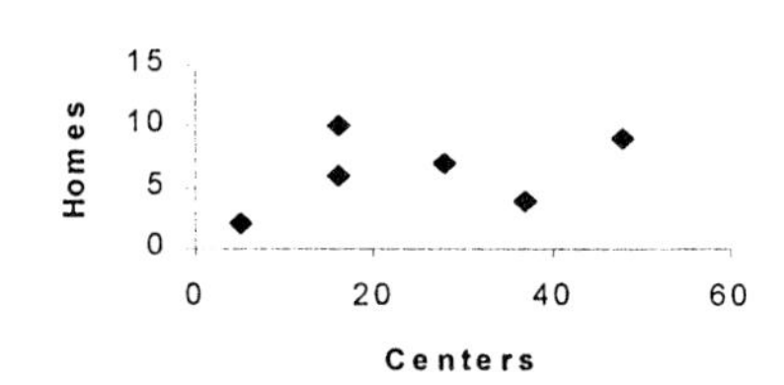

$\sum x = 150$
$\sum y = 38$
$\sum x^2 = 4994$
$\sum y^2 = 286$
$\sum xy = 1042$
$n = 6$

$$r = \frac{6(1042)-(150)(38)}{\sqrt{[6(4994)-(150)^2][6(286)-(38)^2]}}$$

$r = 0.387$

H_0: $\rho = 0$
H_1: $\rho \neq 0$
C. V. = ± 0.917 d. f. = 4

Decision: Do not reject. There is no significant relationship between the number of day care centers and the number of group day care homes. No regression should be done.

3.

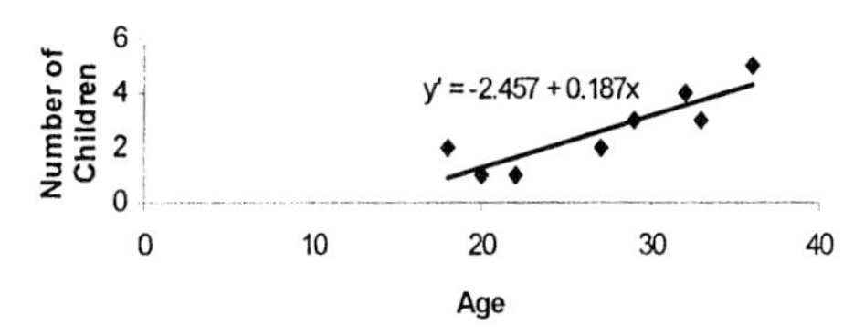

$\sum x = 217$
$\sum y = 21$
$\sum x^2 = 6187$
$\sum y^2 = 69$
$\sum xy = 626$
$n = 8$

$$r = \frac{n(\sum xy)-(\sum x)(\sum y)}{\sqrt{[n(\sum x^2)-(\sum x)^2]\,[n(\sum y^2)-(\sum y)^2]}}$$

$$r = \frac{8(626)-(217)(21)}{\sqrt{[8(6187)-(217)^2][8(69)-(21)^2]}}$$

$r = 0.873$

H_0: $\rho = 0$
H_1: $\rho \neq 0$
C. V. = ± 0.834 d. f. = 6

3. continued
Decision: Reject. There is a significant relationship between the mother's age and the number of children she has.

$$a = \frac{(\sum y)(\sum x^2)-(\sum x)(\sum xy)}{n(\sum x^2)-(\sum x)^2}$$

$$a = \frac{(21)(6187)-(217)(626)}{8(6187)-(217)^2} = -2.457$$

$$b = \frac{n(\sum xy)-(\sum x)(\sum y)}{n(\sum x^2)-(\sum x)^2}$$

$$b = \frac{8(626)-(217)(21)}{8(6187)-(217)^2} = 0.187$$

$y' = a + bx$
$y' = -2.457 + 0.187x$
$y' = -2.457 + 0.187(34) = 3.9$

4.

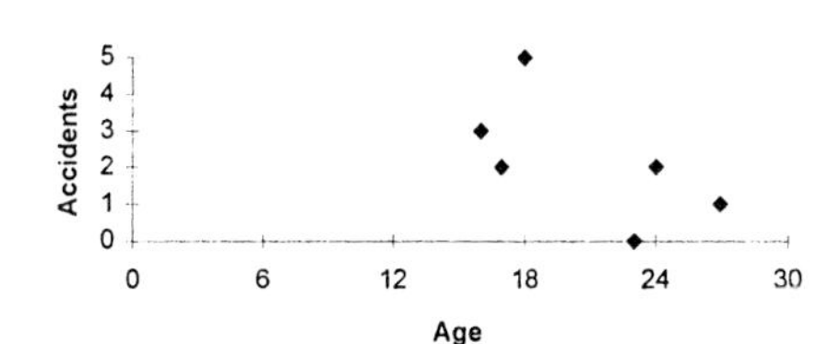

$\sum x = 157$
$\sum y = 14$
$\sum x^2 = 3727$
$\sum y^2 = 44$
$\sum xy = 279$
$n = 7$

$$r = \frac{7(279)-(157)(14)}{\sqrt{[7(3727)-(157)^2][7(44)-(14)^2]}}$$

$r = -0.610$

H_0: $\rho = 0$
H_1: $\rho \neq 0$
C. V. = ± 0.875 d. f. = 5

Decision: Do not reject. There is not a significant relationship between a person's age and the number of accidents a person has.

No regression analysis should be done since the null hypothesis has not been rejected.

5.

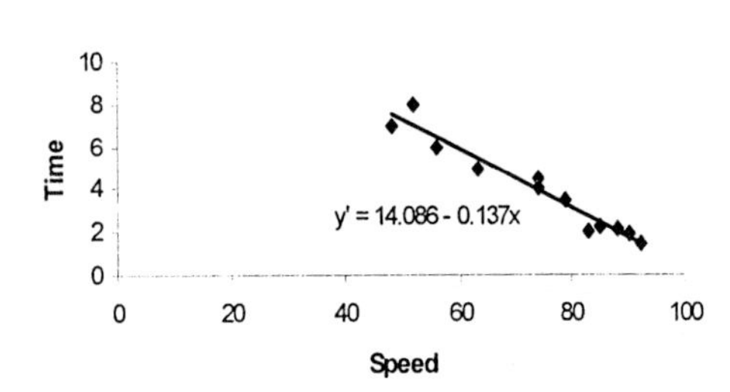

$\sum x = 884$
$\sum y = 47.8$
$\sum x^2 = 67728$
$\sum y^2 = 242.06$
$\sum xy = 3163.8$

$n = 12$

$$r = \frac{n(\sum xy) - (\sum x)(\sum y)}{\sqrt{[n(\sum x^2) - (\sum x)^2][n(\sum y^2) - (\sum y)^2]}}$$

$$r = \frac{12(3163.8) - (884)(47.8)}{\sqrt{[12(67728) - (884)^2][12(242.06) - (47.8)^2]}}$$

$r = -0.974$

H_0: $\rho = 0$
H_1: $\rho \neq 0$
C. V. = ± 0.708 d. f. = 10

Decision: Reject the null. There is a significant relationship between speed and time.

$$a = \frac{(\sum y)(\sum x^2) - (\sum x)(\sum xy)}{n(\sum x^2) - (\sum x)^2}$$

$$a = \frac{(47.8)(67728) - (884)(3163.8)}{12(67728) - (884)^2}$$

$a = 14.086$

$$b = \frac{n(\sum xy) - (\sum x)(\sum y)}{n(\sum x^2) - (\sum x)^2}$$

$$b = \frac{12(3163.8) - (884)(47.8)}{12(67728) - (884)^2}$$

$b = -0.137$

$y' = a + bx$
$y' = 14.086 - 0.137x$
$y' = 14.086 - 0.137(72) = 4.222$

6.

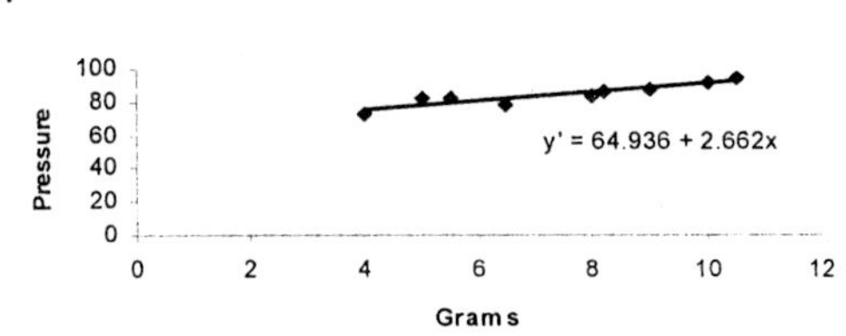

$\sum x = 66.7$
$\sum y = 762$
$\sum x^2 = 535.99$
$\sum y^2 = 64868$
$\sum xy = 5758.2$
$n = 9$

$$r = \frac{9(5758.2) - (66.7)(762)}{\sqrt{[9(535.99) - (66.7)^2][9(64868) - (762)^2]}}$$

$r = 0.916$

H_0: $\rho = 0$
H_1: $\rho \neq 0$
C. V. = ± 0.798 d. f. = 7

Decision: Reject the null.
There is a significant relationship between grams and pressure.

$$a = \frac{(762)(535.99) - (66.7)(5758.2)}{9(535.99) - 66.7^2} = 64.936$$

$$b = \frac{9(5758.2) - (66.7)(762)}{9(535.99) - 66.7^2} = 2.662$$

$y' = 64.936 + 2.662x$
$y' = 64.936 + 2.662(8) = 86.232$

7.

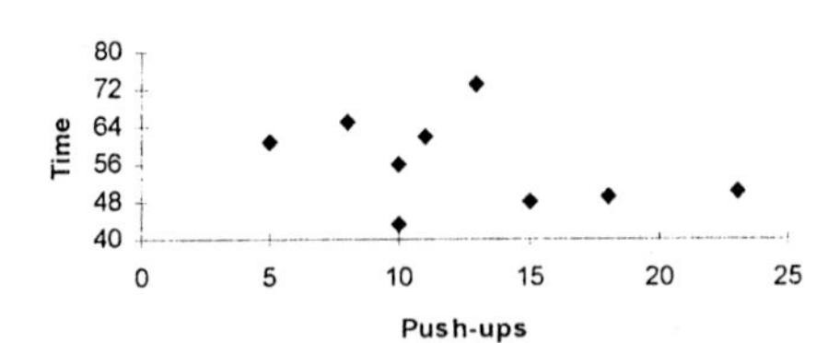

$\sum x = 113$
$\sum y = 507$
$\sum x^2 = 1657$
$\sum y^2 = 29309$
$\sum xy = 6198$
$n = 9$

$$r = \frac{n(\sum xy) - (\sum x)(\sum y)}{\sqrt{[n(\sum x^2) - (\sum x)^2][n(\sum y^2) - (\sum y)^2]}}$$

$$r = \frac{9(6198) - (113)(507)}{\sqrt{[9(1657) - (113)^2][9(29309) - (507)^2]}}$$

$r = -0.397$

7. continued
H_0: $\rho = 0$
H_1: $\rho \neq 0$
C. V. = ± 0.798 d. f. = 7

Decision: Do not reject. Since the null hypothesis is not rejected, no regression should be done.

8.
$$S_{est} = \sqrt{\frac{\sum y^2 - a\sum y - b\sum xy}{n-2}}$$

$$S_{est} = \sqrt{\frac{44-(5.816)(14)-(-0.1701)(279)}{7-2}} = 1.417$$

$$S_{est} = \sqrt{\frac{10.0339}{5}} = \sqrt{2.00678} = 1.417$$

9.
$$S_{est} = \sqrt{\frac{\sum y^2 - a\sum y - b\sum xy}{n-2}}$$

$$S_{est} = \sqrt{\frac{242.06-14.086(47.8)+0.137(3163.8)}{12-2}}$$

$$S_{est} = \sqrt{\frac{2.1898}{10}} = \sqrt{0.21898} = 0.468$$

10.
$$S_{est} = \sqrt{\frac{\sum y^2 - a\sum y - b\sum xy}{n-2}}$$

$$S_{est} = \sqrt{\frac{64868-(64.936)(762)-(2.662)(5758.2)}{9-2}}$$

$$S_{est} = \sqrt{\frac{64868-49481.232-15328.328}{7}}$$

$$S_{est} = \sqrt{\frac{58.44}{7}} = \sqrt{8.349} = 2.89$$

11.
$y' = 14.086 - 0.137x$
$y' = 14.086 - 0.137(72) = 4.222$

$$y' - t_{\frac{\alpha}{2}} \cdot s_{est}\sqrt{1 + \frac{1}{n} + \frac{n(x-\bar{X})^2}{n\Sigma x^2 - (\Sigma x)^2}} < y <$$
$$y' + t_{\frac{\alpha}{2}} \cdot s_{est}\sqrt{1 + \frac{1}{n} + \frac{n(x-\bar{X})^2}{n\Sigma x^2 - (\Sigma x)^2}}$$

$$4.222 - 1.812(0.468)\sqrt{1 + \frac{1}{12} + \frac{12(72-73.667)^2}{12(67{,}728)-884^2}}$$
$$< y < 4.222 + 1.812(0.468)\sqrt{1 + \frac{1}{12} + \frac{12(72-73.667)^2}{12(67{,}728)-884^2}}$$

$4.222 - 1.812(0.468)(1.041) < y <$
$4.222 + 1.812(0.468)(1.041)$
$3.34 < y < 5.10$

12.
$y' = 64.936 + 2.662x$
$y' = 64.936 + 2.662(8) = 86.232$

$$86.232 - 2.365(2.89)\sqrt{1 + \frac{1}{9} + \frac{9(8-7.411)^2}{9(535.99)-66.7^2}}$$

$$< y < 86.232 + 2.365(2.89)\sqrt{1 + \frac{1}{9} + \frac{9(8-7.411)^2}{9(535.99)-66.7^2}}$$

$86.232 - 2.365(2.89)(1.058) < y <$
$86.232 + 2.365(2.89)(1.058)$

$79.1 < y < 93.4$ or $79 < y < 93$

13.
$y' = 12.8 + 2.09x_1 + 0.423x_2$
$y' = 12.8 + 2.09(4) + 0.423(2) = 22.006$

14.
$$R = \sqrt{\frac{(0.681)^2 + (0.872)^2 - 2(0.681)(0.872)(0.746)}{1-(0.746)^2}}$$

$$R = \sqrt{0.7624799} = 0.873$$

15.
$$R^2_{adj} = 1 - \left[\frac{(1-R^2)(n-1)}{n-k-1}\right]$$

$$R^2_{adj} = 1 - \left[\frac{(1-0.873^2)(10-1)}{10-3-1}\right]$$

$$R^2_{adj} = 1 - \left[\frac{2.1408}{6}\right] = 0.643$$

CHAPTER 10 QUIZ

1. False, the y variable would decrease.
2. True
3. True
4. False, the relationship may be affected by another variable, or by chance.
5. False, a relationship may be caused by chance.
6. False, there are several independent variables and one dependent variable.
7. a.
8. a.
9. d.
10. c.
11. b.
12. scatter diagram
13. independent
14. -1, $+1$
15. b.
16. line of best fit
17. $+1$, -1

18.

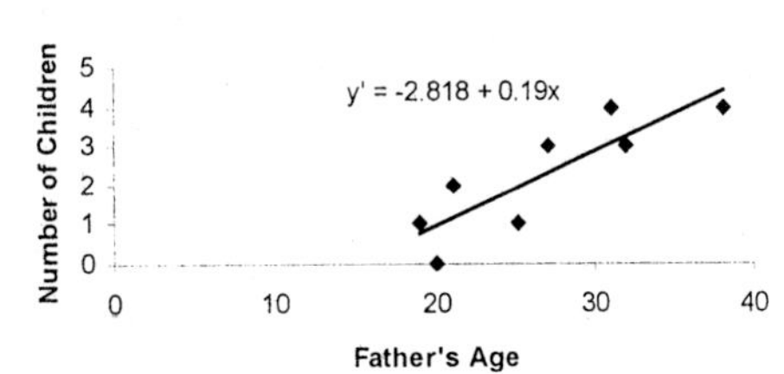

$\sum x = 213$
$\sum x^2 = 5985$
$\sum y = 18$
$\sum y^2 = 56$
$\sum xy = 539$
$n = 8$
$r = 0.857$
H_0: $\rho = 0$ and H_1: $\rho \neq 0$. C.V. = ± 0.707 and d. f. = 6. Reject. There is a significant relationship between father's age and number of children.
$a = -2.818$ $\quad b = 0.19$
$y' = -2.818 + 0.19x$
For x = 35 years old:
$y' = -2.818 + 0.19(35) = 3.8$ or 4

19.

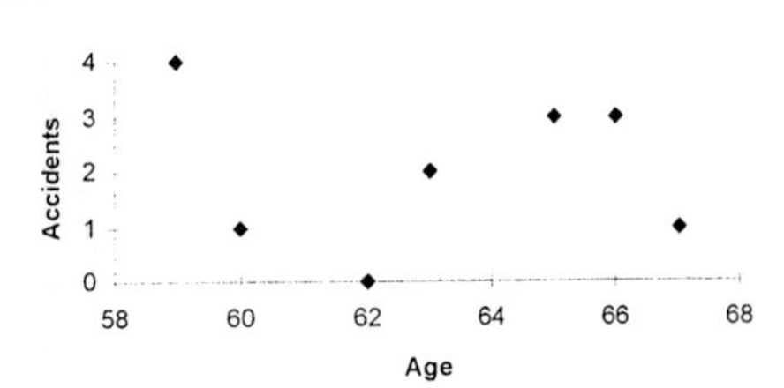

$\sum x = 442$
$\sum x^2 = 27{,}964$
$\sum y = 14$
$\sum y^2 = 40$
$\sum xy = 882$
$n = 7$
$r = -0.078$
H_0: $\rho = 0$ and H_1: $\rho \neq 0$. d. f. = 5 and C. V. = ± 0.764. Decision: do not reject. There is not a significant relationship between age and number of accidents.

20.

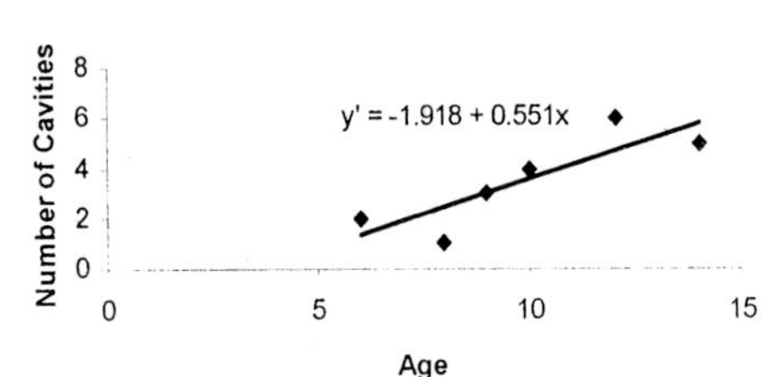

$\sum x = 59$
$\sum x^2 = 621$
$\sum y = 21$
$\sum y^2 = 91$
$\sum xy = 229$
$n = 6$
$r = 0.842$
H_0: $\rho = 0$ and H_1: $\rho \neq 0$; d. f. = 4 and C. V. = ± 0.811. Decision: Reject. There is a significant relationship between age and number of cavities.
$a = -1.918$ $\quad b = 0.551$
$y' = -1.918 + 0.551x$
When x = 11: $y' = -1.918 + 0.551(11)$
$y' = 4.14 \approx 4$ cavities

21.

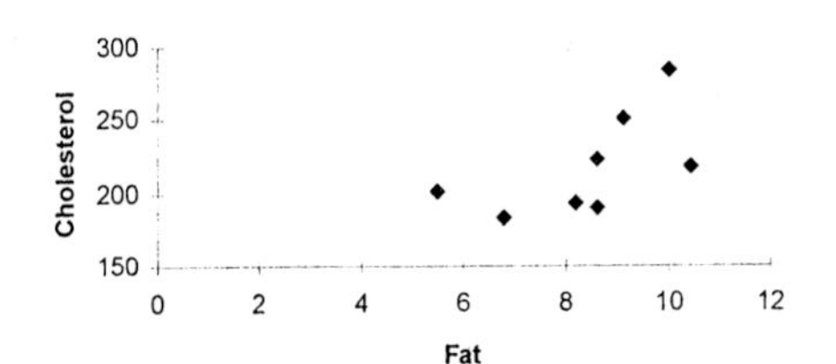

$\sum x = 67.2$
$\sum x^2 = 582.62$
$\sum y = 1740$
$\sum y^2 = 386{,}636$
$\sum xy = 14847.9$
$n = 8$
$r = 0.602$
H_0: $\rho = 0$ and H_1: $\rho \neq 0$; d. f. = 6 and C. V. = ± 0.707. Decision: Do not reject. There is no significant relationship between fat and cholesterol.

22.

$S_{est} = \sqrt{\frac{91-(-1.918)(21)-0.551(229)}{6-2}}$
$S_{est} = 1.129$*

23.
(For calculation purposes only, since no regression should be done.)

$$S_{est} = \sqrt{\frac{386{,}636 - 110.12(1740) - 12.784(14{,}847.9)}{8-2}}$$

$$S_{est} = 29.47^*$$

24.
$y' = -1.918 + 0.551(7) = 1.939$ or 2

$$2 - 2.132(1.129)\sqrt{1 + \frac{1}{6} + \frac{6(11-9.833)^2}{6(621)-59^2}} < y$$

$$< 2 + 2.132(1.129)\sqrt{1 + \frac{1}{6} + \frac{6(11-9.833)^2}{6(621)-59^2}}$$

$$2 - 2.132(1.129)(1.095) < y < 2 + 2.132(1.129)(1.095)$$

$$-0.6 < y < 4.6 \Rightarrow 0 < y < 5^*$$

25.
Since no regression should be done, the prediction interval is $\bar{y} = 217.5$.

26.
$y' = 98.7 + 3.82(3) + 6.51(1.5) = 119.9^*$

27.
$$R = \sqrt{\frac{(0.561)^2 + (0.714)^2 - 2(0.561)(0.714)(0.625)}{1-(0.625)^2}}$$

$R = 0.729^*$

28.
$$R^2_{adj} = 1 - \left[\frac{(1-0.774^2)(8-1)}{(8-2-1)}\right]$$

$R^2_{adj} = 0.439^*$

*These answers may vary due to method of calculation or rounding.

Note: Graphs are not to scale and are intended to convey a general idea.

Answers may vary due to rounding, TI-83's, or computer programs.

EXERCISE SET 11-2

1.
The variance test compares a sample variance to a hypothesized population variance, while the goodness of fit test compares a distribution obtained from a sample with a hypothesized distribution.

2.
The degrees of freedom is the number of categories minus one.

3.
The expected values are computed based on what the null hypothesis states about the distribution.

4.
The categories should be combined with other categories.

5.
H_0: The number of accidents is equally distributed throughout the week. (claim)
H_1: The distribution is not the same as stated in the null hypothesis.
C. V. = 12.592 d. f. = 6 $\alpha = 0.05$

$E = \frac{189}{7} = 27$

$\chi^2 = \sum\frac{(O-E)^2}{E} = \frac{(28-27)^2}{27} + \frac{(32-27)^2}{27} +$

$\frac{(15-27)^2}{27} + \frac{(14-27)^2}{27} + \frac{(38-27)^2}{27} + \frac{(43-27)^2}{27}$

$+ \frac{(19-27)^2}{27} = 28.887$

Alternate Solution:

O	E	O – E	$(O-E)^2$	$\frac{(O-E)^2}{E}$
28	27	1	1	0.037
32	27	5	25	0.926
15	27	-12	144	5.333
14	27	-13	169	6.259
38	27	11	121	4.481
43	27	16	256	9.481
19	27	-8	64	2.370
				28.887

5. continued

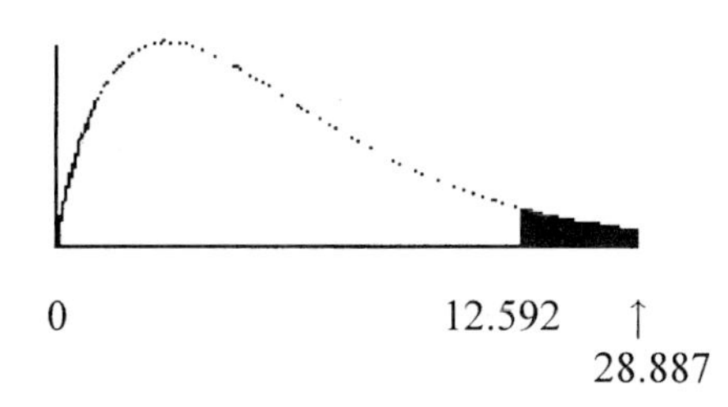

Reject the null hypothesis. There is enough evidence to reject the claim that the number of accidents is equally distributed during the week.

6.
H_0: The customers show no preference for the color of a raincoat.
H_1: The distribution is not the same as stated in the null hypothesis. (claim)
C. V. = 6.251 d. f. = 3 $\alpha = 0.10$
$E = \frac{50}{4} = 12.5$

$\chi^2 = \frac{(17-12.5)^2}{12.5} + \frac{(13-12.5)^2}{12.5}$

$+ \frac{(8-12.5)^2}{12.5} + \frac{(12-12.5)^2}{12.5}$

$\chi^2 = 3.28$

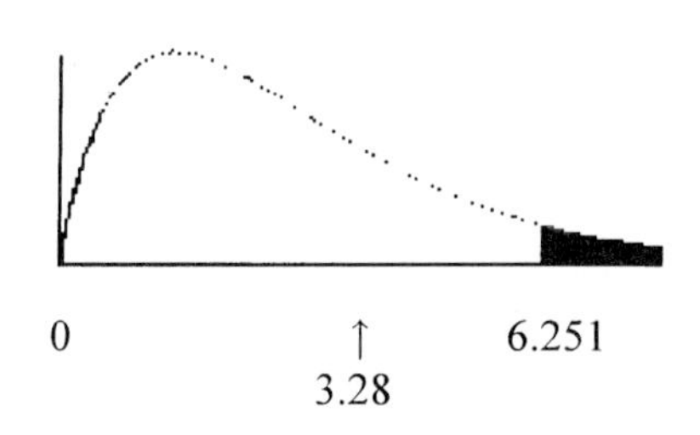

Do not reject the null hypothesis. There is not enough evidence to support the claim that the customers show a preference for the color of a raincoat.

7.
H_0: The proportions are distributed as follows: 28.1% purchased a small car, 47.8% purchased a mid-sized car, 7% purchased a large car, and 17.1% purchased a luxury car.
H_1: The distribution is not the same as stated in the null hypothesis. (claim)
C. V. = 7.815 d. f. = 3 $\alpha = 0.05$

7. continued

$\chi^2 = \sum \frac{(O-E)^2}{E} = \frac{(25-28.1)^2}{28.1} + \frac{(50-47.8)^2}{47.8}$

$+ \frac{(10-7)^2}{7} + \frac{(15-17.1)^2}{17.1} = 1.9869$

Alternate Solution:

O	E	O − E	(O − E)²	$\frac{(O-E)^2}{E}$
25	28.1	-3.1	9.61	0.3420
50	47.8	2.2	4.84	0.1013
10	7	3	9	1.2857
15	17.1	-2.1	4.41	0.2579
				1.9869

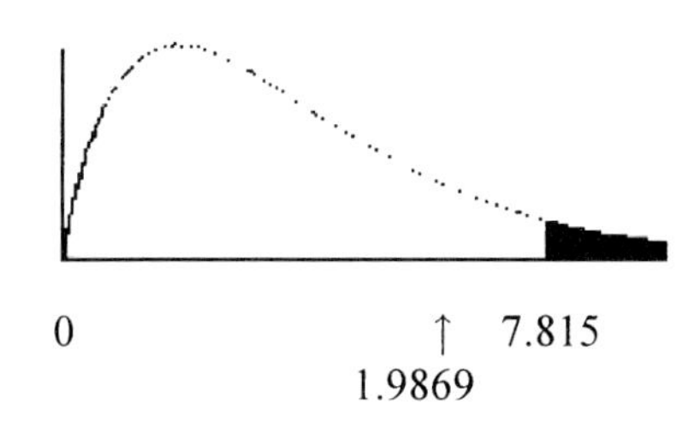

0 ↑ 7.815
1.9869

Do not reject the null hypothesis. There is not enough evidence to support the claim that the proportions are different.

8.
H_0: The proportions are distributed as follows: 68% of families have two parents present, 23% have a mother only present, 5% have a father only present, and 4% have no parent present.
H_1: The distribution is not the same as stated in the null hypothesis. (claim)
C. V. = 7.815 d. f. = 3 $\alpha = 0.05$

$\chi^2 = \sum \frac{(O-E)^2}{E} = \frac{(120-136)^2}{136} + \frac{(40-46)^2}{46}$

$+ \frac{(30-10)^2}{10} + \frac{(10-8)^2}{8} = 43.165$

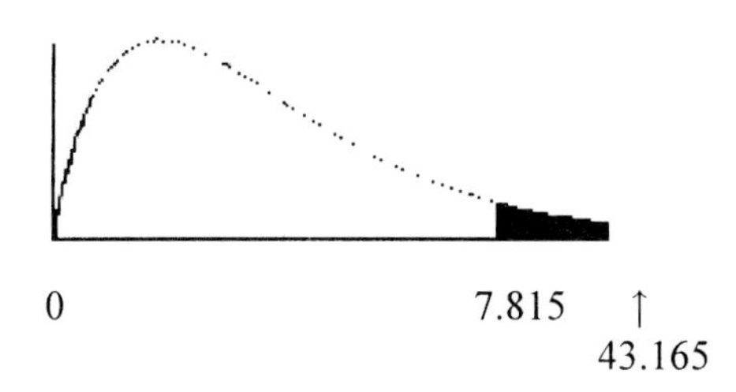

0 7.815 ↑
43.165

Reject the null hypothesis. There is enough evidence to support the claim that the proportions are different.

9.
H_0: The proportions are distributed as follows: safe - 35%, not safe - 52%, no opinion - 13%.
H_1: The distribution is not the same as stated in the null hypothesis. (claim)
C. V. = 9.210 d. f. = 2 $\alpha = 0.01$

$\chi^2 = \frac{(40-42)^2}{42} + \frac{(60-62.4)^2}{62.4} + \frac{(20-15.6)^2}{15.6}$

$\chi^2 = 1.4285$

Alternate Solution:

O	E	O − E	(O − E)²	$\frac{(O-E)^2}{E}$
40	42	-2	4	0.0952
60	62.4	-2.4	5.76	0.0923
20	15.6	4.4	19.36	1.2410
				1.4285

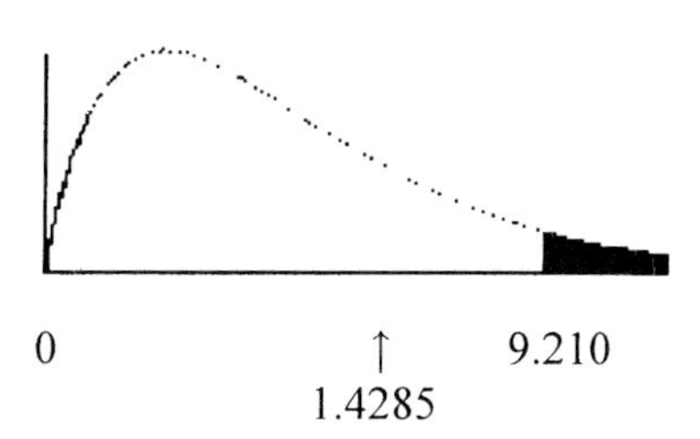

0 ↑ 9.210
1.4285

Do not reject the null hypothesis. There is not enough evidence to support the claim that the proportions are different.

10.
H_0: The proportions are distributed as follows: should continue - 56%, should not continue - 40%, no opinion - 4%.
H_1: The distribution is not the same as stated in the null hypothesis. (claim)
C. V. = 5.991 $\alpha = 0.05$ d. f. = 2

$\chi^2 = \frac{(126-112)^2}{112} + \frac{(65-80)^2}{80} + \frac{(9-8)^2}{8}$

$\chi^2 = 4.6875$

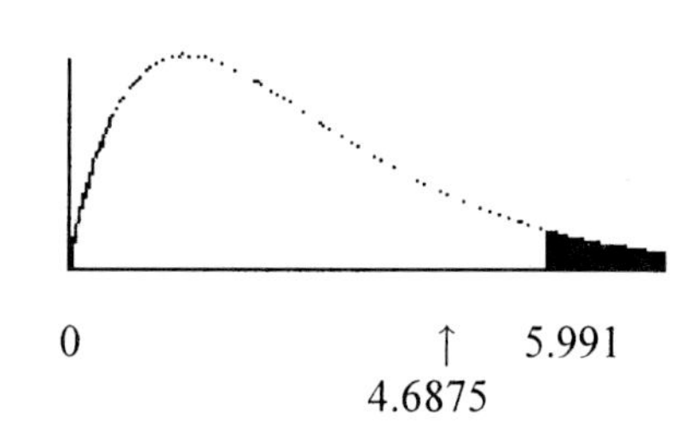

0 ↑ 5.991
4.6875

10. continued
Do not reject the null hypothesis. There is not enough evidence to support the claim that the proportions are different.

11.
H_0: The distribution of loans is as follows: 21% - mortgages, 39% - autos, 20% - unsecured, 12% - real estate, and 8% - miscellaneous. (claim)
H_1: The distribution is not the same as stated in the null hypothesis.
C. V. = 9.488 d. f. = 4 $\alpha = 0.05$

$$\chi^2 = \frac{(25-21)^2}{21} + \frac{(44-39)^2}{39} + \frac{(18-20)^2}{20} +$$

$$\frac{(8-12)^2}{12} + \frac{(4-8)^2}{8} = 4.9362$$

Alternate Solution:

O	E	O − E	(O − E)²	$\frac{(O-E)^2}{E}$
25	21	4	16	0.7619
44	39	5	25	0.6410
18	20	-2	4	0.2
8	12	-4	16	1.3333
4	8	-4	16	2.0000
				4.9362

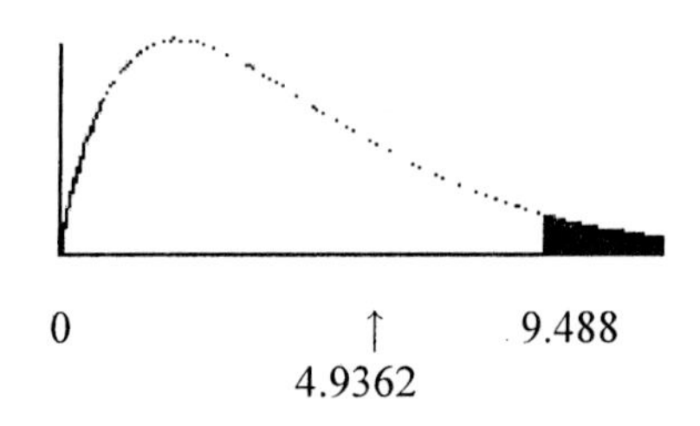

0 ↑ 4.9362 9.488

Do not reject the null hypothesis. There is not enough evidence to support the claim that the distribution is different.

12.
H_0: The proportions are as follows: transportation - 33%, industry - 30%, residential - 20%, and commercial - 17%. (claim)
H_1: The distribution is not the same as stated in the null hypothesis.
C. V. = 7.815 d. f. = 3 $\alpha = 0.05$

$$\chi^2 = \frac{(108-99)^2}{99} + \frac{(93-90)^2}{90} + \frac{(51-60)^2}{60} + \frac{(48-51)^2}{51}$$

$$\chi^2 = 2.4447$$

12. continued

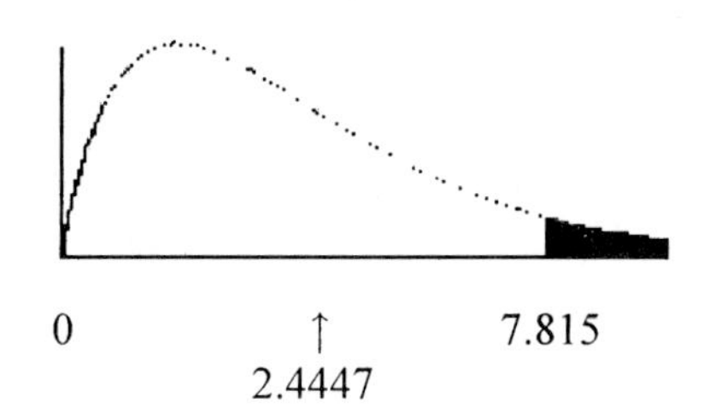

0 ↑ 2.4447 7.815

Do not reject the null hypothesis. There is enough evidence to support the claim that the proportions are as stated.

13.
H_0: The method of payment for purchases is distributed as follows: 53% cash, 30% checks, 16% credit cards, and 1% no preference. (claim)
H_1: The distribution is not the same as stated in the null hypothesis.
C. V. = 11.345 d. f. = 3 $\alpha = 0.01$

$$\chi^2 = \frac{(400-424)^2}{424} + \frac{(210-240)^2}{240} + \frac{(170-128)^2}{128}$$

$$+ \frac{(20-8)^2}{8} = 36.8898$$

Alternate Solution:

O	E	O − E	(O − E)²	$\frac{(O-E)^2}{E}$
400	424	-24	576	1.3585
210	240	-30	900	3.7500
170	128	42	1764	13.7813
20	8	12	144	18.0000
				36.8898

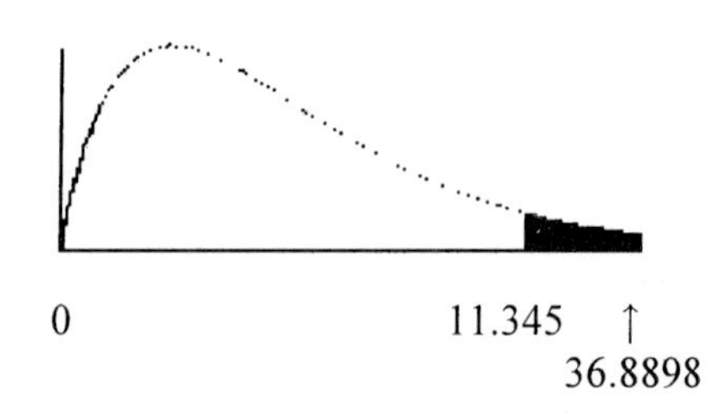

0 11.345 ↑ 36.8898

Reject the null hypothesis. There is enough evidence to reject the claim that the distribution is the same as reported in the survey.

14.
H_0: The distribution is as follows: violent offenses - 12.6%, property offenses - 8.5%,

14. continued
drug offenses - 60.2%, weapons offenses - 8.2%, immigration offenses - 4.9%, and other offenses - 5.6%. (claim)
H_1: The distribution is not the same as stated in the null hypothesis.
C. V. = 11.071 d. f. = 5 $\alpha = 0.05$

$$\chi^2 = \frac{(64-63)^2}{63} + \frac{(40-42.5)^2}{42.5} + \frac{(326-301)^2}{301} +$$

$$\frac{(42-41)^2}{41} + \frac{(25-24.5)^2}{24.5} + \frac{(3-28)^2}{28} = 24.58$$

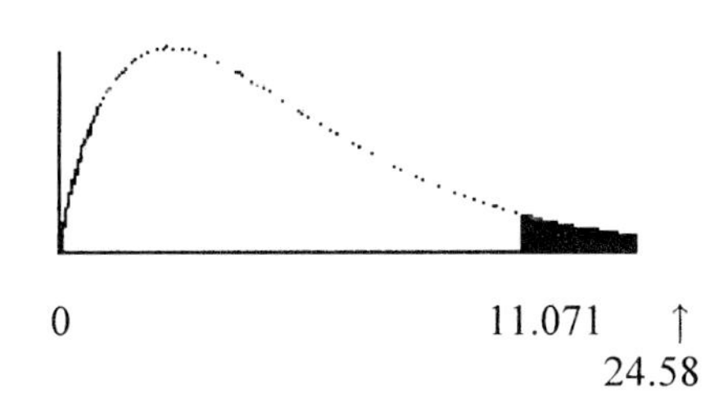

Reject the null hypothesis. There is enough evidence to reject the claim that the proportions are as stated.

15.
H_0: The distribution is as follows: violent offenses - 29.5%, property offenses - 29%, drug offenses - 30.2%, weapons offenses - 10.6%, other offenses - 0.7%. (claim)
H_1: The distribution is not the same as stated in the null hypothesis.
C. V. = 9.488 d. f. = 4 $\alpha = 0.05$

$$\chi^2 = \sum\frac{(O-E)^2}{E} = \frac{(298-295)^2}{295} + \frac{(275-290)^2}{290}$$

$$+ \frac{(344-302)^2}{302} + \frac{(80-106)^2}{106} + \frac{(3-7)^2}{7}$$

$\chi^2 = 15.3106$

Alternate Solution:

O	E	O − E	(O − E)^2	$\frac{(O-E)^2}{E}$
298	295	3	9	0.0305
275	290	-15	225	0.7759
344	302	42	1764	5.8411
80	106	-26	676	6.3774
3	7	-4	16	2.2857
				15.3106

15. continued

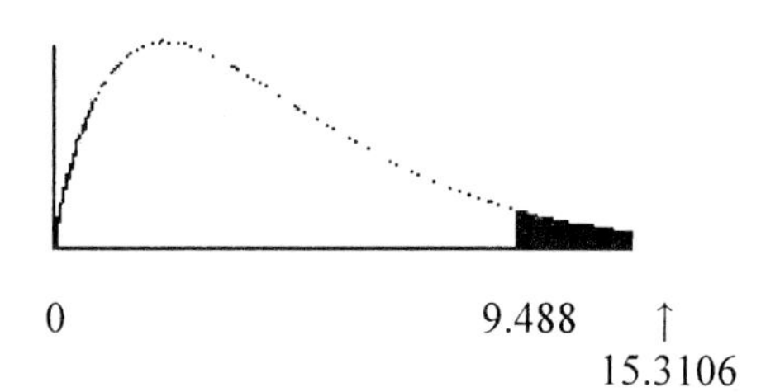

Reject the null hypothesis. There is not enough evidence to support the claim that the proportions are as stated.

16.
H_0: The number of defective parts manufactured each day is the same. (claim)
H_1: The distribution is not the same as stated in the null hypothesis.
$\alpha = 0.05$ d. f. = 4
P-value < 0.005 (0.0047)

$E = \frac{130}{5} = 26$

$$\chi^2 = \frac{(32-26)^2}{26} + \frac{(16-26)^2}{26} + \frac{(23-26)^2}{26} +$$

$$\frac{(19-26)^2}{26} + \frac{(40-26)^2}{26} = 15$$

Reject the null hypothesis since P-value < 0.05. There is enough evidence to reject the claim that the number of defective parts manufactured each day is the same.

17.
H_0: 50% of customers purchase word processing programs, 25% purchase spread sheet programs, and 25% purchase data base programs. (claim)
H_1: The distribution is not the same as stated in the null hypothesis.
$\alpha = 0.05$ d. f. = 2
P-value > 0.10 (0.741)

$$\chi^2 = \sum\frac{(O-E)^2}{E} = \frac{(38-40)^2}{40} + \frac{(23-20)^2}{20} + \frac{(19-20)^2}{20} = 0.6$$

Alternate Solution:

O	E	O − E	(O − E)^2	$\frac{(O-E)^2}{E}$
38	40	-2	4	0.1
23	20	3	9	0.45
19	20	-1	1	0.05
				0.60

17. continued
Do not reject the null hypothesis since P-value > 0.05. There is not enough evidence to reject the store owner's assumption.

18.
H_0: The coins are balanced and randomly tossed. (claim)
H_1: The distribution is not the same as stated in the null hypothesis.
C. V. = 7.815 d. f. = 3
E(0) = 0.125(72) = 9
E(1) = 0.375(72) = 27
E(2) = 0.375(72) = 27
E(3) = 0.125(72) = 9
(use the binomial distribution with n = 3 and p = 0.05)

$\chi^2 = \frac{(3-9)^2}{9} + \frac{(10-27)^2}{27} + \frac{(17-27)^2}{27} +$

$\frac{(42-9)^2}{9} = 139.4$

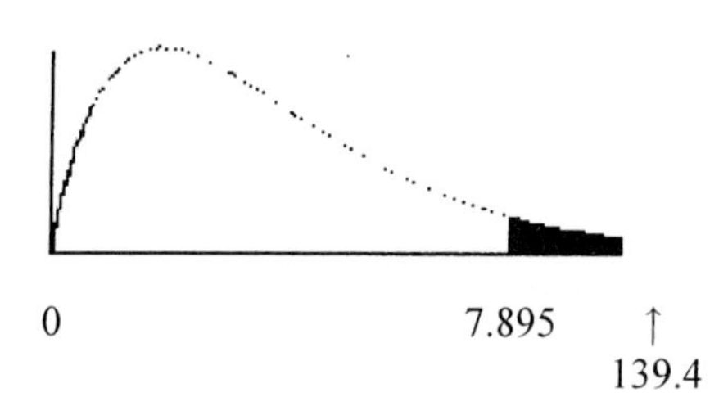

Reject the null hypothesis. There is enough evidence to reject the claim that the coins are balanced and randomly tossed.

19.
Answers will vary.

EXERCISE SET 11-3

1.
The independence test and the goodness of fit test both use the same formula for computing the test-value; however, the independence test uses a contingency table whereas the goodness of fit test does not.

2.
d. f. = (rows − 1)(columns − 1)

3.
H_0: The variables are independent or not related.
H_1: The variables are dependent or related.

4.
Contingency table.

5.
The expected values are computed as (row total · column total) ÷ grand total.

6.
The test of independence is used to determine whether two variables selected from a single sample are related. The test of homogeneity of proportions is used to determine whether proportions are equal.

7.
H_0: $p_1 = p_2 = p_3 = \cdots = p_n$
H_1: At least one proportion is different from the others.

8.
H_0: Blood pressure of individuals is independent of jogging. (claim)
H_1: Blood pressure of individuals is dependent on jogging.
C. V. = 5.991 d. f. = 26 $\alpha = 0.05$

Blood Pressure

	Low	Moderate	High
Joggers	34(26.133)	57(64)	21(21.867)
Non-joggers	15(22.867)	63(56)	20(19.133)

$\chi^2 = \sum\frac{(O-E)^2}{E} = \frac{(34-26.133)^2}{26.133} + \frac{(57-64)^2}{64} +$

$\frac{(21-21.867)^2}{21.867} + \frac{(15-22.867)^2}{22.867} + \frac{(63-56)^2}{56}$

$+ \frac{(20-19.133)^2}{19.133}$

$\chi^2 = 6.789$

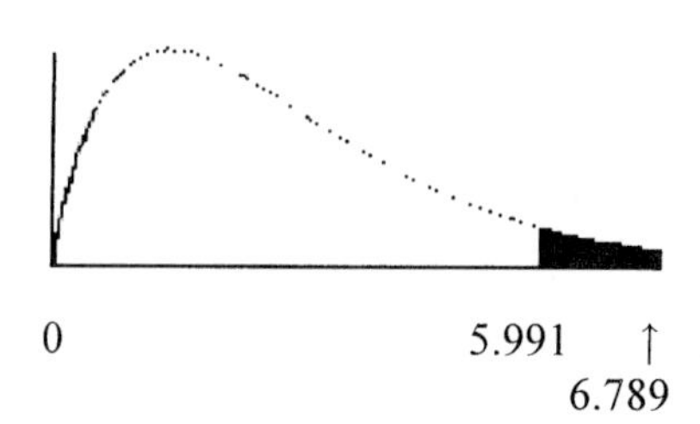

Reject the null hypothesis. There is enough evidence to reject the claim that the blood pressure of individuals is independent of jogging.

9.
H_0: Type of pet owned is independent of annual household income.
H_1: Type of pet owned is dependent on annual household income. (claim)

C. V. = 21.026 d. f. = 12 $\alpha = 0.05$

$E = \frac{\text{(row sum)(column sum)}}{\text{grand total}}$

$E_{1,1} = \frac{(534)(1003)}{4004} = 133.7667$

$E_{1,2} = \frac{(534)(1000)}{4004} = 133.3666$

$E_{1,3} = \frac{(534)(1000)}{4004} = 133.3666$

$E_{1,4} = \frac{(534)(1001)}{4004} = 133.5$

$E_{2,1} = \frac{(800)(1003)}{4004} = 200.3996$

$E_{2,2} = \frac{(800)(1000)}{4004} = 199.8002$

$E_{2,3} = \frac{(800)(1000)}{4004} = 199.8002$

$E_{2,4} = \frac{(800)(1001)}{4004} = 200.0$

$E_{3,1} = \frac{(869)(1003)}{4004} = 208.6661$

$E_{3,2} = \frac{(833)(1000)}{4004} = 208.0420$

$E_{3,3} = \frac{(833)(1000)}{4004} = 208.0420$

$E_{3,4} = \frac{(833)(1001)}{4004} = 208.25$

$E_{4,1} = \frac{(968)(1003)}{4004} = 242.2835$

$E_{4,2} = \frac{(968)(1000)}{4004} = 241.7582$

$E_{4,3} = \frac{(968)(1000)}{4004} = 241.7582$

$E_{4,4} = \frac{(968)(1001)}{4004} = 242.0$

9. continued

Type of Pet

Income	Dog	Cat
< $12,500	127(133.7667)	139(133.3666)
$12,500 - $24,999	191(199.8002)	197(199.8002)
$25,000 - $39,999	216(217.6841)	215(217.0330)
$40,000 - $59,999	215(208.6661)	212(208.0420)
$60,000 & over	254(242.4835)	237(241.7582)

Income	Bird	Horse
< $12,500	173(133.3666)	95(133.5)
$12,500 - $24,999	209(199.8002)	203(200.0)
$25,000 -$39,999	220(217.0330)	218(217.25)
$40,000 - $59,999	175(208.0420)	231(208.25)
$60,000 & over	223(241.7582)	254(242.0)

$$\chi^2 = \sum \frac{(O-E)^2}{E} = \frac{(127-133.7667)^2}{133.7667} + \frac{(139-133.3666)^2}{133.3666}$$

$$+ \frac{(173-133.3666)^2}{133.3666} + \frac{(95-133.5)^2}{133.5} + \frac{(191-199.8002)^2}{199.8002}$$

$$+ \frac{(197-199.8002)^2}{199.8002} + \frac{(209-199.8002)^2}{199.8002} + \frac{(203-200.0)^2}{200}$$

$$+ \frac{(216-217.6841)^2}{217.6841} + \frac{(215-217.0330)^2}{217.0330} + \frac{(220-217.0330)^2}{217.0330}$$

$$+ \frac{(218-217.25))^2}{217.25} + \frac{(215-208.6661)^2}{208.6661} + \frac{(212-208.0420)^2}{208.0420}$$

$$+ \frac{(175-208.0420)^2}{208.0420} + \frac{(231-208.25)^2}{208.25} + \frac{(254-242.4835)^2}{242.4835}$$

$$+ \frac{(237-241.7582)^2}{241.7582} + \frac{(223-241.7582)^2}{241.7582} + \frac{(254-242.0)^2}{242}$$

$\chi^2 = 35.177$

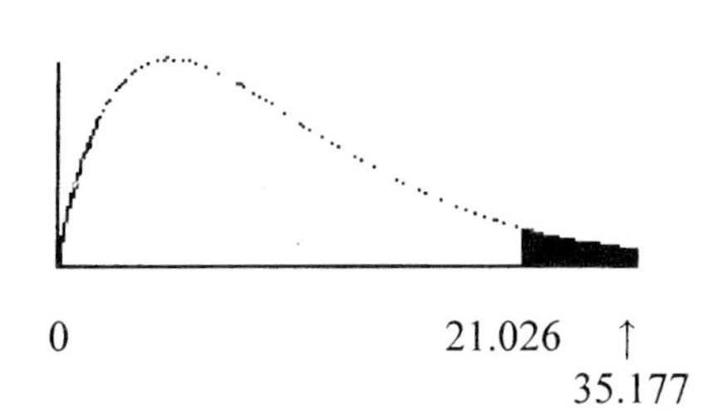

Reject the null hypothesis. There is enough evidence to support the claim that the type of pet is dependent upon the income of the owner.

10.
H_0: The rank of women personnel is independent of the military branch of service.
H_1: The rank of women personnel is dependent upon the military branch of service. (claim)

C. V. = 7.815 d. f. = 3 $\alpha = 0.05$

10. continued

Rank

Branch	Officers	Enlisted
Army	10,791(11,463.0612)	62,491(61,818.9388)
Navy	7816(7909.7344)	42,750(42,656.2656)
Marine Corps	932(1635.7254)	9525(8821.2746)
Air Force	11,819(10,349.4790)	54,344(55,813.5210)

$$\chi^2 = \frac{(10791-11463.0612)^2}{11463.0612} + \frac{(62491-61818.9388)^2}{61818/9388}$$

$$+ \frac{(7816-7909.7344)^2}{7909.7344} + \frac{(42750-42656.2656)^2}{42656.2656}$$

$$+ \frac{(932-1635.7254)^2}{1635.7254} + \frac{(9525-8821.2746)^2}{8821.2746}$$

$$+ \frac{(11819-10349.4790)^2}{10349.4790} + \frac{(54344-55813.5210)^2}{55813.5210}$$

$\chi^2 = 654.2719$

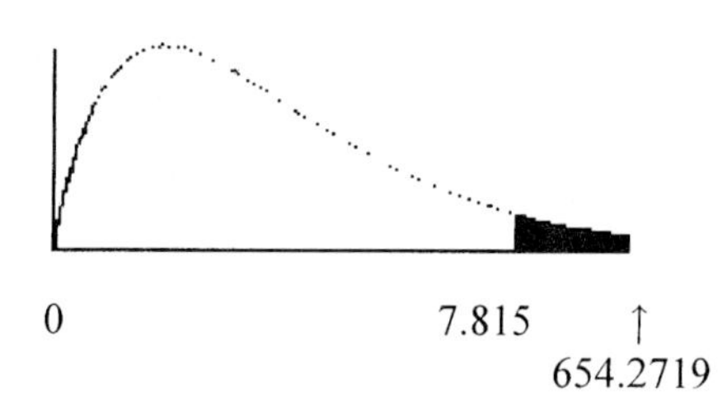

0 7.815 ↑ 654.2719

Reject the null hypothesis. There is enough evidence to support the claim that rank is dependent upon the military branch of service.

11.
H_0: The composition of the House of Representatives is independent of the state.
H_1: The composition of the House of Representatives is dependent upon the state. (claim)

C. V. = 7.815 d. f. = 3 $\alpha = 0.05$

$$E = \frac{\text{(row sum)(column sum)}}{\text{grand total}}$$

$$E_{1,1} = \frac{(203)(320)}{542} = 119.8524$$

$$E_{1,2} = \frac{(203)(222)}{542} = 83.1476$$

$$E_{2,1} = \frac{(98)(320)}{542} = 57.8598$$

$$E_{2,2} = \frac{(98)(222)}{542} = 40.1402$$

$$E_{3,1} = \frac{(100)(320)}{542} = 59.0406$$

$$E_{3,2} = \frac{(100)(222)}{542} 40.9594$$

11. continued

$$E_{4,1} = \frac{(141)(320)}{542} = 83.2472$$

$$E_{4,2} = \frac{(141)(222)}{542} = 57.7528$$

State	Democrats	Republicans
PA	100(119.8524)	103(83.1476)
OH	39(57.8598)	59(40.1402)
WV	75(59.0406)	25(40.9594)
MD	106(83.2472)	35(57.7528)

$$\chi^2 = \sum\frac{(O-E)^2}{E} = \frac{(100-119.8524)^2}{119.8524} + \frac{(103-83.1476)^2}{83.1476}$$

$$+ \frac{(39-57.8598)^2}{57.8598} + \frac{(59-40.1402)^2}{40.1402} + \frac{(75-59.0406)^2}{59.0406}$$

$$+ \frac{(25-40.9594)^2}{40.9594} + \frac{(106-83.2472)^2}{83.2472} + \frac{(35-57.7528)^2}{57.7528}$$

$\chi^2 = 48.7521$

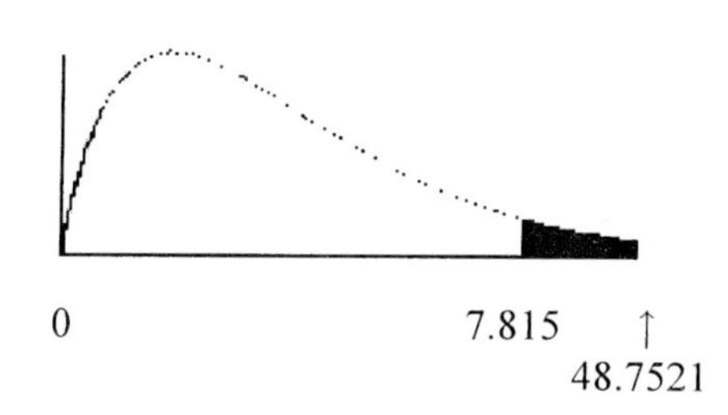

0 7.815 ↑ 48.7521

Reject the null hypothesis. There is enough evidence to support the claim that the composition is dependent upon the state.

12.
H_0: The size of the population (by age) is independent of the state.
H_1: The size of the population (by age) is dependent upon the state. (claim)

C. V. = 11.071 d. f. = 5 $\alpha = 0.05$

State	Under 5	5 - 17	18 - 24
PA	721(753.9308)	2140(2190.0631)	1025(1078.5184)
OH	740(707.0692)	2104(2053.9369)	1065(1011.4816)

State	25 - 44	45 - 64	65 +
PA	3515(3547.2417)	2702(2677.7185)	1899(1754.5275)
OH	3359(3326.7583)	2487(2511.2815)	1501(1645.4725)

$$\chi^2 = \frac{(721-753.9308)^2}{753.9308} + \frac{(2140-2190.0631)^2}{2190.0631}$$

$$+ \frac{(1025-1078.5184)^2}{1078.5184} + \frac{(3515-3547.2417)^2}{3547.2417}$$

$$+ \frac{(2702-2677.7185)^2}{2677.7185} + \frac{(1899-1754.5275)^2}{1754.5275}$$

12. continued

$$+\frac{(740-707.0692)^2}{707.0692}+\frac{(2104-2053.9369)^2}{2053.9369}$$

$$+\frac{(1065-1011.4816)^2}{1011.4816}+\frac{(3359-3326.7583)^2}{3326.7583}$$

$$+\frac{(2487-2511.2815)^2}{2511.2815}+\frac{(1501-1645.4725)^2}{1645.4725}$$

$\chi^2 = 36.4656$

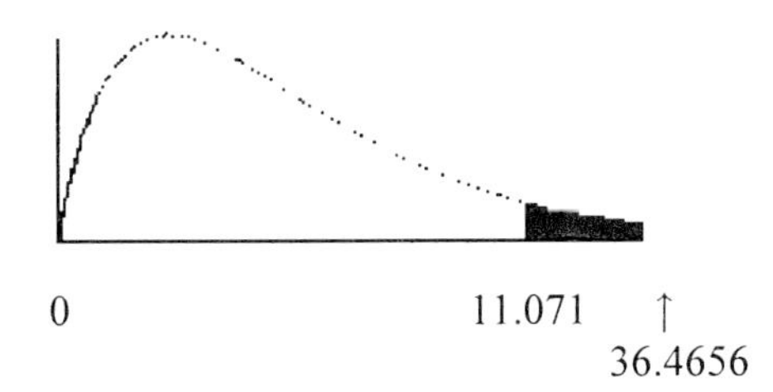

0　　11.071　↑ 36.4656

Reject the null hypothesis. There is enough evidence to support the claim that the size of the population (by age) is dependent upon the state.

13.
H_0: The number of ads people think they've seen or heard in the media is independent of the gender of the individual.
H_1: The number of ads people think they've seen or heard in the media is dependent upon the gender of the individual. (claim)
C. V. = 13.277　d. f. = 4　$\alpha = 0.01$

$E_{1,1} = \frac{(300)(95)}{510} = 55.882$

$E_{1,2} = \frac{(300)(110)}{510} = 64.706$

$E_{1,3} = \frac{(300)(144)}{510} = 84.706$

$E_{1,4} = \frac{(300)(84)}{510} = 49.412$

$E_{1,5} = \frac{(300)(77)}{510} = 45.294$

$E_{2,1} = \frac{(210)(95)}{510} = 39.118$

$E_{2,2} = \frac{(210)(110)}{510} = 45.294$

$E_{2,3} = \frac{(210)(144)}{510} = 59.294$

$E_{2,4} = \frac{(210)(84)}{510} = 34.588$

$E_{2,5} = \frac{(210)(77)}{510} = 31.706$

13. continued

Gender	1 - 30	31 - 50	51 - 100
Men	45(55.882)	60(64.706)	90(84.706)
Women	50(39.118)	50(45.294)	54(59.294)
Total	95	110	144

Gender	101 - 300	301 or more	Total
Men	54(49.412)	51(45.294)	300
Women	30(34.588)	26(31.706)	210
Total	84	77	510

$$\chi^2 = \sum\frac{(O-E)^2}{E} = \frac{(45-55.882)^2}{55.882}+\frac{(60-64.706)^2}{64.706}$$

$$+\frac{(90-84.706)^2}{84.706}+\frac{(54-49.412)^2}{49.412}+\frac{(51-45.294)^2}{45.294}$$

$$+\frac{(50-39.118)^2}{39.118}+\frac{(50-45.294)^2}{45.294}+\frac{(54-59.294)^2}{59.294}$$

$$+\frac{(30-34.588)^2}{34.588}+\frac{(26-31.706)^2}{31.706} = 9.562$$

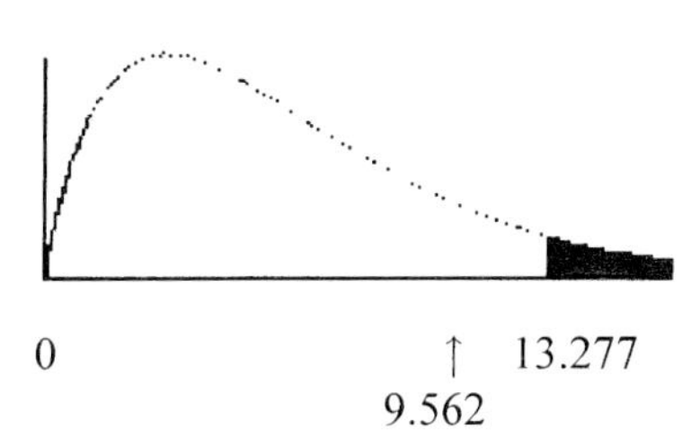

0　↑ 9.562　13.277

Do not reject the null hypothesis. There is not enough evidence to support the claim that the number of ads people think they've seen or heard is related to the gender of the individual.

14.
H_0: The way people obtain information is independent of their educational background. (claim)
H_1: The way people obtain information is dependent on their educational background.
C. V. = 5.991　d. f. = 2

$$\chi^2 = \frac{(159-139.5)^2}{139.5}+\frac{(90-99)^2}{99}+\frac{(51-61.5)^2}{61.5}$$

$$+\frac{(27-46.5)^2}{46.5}+\frac{(42-33)^2}{33}+\frac{(31-20.5)^2}{20.5}$$

$\chi^2 = 21.347$

	TV	Paper	Other	Total
High School	159(139.5)	90(99)	51(61.5)	300
College	27(46.5)	42(33)	31(20.5)	100
Total	186	132	82	400

14. continued

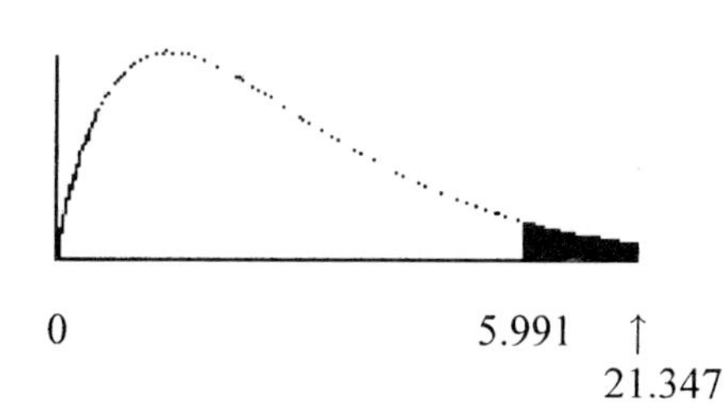

Reject the null hypothesis. There is not enough evidence to support the claim that people obtain information is independent of their educational background.

15.
H_0: The grade a student receives is independent of the number of hours the student works.
H_1: The grade a student receives is dependent upon the number of hours the student works. (claim)

C. V. = 12.592 d. f. = 6 $\alpha = 0.05$

$E_{1,1} = \frac{(21)(18)}{92} = 4.1087$

$E_{1,2} = \frac{(21)(40)}{92} = 9.1304$

$E_{1,3} = \frac{(21)(15)}{92} = 3.4239$

$E_{1,4} = \frac{(21)(19)}{92} = 4.3370$

$E_{2,1} = \frac{(27)(18)}{92} = 5.2826$

$E_{2,2} = \frac{(27)(40)}{92} 11.7391$

$E_{2,3} = \frac{(27)(15)}{92} = 4.4022$

$E_{2,4} = \frac{(27)(19)}{92} = 5.5761$

$E_{3,1} = \frac{(44)(18)}{92} = 8.6087$

$E_{3,2} = \frac{(44)(40)}{92} = 19.1304$

$E_{3,3} = \frac{(44)(15)}{92} = 7.1739$

$E_{3,4} = \frac{(44)(19)}{92} = 9.0870$

15. continued

Hours Working	A	B
20 +	5(4.1087)	8(9.1304)
10 - 20	5(5.2826)	12(11.7391)
< 10	8(8.6087)	20(19.1304)

Hours Working	C	D/F
20 +	3(3.4239)	5(4.3370)
10 - 20	6(4.4022)	4(5.5761)
< 10	6(7.1739)	10(9.0870)

$$\chi^2 = \sum\frac{(O-E)^2}{E} = \frac{(5-4.1087)^2}{4.1087} + \frac{(8-9.1304)^2}{9.1304}$$

$$+ \frac{(3-3.4239)^2}{3.4239} + \frac{(5-4.3370)^2}{4.3370} + \frac{(5-5.2826)^2}{5.2826}$$

$$+ \frac{(12-11.7391)^2}{11.7391} + \frac{(6-4.4022)^2}{4.4022} + \frac{(4-5.5761)^2}{5.5761}$$

$$+ \frac{(8-8.6087)^2}{8.6087} + \frac{(20-19.1304)^2}{19.1304} + \frac{(6-7.1739)^2}{7.1739}$$

$$+ \frac{(10-9.0870)^2}{9.0870} = 1.9$$

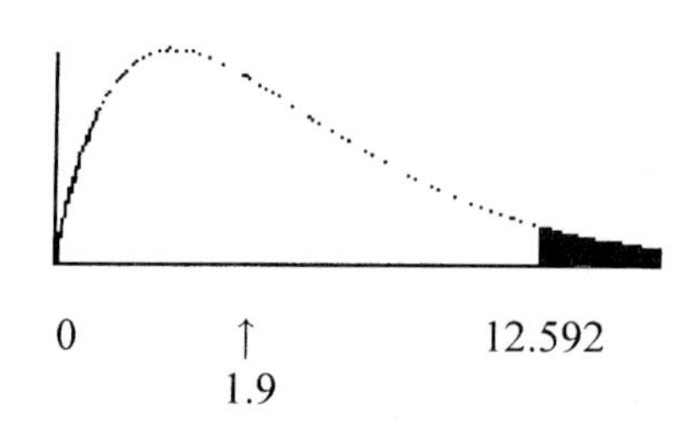

Do not reject the null hypothesis. There is not enough evidence to support the claim that grades are dependent upon the number of hours a student works.

16.
H_0: The number of speeders is independent of the state where the drivers reside.
H_1: The number of speeders is dependent upon the state where the driver resides. (claim)

C. V. = 3.841 d. f. = 1 $\alpha = 0.05$

	Ohio	Michigan
66 mph or more	18(22.7647)	25(20.2353)
65 mph or less	27(22.2353)	15(19.7647)

$$\chi^2 = \frac{(18-22.7647)^2}{22.7647} + \frac{(25-20.2353)^2}{20.2353}$$

$$+ \frac{(27-22.2353)^2}{22.2353} + \frac{(15-19.7647)^2}{19.7647} = 4.289$$

16. continued

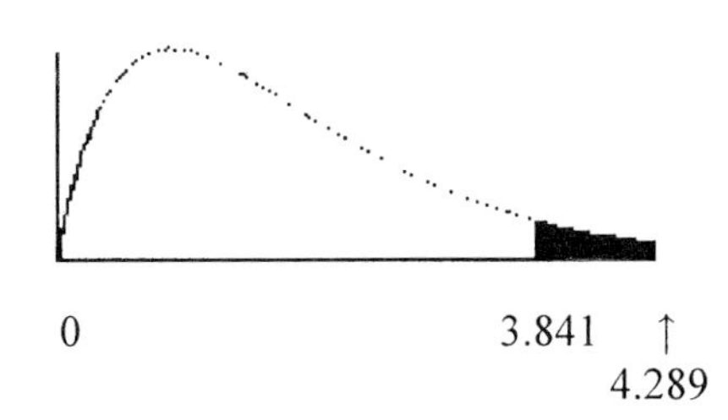

Reject the null hypothesis. There is enough evidence to support the claim that the number of speeders is dependent upon the state where the driver resides.

17.
H_0: The type of video rented by a person is independent of the person's age.
H_1: The type of video a person rents is dependent on the person's age. (claim)
C. V. = 13.362 d. f. = 8 $\alpha = 0.10$

Age	Doc.	Comedy	Mystery
12-20	14(6.588)	9(13.433)	8(10.979)
21-29	15(8.075)	14(16.467)	9(13.458)
30-38	9(14.663)	21(29.9)	39(24.438)
39-47	7(9.775)	22(19.933)	17(16.292)
48 +	6(11.9)	38(24.267)	12(19.833)

$$\chi^2 = \frac{(14-6.588)^2}{6.588} + \frac{(9-13.433)^2}{13.433} + \frac{(8-10.979)^2}{10.979}$$

$$+ \frac{(15-8.075)^2}{8.075} + \frac{(14-16.467)^2}{16.467} + \frac{(9-13.458)^2}{13.458}$$

$$+ \frac{(9-14.663)^2}{14.663} + \frac{(21-29.9)^2}{29.9} + \frac{(39-24.438)^2}{24.438}$$

$$+ \frac{(7-9.775)^2}{9.775} + \frac{(22-19.933)^2}{19.933} + \frac{(17-16.292)^2}{16.292}$$

$$+ \frac{(6-11.9)^2}{11.9} + \frac{(38-24.267)^2}{24.267} + \frac{(12-19.833)^2}{19.833}$$

$\chi^2 = 46.733$

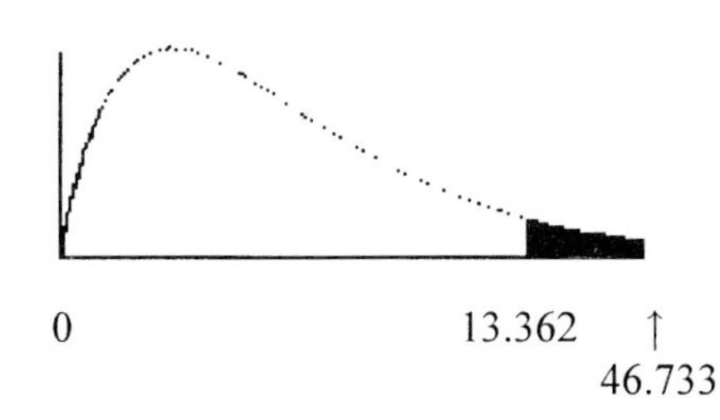

Reject the null hypothesis. There is enough evidence to support the claim that the type of movie selected is related to the age of the customer.

18.
H_0: The preference for two or four wheel drive is independent of the gender of the buyer. (claim)
H_1: The preference for two or four wheel drive is dependent on the gender of the buyer.
C. V. = 3.841 d. f. = 1

Gender	Two-Wheel	Four-Wheel	Total
Male	23(30.067)	43(35.933)	66
Female	18(10.933)	6(13.067)	24
Total	41	49	90

$$\chi^2 = \frac{(23-30.067)^2}{30.067} + \frac{(43-35.933)^2}{35.933} +$$

$$\frac{(18-10.933)^2}{10.933} + \frac{(6-13.067)^2}{13.067} = 11.441$$

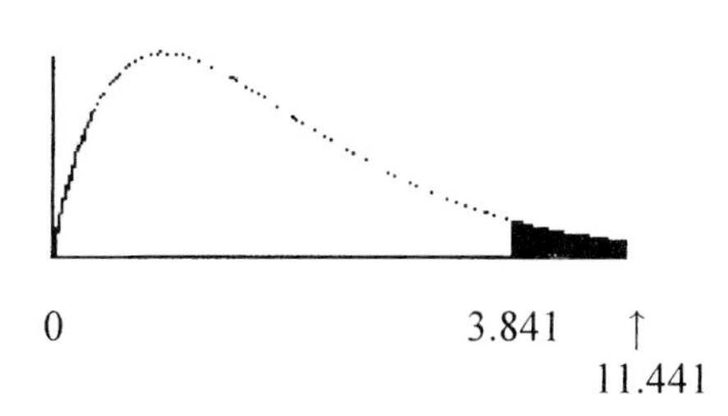

Reject the null hypothesis. There is not enough evidence to support the claim that preference for two or four-wheel drive is independent of gender of the buyer.

19.
H_0: The type of snack purchased is independent of the gender of the consumer. (claim)
H_1: The type of snack purchased is dependent upon the gender of the consumer.

C. V. = 4.605 d. f. = 2

Gender	Hot Dog	Peanuts	Popcorn	Total
Male	12(13.265)	21(15.388)	19(23.347)	52
Female	13(11.735)	8(13.612)	25(20.653)	46
Total	25	29	44	98

$$\chi^2 = \sum\frac{(O-E)^2}{E} = \frac{(12-13.265)^2}{13.265} + \frac{(21-15.388)^2}{15.388}$$

$$+ \frac{(19-23.347)^2}{23.347} + \frac{(13-11.735)^2}{11.735} + \frac{(8-13.612)^2}{13.612}$$

$$+ \frac{(25-20.653)^2}{20.653} = 6.342$$

19. continued

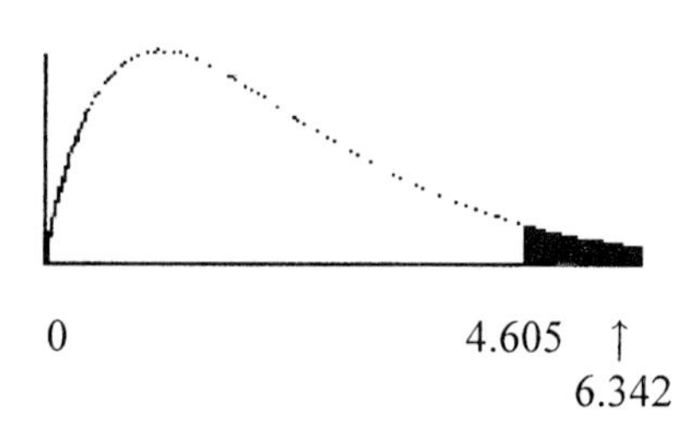

Reject the null hypothesis. There is enough evidence to reject the claim that the type of snack chosen is independent of the gender of the individual.

20.
H_0: The drug is not effective.
H_1: The drug is effective. (claim)
$\alpha = 0.10$ d. f. = 1

	Effective	Not effective	Total
Drug	32(25.408)	9(15.592)	41
Placebo	12(18.592)	18(11.408)	30
Total	44	27	71

$$\chi^2 = \frac{(32-25.408)^2}{25.408} + \frac{(9-15.592)^2}{15.592} +$$

$$\frac{(12-18.592)^2}{18.592} + \frac{(18-11.408)^2}{11.408} = 10.643$$

P-value < 0.005 (0.001)
Reject the null hypothesis since P-value < 0.10. There is enough evidence to support the claim that the drug is effective.

21.
H_0: The type of book selected by the individual is independent of the gender of the individual. (claim)
H_1: The type of book selected by the individual is dependent on the gender of the individual.
$\alpha = 0.05$ d. f. = 2

Gender	Mystery	Romance	Self-help	Total
Male	243(214.121)	201(198.260)	191(222.618)	635
Female	135(163.879)	149(151.740)	202(170.382)	486
Total	378	350	393	1121

$$\chi^2 = \sum\frac{(O-E)^2}{E} = \frac{(243-214.121)^2}{214.121} + \frac{(201-198.260)^2}{198.260}$$

$$+ \frac{(191-222.618)^2}{222.618} + \frac{(135-163.879)^2}{163.879}$$

$$+ \frac{(149-151.740)^2}{151.740} + \frac{(202-170.382)^2}{170.382} = 19.429$$

21. continued
P-value < 0.005 (0.00006)
Reject the null hypothesis since P-value < 0.05. There is enough evidence to reject the claim that the type of book purchased is independent of gender.

22.
H_0: $p_1 = p_2 = p_3$ (claim)
H_1: At least one proportion is different.
C. V. = 9.210 d. f. = 2

$$E(\text{will donate}) = \frac{50(63)}{150} = 21$$

$$E(\text{will not donate}) = \frac{50(87)}{150} = 29$$

	A	B	C
will donate	28(21)	14(21)	21(21)
won't donate	22(29)	36(29)	29(29)

$$\chi^2 = \frac{(28-21)^2}{21} + \frac{(14-21)^2}{21} + \frac{(21-21)^2}{21} +$$

$$\frac{(22-29)^2}{29} + \frac{(36-29)^2}{29} + \frac{(29-29)^2}{29} = 8.046$$

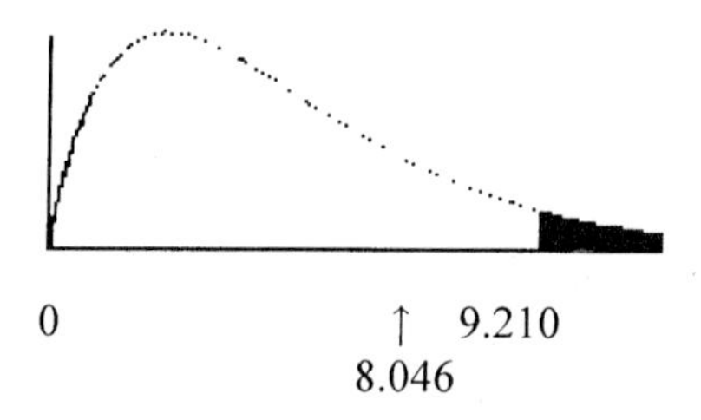

Do not reject the null hypothesis. There is not enough evidence to reject the claim that the proportions are equal.

23.
H_0: $p_1 = p_2 = p_3 = p_4$ (claim)
H_1: At least one proportion is different.

C. V. = 7.815 d. f. = 3

$$E(\text{passed}) = \frac{120(167)}{120} = 41.75$$

$$E(\text{failed}) = \frac{120(313)}{120} = 78.25$$

	Southside	West End	East Hills	Jefferson
Passed	49(41.75)	38(41.75)	46(41.75)	34(41.75)
Failed	71(78.25)	82(78.25)	74(78.25)	86(78.25)

23. continued

$$\chi^2 = \frac{(49-41.75)^2}{41.75} + \frac{(38-41.75)^2}{41.75} + \frac{(46-41.75)^2}{41.75}$$

$$+ \frac{(34-41.75)^2}{41.75} + \frac{(71-78.25)^2}{78.25} + \frac{(82-78.25)^2}{78.25}$$

$$+ \frac{(74-78.25)^2}{78.25} + \frac{(86-78.25)^2}{78.25} = 5.317$$

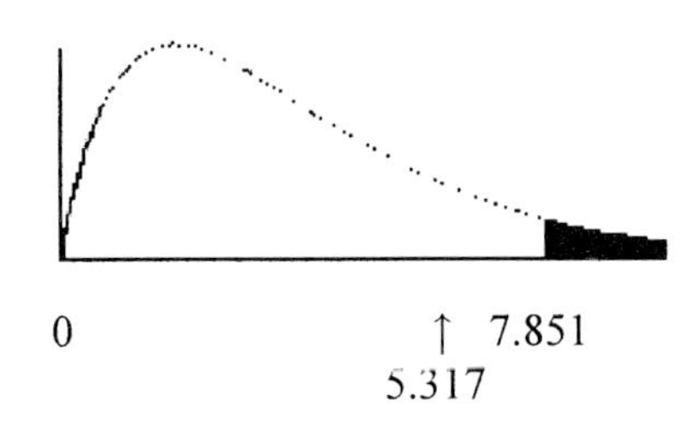

0 ↑ 7.851
5.317

Do not reject the null hypothesis. There is not enough evidence to reject the claim that the proportions are equal.

24.
H_0: $p_1 = p_2 = p_3$ (claim)
H_1: At least one proportion is different.
C. V. = 9.210 d. f. = 2

$$E(\text{yes}) = \frac{92(133)}{276} = 43.47$$

$$E(\text{no}) = \frac{92(143)}{276} = 47.67$$

	Mall A	Mall B	Mall C
Yes	52(43.37)	45(43.37)	36(43.37)
No	40(47.67)	47(47.67)	56(47.67)

$$\chi^2 = \frac{(52-43.37)^2}{43.37} + \frac{(45-43.37)^2}{43.37} + \frac{(36-43.37)^2}{43.37}$$

$$\frac{(40-47.67)^2}{47.67} + \frac{(47-47.67)^2}{47.67} + \frac{(56-47.67)^2}{47.67}$$

$$\chi^2 = 5.73$$

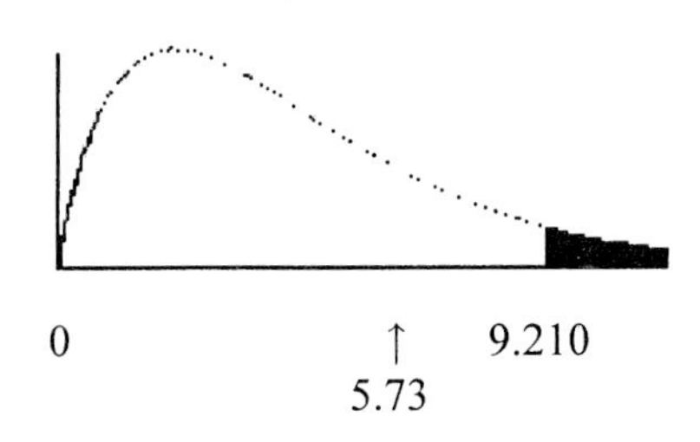

0 ↑ 9.210
5.73

Do not reject the null hypothesis. There is not enough evidence to reject the claim that the proportions are equal.

25.
H_0: $p_1 = p_2 = p_3 = p_4$ (claim)
H_1: At least one proportion is different.
C. V. = 7.815 d. f. = 3

$$E(\text{yes}) = \frac{107(86)}{344} = 26.75$$

$$E(\text{no}) = \frac{237(86)}{344} = 59.25$$

	21-29	30-39	40-49	50 +
Yes	32(26.75)	28(26.75)	26(26.75)	21(26.75)
No	54(59.25)	58(59.25)	60(59.25)	65(59.25)

$$\chi^2 = \frac{(32-26.75)^2}{26.75} + \frac{(28-26.75)^2}{26.75} + \frac{(26-26.75)^2}{26.75} +$$

$$\frac{(21-26.75)^2}{26.75} + \frac{(54-59.25)^2}{59.25} + \frac{(58-59.25)^2}{59.25} + \frac{(60-59.25)^2}{59.25}$$

$$+ \frac{(65-59.25)^2}{59.25} = 3.405$$

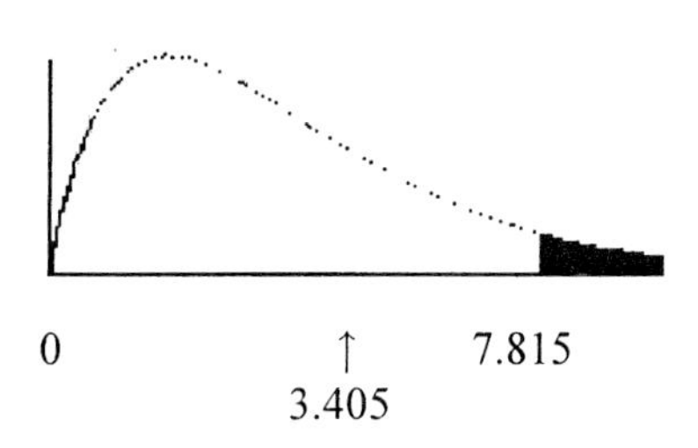

0 ↑ 7.815
3.405

Do not reject the null hypothesis. There is not enough evidence to reject the claim that the proportions are the same.

26.
H_0: $p_1 = p_2 = p_3$ (claim)
H_1: At least one proportion is different.
C. V. = 4.605 d. f. = 2

$$E(\text{mother works}) = \frac{60(118)}{180} = 39.33$$

$$E(\text{mother doesn't work}) = \frac{60(62)}{180} = 20.67$$

	Elem.	Middle	High
work	29(39.33)	38(39.33)	51(39.33)
no work	31(20.67)	22(20.67)	9(20.67)

$$\chi^2 = \frac{(29-39.33)^2}{39.33} + \frac{(38-39.33)^2}{39.33} + \frac{(51-39.33)^2}{39.33}$$

$$+ \frac{(31-20.67)^2}{20.67} + \frac{(22-20.67)^2}{20.67} + \frac{(9-20.67)^2}{20.67}$$

$$\chi^2 = 18.06$$

26. continued

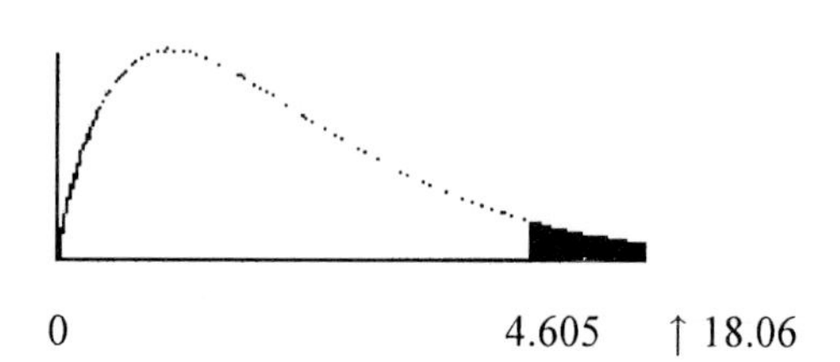

Reject the null hypothesis. There is enough evidence to conclude that at least one proportion is different.

27.
H_0: $p_1 = p_2 = p_3 = p_4$ (claim)
H_1: At least one proportion is different.
C. V. = 6.251 d. f. = 3

$E(\text{yes}) = \frac{(100)(132)}{400} = 33$

$E(\text{no}) = \frac{(100)(268)}{400} = 67$

	North	South	East	West
Yes	43(33)	39(33)	22(33)	28(33)
No	57(67)	61(67)	78(67)	72(67)

$\chi^2 = \frac{(43-33)^2}{33} + \frac{(39-33)^2}{33} + \frac{(22-33)^2}{33} +$

$\frac{(28-33)^2}{33} + \frac{(57-67)^2}{67} + \frac{(61-67)^2}{67} + \frac{(78-67)^2}{67}$

$+ \frac{(72-67)^2}{67} = 12.755$

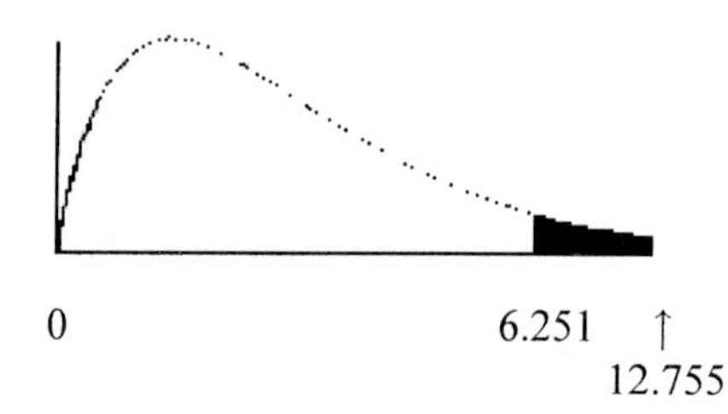

Reject the null hypothesis. There is enough evidence to reject the claim that the proportions are the same.

28.
H_0: $p_1 = p_2 = p_3 = p_4$ (claim)
H_1: At least one proportion is different.
C. V. = 7.815 d. f. = 3

$E(\text{present}) = \frac{239(75)}{300} = 59.75$

$E(\text{not present}) = \frac{61(75)}{300} = 15.25$

28. continued

	A	B	C	D
Present	66(59.75)	60(59.75)	57(59.75)	56(59.75)
Not present	9(15.25)	15(15.25)	18(15.25)	19(15.25)

$\chi^2 = \frac{(66-59.75)^2}{59.75} + \frac{(60-59.75)^2}{59.75} + \frac{(57-59.75)^2}{59.75}$

$+ \frac{(56-59.75)^2}{59.75} + \frac{(9-15.25)^2}{15.25} + \frac{(15-15.25)^2}{15.25} +$

$\frac{(18-15.25)^2}{15.25} + \frac{(19-15.25)^2}{15.25} = 5$

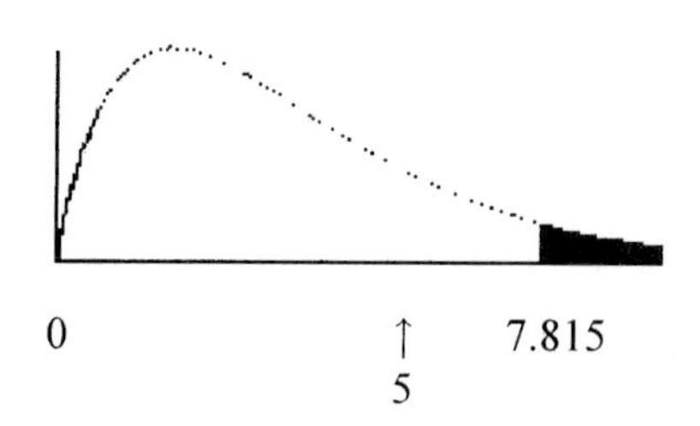

Do not reject the null hypothesis. There is not enough evidence to reject the claim that the proportions are the same.

29.
H_0: $p_1 = p_2 = p_3 = p_4$ (claim)
H_1: At least one proportion is different.
$\alpha = 0.05$ d. f. = 3

$E(\text{on bars}) = \frac{30(62)}{120} = 15.5$

$E(\text{not on bars}) = \frac{30(58)}{120} = 14.5$

	N	S	E	W
on	15(15.5)	18(15.5)	13(15.5)	16(15.5)
off	15(14.5)	12(14.5)	17(14.5)	14(14.5)

$\chi^2 = \frac{(15-15.5)^2}{15.5} + \frac{(18-15.5)^2}{15.5} + \frac{(13-15.5)^2}{15.5} +$

$\frac{(16-15.5)^2}{15.5} + \frac{(15-14.5)^2}{14.5} + \frac{(12-14.5)^2}{14.5} +$

$\frac{(17-14.5)^2}{14.5} + \frac{(14-14.5)^2}{14.5} = 1.734$

P-value > 0.10 (0.629)
Do not reject the null hypothesis. There is not enough evidence to reject the claim that the proportions are the same.

30.
H_0: $p_1 = p_2 = p_3 = p_4$ (claim)
H_1: At least one proportion is different.
$\alpha = 0.10$ d. f. = 3

30. continued

E(will travel) $= \frac{125(184)}{500} = 46$

E(will not travel) $= \frac{125(316)}{500} = 79$

	A	B	C	D
will	37(46)	52(46)	46(46)	49(46)
will not	88(79)	73(79)	79(79)	76(79)

$\chi^2 = \frac{(37-46)^2}{46} + \frac{(52-46)^2}{46} + \frac{(46-46)^2}{46} +$

$\frac{(49-46)^2}{46} + \frac{(88-79)^2}{79} + \frac{(73-79)^2}{79} +$

$\frac{(79-79)^2}{79} + \frac{(76-79)^2}{79} = 4.334$

P-value > 0.10 (0.228)

Do not reject the null hypothesis since P-value > 0.10. There is not enough evidence to reject the claim that the proportions are the same.

31.

H_0: $p_1 = p_2 = p_3$ (claim)

H_1: At least one proportion is different.

C. V. = 4.605 d. f. = 2

E(list) $= \frac{96(219)}{288} = 73$

E(no list) $= \frac{96(69)}{288} = 23$

	A	B	C
list	77(73)	74(73)	68(73)
no list	19(23)	22((23)	28(23)

$\chi^2 = \frac{(77-73)^2}{73} + \frac{(74-73)^2}{73} + \frac{(68-73)^2}{73}$

$+ \frac{(19-23)^2}{23} + \frac{(22-23)^2}{23} + \frac{(28-23)^2}{23}$

$\chi^2 = 2.401$

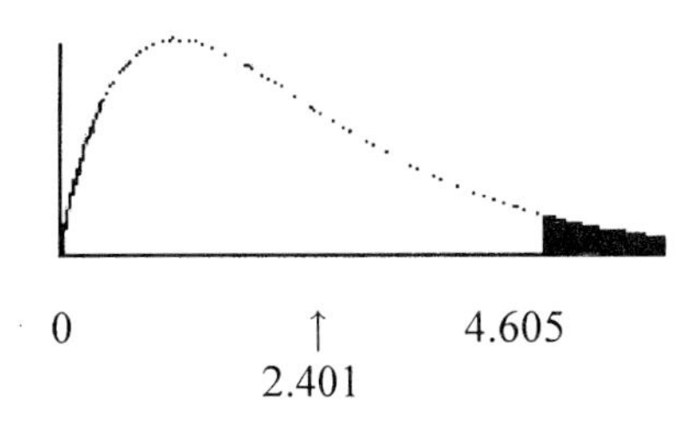

0 ↑ 4.605

2.401

Do not reject the null hypothesis. There is not enough evidence to reject the claim that the proportions are the same.

32.

			Total
	12(9.61)	15(17.39)	27
	9(11.39)	23(20.61)	32
Total	21	38	59

$\chi^2 = \frac{(12-9.61)^2}{9.61} + \frac{(15-17.39)^2}{17.39}$

$+ \frac{(9-11.39)^2}{11.39} + \frac{(23-20.61)^2}{20.61} = 1.70$

$\chi^2 = \frac{59(12\cdot23-15\cdot9)^2}{(12+15)(12+9)(9+23)(15+23)}$

$= \frac{1172979}{689472} = 1.70$

Alternate Method:

$\chi^2 = \frac{n(ad-bc)^2}{(a+b)(a+c)(c+d)(b+d)}$

$\chi^2 = \frac{59(12\cdot23-15\cdot9)^2}{(12+15)(12+9)(9+23)(15+23)} = 1.70$

Both answers are the same.

33.

$\chi^2 = \frac{(|O-E|-0.5)^2}{E} = \frac{(|12-9.6|-0.5)^2}{9.6}$

$+ \frac{(|15-17.4|-0.5)^2}{17.4} + \frac{(|9-11.4|-0.5)^2}{11.4}$

$+ \frac{(|23-20.6|-0.5)^2}{20.6}$

$= \frac{3.61}{9.6} + \frac{3.61}{17.4} + \frac{3.61}{11.4} + \frac{3.61}{20.6}$

$= 0.376 + 0.207 + 0.317 + 0.175 = 1.075$

34.

For 8:

$\chi^2 = 6.789$

$n = 34 + 57 + 21 + 15 + 63 + 20 = 210$

$C = \sqrt{\frac{\chi^2}{\chi^2+n}} = \sqrt{\frac{6.789}{6.789 + 210}} = 0.177$

For 20:

$\chi^2 = 10.643$

$n = 32 + 9 + 12 + 18 = 71$

$C = \sqrt{\frac{10.643}{10.643 + 71}} = 0.361$

REVIEW EXERCISES - CHAPTER 11

1.

H_0: The number of sales is equally distributed over five regions. (claim)

H_1: The null hypothesis is not true.

1. continued
C. V. = 9.488 d. f. = 4
$E = \frac{1328}{5} = 265.6$

$\chi^2 = \sum \frac{(O-E)^2}{E} = \frac{(236-265.6)^2}{265.6}$

$+ \frac{(324-265.6)^2}{265.6} + \frac{(182-265.6)^2}{265.6}$

$+ \frac{(221-265.6)^2}{265.6} + \frac{(365-265.6)^2}{265.6} = 87.14$

Alternate Solution:

O	E	O − E	$(O-E)^2$	$\frac{(O-E)^2}{E}$
236	265.6	-29.6	876.18	3.299
324	265.6	58.4	3410.56	12.841
182	265.6	-83.6	6988.96	26.314
221	265.6	-44.6	1989.16	7.489
365	265.6	99.4	9880.36	37.200
				87.143

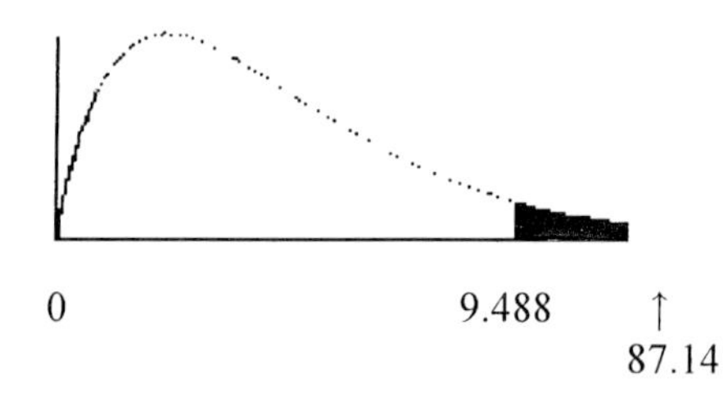

0 9.488 ↑
87.14

Reject the null hypothesis. There is enough evidence to reject the claim that the number of items sold in each region is the same.

2.
H_0: The ad produced the same number of responses in each county. (claim)
H_1: The null hypothesis is not true.
C. V. = 11.345 d. f. = 3

$E = \frac{298}{4} = 74.5$

$\chi^2 = \frac{(87-74.5)^2}{74.5} + \frac{(62-74.5)^2}{74.5}$

$+ \frac{(56-74.5)^2}{74.5} + \frac{(93-74.5)^2}{74.5} = 13.38$

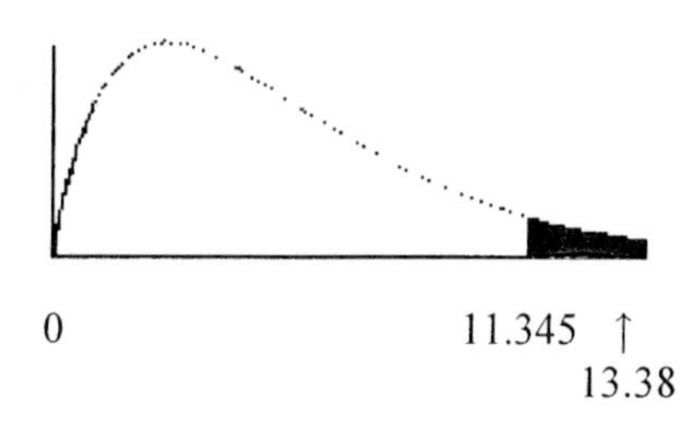

0 11.345 ↑
13.38

2. continued
Reject the null hypothesis. There is enough evidence to reject the claim that the ad produced the same number of responses in each county.

3.
H_0: The gender of the individual is not related to whether or not a person would use the labels. (claim)
H_1: The gender is related to use of the labels.
C. V. = 4.605 d. f. = 2

Gender	Yes	No	Undecided
Men	114(120.968)	30(22.258)	6(6.774)
Women	136(129.032)	16(23.742)	8(7.226)

$\chi^2 = \frac{(114-120.968)^2}{120.968} + \frac{(30-22.258)^2}{22.258} + \frac{(6-6.774)^2}{6.774}$

$+ \frac{(136-129.032)^2}{129.032} + \frac{(16-23.742)^2}{23.742} + \frac{(8-7.226)^2}{7.226}$

$\chi^2 = 6.16$

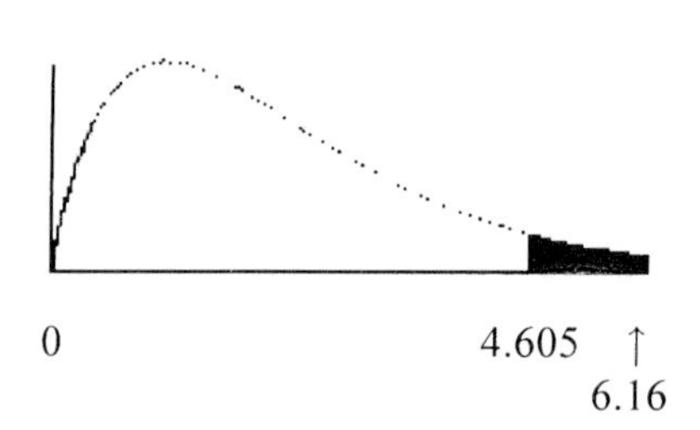

0 4.605 ↑
6.16

Reject the null hypothesis. There is not enough evidence to support the claim that opinion is independent of gender.

4.
H_0: The condiment preference is independent of the sex of the purchaser. (claim)
H_1: The condiment preference is dependent on the sex of the purchaser.
C. V. = 4.605 d. f. = 2

	Relish	Catsup	Mustard
Men	15(19.11)	18(15.29)	10(8.60)
Women	25(20.89)	14(16.71)	8(9.60)

$\chi^2 = \sum \frac{(O-E)^2}{E} = \frac{(15-19.11)^2}{19.11} + \frac{(18-15.29)^2}{15.29}$

$+ \frac{(10-8.60)^2}{8.60} + \frac{(25-20.89)^2}{20.89} + \frac{(14-16.71)^2}{16.71}$

$+ \frac{(8-9.40)^2}{9.40} = 3.050$

4. continued

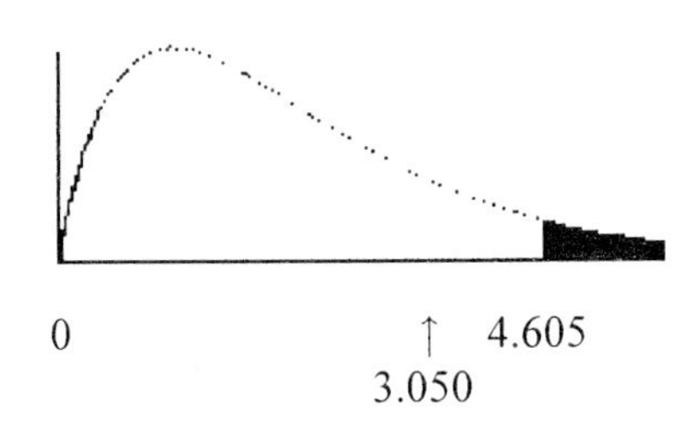

Do not reject the null hypothesis. There is not enough evidence to reject the claim that the condiment chosen is independent of the gender of the individual.

5.
H_0: The type of investment is independent of the age of the investor.
H_1: The type of investment is dependent upon the age of the investor. (claim)
C. V. = 9.488 d. f. = 4

Age	Large	Small	Inter.
45	20(20.18)	10(15.45)	10(15.45
65	42(33.82)	24(18.55)	24(18.55)

Age	CD	Bond
45	15(9.55)	45(31.36)
65	6(11.45)	24(37.64)

$$\chi^2 = \frac{(20-20.18)^2}{20.18} + \frac{(10-15.45)^2}{15.45} + \frac{(10-15.45)^2}{15.45}$$

$$+ \frac{(15-9.55)^2}{9.55} + \frac{(45-31.36)^2}{31.36} + \frac{(42-33.82)^2}{33.82} +$$

$$\frac{(24-18.55)^2}{18.55} + \frac{(24-18.55)^2}{18.55} + \frac{(6-11.45)^2}{11.45} +$$

$$\frac{(24-37.64)^2}{37.64} = 25.6$$

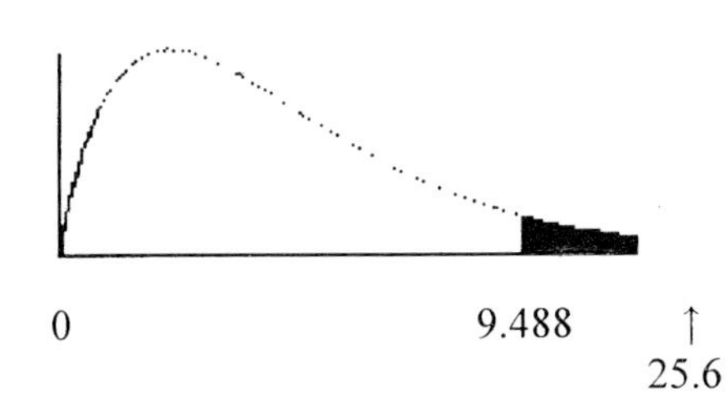

Reject the null hypothesis. There is enough evidence to support the claim that the type of investment is dependent on age.

6.
H_0: The type of car a person purchases is independent of the gender of the purchaser.

6. continued
H_1: The type of car a person purchases is dependent upon the gender of the purchaser. (claim)
C. V. = 11.345 d. f. = 3 $\alpha = 0.01$

	Sedan	Compact
Male	33(25.234)	27(28.505)
Female	21(28.766)	34(32.495)

	Station Wagon	SUV
Male	23(29.907)	17(16.355)
Female	41(34.093)	18(18.645)

$$\chi^2 = \sum\frac{(O-E)^2}{E} = \frac{(33-25.234)^2}{25.234} + \frac{(27-28.505)^2}{28.505}$$

$$+ \frac{(23-29.907)^2}{29.907} + \frac{(17-16.355)^2}{16.355} + \frac{(21-28.766)^2}{28.766}$$

$$+ \frac{(34-32.495)^2}{32.495} + \frac{(41-34.093)^2}{34.093} + \frac{(18-18.645)^2}{18.645}$$

$$\chi^2 = 7.678$$

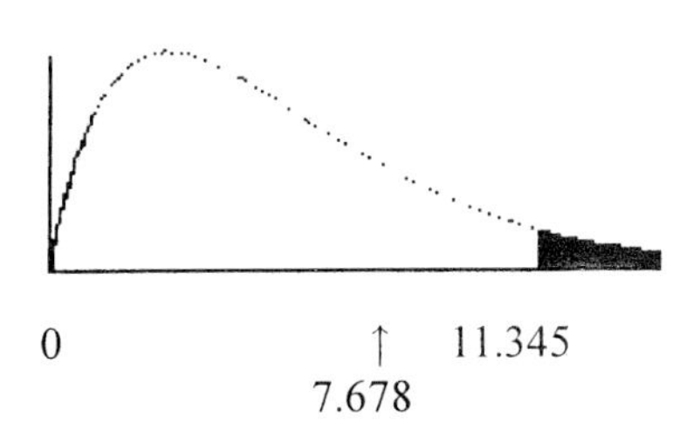

Do not reject the null hypothesis. There is not enough evidence to support the claim that the type of car purchased is dependent on the gender of the purchaser.

7.
H_0: $p_1 = p_2 = p_3$ (claim)
H_1: At least one proportion is different.
$\alpha = 0.01$ d. f. = 2

$$E(\text{work}) = \frac{80(114)}{240} = 38$$

$$E(\text{don't work}) = \frac{80(126)}{240} = 42$$

	16	17	18
work	45(38)	31(38)	38(38)
don't work	35(42)	49(42)	42(42)

$$\chi^2 = \frac{(45-38)^2}{38} + \frac{(31-38)^2}{38} + \frac{(38-38)^2}{38}$$

$$+ \frac{(35-42)^2}{42} + \frac{(49-42)^2}{42} + \frac{(42-42)^2}{42} = 4.912$$

$0.05 < \text{P-value} < 0.10$ (0.086)

7. continued
Do not reject the null hypothesis since P-value > 0.01. There is not enough evidence to reject the claim that the proportions are the same.

8.
H_0: $p_1 = p_2 = p_3 = p_4$ (claim)
H_1: At least one proportion is different.
C. V. = 7.815 d. f. = 3

$E(\text{male}) = \frac{100(219)}{400} = 54.75$

$E(\text{female}) = \frac{100(181)}{400} = 45.25$

	May	June	July	Aug
Male	51(54.75)	47(54.75)	58(54.75)	63(54.75)
Female	49(45.25)	53(45.25)	42(45.25)	37(45.25)

$\chi^2 = \frac{(51-54.75)^2}{54.75} + \frac{(47-54.75)^2}{54.75} + \frac{(58-54.75)^2}{54.75}$

$+ \frac{(63-54.75)^2}{54.75} + \frac{(49-45.25)^2}{45.25} + \frac{(53-45.25)^2}{45.25} +$

$\frac{(42-45.25)^2}{45.25} + \frac{(37-45.25)^2}{45.25} = 6.17$

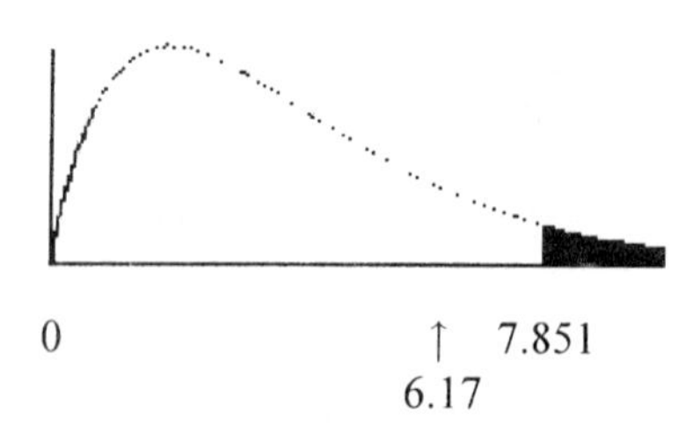

Do not reject the null hypothesis. There is not enough evidence to reject the claim that the proportions are the same.

9.
H_0: $p_1 = p_2 = p_3 = p_4$
H_1: At least one proportion is different.
C. V. = 6.251 d. f. = 3

$E(\text{yes}) = \frac{50(58)}{200} = 14.5$

$E(\text{no}) = \frac{50(142)}{200} = 35.5$

	A	B	C	D
Yes	12(14.5)	15(14.5)	10(14.5)	21(14.5)
No	38(35.5)	35(35.5)	40(35.5)	29(35.5)

$\chi^2 = \frac{(12-14.5)^2}{14.5} + \frac{(15-14.5)^2}{14.5} + \frac{(10-14.5)^2}{14.5} +$

$\frac{(21-14.5)^2}{14.5} + \frac{(38-35.5)^2}{35.5} + \frac{(35-35.5)^2}{35.5}$

9. continued

$+ \frac{(40-35.5)^2}{35.5} + \frac{(29-35.5)^2}{35.5} = 6.702$

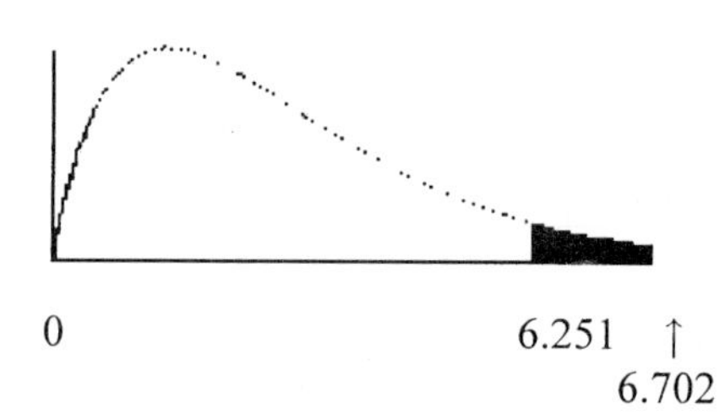

Reject the null hypothesis. There is enough evidence to reject the claim that the proportions are the same.

10.
H_0: $p_1 = p_2 = p_3$ (claim)
H_1: At least one proportion is different.
C. V. = 9.210 d. f. = 2

$E(\text{yes}) = \frac{200(186)}{600} = 62$

$E(\text{no}) = \frac{200(414)}{600} = 138$

	#1	#2	#3
Yes	87(62)	56(62)	43(62)
No	113(138)	144(138)	157(138)

$\chi^2 = \frac{(87-62)^2}{62} + \frac{(56-62)^2}{62} + \frac{(43-62)^2}{62}$

$+ \frac{(113-138)^2}{138} + \frac{(144-138)^2}{138} + \frac{(157-138)^2}{138}$

$\chi^2 = 23.89$

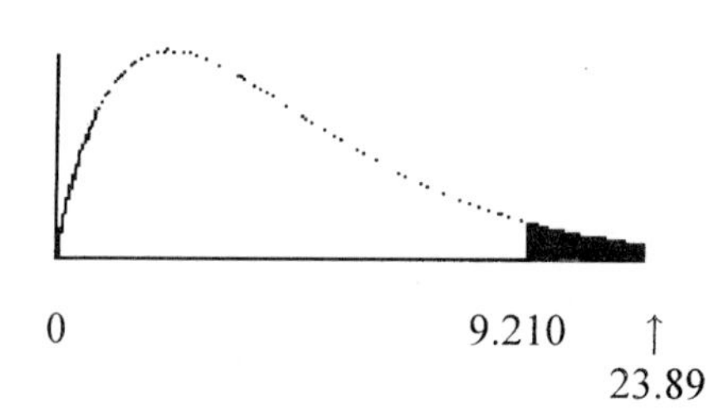

Reject the null hypothesis. There is not enough evidence to support the claim that the proportions are the same.

CHAPTER 11 QUIZ

1. False, it is one-tailed right.
2. True
3. False, there is little agreement between observed and expected frequencies.

4. c.
5. b.
6. d.
7. 6
8. independent
9. right
10. at least five

11. H_0: The number of advertisements is equally distributed over five geographic regions. (claim)
H_1: The number of advertisements is not equally distributed over five regions.
C. V. = 9.488 d. f. = 4 E = 240.4
$\chi^2 = \sum\frac{(O-E)^2}{E} = 45.4$
Reject the null hypothesis. There is enough evidence to reject the claim that the number of advertisements is equally distributed.

12. H_0: The ads produced the same number of responses. (claim)
H_1: The ads produced different numbers of responses.
C. V. = 13.277 d. f. = 4 E = 64.6
$\chi^2 = \sum\frac{(O-E)^2}{E} = 12.6$
Do not reject the null hypothesis. There is not enough evidence to reject the claim that the ads produced the same number of responses.

13. H_0: 62% of respondents never watch shopping channels, 23% watch rarely, 11% watch occasionally, and 4% watch frequently.
H_1: College students show a different preference for shopping channels. (claim)
C. V. = 7.815 d. f. = 3
$\chi^2 = 21.789$
Reject the null hypothesis. There is enough evidence to support the claim that college students show a different preference for shopping channels.

14. H_0: The number of commuters is distributed as follows: alone - 75.7%, carpooling - 12.2%, public transportation - 4.7%, walking - 2.9%, other - 1.2%, and working at home - 3.3%.
H_1: The proportions are different from the null hypothesis. (claim)
C. V. = 11.071 d. f. = 5
$\chi^2 = 41.692$
Reject the null hypothesis. There is enough evidence to support the claim that the

14. continued
distribution is different from the one stated in the null hypothesis.

15. H_0: The type of novel purchased is independent of the gender of the purchaser. (claim)
H_1: The type of novel purchased is dependent on the gender of the purchaser.
C. V. = 5.991 d. f. = 2
$\chi^2 = 132.9$
Reject the null hypothesis. There is enough evidence to reject the claim that the novel purchased is independent of the gender of the purchaser.

16. H_0: The type of pizza ordered is independent of the age of the purchaser. (claim)
H_1: The type of pizza ordered is dependent on the age of the purchaser.
$\alpha = 0.10$ d. f. = 9
$\chi^2 = 107.3$
P-value < 0.005
Reject the null hypothesis since P-value $<$ 0.10. There is enough evidence to reject the claim that the type of pizza is independent of the age of the purchaser.

17. H_0: The color of the pennant purchased is independent of the gender of the purchaser. (claim)
H_1: The color of the pennant purchased is dependent on the gender of the purchaser.
C. V. = 4.605 d. f. = 2
$\chi^2 = 5.6$
Reject the null hypothesis. There is enough evidence to reject the claim that the color of the pennant purchased is independent of the gender of the purchaser.

18. H_0: The opinion of the children on the use of the tax credit is independent of the gender of the children.
H_1: The opinion of the children on the use of the tax credit is dependent upon the gender of the children. (claim)
C. V. = 4.605 d. f. = 2
$\chi^2 = 1.534$
Do not reject the null hypothesis. There is not enough evidence to support the claim that the opinion of the children is dependent upon their gender.

Note: Graphs are not to scale and are intended to convey a general idea. Answers may vary due to rounding.

EXERCISE SET 12-3

1.
The analysis of variance using the F-test can be used to compare 3 or more means.

2.
1. Comparing two means at a time ignores all other means.
2. The probability of type I error is actually larger than α when multiple t-tests are used.
3. The more sample means, the more t-tests are needed.

3.
The populations from which the samples were obtained must be normally distributed. The samples must be independent of each other. The variances of the populations must be equal.

4.
The between group variance estimates the population variance using the means. The within group variance estimates the population variance using all the data values.

5.
$F = \frac{s_B^2}{s_W^2}$

6.
H_0: $\mu_1 = \mu_2 = ... = \mu_n$
H_1: At least one mean is different from the others.

7.
Scheffe´ Test and the Tukey Test

8.
H_0: $\mu_1 = \mu_2 = \mu_3$
H_1: At least one mean is different from the others. (claim)
C. V. = 3.52 $\alpha = 0.05$
d. f. N = 2 d. f. D = 19

$\bar{X}_1 = 165.714$ $\bar{X}_2 = 245.714$ $\bar{X}_3 = 237.5$

$s_1^2 = 5695.238$ $s_2^2 = 3928.571$ $s_3^2 = 7335.714$

8. continued
$\bar{X}_{GM} = \frac{4780}{22} = 217.273$

$s_B^2 = \frac{\sum n_i(\bar{X}_i - \bar{X}_{GM})^2}{k - 1} =$

$\frac{7(165.714-217.273)^2+7(245.714-217.273)^2+8(237.5-217.273)^2}{2}$

$= 13,771.799$

$s_W^2 = \frac{\sum(n_i - 1)s_i^2}{\sum(n_i - 1)}$

$= \frac{6(5695.238) + 6(3928.571) + 7(7335.714)}{6 + 6 + 7} = 5741.729$

$F = \frac{s_B^2}{s_W^2} = \frac{13771.799}{5741.729} = 2.3985$

Do not reject the null hypothesis. There is not enough evidence to support the claim that at least one mean is different.

9.
H_0: $\mu_1 = \mu_2 = \mu_3$
H_1: At least one mean is different from the others. (claim)
C. V. = 3.47 $\alpha = 0.05$
d. f. N. = 2 d. f. D. = 21

$\bar{X}_{GM} = 4.554$ $s_B^2 = 9.82113$ $s_W^2 = 4.93225$

$F = \frac{9.82113}{4.93225} = 1.9912$

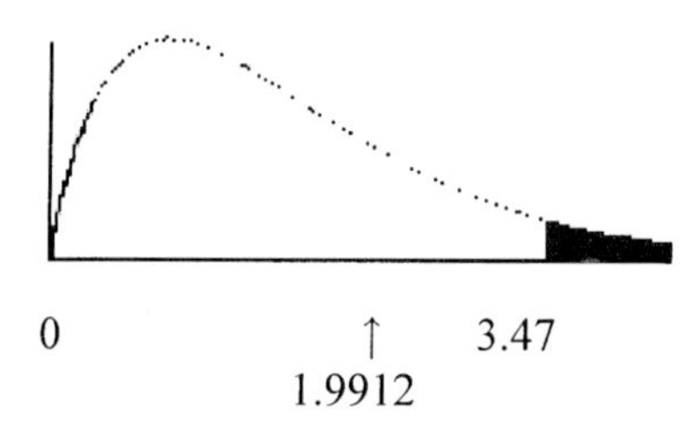

Do not reject the null hypothesis. There is not enough evidence to support the claim that at least one mean is different from the others.

10.
H_0: $\mu_1 = \mu_2 = \mu_3 = \mu_4$ (claim)
H_1: At least one mean is different from the others.
C. V. = 2.30 $\alpha = 0.10$
d. f. N = 3 d. f. D = 27

$\bar{X}_1 = 3689$ $s_1^2 = 904,361.667$

10. continued

$\bar{X}_2 = 5183.143 \quad s_2^2 = 2{,}387{,}825.81$

$\bar{X}_3 = 5889.125 \quad s_3^2 = 6{,}446{,}701.554$

$\bar{X}_4 = 4671.222 \quad s_4^2 = 4{,}951{,}768.194$

$\bar{X}_{GM} = \frac{\sum X}{n} = \frac{151259}{31} = 4879.323$

$s_B^2 = \frac{\sum n_i(\bar{X}_i - \bar{X}_{GM})^2}{k-1} =$

$\frac{7(3689-4879.323)^2+7(5183.143-4879.323)^2}{4-1}$

$+ \frac{8(5889.125-4879.323)^2+9(4671.222-4879.323)^2}{4-1}$

$= 6{,}370{,}527.64$

$s_W^2 = \frac{\sum(n_i-1)s_i^2}{\sum(n_i-1)}$

$= \frac{6(904361.667) + 6(2387825.81)}{6+6+7+8}$

$+ \frac{7(6446701.554) + 8(4951768.194)}{6+6+7+8}$

$= 3{,}870{,}154.86$

$F = \frac{6370527.64}{3870154.86} = 1.646$

Do not reject the null hypothesis. There is not enough evidence to reject the claim that the means are the same.

11.

H_0: $\mu_1 = \mu_2 = \mu_3$

H_1: At least one mean is different from the others. (claim)

C. V. = 3.98 $\quad \alpha = 0.05$

d. f. N = 2 $\quad$ d. f. D = 11

$\bar{X}_{GM} = \frac{52414}{14} = 3743.857$

$s_B^2 = 3{,}633{,}540.88$

$s_W^2 = 1{,}330{,}350$

$F = \frac{3633540.88}{1330350} = 2.7313$

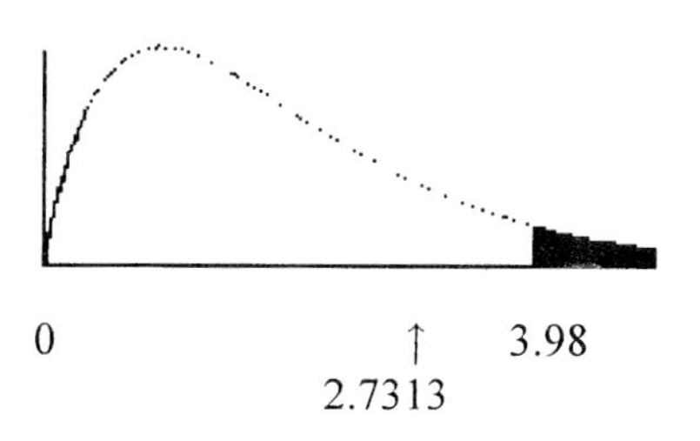

11. continued

Do not reject the null hypothesis. There is not enough evidence to support the claim that at least one mean is different from the others.

12.

H_0: $\mu_1 = \mu_2 = \mu_3$

H_1: At least one mean is different from the others. (claim)

$\alpha = 0.05$

d. f. N = 2 $\quad$ d. f. D = 11

$\bar{X}_{GM} = \frac{\sum X}{n} = \frac{99}{14} = 7.07$

$s_B^2 = \frac{101.095}{2} = 50.548$

$s_W^2 = \frac{71.833}{11} = 6.530$

$F = \frac{s_B^2}{s_W^2} = \frac{50.548}{6.530} = 7.74$

P-value = 0.00797

Reject since P-value < 0.05.

Scheffe´ Test

C. V. = 7.96

$F_s = \frac{(\bar{X}_i - \bar{X}_j)^2}{s_W^2\left(\frac{1}{n_i} + \frac{1}{n_j}\right)}$

For $\bar{X}_1$ vs $\bar{X}_2$

$F_S = \frac{(5-10.167)^2}{6.530\left(\frac{1}{4}+\frac{1}{6}\right)} = \frac{26.698}{2.721} = 9.812$

For $\bar{X}_1$ vs $\bar{X}_3$

$F_S = \frac{(5-4.5)^2}{6.530\left(\frac{1}{4}+\frac{1}{4}\right)} = \frac{0.25}{3.265} = 0.077$

For $\bar{X}_2$ vs $\bar{X}_3$

$F_S = \frac{(10.167-4.5)^2}{6.530\left(\frac{1}{6}+\frac{1}{4}\right)} = \frac{32.115}{2.721} = 11.803$

There is a significant difference between $\bar{X}_1$ and $\bar{X}_2$ and between $\bar{X}_2$ and $\bar{X}_3$.

13.

H_0: $\mu_1 = \mu_2 = \mu_3$ (claim)

H_1: At least one mean is different.

k = 3 $\quad$ N = 17 $\quad$ d.f.N. = 2 $\quad$ d.f.D. = 14

CV = 3.74

17. continued

$\overline{X}_2$ vs $\overline{X}_3$:

$$F_s = \frac{(203.125-155.625)^2}{1073.776(\frac{1}{8}+\frac{1}{8})} = 8.40$$

There is a significant difference between $\overline{X}_1$ and $\overline{X}_3$ and between $\overline{X}_2$ and $\overline{X}_3$.

18.
H_0: $\mu_1 = \mu_2 = \mu_3$
H_1: At least one mean is different from the others.

C. V. = 3.63 $\alpha = 0.05$
d. f. N = 2 d. f. D = 16

$\overline{X}_1 = 4.5214$ $s_1^2 = 1.3412$

$\overline{X}_2 = 4.5667$ $s_2^2 = 1.1949$

$\overline{X}_3 = 4.2444$ $s_3^2 = 0.8880$

$$\overline{X}_{GM} = \frac{\sum X}{N} = 4.4483$$

$$s_B^2 = \frac{\sum n_i(\overline{X}_i - \overline{X}_{GM})^2}{k-1}$$

$$= \frac{7(4.5214-4.4483)^2 + 6(4.5667-4.4483)^2}{3-1}$$

$$+ \frac{6(4.2444-4.4483)^2}{3-1} = 0.1855$$

$$s_W^2 = \frac{\sum(n_i-1)s_i^2}{\sum(n_i-1)}$$

$$= \frac{6(1.3412)+5(1.1949)+5(0.888)}{6+5+5} = 1.1538$$

$$F = \frac{s_B^2}{s_W^2} = \frac{0.1855}{1.1538} = 0.1607$$

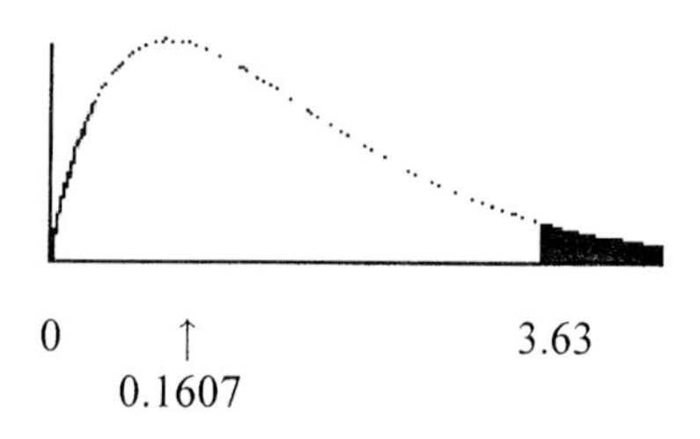

0 ↑ 3.63
0.1607

Do not reject the null hypothesis. There is not enough evidence to support the claim that at least one mean is different from the others.

19.
H_0: $\mu_1 = \mu_2 = \mu_3 = \mu_4$
H_1: At least one mean is different. (claim)
C. V. = 4.13 $\alpha = 0.01$
d. f. N = 3 d. f. D = 64

$$\overline{X}_{GM} = \frac{36,254}{68} = 533.147$$

$$s_B^2 = \frac{40,968.145}{3} = 13,656.05$$

$$s_w^2 = \frac{787,506.385}{64} = 12,304.787$$

$$F = \frac{13,656.05}{12,304.787} = 1.11$$

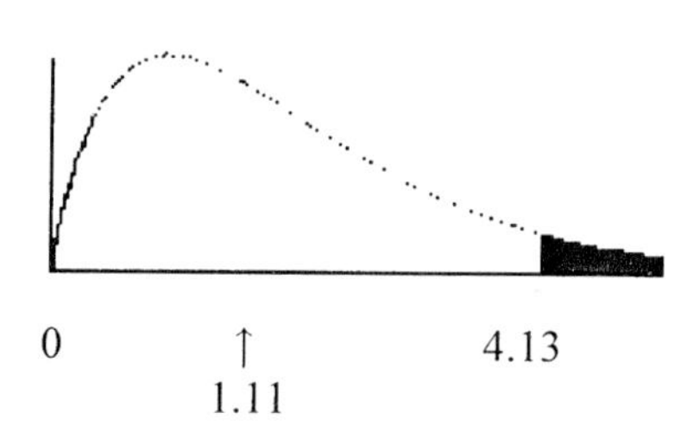

0 ↑ 4.13
1.11

Do not reject the null hypothesis. There is not enough evidence to support the claim that at least one mean is different.

EXERCISE SET 12-4

1.
The two-way ANOVA allows the researcher to test the effects of two independent variables and a possible interaction effect. The one-way ANOVA can test the effects of one independent variable only.

2.
The main effects are the effects of the independent variables taken separately. The interaction effect occurs when one independent variable effects the dependent variable differently at different levels of the other independent variable.

3.
The mean square values are computed by dividing the sum of squares by the corresponding degrees of freedom.

4.
The F test value is computed by dividing the Mean Square for the variable by the Mean Square for the error term.

5.
a. d. $f._A = (3 - 1) = 2$ for factor A
b. d. $f._B = (2 - 1) = 1$ for factor B
c. d. $f._{AxB} = (3 - 1)(2 - 1) = 2$
d. d. $f._{within} = 3 \cdot 2(5 - 1) = 24$

6.
a. d. $f._A = (6 - 1) = 5$
b. d. $f._B = (5 - 1) = 4$
c. d. $f._{AxB} = (6 - 1)(5 - 1) = 20$
d. d. $f._{within} = 6 \cdot 5(7 - 1) = 180$

7.
The two types of interactions that can occur are ordinal and disordinal.

8.
The main effects can be interpreted independently when the interaction effect is not significant or the interaction is an ordinal interaction.

9.
a. The lines will be parallel or approximately parallel. They could also coincide.
b. The lines will not intersect and they will not be parallel.
c. The lines will intersect.

10.
H_0: There is no interaction effect between the type of ad and the medium on the effectiveness of the ad.
H_1: There is an interaction effect between the type of ad and the medium on the effectiveness of the ad.

H_0: There is no difference between the means of the ratings for the type of ad used.
H_1: There is a difference between the means of the rating for the type of ad used.

H_0: There is no difference between the means of the ratings for the medium used.
H_1: There is a difference in the means of the ratings for the medium used.

10. continued

ANOVA SUMMARY TABLE

Source	SS	d. f.	M. S.	F
Type	10.563	1	10.563	1.913
Medium	175.563	1	175.563	31.800
Interaction	0.063	1	0.063	0.011
Within	66.250	12	5.521	
Total	252.439	15		

The critical value at $\alpha = 0.01$ with d. f. N = 1 and d. f. D = 12 is 9.33 for F_A, F_B and F_{AxB}.

Since the F test value for the medium, 31.800, is greater than the critical value, 9.33, the decision is to reject the null hypothesis for the medium. It can be concluded that there is a significant difference for the ratings for the medium used. Since each of 1.913 and 0.011 is less than the C. V. of 9.33, do not reject the null hypothesis for type and interaction.

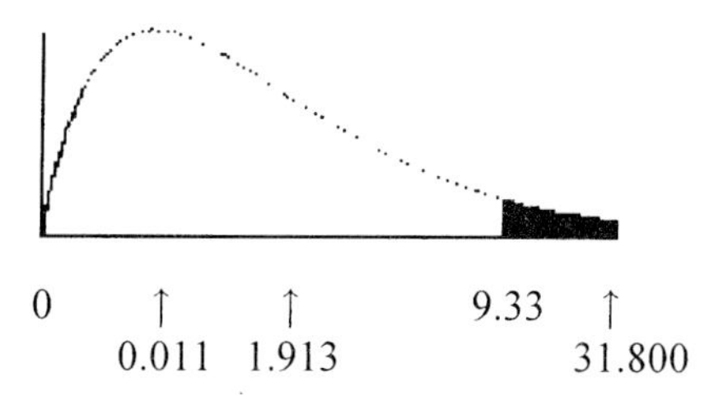

11.
H_0: There is no interaction effect between the time of day and the type of diet on a person's sodium level.
H_1: There is an interaction effect between the time of day and the type of diet on a person's sodium level.

H_0: There is no difference between the means for the sodium level for the times of day.
H_1: There is a difference between the means for the sodium level for the times of day.

H_0: There is no difference between the means for the sodium level for the type of diet.
H_1: There is a difference between the means for the sodium level for the type of diet.

11. continued

ANOVA SUMMARY TABLE

Source	SS	d. f.	MS	F
Time	1800	1	1800	25.806
Diet	242	1	242	3.470
Interaction	264.5	1	264.5	3.792
Within	279	4	69.75	
Total	2585.5	7		

The critical value at $\alpha = 0.05$ with d. f. N = 1 and d. f. D = 4 is 7.71 for F_A, F_B, and F_{AxB}.

Since the only F test value that exceeds 7.71 is the one for the time, 25.806, it can be concluded that there is a difference in the means for the sodium level taken at two different times.

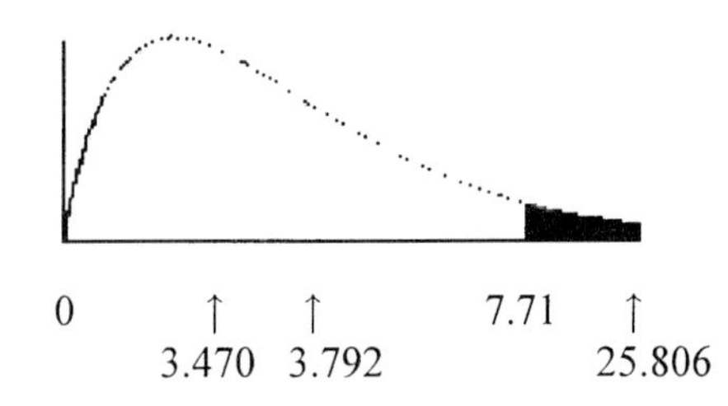

0 ↑ ↑ 7.71 ↑
3.470 3.792 25.806

12.
H_0: There is no interaction effect between the subcontractors and the types of homes they build or the times it takes to build the homes.
H_1: There is an interaction effect between the subcontractors and the types of homes they build on the times it takes to build the homes.

H_0: There is no difference in the means of the times it takes the subcontractors to build the homes.
H_1: There is a difference in the means of the times it takes the subcontractors to build the homes.

H_0: There is no difference among the means of the times for the types of homes built.
H_1: There is a difference among the means of the times for the types of homes built.

ANOVA SUMMARY TABLE

Source	SS	d. f.	MS	F
Subcontractor	1672.553	1	1672.553	122.084
Home Type	444.867	2	222.434	16.236
Interaction	313.267	2	156.634	11.433
Within	328.8	24	13.700	
Total	2579.487	29		

12. continued
The critical values at $\alpha = 0.05$:
For the subcontractor with d. f. N = 1, d. f. D = 24, C. V. = 4.26.

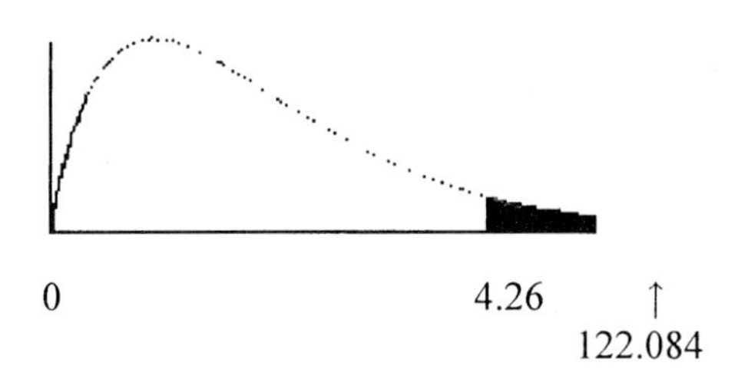

0 4.26 ↑
122.084

For the home type and interaction d. f. N = 2 and d. f. D = 24, C. V. = 3.40.

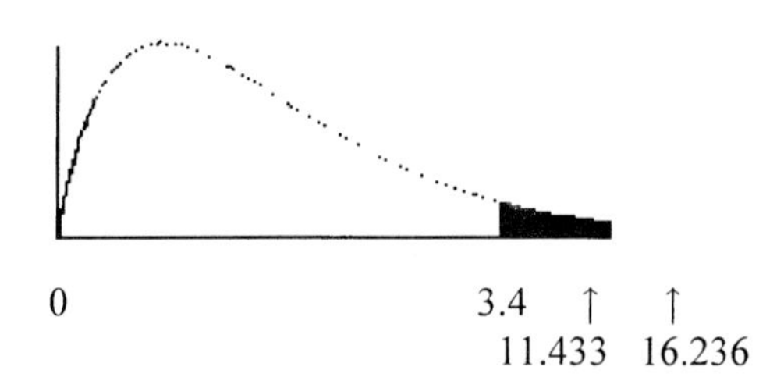

0 3.4 ↑ ↑
11.433 16.236

All F test values exceed the critical values and all of the null hypotheses are rejected. Since there is a significant interaction effect the means of the cells must be computed and graphed to determine the type of interaction.

The cell means are:

	HOME TYPE		
Contractor	I	II	III
A	28	31.4	44.4
B	18.6	20.000	20.400

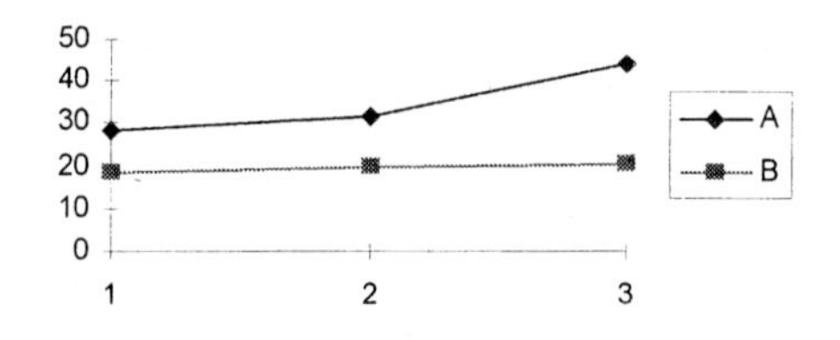

Since all of the three means for the home types for contractor A are greater than the three means for contractor B, and the differences are not equal, there is an ordinal interaction. Hence it can be concluded that there is a difference in means for the subcontractors, home types, and also an interaction effect is present.

13.
H_0: There is no interaction effect between the gender of the individual and the duration of the training on the test scores.
H_1: There is an interaction effect between the gender of the individual and the duration of the training on the test scores.

H_0: There is no difference between the means of the test scores for the males and females.
H_1: There is a difference between the means of the test scores for the males and females.

H_0: There is no difference between the means of the test scores for the two different durations.
H_1: There is a difference between the means of the test scores for the two different durations.

ANOVA SUMMARY TABLE

Source	SS	d. f.	MS	F
Gender	57.042	1	57.042	0.835
Duration	7.042	1	7.042	0.103
Interaction	3978.375	1	3978.375	58.270
Within	1365.5	20	68.275	
Total	5407.959	23		

The critical value at $\alpha = 0.10$ with d. f. N = 1 and d. f. D = 20 is 2.97. Since the F test value for the interaction is greater than the critical value, it can be concluded that the gender affects the test scores differently for the duration levels.

14.
H_0: There is no interaction effect between the type of paint and the geographic location on the lifetimes of the paint.
H_1: There is an interaction effect between the type of paint and the geographic location on the lifetimes of the paint.

H_0: There is no difference between the means of the lifetimes of the two types of paints.
H_1: There is a difference between the means of the lifetimes of the two types of paints.

H_0: There is no difference among the means of the lifetimes of the paints used in different geographic locations.
H_1: There is a difference in the means of the lifetimes of the paints used in different geographic locations.

14. continued

ANOVA SUMMARY TABLE

Source	SS	d. f.	MS	F
Paint Type	12.1	1	12.1	0.166
Location	2501	3	833.667	11.465
Interaction	268.1	3	89.367	1.229
Within	2326.8	32	72.713	
Total	5108	39		

For $\alpha = 0.01$ the critical values are:
For the paint type d. f. N = 1, d. f. D = 32 (use 30), and C. V. = 7.56

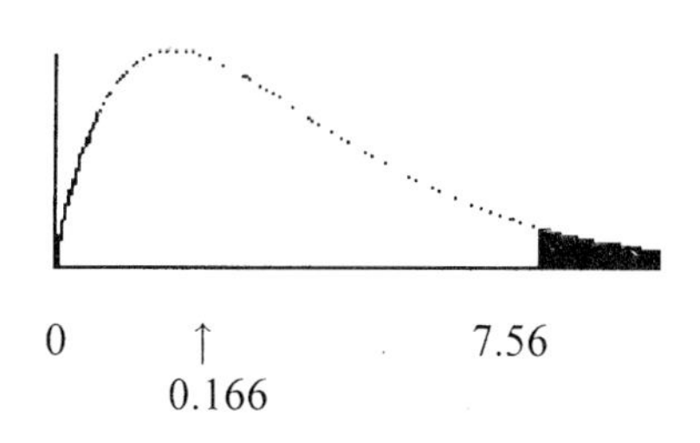

For the location and interaction
d. f. N = 3,
d. f. D = 32 (use 30), and C. V. = 4.51

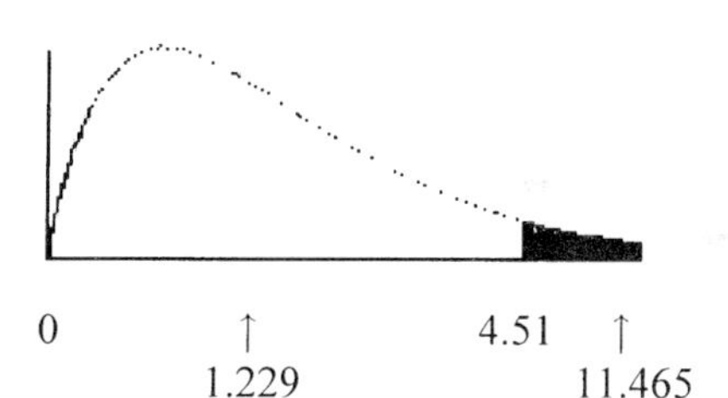

Since the only F test value that exceeds the critical value is the one for the location, it can be concluded that there is a difference in the means for the geographic locations.

15.
H_0: There is no interaction effect between the ages of the salespersons and the products they sell on the monthly sales.
H_1: There is an interaction effect between the ages of the salespersons and the products they sell on the monthly sales.

H_0: There is no difference in the means of the monthly sales of the two age groups.
H_1: There is a difference in the means of the monthly sales of the two age groups.

H_0: There is no difference among the means of the sales for the different products.
H_1: There is a difference among the means of the sales for the different products.

15. continued

ANOVA SUMMARY TABLE

Source	SS	d. f.	MS	F
Age	168.033	1	168.033	1.567
Product	1762.067	2	881.034	8.215
Interaction	7955.267	2	3877.634	37.087
Error	2574	24	107.250	
Total	12459.367	29		

At $\alpha = 0.05$, the critical values are:

For age, d. f. N = 1, d. f. D = 24, C. V. = 4.26

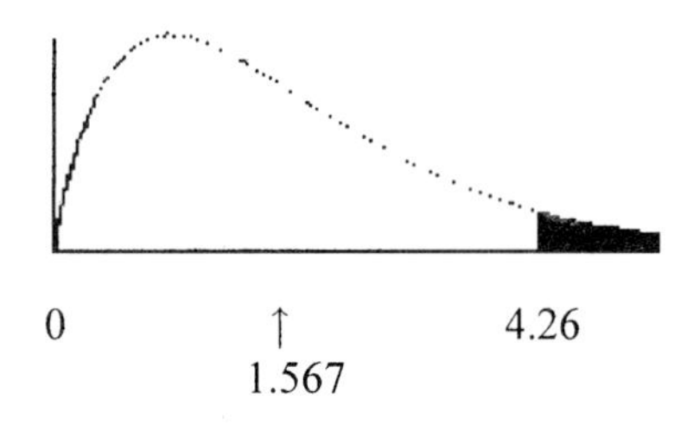

0 ↑ 1.567 4.26

For product and interaction, d. f. N = 2, d. f. D = 24, and C. V. = 3.40

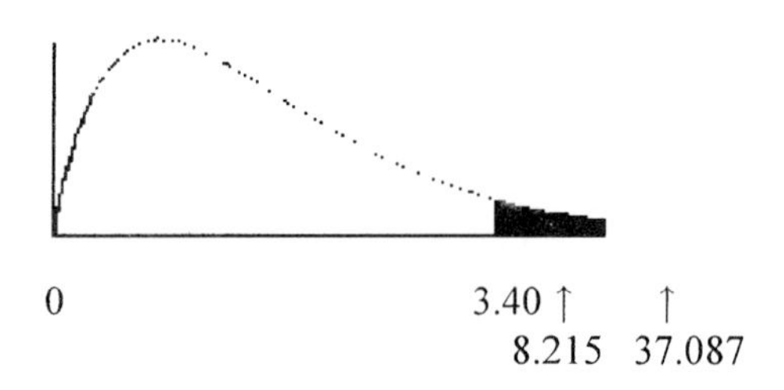

0 3.40 ↑ 8.215 ↑ 37.087

The null hypotheses for the interaction effect and for the type of product sold are rejected since the F test values exceed the critical value, 3.40. The cell means are:

Age	Pools	Spas	Saunas
over 30	38.8	28.6	55.4
30 & under	21.2	68.6	18.8

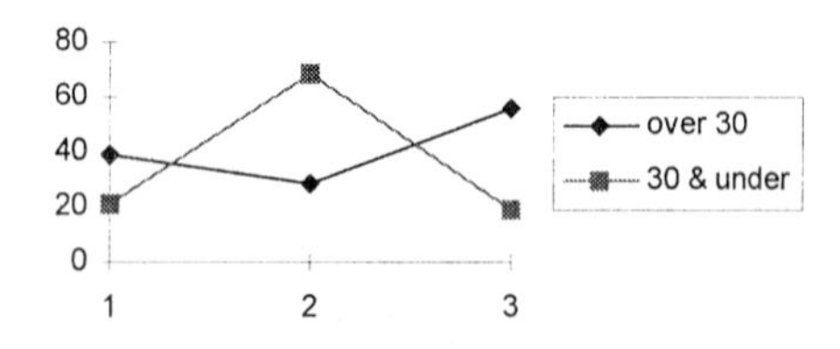

Since the lines cross, there is a disordinal interaction hence there is an interaction effect between the age of the sales person and the type of products sold on the sales.

REVIEW EXERCISES - CHAPTER 12

1.

H_0: $\mu_1 = \mu_2 = \mu_3$ (claim)

H_1: At least one mean is different from the others.

C. V. = 5.39 $\alpha = 0.01$

d. f. N = 2 d. f. D = 33

$\bar{X}_1 = 620.5$ $s_1^2 = 5445.91$

$\bar{X}_2 = 610.17$ $s_2^2 = 22{,}108.7$

$\bar{X}_3 = 477.83$ $s_3^2 = 5280.33$

$\bar{X}_{GM} = \frac{20{,}502}{36} = 569.5$

$s_B^2 = \frac{151{,}890.667}{2} = 75{,}945.333$

$s_W^2 = \frac{361{,}184.333}{33} = 10{,}944.9798$

$F = \frac{s_B^2}{s_W^2} = \frac{75{,}945.333}{10{,}944.9798} = 6.94$

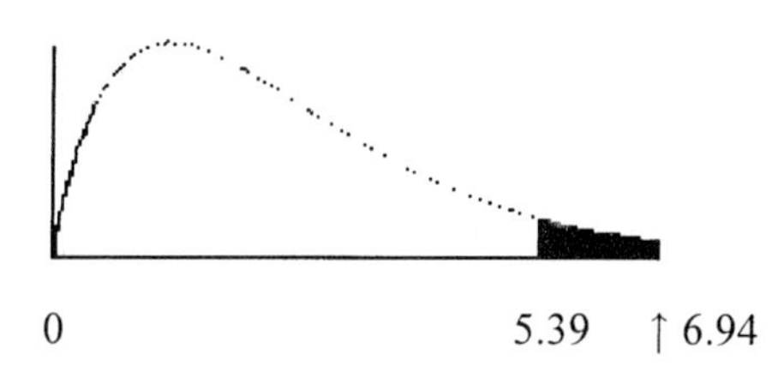

0 5.39 ↑ 6.94

Reject. At least one mean is different.

Tukey Test C. V. = 4.45 using (3, 33)

$\bar{X}_1$ vs $\bar{X}_2$

$q = \frac{\bar{X}_1 - \bar{X}_2}{\sqrt{\frac{s_W^2}{n}}} = \frac{620.5 - 610.17}{\sqrt{\frac{10{,}944.98}{12}}} = 0.342$

$\bar{X}_1$ vs $\bar{X}_3$

$q = \frac{620.5 - 477.83}{\sqrt{\frac{10{,}944.98}{12}}} = 4.72$

$\bar{X}_2$ vs $\bar{X}_3$

$q = \frac{610.17 - 477.83}{\sqrt{\frac{10{,}944.98}{12}}} = 4.38$

There is a significant difference between $\bar{X}_1$ and $\bar{X}_3$.

2.
H_0: $\mu_1 = \mu_2 = \mu_3$ (claim)
H_1: At least one mean is different from the others.
C. V. = 6.23 $\alpha = 0.01$
d. f.N = 2 d. f. D = 16

$\overline{X}_1 = 103.3333$ $s_1^2 = 41.0667$

$\overline{X}_2 = 115.8571$ $s_2^2 = 134.8095$

$\overline{X}_3 = 88.5$ $s_3^2 = 180.3$

$X_{GM} = 103.2632$

$$s_B^2 = \frac{\sum n_i(\overline{X}_i - \overline{X}_{GM})^2}{k-1}$$

$$s_B^2 = \frac{6(103.3333-103.2632)^2}{2} + \frac{7(115.8571-103.2632)^2}{2}$$

$$+ \frac{6(88.5-103.2632)^2}{2} = 1208.993$$

$$s_W^2 = \frac{\sum(n_i-1)s_i^2}{\sum(n_i-1)}$$

$$= \frac{5(41.0667) + 6(134.8095) + 5(180.3)}{5+6+5}$$

$$= 119.7307$$

$$F = \frac{s_B^2}{s_W^2} = \frac{1208.993}{119.7307} = 10.0976$$

Reject the null hypothesis. There is not enough evidence to support the claim that the means are the same.

Scheffe Test:
C. V. = 12.56
$\overline{X}_1$ vs $\overline{X}_2$:

$$F_s = \frac{(\overline{X}_i - \overline{X}_j)^2}{s_W^2(\frac{1}{n_i} + \frac{1}{n_j})} = \frac{(103.333-115.8571)^2}{119.7307(\frac{1}{6}+\frac{1}{7})}$$

$$= 4.232$$

$\overline{X}_1$ vs $\overline{X}_3$:

$$F_s = \frac{(103.333-88.5)^2}{119.7307(\frac{1}{6}+\frac{1}{6})} = 5.5128$$

$\overline{X}_2$ vs $\overline{X}_3$:

$$F_s = \frac{(115.8571-88.5)^2}{119.7307(\frac{1}{7}+\frac{1}{6})} = 20.195$$

There is a significant difference between $\overline{X}_2$ and $\overline{X}_3$.

3.
H_0: $\mu_1 = \mu_2 = \mu_3$
H_1: At least one mean is different from the others. (claim)

C. V. = 3.55 $\alpha = 0.05$
d. f. N = 2 d. f. D = 18

$\overline{X}_1 = 29.625$ $s_1^2 = 59.125$

$\overline{X}_2 = 29$ $s_2^2 = 63.333$

$\overline{X}_3 = 28.5$ $s_3^2 = 37.1$

$\overline{X}_{GM} = 29.095$

$$s_B^2 = \frac{\sum n_i(\overline{X}_i - \overline{X}_{GM})^2}{k-1}$$

$$s_B^2 = \frac{8(29.625-29.095)^2}{2} + \frac{7(29-29.095)^2}{2}$$

$$+ \frac{6(28.5-29.095)^2}{2} = 2.21726$$

$$s_W^2 = \frac{\sum(n_i-1)s_i^2}{\sum(n_i-1)}$$

$$s_W^2 = \frac{7(59.125)+6(63.333)+5(37.1)}{7+6+5}$$

$$s_W^2 = 54.509611$$

$$F = \frac{s_B^2}{s_W^2} = \frac{2.21726}{54.509611} = 0.04075$$

Do not reject the null hypothesis. There is not enough evidence to support the claim that at least one mean is different from the others.

4.
H_0: $\mu_1 = \mu_2 = \mu_3$
H_1: At least one mean is different from the others. (claim)

C. V. = 6.01 $\alpha = 0.01$
d. f. N = 2 d. f. D = 18

$\overline{X}_1 = 17.28571$ $s_1^2 = 11.2381$

$\overline{X}_2 = 19.57143$ $s_2^2 = 9.28571$

$\overline{X}_3 = 19.42857$ $s_3^2 = 32.28571$

$\overline{X}_{GM} = \frac{\sum X}{N} = 18.7619$

4. continued

$s_B^2 = \frac{\sum n_i(\bar{X}_i - \bar{X}_{GM})^2}{k-1}$

$= \frac{7(17.28571 - 18.7619)^2 + 7(19.57143 - 18.7619)^2}{3-1}$

$+ \frac{7(19.42857 - 18.7619)^2}{3-1} = 11.4762$

$s_W^2 = \frac{\sum(n_i - 1)s_i^2}{\sum(n_i - 1)}$

$= \frac{6(11.2381) + 6(9.28571) + 6(32.28571)}{6+6+6}$

$= 17.6032$

$F = \frac{s_B^2}{s_W^2} = \frac{11.4762}{17.6032} = 0.6519$

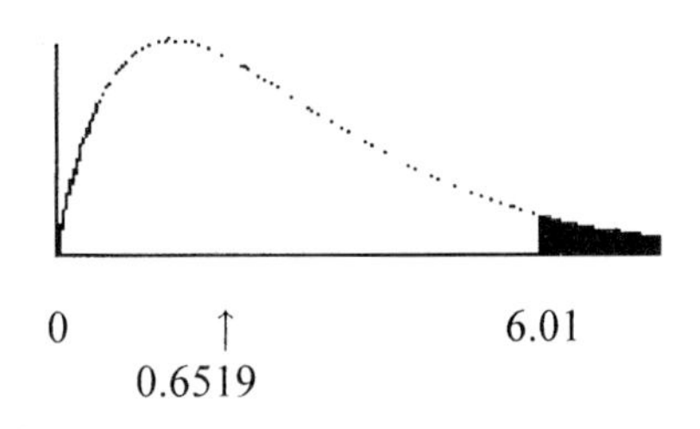

0 ↑ 6.01
0.6519

Do not reject the null hypothesis. There is not enough evidence to support the claim that at least one mean is different from the others.

5.
H_0: $\mu_1 = \mu_2 = \mu_3$
H_1: At least one mean is different. (claim)
C. V. = 2.61 $\alpha = 0.10$
d. f. N = 2 d. f. D = 19

$\bar{X}_{GM} = 3.8591$

$s_B^2 = 1.65936$

$s_W^2 = 3.40287$

$F = \frac{1.65936}{3.40287} = 0.4876$

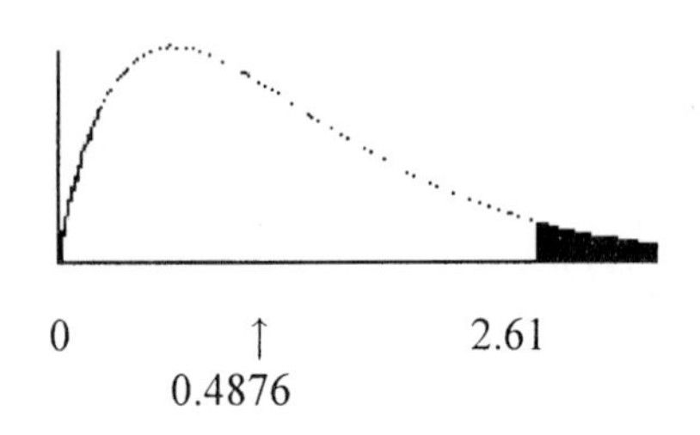

0 ↑ 2.61
0.4876

5. continued
Do not reject. There is not enough evidence to support the claim that at least one mean is different from the others.

6.
H_0: $\mu_1 = \mu_2 = \mu_3 = \mu_4$
H_1: At least one mean is different from the others.
C. V. = 3.10 $\alpha = 0.05$
d. f. N = 3 d. f. D = 20

$\bar{X}_{GM} = \frac{125}{24} = 5.208$

$s_B^2 = \frac{60.7917}{3} = 20.264$

$s_W^2 = \frac{255.167}{20} = 12.758$

$F = \frac{20.264}{12.758} = 1.59$

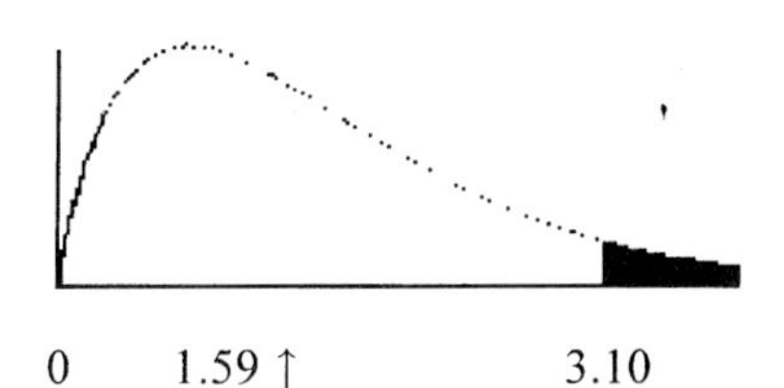

0 1.59 ↑ 3.10

Do not reject the null hypothesis. There is not enough evidence to conclude that there is a difference in the means.

7.
H_0: $\mu_1 = \mu_2 = \mu_3$
H_1: At least one mean is different from the others.
C. V. = 3.68 $\alpha = 0.05$
d. f. N = 2 d. f. D = 15

$\bar{X}_{GM} = 34.611$

$s_B^2 = \frac{1445.7778}{2} = 722.8889$

$s_W^2 = \frac{1376.5}{15} = 91.7667$

$F = \frac{722.8889}{91.7667} = 7.8775$

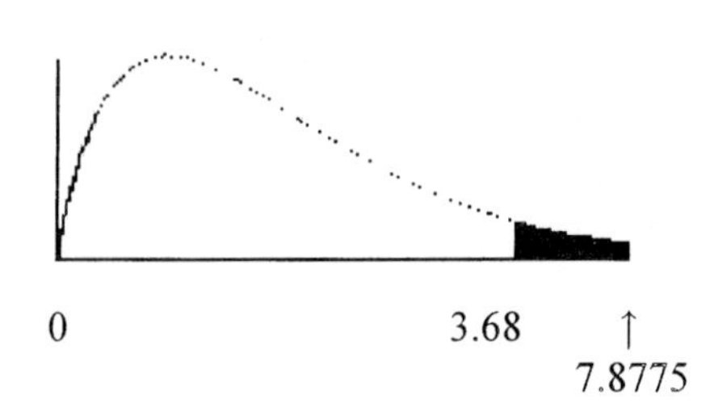

0 3.68 ↑
7.8775

7. continued
Reject the null hypothesis. There is enough evidence to support the claim that at least one mean is different from the others.

Tukey Test:
C. V. = 3.67

$\overline{X}_1$ vs $\overline{X}_2$:

$$q = \frac{47.167 - 26.833}{\sqrt{\frac{91.767}{6}}} = 5.1994$$

$\overline{X}_1$ vs $\overline{X}_3$:

$$q = \frac{47.167 - 29.833}{\sqrt{\frac{91.767}{6}}} = 4.4323$$

$\overline{X}_2$ vs $\overline{X}_3$:

$$q = \frac{26.833 - 29.833}{\sqrt{\frac{91.767}{6}}} = -0.7671$$

There is a significant different between $\overline{X}_1$ and $\overline{X}_2$ and between $\overline{X}_1$ and $\overline{X}_3$.

8.
H_0: There is no interaction effect between the ages of the students and the subjects on the anxiety scores of the students.
H_1: There is an interaction effect between the ages of the students and the subject on the anxiety scores of the students.

H_0: There is no difference between the means of the anxiety scores of the students in the two age groups.
H_1: There is a difference between the means of the anxiety scores of the students in the two age groups.

H_0: There is no difference between the means of the anxiety scores of the students in the two subjects.
H_1: There is a difference between the means of the anxiety scores of the students in the two subjects.

8. continued

ANOVA SUMMARY TABLE

Source	SS	d. f.	MS	F
Age	2376.2	1	2376.2	49.816
Class	105.8	1	105.8	2.218
Interaction	2645	1	2645	55.451
Within	763.2	16	47.7	
Total	5890.2	19		

At $\alpha = 0.10$ the d. f. N = 1 and the d. f. D = 16. The critical value is 3.05.

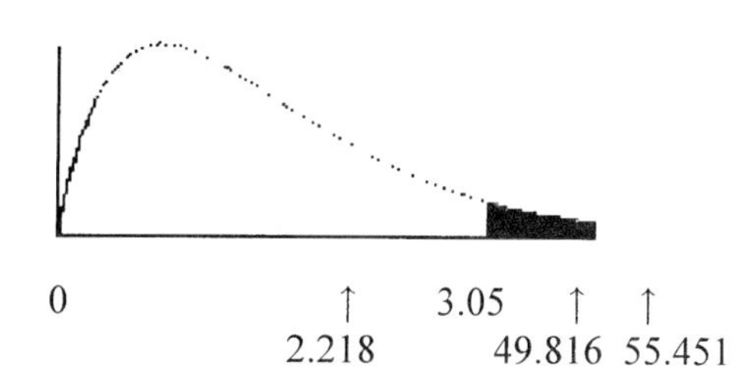

Since the interaction hypothesis is rejected it is necessary to calculate and graph the cell means. Note: H_0 for age is also rejected.

The cell means are:

	CLASS	
Age	Calculus	Statistics
under 20	53.6	26.000
20 & over	52.4	70.800

The graph is:

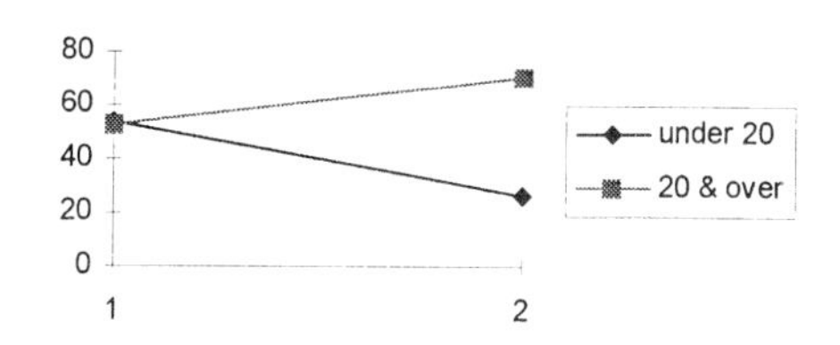

There is a significant interaction effect between the age of the student and the subject on the anxiety test score.

9.
H_0: There is no interaction effect between the type of exercise program and the type of diet on a person's glucose level.
H_1: There is an interaction effect between the type of exercise program and the type of diet on a person's glucose level.

9. continued

H_0: There is no difference in the means for the glucose levels of the persons in the two exercise programs.

H_1: There is a difference in the means for the glucose levels of the persons in the two exercise programs.

H_0: There is no difference in the means for the glucose levels of the persons in the two diet programs.
H_1: There is a difference in the means for the glucose levels of the persons in the two diet programs.

ANOVA SUMMARY TABLE

Source	SS	d. f.	MS	F
Exercise	816.75	1	816.75	60.5
Diet	102.083	1	102.083	7.562
Interaction	444.083	1	444.083	32.895
Within	108	8	13.5	
Total	1470.916	11		

At $\alpha = 0.05$ and d. f. N = 1 and d. f. D = 8 the critical value is 5.32 for each F_A, F_B, and F_{AxB}.

Hence all three null hypotheses are rejected.

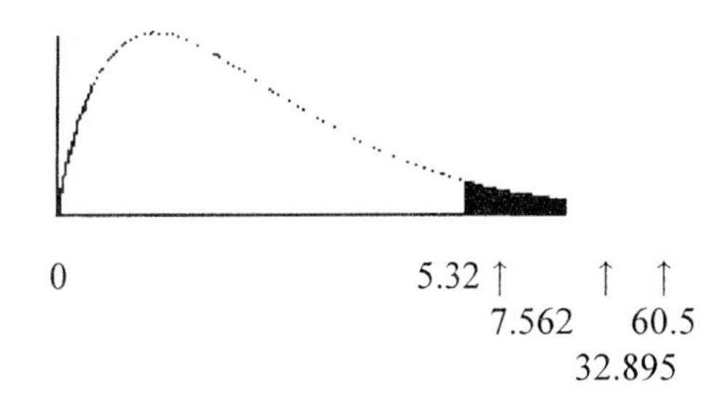

The cell means should be calculated.

	Diet	
Exercise	A	B
I	64	57.667
II	68.333	86.333

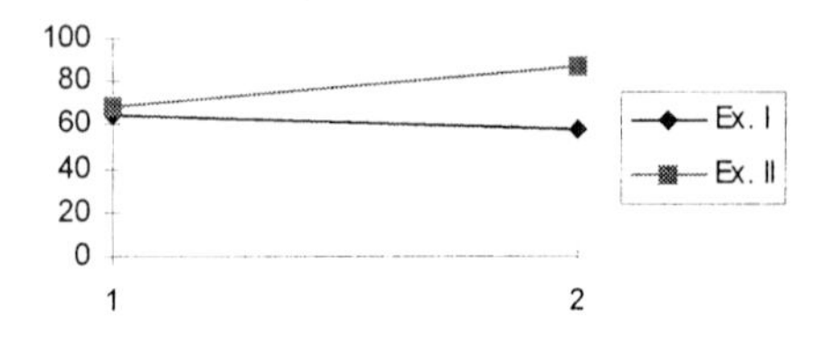

9. continued

Since the means for the Exercise Program I are both smaller than those for Exercise Program II and the vertical differences are not the same, the interaction is ordinal. Hence one can say that there is a difference for exercise, diet; and that an interaction effect is present.

CHAPTER 12 QUIZ

1. False, there could be a significant difference between only some of the means.
2. False, degrees of freedom are used to find the critical value.
3. False, the null hypothesis should not be rejected.
4. True
5. d.
6. a.
7. a.
8. c.
9. ANOVA
10. Tukey
11. two

12. H_0: $\mu_1 = \mu_2 = \mu_3$ (claim)
H_1: At least one mean is different from the others.
C. V. = 3.55

$s_B^2 = 6.0476$ $\quad s_W^2 = 0.8730$
$F = \frac{6.0476}{0.8730} = 6.927$

Reject the null hypothesis. There is not enough evidence to support the claim that the means are the same.

$\overline{X}_1 = 3.29$
$\overline{X}_2 = 5.14$
$\overline{X}_3 = 4.29$

Tukey Test C. V. = 3.61
$\overline{X}_1$ vs $\overline{X}_2$:

$$q = \frac{3.29 - 5.14}{\sqrt{\frac{0.873}{7}}} = -5.24$$

$\overline{X}_1$ vs $\overline{X}_3$:

$$q = \frac{3.29 - 4.29}{\sqrt{\frac{0.873}{7}}} = -2.83$$

12. continued
$\overline{X}_2$ vs $\overline{X}_3$:

$$q = \frac{5.14 - 4.29}{\sqrt{\frac{0.873}{7}}} = 2.41$$

There is a significant difference between $\overline{X}_1$ and $\overline{X}_2$.

13. H_0: $\mu_1 = \mu_2 = \mu_3$ (claim)
H_1: At least one mean is different from the others.
C. V. = 6.01 $\alpha = 0.01$
$s_B^2 = 70.776$ $s_W^2 = 6.094$
$F = \frac{70.776}{6.094} = 11.614$

Reject the null hypothesis. There is not enough evidence to support the claim that the means are the same.

Scheffe Test C. V. = 12.02
$\overline{X}_1$ vs $\overline{X}_2$:

$$F_s = \frac{(27.75 - 26.667)^2}{6.094(\frac{1}{8} + \frac{1}{6})} = 0.660$$

$\overline{X}_1$ vs $\overline{X}_3$:

$$F_s = \frac{(27.75 - 21.857)^2}{6.094(\frac{1}{8} + \frac{1}{7})} = 21.274$$

$\overline{X}_2$ vs $\overline{X}_3$:

$$F_s = \frac{(26.667 - 21.857)^2}{6.094(\frac{1}{6} + \frac{1}{7})} = 12.266$$

There is a significant difference between means 1 and 3 and means 2 and 3.

14. H_0: $\mu_1 = \mu_2 = \mu_3$
H_1: At least one mean is different from the others. (claim)
C. V. = 3.68 $\alpha = 0.05$
$s_B^2 = 617.167$ $s_W^2 = 58.811$
$F = \frac{617.167}{58.811} = 10.494$
Reject the null hypothesis. There is enough evidence to support the claim that at least one mean is different from the others.

Tukey Test:
C. V. = 3.67
$\overline{X}_1 = 47.67$
$\overline{X}_2 = 63$
$\overline{X}_3 = 43.83$

14. continued
$\overline{X}_1$ vs $\overline{X}_2$:

$$q = \frac{47.67 - 63}{\sqrt{\frac{58.811}{6}}} = -4.90$$

$\overline{X}_1$ vs $\overline{X}_3$:

$$q = \frac{47.67 - 43.83}{\sqrt{\frac{58.811}{6}}} = 1.23$$

$\overline{X}_2$ vs $\overline{X}_3$:

$$q = \frac{63 - 43.83}{\sqrt{\frac{58.811}{6}}} = 6.12$$

There is a significant difference between $\overline{X}_1$ and $\overline{X}_2$ and between $\overline{X}_2$ and $\overline{X}_3$.

15. H_0: $\mu_1 = \mu_2 = \mu_3$ (claim)
H_1: At least one mean is different from the others.
C. V. = 2.70 $\alpha = 0.10$
$s_B^2 = 0.1213$ $s_W^2 = 2.3836$
$F = \frac{0.1213}{2.3836} = 0.0509$
Do not reject. There is not enough evidence to reject the claim that the means are the same.

16. H_0: $\mu_1 = \mu_2 = \mu_3 = \mu_4$
H_1: At least one mean is different from the others. (claim)
C. V. = 3.07 $\alpha = 0.05$
$s_B^2 = 15.3016$ $s_W^2 = 33.5283$
$F = \frac{15.3016}{33.5283} = 0.4564$
Do not reject. There is not enough evidence to support the claim that at least one mean is different.

17.
a. two-way ANOVA
b. diet and exercise program
c. 2

d. H_0: There is no interaction effect between the type of exercise program and the type of diet on a person's weight loss.
H_1: There is an interaction effect between the type of exercise program and the type of diet on a person's weight loss.

H_0: There is no difference in the means of the weight losses for those in the exercise programs.

17d. continued
H_1: There is a difference in the means of the weight losses for those in the exercise programs.

H_0: There is no difference in the means of the weight losses for those in the diet programs.
H_1: There is a difference in the means of the weight losses for those in the diet programs.

e. Diet: $F = 21$, significant
Exercise Program: $F = 0.429$, not significant
Interaction: $F = 0.429$, not significant

f. Reject the null hypothesis for the diets.

Note: Graphs are not to scale and are intended to convey a general idea. Answers may vary due to rounding.

EXERCISE SET 13-2

1.
Non-parametric means hypotheses other than those using population parameters can be tested, whereas distribution free means no assumptions about the population distributions have to be satisfied.

2.
When the assumptions for the parametric methods cannot be met, statisticians use non-parametric methods.

3.
The advantages of non-parametric methods are:

1. They can be used to test population parameters when the variable is not normally distributed.
2. They can be used when data is nominal or ordinal in nature.
3. They can be used to test hypotheses other than those involving population parameters.
4. The computations are easier in some cases than the computations of the parametric counterparts.
5. They are easier to understand.

The disadvantages are:

1. They are less sensitive than their parametric counterparts.
2. They tend to use less information than their parametric counterparts.
3. They are less efficient than their parametric counterparts.

4.

DATA	1	2	4	5	7	8	9
RANK	1	2	3	4	5	6	7

5.

DATA	21	31	34	41	41	61	65	72
RANK	1	2	3	4.5	4.5	6	7	8

6.

DATA	73	186	241	320	432
RANK	1	2	3	4	5

7.

DATA	3	5	5	6	7	8	8	9	12	14	15	17
RANK	1	2.5	2.5	4	5	6.5	6.5	8	9	10	11	12

8.

DATA	18	22	25	28	28	32	37	41	41	43
RANK	1	2	3	4.5	4.5	6	7	8.5	8.5	10

9.

DATA	187	190	190	236	321	532	673
RANK	1	2.5	2.5	4	5	6	7

10.

DATA	2.5	3.21	3.6	3.8	4.1	4.1	4.12	7.9	7.9
RANK	1	2	3	4	5.5	5.5	7	8.5	8.5

EXERCISE SET 13-3

1.
The sign test uses only + or − signs.

2.
The median.

3.
The smaller number of + or − signs.

4.
The normal approximation.

5.

+	−	−	−	−
+	+	+	+	+
−	−	+	+	0
−	−	+	+	+

H_0: Median = 2.8 (claim)
H_1: Median $\neq$ 2.8

$\alpha = 0.05 \quad n = 19$
C. V. = 4
Test value = 8

8 ↓

3	4	5	6	7	8	9

Do not reject the null hypothesis. There is not enough evidence to reject the claim that the median is 2.8.

6.

0	+	+	+	+
−	−	0	+	+
−	−	+	+	+
+	−	−	−	−

6. continued
H_0: Median = 81° (claim)
H_1: Median $\neq$ 81°

$\alpha = 0.01$ n = 18
C. V. = 3 Test value = 8

8 ↓

| | | | | | |
3 4 5 6 7 8 9

Do not reject the null hypothesis. There is not enough evidence to reject the claim that the median temperature is 81°.

7.
\+ + + +
− + + +
\+ − − +

H_0: Median = \$325 (claim)
H_1: Median $\neq$ \$325

$\alpha = 0.05$ n = 12
C. V. = 2 Test value = 3

3 ↓

| | | | |
0 1 2 3 4

Do not reject the null hypothesis. There is not enough evidence to reject the claim that the median rent is \$325.

8.
\+ + − − + +
− − − 0 − +
\+ 0 + + + +
\+ + + +

H_0: Median = \$5.00 (claim)
H_1: Median $\neq$ \$5.00

$\alpha = 0.10$ n = 20
C. V. = 5 Test value = 6

6 ↓

| | | | |
3 4 5 6 7

Do not reject the null hypothesis. There is not enough evidence to reject the claim that the median cost of beef is \$5.00 per pound.

9.
H_0: median = 7.30 (claim)
H_1: median $\neq$ 7.30

C. V. = ± 1.96

$$z = \frac{(x+0.5)-\left(\frac{n}{2}\right)}{\frac{\sqrt{n}}{2}} = \frac{(15+0.5)-\frac{37}{2}}{\frac{\sqrt{37}}{2}}$$

$$= \frac{-3}{3.041} = -0.99$$

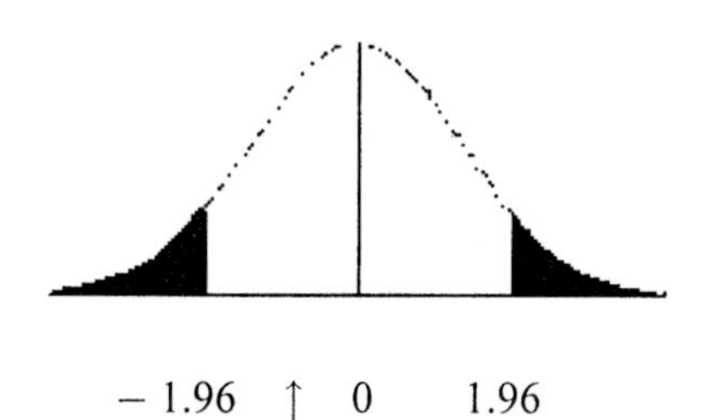

− 1.96 ↑ 0 1.96
− 0.99

Do not reject the null hypothesis. There is not enough evidence to reject the claim that the median is 7.30 minutes.

10.
H_0: Median $\geq$ 50
H_1: Median < 50 (claim)

C. V. = − 1.28

$$z = \frac{(42+0.5)-\frac{100}{2}}{\frac{\sqrt{100}}{2}} = -1.5$$

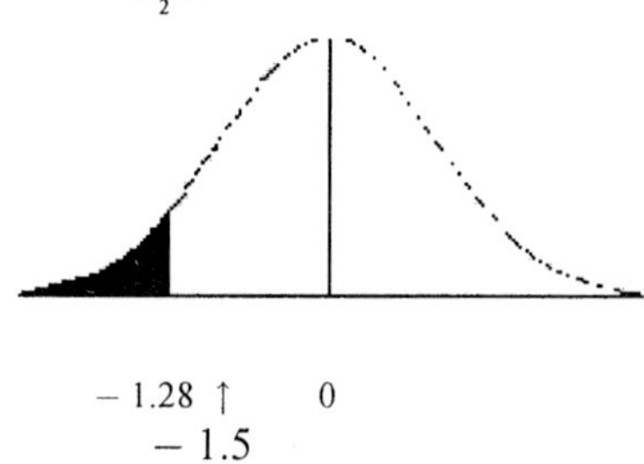

− 1.28 ↑ 0
− 1.5

Reject the null hypothesis. There is enough evidence to support the claim that the diet is effective.

11.
H_0: p $\leq$ 50%
H_1: p > 50% (claim)

$$z = \frac{(x+0.5)-\left(\frac{n}{2}\right)}{\frac{\sqrt{n}}{2}} = \frac{(21+0.5)-\frac{50}{2}}{\frac{\sqrt{50}}{2}}$$

$$= \frac{21.5-25}{3.54} = \frac{-3.5}{3.54} = -0.99$$

11. continued
P-value = 0.1611

Do not reject. There is not enough evidence to support the claim that more than 50% of the students favor single room dormitories.

12.
H_0: Median $\geq$ 37.5
H_1: Median $<$ 37.5 (claim)

C. V. = -2.05

$$z = \frac{(23 + 0.5) - \frac{75}{2}}{\frac{\sqrt{75}}{2}} = -3.23$$

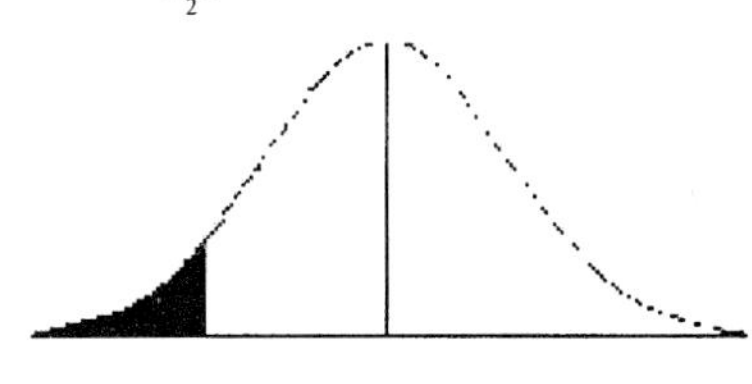

↑ -1.28 0
-3.23

Reject. There is enough evidence to support the claim that less than half of the lifeguards favor water to maintain hydration.

13.
H_0: Median = 50 (claim)
H_1: Median $\neq$ 50

$$z = \frac{(x + 0.5) - \left(\frac{n}{2}\right)}{\frac{\sqrt{n}}{2}} = \frac{(38 + 0.5) - \frac{100}{2}}{\frac{\sqrt{100}}{2}}$$

$$= \frac{38.5 - 50}{5} = \frac{-13.5}{5} = -2.3$$

P-value = 0.0057

Reject. There is enough evidence to reject the claim that 50% of the students are against extending the school year.

14.
H_0: Median = 27 (claim)
H_1: Median $\neq$ 27

C. V. = ± 1.96

$$z = \frac{(x + 0.5) - \left(\frac{n}{2}\right)}{\frac{\sqrt{n}}{2}} = \frac{(12 + 0.5) - \frac{29}{2}}{\frac{\sqrt{29}}{2}}$$

$$= \frac{-2}{2.693} = -0.74$$

14. continued

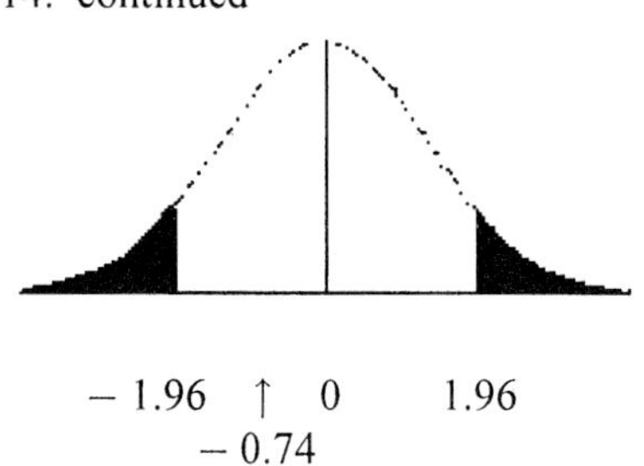

-1.96 ↑ 0 1.96
-0.74

Do not reject the null hypothesis. There is not enough evidence to reject the claim that the median age is 27.

15.

A	B	C	D	E	F	G	H
+	+	+	+	+	−	+	+

H_0: The medication has no effect on weight loss.
H_1: The medication affects weight loss. (claim)
$\alpha = 0.05$ n = 8
C. V. = 0 Test value = 1

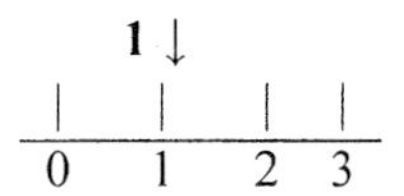

Do not reject the null hypothesis. There is not enough evidence to support the claim that the medication affects weight loss.

16.

1	2	3	4	5	6	7	8	9	10
−	0	+	+	+	−	−	+	+	−

H_0: The two machines give equivalent readings. (claim)
H_1: The two machines give different readings.
$\alpha = 0.01$ n = 9
C. V. = 0 Test value = 4

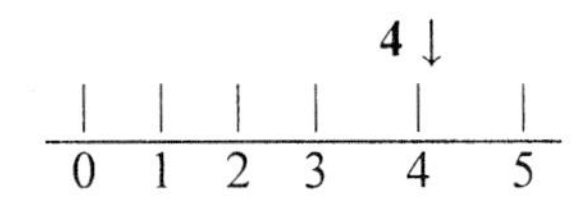

Do not reject the null hypothesis. There is not enough evidence to reject the claim that both machines gave the same readings.

17.

1	2	3	4	5	6	7	8	9
−	−	+	−	−	−	−	−	−

H_0: Reasoning ability will not be affected by the course.
H_1: Reasoning ability increased after the course. (claim)
$\alpha = 0.05$ n = 9
C. V. = 1 Test value = 1

1 ↓
0 1 2 3

Reject the null hypothesis. There is enough evidence to support the claim that the reasoning ability has increased after the course.

18.

1	2	3	4	5	6	7	8	9	10	11	12
+	+	+	−	+	+	−	+	+	+	+	+

H_0: The pill has no effect on the caloric intake of the person eating.
H_1: The pill has an effect on the caloric intake of the person eating. (claim)
$\alpha = 0.02$ n = 12
C. V. = 1 Test value = 2

2 ↓
0 1 2 3

Do not reject the null hypothesis. There is not enough evidence to support the claim that the pill has any effect on the caloric intake of a person.

19.

1	2	3	4	5	6	7	8	9	10
−	+	+	+	+	+	−	0	−	+

H_0: Alcohol has no effect on a person's I. Q. test score.
H_1: Alcohol does effect a person's I. Q. Test score. (claim)

$\alpha = 0.10$ n = 9
C. V. = 1 Test value = 3

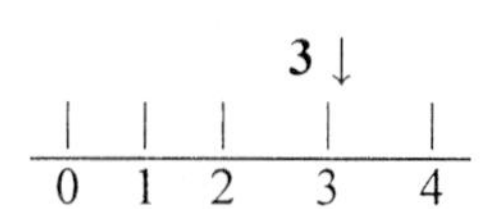

19. continued
Do not reject the null hypothesis. There is not enough evidence to reject the claim that alcohol has no effect on a person's I. Q. score.

20.

1	2	3	4	5	6	7	8	9
+	+	−	0	−	+	+	+	+

H_0: Increased maintenance does not reduce the number of defective parts a machine produces.
H_1: Increased maintenance reduces the number of defective parts a machine produces. (claim)
$\alpha = 0.01$ n = 8
C. V. = 0 Test value = 2

2 ↓
0 1 2 3 4

Do not reject the null hypothesis. There is not enough evidence to support the claim that increased maintenance reduces the number of defective parts manufactured by the machines.

21.
3, 4, 6, 9, 12, 15, 15, 16, 18, 22, 25, 30
At $\alpha = 0.05$, the value from Table J with n = 12 is 2; hence, count in 3 numbers from each end to get $6 \leq MD \leq 22$.

22.
100, 101, 106, 115, 115, 141, 142, 143, 143, 145, 147, 147, 150, 152, 153, 155, 157, 160, 163, 164
MD = 146
$141 \leq MD \leq 153$

23.
4.2, 4.5, 4.7, 4.8, 5.1, 5.2, 5.6, 6.3, 7.1, 7.2, 7.8, 8.2, 9.3, 9.3, 9.5, 9.6
At $\alpha = 0.02$, the value from Table J with n = 16 is 2; hence, count 3 numbers from each end to get $4.7 \leq MD \leq 9.3$

24.
1, 2, 3, 5, 6, 8, 10, 15, 15, 21, 24, 31, 33, 41, 42, 54, 56, 58, 65
MD = 21
$5 \leq MD \leq 54$

25.
12, 14, 14, 15, 16, 17, 18, 19, 19, 21, 23, 25, 27, 32, 33, 35, 39, 41, 42, 47
At $\alpha = 0.05$, the value from Table J with n = 20 is 5; hence, count in 6 numbers from each end to get $17 \leq MD \leq 33$.

EXERCISE SET 13-4

1.
The sample sizes must be greater than or equal to 10.

2.
The t-test for independent samples.

3.
The standard normal distribution.

4.
H_0: There is no difference in the length of the sentences of the males and females. (claim)
H_1: There is a difference in the length of the sentences of the males and females.

C. V. = ± 1.96

2	3	4	5	6	7	8	9	11
1	2	3	4	5	6	7	8	9
F	F	F	F	M	F	M	F	F

12	12	13	14	15	16	17	19	21
10.5	10.5	12	13	14	15	16	17	18
F	M	M	M	M	F	F	M	F

22	23	24	26	26	27	30	32
19	20	21	22.5	22.5	24	25	26
M	F	M	F	M	M	F	M

R = 191

$$\mu_R = \frac{12(12 + 14 + 1)}{2} = 162$$

$$\sigma_R = \sqrt{\frac{12 \cdot 14(12 + 14 + 1)}{12}} = 19.44$$

$$Z = \frac{191 - 162}{19.44} = 1.49$$

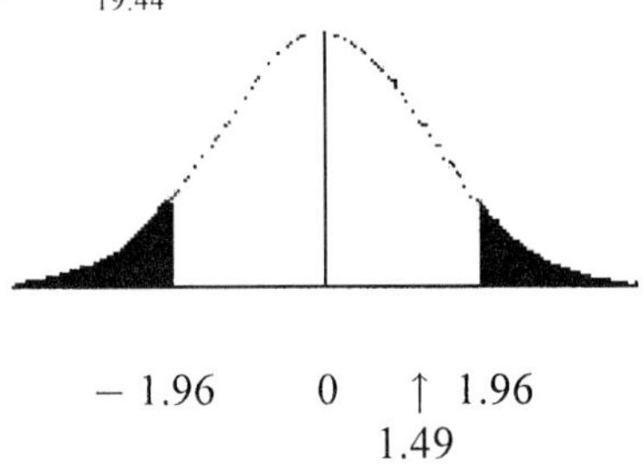

− 1.96 0 ↑ 1.96
1.49

4. continued
Do not reject the null hypothesis. There is not enough evidence to reject the claim that there is no difference in the sentences received by the men and women.

5.
H_0: There is no difference in the number of books each group read. (claim)
H_1: There is a difference in the number of books each group read.
C. V. = ± 1.65

0	1	2	3	3	4	4	5	5
1	2	3	4.5	4.5	6.5	6.5	8.5	8.5
S	S	S	S	S	M	S	S	S

6	7	7	8	9	10	11	11	12
10	11.5	11.5	13	14	15	16.5	16.5	18.5
M	M	M	M	M	M	S	S	M

12	13	15	16	18
18.5	20	21	22	23
S	M	M	S	M

R = 164

$$\mu_R = \frac{n_1(n_1 + n_2 + 1)}{2}$$

$$= \frac{11(11 + 12 + 1)}{2} = \frac{11(24)}{2} = \frac{264}{2} = 132$$

$$\sigma_R = \sqrt{\frac{n_1 \cdot n_2(n_1 + n_2 + 1)}{12}}$$

$$= \sqrt{\frac{11 \cdot 12(11 + 12 + 1)}{12}} = \sqrt{\frac{(11)(12)(24)}{12}}$$

$$= \sqrt{264} = 16.25$$

$$Z = \frac{R - \mu_R}{\sigma_R} = \frac{164 - 132}{16.25} = 1.97$$

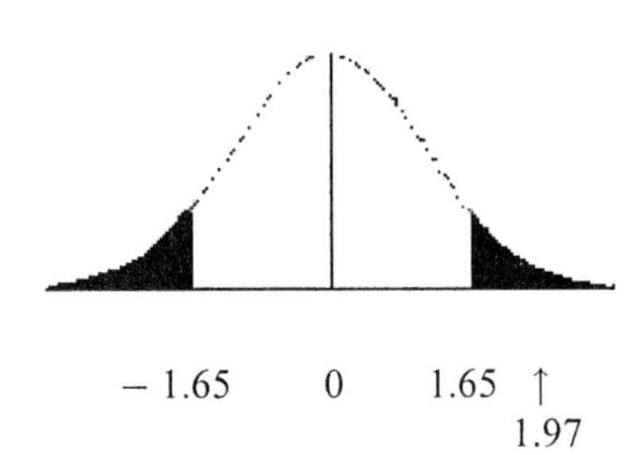

− 1.65 0 1.65 ↑
1.97

Reject the null hypothesis. There is enough evidence to reject the claim that there is no difference in the number of books read by each group.

6.
H_0: There is no difference in the lifetimes of the two brands of video games. (claim)
H_1: There is a difference in the lifetimes of the two brands of video games.

C. V. = ± 2.58

22	28	29	32	34	34	38	38	39	39	39
1	2	3	4	5.5	5.5	7.5	7.5	10	10	10
A	A	B	B	A	A	B	B	A	B	A

41	42	42	43	43	44	45	47	49	51	53
12	13.5	13.5	15.5	15.5	17	18	19	20	21	22
A	A	A	B	B	B	B	A	B	A	B

R = 113 for A $\quad \mu_R = \frac{11(11+11+1)}{2} = 126.5$

$$\sigma_R = \sqrt{\frac{11(11)(23)}{12}} = 15.23$$

$$z = \frac{113 - 126.5}{15.23} = -0.89$$

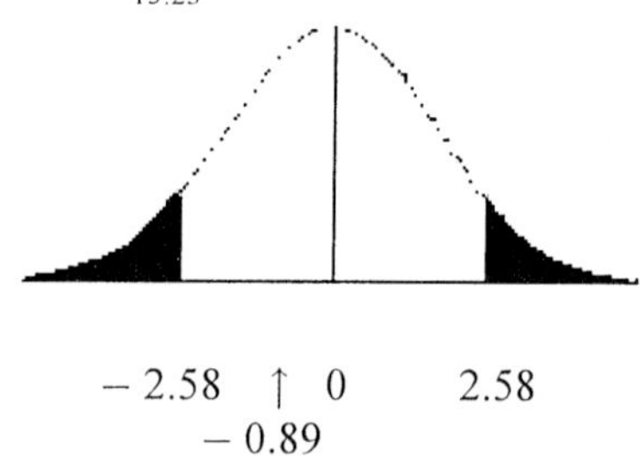

$-2.58 \quad \uparrow \quad 0 \quad 2.58$
-0.89

Do not reject the null hypothesis. There is not enough evidence to reject the claim that there is no difference in the lifetimes of the two brands of video games.

7.
H_0: There is no difference in the number of scholarships for the two high schools.
H_1: There is a difference between the number of scholarships for the two schools. (claim)

C. V. = ± 1.96

1	2	3	3	3	4	4	4	4	5	5	6
1	2	4	4	4	7.5	7.5	7.5	7.5	10.5	10.5	13
V	O	V	O	O	V	V	O	V	O	O	V

6	6	7	7	7	8	8	9	9	10	11	12
13	13	16	16	16	18.5	18.5	20.5	20.5	22	23	24
O	O	V	V	O	V	O	V	O	V	V	V

R = 156.5 for Valley View High School
R = 143.5 for Ocean View High School

7. continued

$$\mu_R = \frac{12(12+12+1)}{2} = 150$$

$$\sigma_R = \sqrt{\frac{12 \cdot 12(12+12+1)}{12}} = 17.32$$

$$z = \frac{156.5-150}{17.32} = 0.38$$

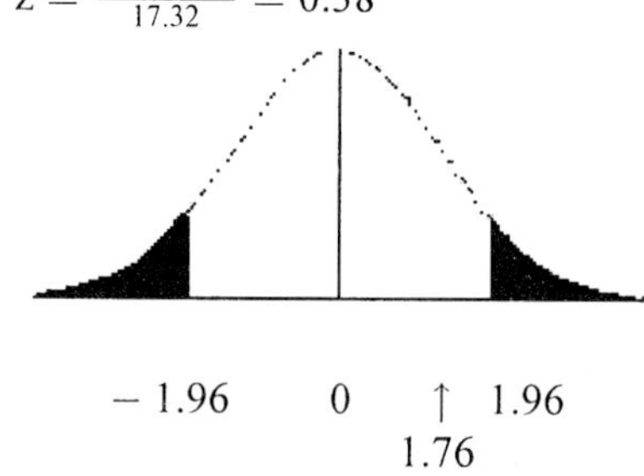

$-1.96 \quad 0 \quad \uparrow \quad 1.96$
1.76

Do not reject the null hypothesis. There is enough evidence to support the claim that there is a difference in the number of scholarships.

8.
H_0: There is no difference in job satisfaction scores for each group. (claim)
H_1: There is a difference in job satisfaction scores for the two groups.

52	56	58	59	63	64	66	68	68
1	2	3	4	5	6	7	8.5	8.5
O	U	O	O	O	O	O	U	O

72	73	75	77	78	79	82	83	85
10	11	12	13	14	15	16	17	18
U	O	U	U	U	O	O	U	O

86	88	93	93	94	97	98	99
19	20	21	22	23	24	25	26
U	O	U	U	O	U	U	U

R = 137.5 (for 5 years and over)

$$\mu_R = \frac{13(13+13+1)}{2} = 175.5$$

$$\sigma_R = \sqrt{\frac{13 \cdot 13(13+13+1)}{12}} = 19.5$$

$$Z = \frac{137.5-175.5}{19.5} = -1.95$$

P-value = 0.026
Reject the null hypothesis. There is enough evidence to reject the claim that there is no difference in job satisfaction scores for each group.

9.

H_0: There is no difference in the amount of money awarded to each city.

H_1: There is a difference in the amount of money awarded to each city. (claim)

C. V. = ± 1.96

232	239	324	431	453	563	601	602
1	2	3	4	5	6	7	8
A	B	B	A	B	A	A	A

605	626	648	687	718	752	769	824
9	10	11	12	13	14	15	16
A	B	A	A	B	B	B	A

832	869	885	918	921	925	927	953
17	18	19	20	21	22	23	24
B	B	B	B	A	A	B	A

R = 141 (for city A)

$$\mu_R = \frac{n_1(n_1 + n_2 + 1)}{2} = \frac{12(12 + 12 + 1)}{2} = 150$$

$$\sigma_R = \sqrt{\frac{n_1 \cdot n_2(n_1 + n_2 + 1)}{12}}$$

$$= \sqrt{\frac{12 \cdot 12(12 + 12 + 1)}{12}} = 17.321$$

$$Z = \frac{R - \mu_R}{\sigma_R} = \frac{141 - 150}{17.321} = -0.520$$

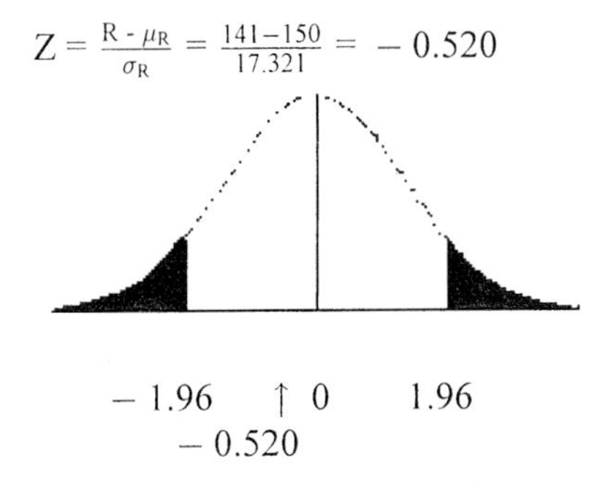

-1.96 ↑ 0 1.96

-0.520

Do not reject the null hypothesis. There is not enough evidence to support the claim that there is a difference in the amount of money awarded to the cities.

10.

H_0: Married men receive the same or lower ratings than do single men.

H_1: Married men receive higher ratings than do single men. (claim)

C. V. = ± 2.58

24	28	29	31	31	32	34	35
1	2	3	4.5	4.5	6	7	8
S	S	M	S	M	M	S	M

10. continued

36	36	37	37	38	40	41	42
9.5	9.5	11.5	11.5	13	14	15	16.5
M	S	M	M	S	S	M	M

42	43	44	46	48	49	50
16.5	18	19	20	21	22	23
S	M	M	S	S	S	S

R = 122.5

$$\mu_R = \frac{11(11 + 12 + 1)}{2} = 132$$

$$\sigma_R = \sqrt{\frac{11 \cdot 12(11 + 12 + 1)}{12}} = 16.25$$

$$Z = \frac{122.5 - 132}{16.25} = -0.5846$$

-2.58 ↑ 0 2.58

-0.5846

Do not reject the null hypothesis. There is not enough evidence to support the claim that married men receive higher ratings.

11.

H_0: There is no difference in the times needed to assemble the product.

H_1: There is a difference in the times needed to assemble the product. (claim)

C. V. = ± 1.96

1.6	1.7	1.9	2.0	2.4	2.6	2.7	2.9	3.0
1	2	3	4	5	6	7	8	9
N	N	N	N	N	N	N	N	G

3.1	3.2	3.4	3.6	3.8	3.9	4.2	4.4	4.7
10	11	12	13	14	15	16	17	18
N	G	N	G	N	N	G	G	G

5.3	5.6	5.8	6.3	6.4	7.1	7.3	8.2
19	20	21	22	23	24	25	26
N	G	G	G	G	G	G	G

R = 245 (for graduates)

11. continued

$\mu_R = \frac{n_1(n_1 + n_2 + 1)}{2} = \frac{13(13 + 13 + 1)}{2}$

$= \frac{13(27)}{2} = 175.5$

$\sigma_R = \sqrt{\frac{n_1 \cdot n_2(n_1 + n_2 + 1)}{12}}$

$= \sqrt{\frac{13 \cdot 13(13 + 13 + 1)}{12}} = 19.5$

$Z = \frac{R - \mu_R}{\sigma_R} = \frac{245 - 175.5}{19.5} = \frac{69.5}{19.5} = 3.56$

− 1.96   0   1.96   ↑ 3.56

Reject the null hypothesis. There is enough evidence to support the claim that there is a difference in the productivity of the two groups.

EXERCISE SET 13-5

1.
The t-test for dependent samples.

2.

B	A	B − A	|B − A|	Rank	Signed Rank
18	20	− 2	2	2	− 2
32	21	11	11	8	8
35	26	9	9	7	7
37	37	0			
25	29	− 4	4	4	− 4
41	40	1	1	1	1
52	31	21	21	9	9
43	37	6	6	6	6
56	51	5	5	5	5
62	65	− 3	3	3	− 3

Sum of the − ranks: $(-2) + (-4) + (-3) = -9$.
Sum of the + ranks: $8 + 7 + 1 + 9 + 6 + 5 = 36$.
$w_s = 9$

3.

B	A	B − A	|B − A|	Rank	Signed Rank
108	110	− 2	2	1	− 1
97	97	0			
115	103	12	12	4.5	4.5
162	168	− 6	6	2	− 2
156	143	13	13	6	6
105	112	− 7	7	3	− 3
153	141	12	12	4.5	4.5

Sum of the − ranks: $(-1) + (-2) + (-3) = -6$.
Sum of the + ranks: $4.5 + 6 + 4.5 = 15$
$w_s = 6$

4.
C. V. = 59 $w_s = 62$
Since $62 > 59$, do not reject the null hypothesis.

5.
C. V. = 20 $w_s = 18$
Since $18 \leq 20$, reject the null hypothesis.

6.
C. V. = 101 $w_s = 53$
Since $53 \leq 101$, reject the null hypothesis.

7.
C. V. = 130 $w_s = 142$
Since $142 > 130$, do not reject the null hypothesis.

8.
C. V. = 107 $w_s = 109$
Since $109 > 107$, do not reject the null hypothesis.

9.
H_0: The workshop did not reduce anxiety.
H_1: The workshop reduced anxiety. (claim)

B	A	B − A	|B − A|	Rank	Signed Rank
23	22	1	1	1.5	1.5
26	29	− 3	3	6	− 6
30	27	3	3	6	6
31	29	2	2	3.5	3.5
39	33	6	6	8	8
23	21	2	2	3.5	3.5
28	25	3	3	6	6
27	28	− 1	1	1.5	− 1.5

Sum of the − ranks:
$(-6) + (-1.5) = -7.5$
Sum of the + ranks:
$1.5 + 6 + 3.5 + 8 + 3.5 + 6 = 28.5$

9. continued

n = 8 C. V. = 6

$w_s = 7.5$

Since 7.5 > 6, do not reject the null hypothesis. There is not enough evidence to support the claim that the workshop reduced the anxiety of the subjects.

10.

H_0: There is no difference in the salaries of the males and the females.

H_1: There is a difference in the salaries of the males and females. (claim)

n = 7 $\alpha = 0.10$ C. V. = 4

M	F	M – F	|M – F|	Rank	Signed Rank
18	16	2	2	2.5	2.5
43	38	5	5	6	6
32	35	– 3	3	4	– 4
27	29	– 2	2	2.5	2.5
15	15	0			
45	46	– 1	1	1	– 1
21	25	– 4	4	5	– 5
22	28	– 6	6	7	– 7

Sum of the – ranks:

$(-4) + (-2.5) + (-1) + (-5) + (-7)$
$= -19$

Sum of the + ranks: $2.5 + 6 = 8.5$

$w_s = 8.5$

Since 8.5 > 4, do not reject the null hypothesis. There is not enough evidence to support the claim that there is a difference in the salaries for the two groups.

11.

H_0: The sizes of police forces have decreased or remained the same.

H_1: The sizes of police forces have increased. (claim)

n = 10 $\alpha = 0.05$ C. V. = 11

B	A	B – A	|B – A|	Rank	Signed Rank
23,339	29,327	– 5988	5988	10	– 10
6886	7637	– 751	751	7	– 7
12,353	12,093	260	260	2	2
3716	4734	– 1018	1018	9	– 9
7218	6225	993	993	8	8
1376	1861	– 485	485	4	– 4
3808	3860	– 52	52	1	– 1
2084	2807	– 723	723	6	– 6
1635	1978	– 343	343	3	– 3
1159	1662	– 503	503	5	– 5

11. continued

Sum of the – ranks:

$(-10) + (-7) + (-9) + (-4) + (-1)$
$+ (-6) + (-3) + (-5) = -45$

Sum of the + ranks: $2 + 8 = 10$

$w_s = 10$

Since $10 \leq 11$, reject the null hypothesis. There is enough evidence to support the claim that the police forces have increased.

12.

H_0: There is no difference in the attitudes of the couples.

H_1: There is a difference in the attitudes of the couples. (claim)

n = 7 $\alpha = 0.10$ C. V. = 4

B	A	B – A	|B – A|	Rank	Signed Rank
43	48	– 5	5	3	– 3
52	59	– 7	7	5	– 5
37	36	1	1	1	1
29	29	0			
51	60	– 9	9	6.5	– 6.5
62	68	– 6	6	4	– 4
57	59	– 2	2	2	– 2
61	72	– 9	9	6.5	– 6.5

Sum of the – ranks:

$(-3) + (-5) + (-6.5) + (-4) + (-2)$
$+ (-6.5) = 27$

Sum of the + ranks: 1

$w_s = 1$

Since $1 \leq 4$, reject the null hypothesis. There is enough evidence to support the claim that there is a difference in the attitude of the couples.

13.

H_0: There is no change in the size of the rosters. (claim)

H_1: The size of the rosters has changed.

n = 8 $\alpha = 0.10$ C. V. = 6

13. continued

B	A	B – A	|B – A|	Rank	Signed Rank
33	33	0			
29	37	– 8	8	5.5	– 5.5
42	46	– 4	4	4	– 4
32	34	– 2	2	1	– 1
28	45	– 17	17	7	– 7
55	58	– 3	3	2.5	– 2.5
46	46	0			
26	23	3	3	2.5	2.5
17	35	– 18	18	8	– 8
30	22	8	8	5.5	5.5
45	45	0			
21	21	0			
42	42	0			

Sum of the + ranks: $2.5 + 5.5 = 8$

Sum of the – ranks:
$(-5.5) + (-4) + (-1) + (-7) +$
$(-2.5) + (-8) = -28$

$w_s = 8$

Since $8 > 6$, do not reject the null hypothesis. There is not enough evidence to reject the claim that there is no change in the rosters.

EXERCISE SET 13-6

1.
H_0: There is no difference in the number of calories each brand contains.
H_1: There is a difference in the number of calories each brand contains. (claim)
C. V. = 7.815 $\alpha = 0.05$ d. f. = 3

A	Rank	B	Rank	C	Rank	D	Rank
112	7	110	6	109	5	106	3
120	13	118	12	116	9.5	122	15
135	24	123	16	125	17.5	130	21.5
125	17.5	128	19.5	130	21.5	117	11
108	4	102	2	128	19.5	116	9.5
121	14	101	1	132	23	114	8
R_1=	79.5	R_2=	56.5	R_3=	96	R_4=	68

$$H = \frac{12}{N(N+1)}\left(\frac{R_1^2}{n_1} + \frac{R_2^2}{n_2} + \frac{R_3^2}{n_3} + \frac{R_4^2}{n_4}\right)$$

$$- 3(N + 1)$$

$$H = \frac{12}{12(24+1)}\left(\frac{79.5^2}{6} + \frac{56.5^2}{6} + \frac{96^2}{6}\right.$$

$$\left. + \frac{68^2}{6}\right) = 2.842$$

1. continued

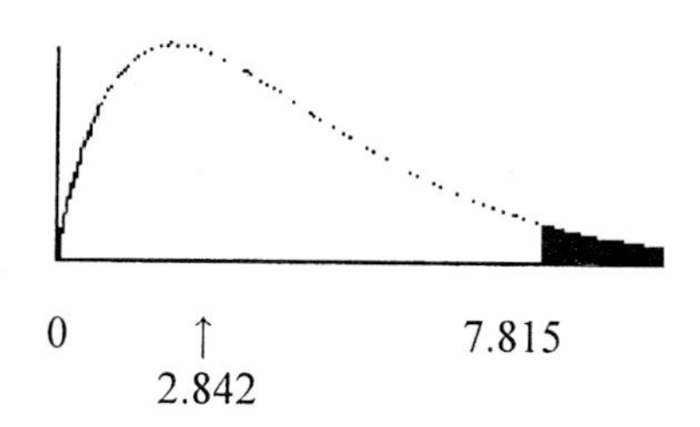

Do not reject the null hypothesis. There is not enough evidence to reject the claim that there is no difference in the calories.

2.
H_0: There is no difference in the scores of the three groups.
H_1: There is a difference in the scores of the three groups. (claim)
C. V. = 5.991 d. f. = 2 $\alpha = 0.05$

Oldest	Rank	Middle	Rank	Youngest	Rank
48	16	50	18	47	15
46	13.5	49	17	45	12
42	9.5	42	9.5	46	13.5
41	7.5	43	11	30	2
37	5	39	6	32	3.5
32	3.5	28	1	41	7.5
R_1=	55	R_2=	62.5	R_3=	53.5

$$H = \frac{12}{18(18+1)}\left(\frac{55^2}{6} + \frac{62.5^2}{6} + \frac{53.5^2}{6}\right)$$

$$- 3(18 + 1) = 0.272$$

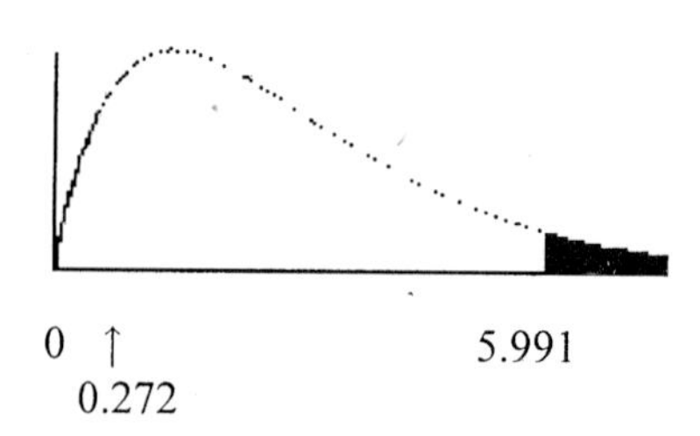

Do not reject the null hypothesis. There is not enough evidence to support the claim that there is a difference in the scores.

3.
H_0: There is no difference in the sales of the stores.
H_1: There is a difference in the sales of the stores. (claim)

C. V. = 4.605 d. f. = 2 $\alpha = 0.10$

3. continued

Radio	Rank	TV	Rank	Paper	Rank
832	13	1024	18	329	2.5
648	11	996	16	437	5
562	10	1011	17	561	9
786	12	853	14	329	2.5
452	6	471	7	382	4
975	15	$R_2=$	72	495	8
$R_1=$	67			262	1
				$R_3=$	32

$$H = \frac{12}{N(N+1)}\left(\frac{R_1^2}{n_1} + \frac{R_2^2}{n_2} + \frac{R_3^2}{n_3} + \frac{R_4^2}{n_4}\right)$$

$$-\,3(N+1)$$

$$= \frac{12}{18(18+1)}\left(\frac{67^2}{6} + \frac{72^2}{5} + \frac{32^2}{7}\right) - 3(18+1)$$

$$= \frac{12}{342}\,(748.167 + 1036.8 + 146.286) - 3(19)$$

$H = 10.8$

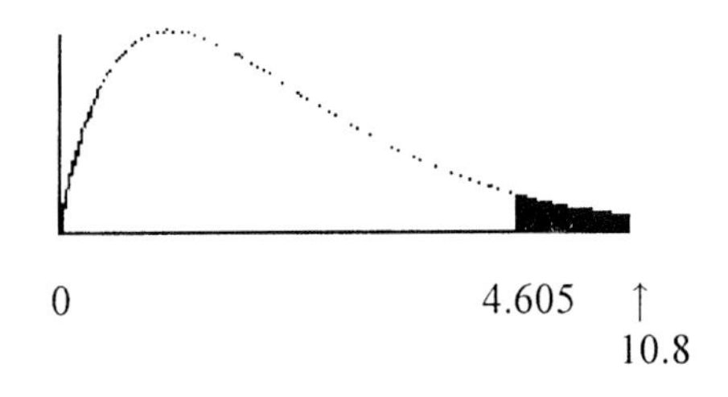

0 4.605 ↑ 10.8

Reject the null hypothesis. There is enough evidence to support the claim that there is a difference in sales.

4.

H_0: There is no difference in the amounts of sodium in the different brands of microwave dinners.

H_1: There is a difference in the amounts of sodium in the different brands of microwave dinners. (claim)

C. V. = 5.991 d. f. = 2 $\alpha = 0.05$

A	Rank	B	Rank	C	Rank
810	12	917	18	893	15.5
702	5	912	17	790	11
853	13	952	19	603	1
703	6	958	20	744	10
892	14	893	15.5	623	4
732	8			743	9
713	7			609	2
613	3				
$R_1=$	68	$R_2=$	89.5	$R_3=$	52.5

4. continued

$$H = \frac{12}{20(20+1)}\left(\frac{68^2}{8} + \frac{89.5^2}{5} + \frac{52.5^2}{7}\right)$$

$$-\,3(20+1) = 10.533$$

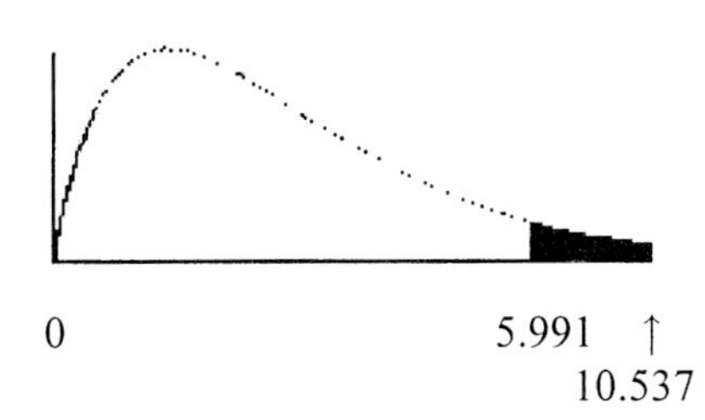

0 5.991 ↑ 10.537

Reject the null hypothesis. There is enough evidence to support the claim that there is a difference in the amounts of sodium in the different brands of microwave dinners.

5.

H_0: There is no difference in the yields of the three plots.

H_1: There is a difference in the yields of the three plots. (claim)

C. V. = 9.210 d. f. = 2 $\alpha = 0.01$

A	Rank	B	Rank	C	Rank
32	2	43	6	50	10
38	3	45	7	56	14
31	1	49	9	58	15
40	5	46	8	54	13
39	4	51	11	52	12
$R_1=$	15	$R_2=$	41	$R_3=$	64

$$H = \frac{12}{N(N+1)}\left(\frac{R_1^2}{n_1} + \frac{R_2^2}{n_2} + \frac{R_3^2}{n_3} + \frac{R_4^2}{n_4}\right) - 3(N+1)$$

$$= \frac{12}{15(15+1)}\left(\frac{15^2}{5} + \frac{41^2}{5} + \frac{64^2}{5}\right) - 3(15+1)$$

$H = 12.020$

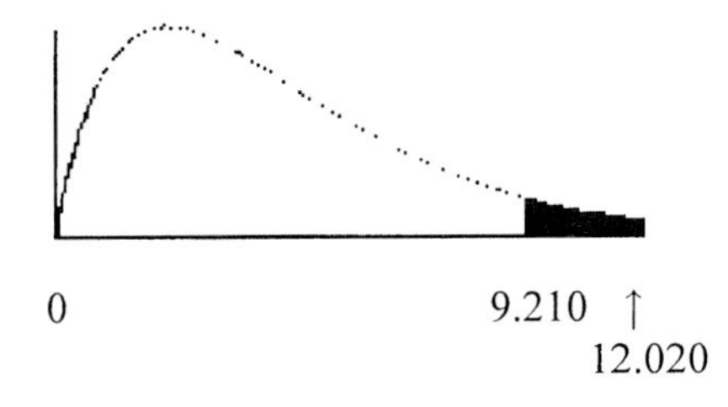

0 9.210 ↑ 12.020

Reject the null hypothesis. There is enough evidence to support the claim that the yields of the three plots are different.

6.
H_0: There is no difference in the number of job offers received by each group.
H_1: There is a difference in the number of job offers received by each group. (claim)
C. V. = 5.991 d. f. = 2 $\alpha = 0.05$

A	Rank	B	Rank	C	Rank
6	8	2	3	10	13
8	11	1	2	12	14
7	10	0	1	9	12
5	6	3	4	13	15
6	8	6	8	4	5
$R_1=$	43	$R_2=$	18	$R_3=$	59

$$H = \frac{12}{15(15+1)}\left(\frac{43^2}{5} + \frac{18^2}{5} + \frac{59^2}{5}\right) - 3(15+1)$$

H = 8.54

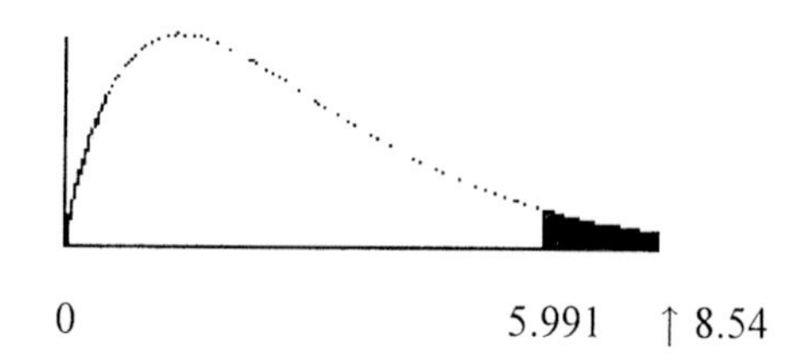

Reject the null hypothesis. There is enough evidence to support the claim that the number of job offers is different.

7.
H_0: There is no difference in the number of deaths due to lightning, tornado, flood or blizzard.
H_1: There is a difference in the number of weather-related deaths. (claim)
C. V. = 6.251

L	Rank	T	Rank	F	Rank	B	Rank
39	6.5	30	1.5	46	12	54	16
41	9	39	6.5	55	17	43	10
73	22	39	6.5	45	11	39	6.5
74	23	53	15	109	24	35	4
67	20	50	14	62	19	56	18
68	21	32	3	30	1.5	48	13
$R_1=$	101.5	$R_2=$	46.5	$R_3=$	85.5	$R_4=$	67.5

$$H = \frac{12}{N(N+1)}\left(\frac{R_1^2}{n_1} + \frac{R_2^2}{n_2} + \frac{R_3^2}{n_3} + \frac{R_4^2}{n_4}\right) - 3(N+1)$$

$$H = \frac{12}{24(24+1)}\left(\frac{101.5^2}{6} + \frac{46.5^2}{6} + \frac{85.5^2}{6} + \frac{67.5^2}{6}\right) - 3(24+1)$$

7. continued
H = 5.537

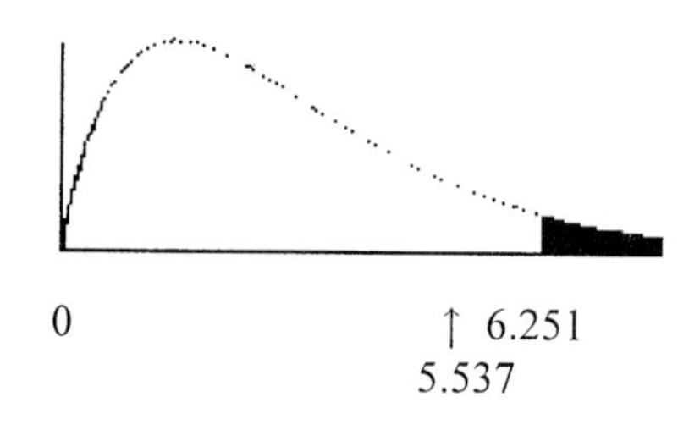

Do not reject the null hypothesis. There is not enough evidence to support the claim that there is a difference in the number of weather-related deaths.

8.
H_0: There is no difference in the monthly maintenance costs for the 3 brands.
H_1: There is a difference in the monthly maintenance costs for the 3 brands. (claim)
C. V. = 5.991 d. f. = 2 $\alpha = 0.05$

I	Rank	II	Rank	III	Rank
56	6	63	7	82	15
42	1	72	9	81	14
48	2	71	8	79	13
53	4	74	10	77	12
51	3	76	11	55	8
$R_1=$	16	$R_2=$	45	$R_3=$	59

$$H = \frac{12}{15(15+1)}\left(\frac{16^2}{5} + \frac{45^2}{5} + \frac{59^2}{5}\right) - 3(15+1)$$

H = 9.620

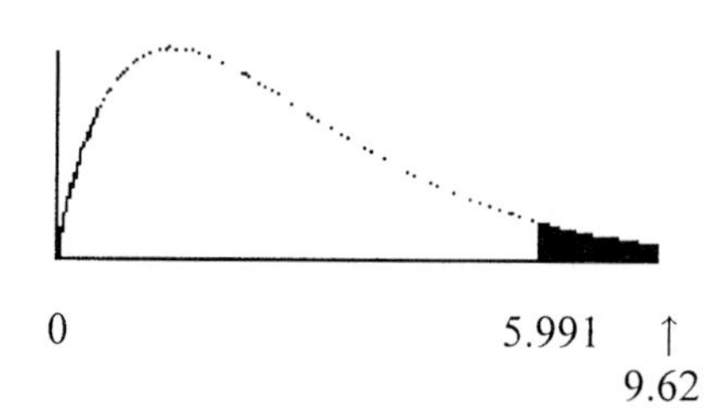

Reject the null hypothesis. There is enough evidence to support the claim that there is a difference in the monthly maintenance costs for the 3 brands.

9.
H_0: There is no difference in the number of crimes in the 5 precincts.
H_1: There is a difference in the number of crimes in the 5 precincts. (claim)

9. continued
C. V. = 13.277 d. f. = 4 $\alpha = 0.01$

1	Rank	2	Rank	3	Rank
105	24	87	13	74	7.5
108	25	86	12	83	11
99	22	91	16	78	9
97	20	93	18	74	7.5
92	17	82	10	60	5
R_1=	108	R_2=	69	R_3=	40

4	Rank	5	Rank
56	3	103	23
43	1	98	21
52	2	94	19
58	4	89	15
62	6	88	14
R_4=	16	R_5=	92

$$H = \frac{12}{N(N+1)}\left(\frac{R_1^2}{n_1} + \frac{R_2^2}{n_2} + \frac{R_3^2}{n_3} + \frac{R_4^2}{n_4} + \frac{R_5^2}{n_5}\right) - 3(N+1)$$

$$= \frac{12}{25(25+1)}\left(\frac{108^2}{5} + \frac{69^2}{5} + \frac{40^2}{5} + \frac{16^2}{5} + \frac{92^2}{5}\right) - 3(25+1) = 20.753$$

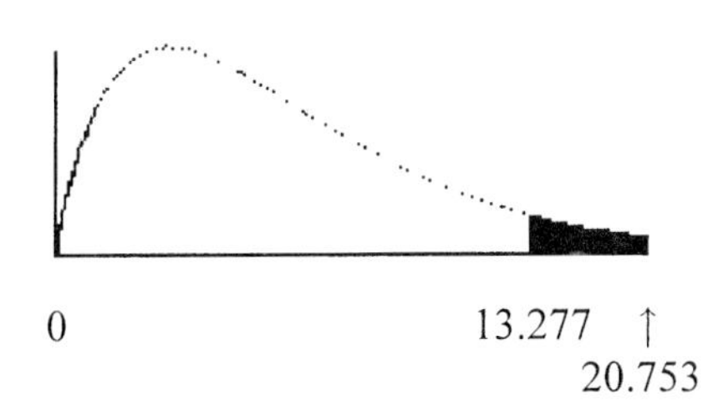

Reject the null hypothesis. There is enough evidence to support the claim that there is a difference in the number of crimes for the precincts.

10.
H_0: There is no difference in the number of unemployed people seeking employment based on education.
H_1: There is a difference in the number of unemployed people seeking employment based on education. (claim)
d. f. = 2 $\alpha = 0.05$
P-value > 0.10 (0.119)

10. continued

HS	Rank	CD	Rank	PGD	Rank
49	8.5	23	2.5	17	1
43	7	49	8.5	38	6
51	10	54	12	23	2.5
108	15	87	14	52	11
68	13	28	5	26	4
R_1=	53.5	R_2=	42	R_3=	24.5

$$H = \frac{12}{15(15+1)}\left(\frac{53.5^2}{5} + \frac{42^2}{5} + \frac{24.5^2}{5}\right) - 3(15+1) = 4.265$$

Since P-value > 0.05, do not reject the null hypothesis. There is not enough evidence to support the claim that there is a difference in the number of unemployed people seeking employment based on education.

11.
H_0: There is no difference in the final exam scores of the three groups.
H_1: There is a difference in the final exam scores of the three groups. (claim)
d. f. = 2 $\alpha = 0.10$

A	Rank	B	Rank	C	Rank
98	18	97	17	99	19
100	20	88	11	94	14
95	15	82	8	96	16
92	13	84	9	89	12
86	10	75	5	81	7
76	6	73	3	72	2
71	1	74	4	R_3=	70
R_1=	83	R_2=	57		

$$H = \frac{12}{N(N+1)}\left(\frac{R_1^2}{n_1} + \frac{R_2^2}{n_2} + \frac{R_3^2}{n_3}\right) - 3(N+1)$$

$$= \frac{12}{20(20+1)}\left(\frac{83^2}{7} + \frac{57^2}{7} + \frac{70^2}{6}\right) - 3(20+1)$$

H = 1.710

P-value > 0.10 (0.425)
Do not reject. There is not enough evidence to support the claim that there is a difference in the final exam scores of the three groups.

EXERCISE SET 13-7

1.
0.716

2.
0.488

3.
0.648

4.
0.833

5.
H_0: $\rho = 0$
H_1: $\rho \neq 0$
C. V. = ± 0.564 n = 10 $\alpha = 0.10$

Tornadoes	R_1	Temp	R_2	$R_1 - R_2$	d^2
668	6	112	5	1	1
781	7	118	9	− 2	4
1590	10	109	3	7	49
798	8	117	8	0	0
1198	9	121	10	− 1	1
169	2	108	2	0	0
310	3	111	4	− 1	1
360	4	113	6	− 2	4
21	1	105	1	0	0
625	5	114	7	− 2	4
				$\sum d^2 =$	64

$r_s = 1 - \frac{6 \cdot \sum d^2}{n(n^2-1)} = 1 - \frac{6 \cdot 64}{10\,(10^2-1)}$

$r_s = 0.612$

Reject the null hypothesis. There is a significant relationship between the number of tornadoes and high temperatures.

6.
H_0: $\rho = 0$
H_1: $\rho \neq 0$

C. V. = ± 0.738

$r_s = 1 - \frac{6 \cdot 34}{8(8^2-1)} = 0.595$

Do not reject the null hypothesis. There is no significant relationship between the rankings.

7.
H_0: $\rho = 0$
H_1: $\rho \neq 0$

C. V. = ± 0.886 n = 6 $\alpha = 0.05$

7. continued

Sentence	R_1	Time	R_2	$R_1 - R_2$	d^2
227	6	97	6	0	0
120	5	45	5	0	0
106	4	40	4	0	0
77	3	26	3	0	0
60	2	18	1	1	1
49	1	21	2	− 1	1
				$\sum d^2 =$	2

$r_s = 1 - \frac{6 \cdot \sum d^2}{n(n^2-1)}$

$r_s = 1 - \frac{6 \cdot 2}{6\,(6^2-1)} = 0.943$

Reject. There is a significant relationship between sentence and time served.

8.
H_0: $\rho = 0$
H_1: $\rho \neq 0$

C. V. = ± 0.886

$r_s = 1 - \frac{6 \cdot 18}{6(6^2-1)} = 0.486$

Do not reject the null hypothesis. There is not enough evidence to say that a significant relationship exists.

9.
H_0: $\rho = 0$
H_1: $\rho \neq 0$

C. V. = ± 0.738

Teen	R_1	Parent	R_2	$R_1 - R_2$	d^2
4	4	1	1	3	9
6	6	7	7	-1	1
2	2	5	5	-3	9
8	8	4	4	4	16
1	1	3	3	-2	4
7	7	8	8	-1	1
3	3	2	2	1	1
5	5	6	6	-1	3
				$\sum d^2 =$	42

$r_s = 1 - \frac{6 \cdot 42}{8(8^2-1)} = 1 - \frac{252}{504} = 0.5$

Do not reject the null hypothesis. There is not enought evidence to say that a significant relationship exists between the rankings.

10.
H_0: $\rho = 0$
H_1: $\rho \neq 0$

C. V. = ± 0.786
$r_s = 1 - \frac{6 \cdot 2}{7(7^2 - 1)} = 0.964$

Reject the null hypothesis. There is enough evidence to say that a significant relationship exists between the rankings.

11.
H_0: $\rho = 0$
H_1: $\rho \neq 0$
C. V. = ± 0.886

Designer	R_1	Customer	R_2	$R_1 - R_2$	d^2
48	2	35	2	0	0
76	4	44	3	1	1
30	1	28	1	0	0
88	5	50	4	1	1
61	3	75	5	-2	4
93	6	85	6	0	0
				$\sum d^2 =$	6

$r_s = 1 - \frac{6\sum d^2}{n(n^2-1)} = 1 - \frac{6 \cdot 6}{6(6^2-1)}$

$r_s = 0.829$

Do not reject the null hypothesis. There is not enough evidence to say that a significant relationship exists between the rankings.

12.
H_0: $\rho = 0$
H_1: $\rho \neq 0$

C. V. = ± 0.700

$r_s = 1 - \frac{6 \cdot 24}{9(9^2-1)} = 0.80$

Reject the null hypothesis. There is enough evidence to say that a significant relationship exists between the rankings.

13.
H_0: $\rho = 0$
H_1: $\rho \neq 0$

C. V. = ± 0.591

13. continued

Engineer	R_1	Customer	R_2	$R_1 - R_2$	d^2
81	8	85	9	-1	1
70	7	75	7	0	0
65	6	68	6	0	0
54	5	50	4	1	1
43	4	52	5	-1	1
90	12	95	11	1	1
41	3	48	3	0	0
88	11	100	12	-1	1
40	2	44	2	0	0
85	10	90	10	0	0
82	9	83	8	1	1
35	1	20	1	0	0
				$\sum d^2 =$	6

$r_s = 1 - \frac{6\sum d^2}{n(n^2-1)} = 1 - \frac{6 \cdot 6}{12(12^2-1)}$

$r_s = 0.979$

Reject the null hypothesis. There is enough evidence to say that a significant relationship exists between the rankings.

14.
H_0: $\rho = 0$
H_1: $\rho \neq 0$

C. V. = ± 0.886

$r_s = 1 - \frac{6 \cdot 8}{6(6^2-1)} = 0.771$

Do not reject the null hypothesis. There is no significant relationship between the rankings.

15.
H_0 = The occurrance of cavities is random.
H_1= The null hypothesis is not true.

The median of the data set is two. Using A = above and B = below the runs (going across) are shown:

B AA B AAA B A BB AAAA B A B A B A
B A B AAA B A BB

There are 21 runs. The expected number of runs is between 10 and 22; therefore, the null hypothesis should not be rejected. The number of cavities occurs at random.

16.
H_0: The numbers occur at random. (claim)
H_1: The null hypothesis is not true.

16. continued
O E O EE O E OO E OO EE OO EE OOO E O EE OOO EE

There are 18 runs and since this value is between 10 and 22; the decision is do not reject the null hypothesis. The numbers occur at random.

17.
H_0: The lotto numbers occur at random.
H_1: The null hypothesis is not true.

OO E OO EE O EE OOO EE OO E O E O EE

There are 14 runs and this is between 7 and 18; hence, do not reject the null hypothesis. The numbers occur at random.

18.
H_0: The answers to the test questions are random. (claim)
H_1: The null hypothesis is not true.

TTT FF TTT FFFFFF TTT FFF TTT F T FF TT F

There are 12 runs and since this is between 10 and 22, the null hypothesis is not rejected. The answers are random.

19.
H_0: The number of defective cigarettes manufactured by a machine occurs at random.
H_1: The null hypothesis is not true.

D AAAAAA DD A DD AAA DD AAAAAAA DDD AAA

There are 10 runs and since this is between 9 and 20, the null hypothesis is not rejected. The defective cigarettes occur at random.

20.
H_0: The gender occurs at random.
H_1: The null hypothesis is not true.

F MM FF M F MM FF MMM FFFFF M
There are 10 runs. The expected number of runs is between 6 and 16, hence the null hypothesis is not rejected. The gender occurs at random.

21.
H_0: The number of absences of employees occur at random.
H_1: The null hypothesis is not true.

The median of the data is 12. Using A = above and B = below, the runs are shown.

A B AAAAAAA BBBBBBBBB AAAAAA BBBB
There are six runs. The expected number of runs is between 9 and 21, hence the null hypothesis is rejected since six is not between 9 and 21. The number of absences do not occur at random.

22.
H_0: The days customers are able to ski occur at random. (claim)
H_1: The null hypothesis is false.

SSSSSS NNNNNNNNN SSS NN SSSSSSSS
There are 5 runs. The expected number of runs is between 9 and 20, hence the null hypothesis should be rejected. The days customers are able to ski do not occur at random.

23.
H_0: The I. Q.'s are random.
H_1: The null hypothesis is not true.

The median of the data is 98. Using A = above and B = below, the runs are shown.

AAAAAAA BBBBB A BB A BB AA
Since there are 7 runs, the null hypothesis is not rejected because the 7 is within the 6 to 16 range. The I. Q.'s are random.

24.
$r = \frac{\pm 1.96}{\sqrt{50-1}} = \pm 0.28$

25.
$r = \frac{\pm z}{\sqrt{n-1}} = \frac{\pm 2.58}{\sqrt{30-1}} = \pm 0.479$

26.
$r = \frac{\pm 2.33}{\sqrt{35-1}} = \pm 0.400$

27.
$r = \frac{\pm z}{\sqrt{n-1}} = \frac{\pm 1.65}{\sqrt{60-1}} = \pm 0.215$

28.
$r = \frac{\pm 2.58}{\sqrt{40-1}} = \pm 0.413$

REVIEW EXERCISES - CHAPTER 13

1.
+ + + + − − − − − − − + +

H_0: median = 5 (claim)
H_1: median ≠ 5

C. V. = 2 Test Value = 6

Do not reject. There is not enough evidence to reject the claim that the median is 5.

2.
H_0: median = 40,000 miles (claim)
H_1: median ≠ 40,000 miles

C. V. = ± 1.96

$$Z = \frac{(12 + 0.5) - \left(\frac{30}{2}\right)}{\frac{\sqrt{30}}{2}} = -0.913$$

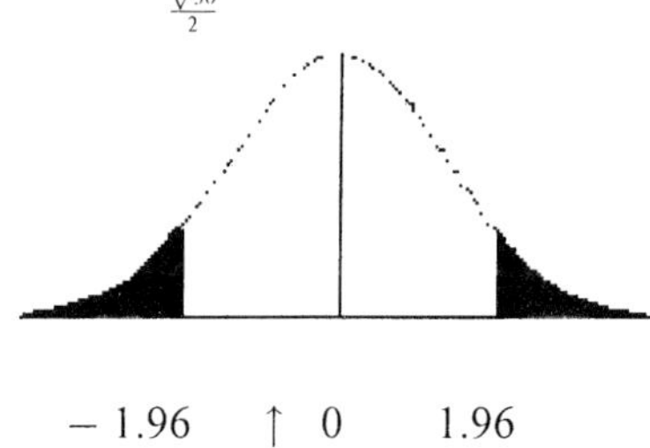

− 1.96 ↑ 0 1.96
− 0.913

Do not reject the null hypothesis. There is not enough evidence to reject the claim that the median is 40,000 miles.

3.
− + − − − − − − + +

H_0: The special diet has no effect on weight.
H_1: The diet increases weight. (claim)

C. V. = 1 Test value = 3

Do not reject the null hypothesis. There is not enough evidence to support the claim that there was an increase in weight.

4.
H_0: There is no difference in the record high temperatures.
H_1: There is a difference in the record high temperatures. (claim)
$\alpha = 0.05$

4. continued

47	47	50	51	51	52	52	57
W	W	W	W	W	D	D	D
1.5	1.5	3	4.5	4.5	6.5	6.5	8

60	66	66	69	71	80	80	86
D	D	W	W	D	W	D	W
9	10.5	10.5	12	13	14.5	14.5	17

86	86	88	89	89	91	93	94
W	D	D	D	W	W	D	D
17	17	19	20.5	20.5	22	23	24

R = 128.5 for Whitehorse

$$\mu_R = \frac{12(12 + 12 + 1)}{2} = 150$$

$$\sigma_R = \sqrt{\frac{12 \cdot 12(12 + 12 + 1)}{12}} = 17.32$$

$$z = \frac{R - \mu_R}{\sigma_R} = \frac{128.5 - 150}{17.32} = -1.24$$

P-value = 0.1075
Do not reject the null hypothesis. There is not enough evidence to support the claim that there is a difference in temperatures.

5.
H_0: There is no difference in the amounts of money each group spent for the textbook.
H_1: There is a difference in the amounts of money each group spent for the textbook. (claim)
C. V. = ± 1.65

36	46	48	49	50	51	52	53	55
1	2	3	4	5	6	7	8	9
B	B	B	B	B	B	B	B	E

58	58	62	63	63	64	72	73
10.5	10.5	12	13.5	13.5	15	16	17
B	E	B	B	E	E	E	E

74	78	78	85	86	88	93	98
18	19.5	19.5	21	22	23	24	25
B	E	E	E	E	E	E	E

R = 90

$$\mu_R = \frac{n_1(n_1 + n_2 + 1)}{2} = \frac{12(12 + 13 + 1)}{2} = 156$$

$$\sigma_R = \sqrt{\frac{n_1 n_2(n_1 + n_2 + 1)}{12}}$$

5. continued

$$\sigma_R = \sqrt{\frac{12 \cdot 13(12 + 13 + 1)}{12}} = 18.38$$

$$Z = \frac{R - \mu_R}{\sigma_R} = \frac{90 - 156}{18.38} = \frac{-66}{18.38} = -3.59$$

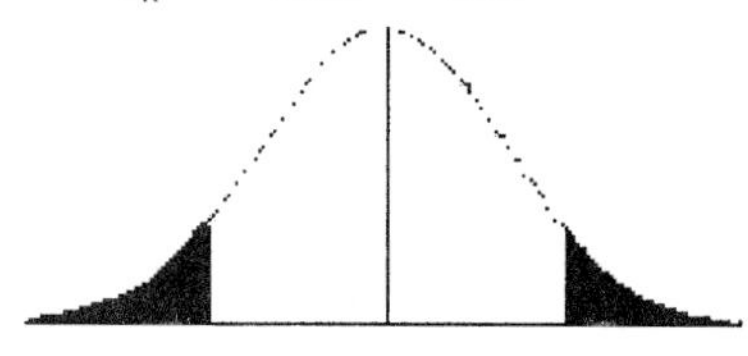

↑ − 1.65 0 1.65
− 3.59

Reject the null hypothesis. There is enough evidence to support the claim that there is a difference in the amount spent on the textbooks.

6.
H_0: The additive did not improve the gas mileage.
H_1: The additive did improve the gas mileage. (claim)

B	A	B − A	|B − A|	Rank	Signed Rank
13.6	18.3	− 4.7	4.7	11	− 11
18.2	19.5	− 1.3	1.3	3	− 3
16.1	18.2	− 2.1	2.1	8.5	− 8.5
15.3	16.7	− 1.4	1.4	4.5	− 4.5
19.2	21.3	− 2.1	2.1	8.5	− 8.5
18.8	17.2	1.6	1.6	6	6
22.6	23.7	− 1.1	1.1	1.5	− 1.5
21.9	20.8	1.1	1.1	1.5	1.5
25.3	25.3	0			
28.6	27.2	1.4	1.4	4.5	4.5
15.2	17.2	− 2.0	2.0	7	− 7
16.3	18.5	− 2.2	2.2	10	− 10

Sum of + ranks = 12
Sum of − ranks = 54
C. V. = 14 $\alpha = 0.05$ n = 11

$w_s = 12$

Since 12 < 14, reject the null hypothesis. There is enough evidence to support the claim that the additive improved the gas mileage.

7.
H_0: The number of sick days workers used was not reduced.
H_1: The number of sick days workers used was reduced. (claim)

7. continued

B	A	B − A	|B − A|	Rank	Signed Rank
6	8	− 2	2	1.5	− 1.5
15	12	3	3	3.5	3.5
18	16	2	2	1.5	1.5
14	9	5	5	6	6
27	23	4	4	5	4
17	14	3	3	3.5	3
9	15	− 6	6	7	− 7

Sum of the + ranks: 18
Sum of the − ranks: − 8.5
$w_s = 8.5$
C. V. = 4 $\alpha = 0.05$ n = 7

Do not reject the null hypothesis. There is not enough evidence to support the claim that the number of sick days was reduced.

8.
H_0: There is no difference in the breaking strengths of the ropes.
H_1: There is a difference in the breaking strengths of the ropes. (claim)

C. V. = 5.991 d. f. = 2 $\alpha = 0.05$

Cotton	Rank	Nylon	Rank	Hemp	Rank
230	1	356	7	506	34
432	18.5	303	2	527	36
505	32.5	361	8.5	581	42
487	27	405	17	497	29
451	20	432	18.5	459	22
390	15	378	14	507	35
462	23.5	361	8.5	562	40
531	37	399	16	571	41
366	11	372	12.5	499	30
372	12.5	363	10	475	26
453	21	306	4	505	32.5
488	28	304	3	561	39
462	23.5	318	5	432	38
467	25	322	6	501	31
$R_1 =$	295.5	$R_2 =$	132	$R_3 =$	475.5

$$H = \frac{12}{42(43)}\left(\frac{295.5^2}{14} + \frac{132^2}{14} + \frac{475.5^2}{14}\right)$$

$$- 3(42 + 1) = 28.02$$

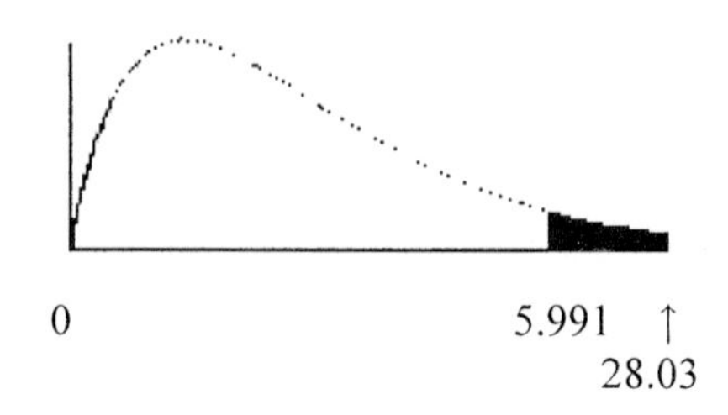

0 5.991 ↑
28.03

8. continued
Reject the null hypothesis. There is enough evidence to support the claim that there is a difference in the breaking strengths of the ropes.

9.
H_0: The diet has no effect on learning.
H_1: The diet affects learning. (claim)

C. V. = 5.991 d. f. = 2 $\alpha = 0.05$

Diet 1	Rank	Diet 2	Rank	Diet 3	Rank
8	11	2	1	9	13.5
6	6.5	3	2	15	17.5
12	15	6	6.5	17	19
15	17.5	8	11	8	11
9	13.5	7	8.5	4	3.5
7	8.5	4	3.5	13	16
5	5	R_2=	32.5	18	20
R_1=	77			20	21
				R_3=	121.5

$$H = \frac{12}{N(N+1)}\left(\frac{R_1^2}{n_1} + \frac{R_2^2}{n_2} + \frac{R_3^2}{n_3}\right) - 3(N+1)$$

$$= \frac{12}{21(21+1)}\left(\frac{77^2}{7} + \frac{32.5^2}{6} + \frac{121.5^2}{8}\right)$$

$$- 3(21+1) = 8.5$$

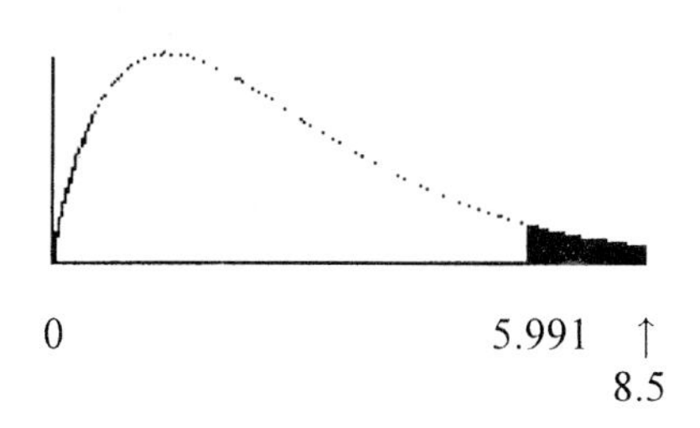

0 5.991 ↑ 8.5

Reject the null hypothesis. There is enough evidence to support the claim that the diets do affect learning.

10.
$r_s = 1 - \frac{6 \cdot 8}{9(9^2-1)} = 0.933$

H_0: $\rho = 0$
H_1: $\rho \neq 0$

C. V. = ± 0.700

Reject. There is a significant relationship in between the rankings.

11.

Boys	R_1	Girls	R_2	$R_1 - R_2$	d^2
3	3	4	4	− 1	1
2	2	5	5	− 3	9
6	6	1	1	5	25
1	1	3	3	− 2	4
5	5	2	2	3	9
4	4	6	6	− 2	4
				$\sum d^2 =$	52

$$r_s = 1 - \frac{6\sum d^2}{6(6^2-1)} = 1 - \frac{6(52)}{6(35)} = -0.486$$

H_0: $\rho = 0$
H_1: $\rho \neq 0$

C. V. = ± 0.886

Do not reject the null hypothesis. There is not enough evidence to say that a relationship exists between the rankings of the boys and girls.

12.
H_0: The students arrive at class at random according to whether or not they work. (claim)
H_1: The null hypothesis is false.

W NNN WWW N W NN WW NN W NN W N

The number of runs is 12. Since the expected number of runs is between 6 and 16, the null hypothesis should not be rejected.

13.
H_0: The grades of the students who finish the exam occur at random.
H_1: The null hypothesis is not true.

The median grade is 73. Using A = above and B = below, the runs are:

AAAA B AAAA BBBB AAAAA BB A BBBBBBB

Since there are eight runs and this does not fall between 9 and 21, the null hypothesis is rejected. The grades do not occur at random.

CHAPTER 13 QUIZ

1. False
2. False, they are less sensitive.
3. True

4. True
5. a.
6. c.
7. d.
8. b.
9. non-parametric
10. nominal, ordinal
11. sign
12. sensitive

13. H_0: Median = 300 (claim)
H_1: Median $\neq$ 300
There are seven + signs. Do not reject since 7 is greater than the critical value 5. There is not enough evidence to reject the claim that the median is 300.

14. H_0: Median = 1200 (claim)
H_1: Median $\neq$ 1200
There are ten − signs. Do not reject since 10 is greater than the critical value 6. There is not enough evidence to reject the claim that the median is 1,200.

15. H_0: There will be no change in the weight of the turkeys after the special diet.
H_1: The turkeys will weigh more after the special diet. (claim)
There is one + sign. Reject the null hypothesis. There is enough evidence to support the claim that the turkeys gained weight on the special diet.

16. H_0: The distributions are the same.
H_1: The distributions are different. (claim)
C. V. = $\pm$ 1.96
z = − 0.05
Do not reject the null hypothesis. There is not enough evidence to reject the claim that the distributions are the same.

17. H_0: The distributions are the same.
H_1: The distributions are different. (claim)
C. V. = $\pm$ 1.65
z = − 0.14
Do not reject the null hypothesis. There is not enough evidence to say there is a difference in the costs of the textbooks.

18. H_0: There is no difference in the GPA's before and after the workshop.
H_1: There is a difference in GPA's before and after the workshop. (claim)
C. .V. = 2 Test statistic = 0
Reject the null hypothesis. There is enough evidence to support the claim that there is a difference in the GPA's of the students.

19. H_0: There is no difference in the breaking strengths of the tapes.
H_1: There is a difference in breaking strengths. (claim)
H = 29.25
$\chi^2 = 5.991$
Reject the null hypothesis. There is enough evidence to support the claim that there is a difference in the breaking strengths of the tapes.

20. H_0: There is no difference in reaction times.
H_1: There is a difference in reaction times. (claim)
H = 6.9
$0.025 < \text{P-value} < 0.05$ (0.032)
Reject the null hypothesis. There is enough evidence to support the claim that there is a difference in the reaction times of the monkeys.

21. H_0: $\rho = 0$
H_1: $\rho \neq 0$
C. V. = $\pm$ 0.700
r = 0.846
Reject the null hypothesis. There is a significant relationship between the number of homework exercises and exam scores.

22. H_0: $\rho = 0$
H_1: $\rho \neq 0$
r = − 0.400
C. V. = 0.900
Do not reject the null hypothesis. There is not enough evidence ot say that a significant relationship exists between the rankings of the brands by males and females.

23. H_0: The gender of babies occurs at random.
H_1: The null hypothesis is false.
$\alpha = 0.05$ C. V. = 8, 19
There are 10 runs, which is between 8 and 19. Do not reject the null hypothesis. There is not enough evidence to reject the null hypothesis that the gender occurs at random.

24. H_0: There is no difference in output ratings.
H_1: There is a difference in output ratings after reconditioning. (claim)
$\alpha = 0.05$ n = 9 C. V. = 6
test statistic = 0
Do not reject the null hypothesis. There is not enough evidence to support the claim that there is a difference in output ratings before and after reconditioning.

25. H_0: The numbers occur at random
H_1: The null hypothesis is false.

25. continued

$\alpha = 0.05$ C. V. = 9, 21
The median number is 538.
There are 16 runs, reading from left to right, and since this is between 9 and 21, the null hypothesis is not rejected. There is not enough evidence to reject the null hypothesis that the numbers occur at random.

EXERCISE SET 14-2

1.
Random, systematic, stratified, cluster.

2.
Samples can save the researcher time and money. They are used when the population is large or infinite. They are used when the units are to be destroyed such as in testing the breaking strength of ropes.

3.
A sample must be randomly selected.

4.
Random numbers are used to ensure every element of the population has the same chance of being selected.

5.
Talking to people on the street, calling people on the phone, and asking one's friends are three incorrect ways of obtaining a sample.

6.
Over the long run each digit, zero through nine, will occur with the same probability.

7.
Random sampling has the advantage that each unit of the population has an equal chance of being selected. One disadvantage is that the units of the population must be numbered, and if the population is large this could be somewhat time consuming.

8.
Systematic sampling has an advantage that once the first unit is selected each succeeding unit selected has been determined. This will save time. A disadvantage would be if the list of units were arranged in some manner so that a bias would occur, such as selecting all men when the population consists of both men and women.

9.
An advantage of stratified sampling is that it ensures representation for the groups used in stratification; however, it is virtually impossible to stratify the population so that all groups could be represented.

10.
Clusters are easy to use since they already exist, but it is difficult to justify that the clusters actually represent the population.

11 through 20.
Answers will vary.

EXERCISE SET 14-3

1.
This is a biased question. Change the question to read: "Do you think XYZ Department Store should carry brand-name merchandise?"

2.
This is a biased question. Change the question to read: "Do you think ABC cars should use imported parts when they are manufactured?"

3. This is a biased question. Change the question to read: "Should banks charge a fee to balance their customers' checkbooks?"

4.
This question uses a double negative. Change the question to read: "Do you feel that it is appropriate for shopping malls to have activities for children who cannot read?"

5.
This question has confusing wording. Change the question to read: "How many hours did you study for this exam?"

6.
This is a double-barreled question. Change the question to read: "Do you think children should watch less television?"

7.
This question has confusing wording. Change the question to read: "If a plane were to crash on the border of New York and New Jersey, where should the victims be buried?"

8.
No flaw.

9.
Answers will vary.

10.
Answers will vary.

EXERCISE SET 14-5

1.
Simulation involves setting up probability experiments that mimic the behavior of real life events.

2.
Answers will vary.

3.
John Van Neumann and Stanislaw Ulam.

4.
Using the computer to simulate real life situations can save time since the computer can generate random numbers and keep track of the outcomes very quickly and easily.

5.
The steps are:
1. List all possible outcomes.
2. Determine the probability of each outcome.
3. Set up a correspondence between the outcomes and the random numbers.
4. Conduct the experiment using random numbers.
5. Repeat the experiment and tally the outcomes.
6. Compute any statistics and state the conclusions.

6.
Random numbers can be used to ensure the outcomes occur with appropriate probability.

7.
When the repetitions increase there is a higher probability that the simulation will yield more precise answers.

8.
Use random numbers one through five. Ignore six through nine and zero.

9.
Use random numbers one through eight to make a shot and nine and zero to represent a "miss".

10.
Use one and two for a defective VCR and three through nine and zero to represent a non-defective VCR.

11.
Use random numbers one through seven to represent a "hit" and eight, nine and zero to represent a "miss".

12.
Use the odd digits to represent a match and the even digits to represent a non-match.

13.
Let an odd number represent "heads" and an even number to represent "tails"; then each person selects a digit at random.

14 through 24.
Answers will vary.

REVIEW EXERCISES - CHAPTER 15

1 - 8.
Answers will vary.

9.
Use two digit random numbers 01 through 65 to represent a strike-out and 66 through 99 and 00 to represent anything other than a strike-out.

10.
Select two-digit numbers. The numbers 01 through 05 represent an over-booked flight.

11.
Select two digits between one and six to represent the dice.

12.
Select two digits between one and six for the first person's roll and one digit to represent the second person's roll.

13.
Let the digits 1 – 3 represent "rock"
Let the digits 4 – 6 represent "paper"
Let the digits 7 – 9 represent "scissors"
Omit 0.

14 through 18.
Answers will vary.

19.
This is a biased question. Change the question to read: "Have you ever driven through a red light?"

20.
This question uses a double negative. Change the question to read: "Do you think students who are not failing should be given tutoring if they request it?"

21.
This is a double-barreled question. Change the question to read: "Do you think all automobiles should have heavy-duty bumpers?"

22.
Answers will vary.

CHAPTER 14 QUIZ

1. True
2. True
3. False, only random numbers generated by a random number table are random.
4. True
5. a.
6. c.
7. c.
8. larger
9. biased
10. cluster
11. Answers will vary.
12. Answers will vary.
13. Answers will vary.
14. Answers will vary.
15. Use two-digit random numbers. 01 through 35 constitute a win and 36 through 00 constitute a loss.
16. Use two-digit random numbers. 01 - 03 constitute a cancellation.
17. Use two-digit random numbers. 01 - 13 for cards.
18. Use random numbers 1 - 6 to simulate the roll of a die and random numbers 01 - 13 to simulate the cards.
19. Use random numbers 1 - 6 to simulate a roll of a die.
20 - 24. Answers will vary.

A-1

A-1. $9! = 9 \cdot 8 \cdot 7 \cdot 6 \cdot 5 \cdot 4 \cdot 3 \cdot 2 \cdot 1 =$ 362,880

A-2. $7! = 7 \cdot 6 \cdot 5 \cdot 4 \cdot 3 \cdot 2 \cdot 1 = 5040$

A-3. $5! = 5 \cdot 4 \cdot 3 \cdot 2 \cdot 1 = 120$

A-4. $0! = 1$

A-5. $1! = 1$

A-6. $3! = 3 \cdot 2 \cdot 1 = 6$

A-7. $\frac{12!}{9!} = \frac{12 \cdot 11 \cdot 10 \cdot 9!}{9!} = 1320$

A-8. $\frac{10!}{2!} = \frac{10 \cdot 9 \cdot 8 \cdot 7 \cdot 6 \cdot 5 \cdot 4 \cdot 3 \cdot 2!}{2!}$

$= 1{,}814{,}400$

A-9. $\frac{5!}{3!} = \frac{5 \cdot 4 \cdot 3!}{3!} = 20$

A-10. $\frac{11!}{7!} = \frac{11 \cdot 10 \cdot 9 \cdot 8 \cdot 7!}{7!} = 7920$

A-11. $\frac{9!}{(4!)(5!)} = \frac{9 \cdot 8 \cdot 7 \cdot 6 \cdot 5!}{4 \cdot 3 \cdot 2 \cdot 1 \cdot 5!} = 126$

A-12. $\frac{10!}{(7!)(3!)} = \frac{10 \cdot 9 \cdot 8 \cdot 7!}{3 \cdot 2 \cdot 1 \cdot 7!} = 120$

A-13. $\frac{8!}{4!4!} = \frac{8 \cdot 7 \cdot 6 \cdot 5 \cdot 4!}{4 \cdot 3 \cdot 2 \cdot 1 \cdot 4!} = 70$

A-14. $\frac{15!}{12!3!} = \frac{15 \cdot 14 \cdot 13 \cdot 12!}{3 \cdot 2 \cdot 1 \cdot 12!} = 455$

A-15. $\frac{10!}{(10!)(0!)} = \frac{10!}{10! \cdot 1} = 1$

A-16. $\frac{5!}{3!2!1!} = \frac{5 \cdot 4 \cdot 3!}{3! \cdot 2 \cdot 1 \cdot 1} = 10$

A-17. $\frac{8!}{3!3!2!} = \frac{8 \cdot 7 \cdot 6 \cdot 5 \cdot 4 \cdot 3!}{3! \cdot 3 \cdot 2 \cdot 1 \cdot 2 \cdot 1} = 560$

A-18. $\frac{11!}{7!2!2!} = \frac{11 \cdot 10 \cdot 9 \cdot 8 \cdot 7!}{7! \cdot 2 \cdot 1 \cdot 2 \cdot 1} = 1980$

A-19. $\frac{10!}{3!2!5!} = \frac{10 \cdot 9 \cdot 8 \cdot 7 \cdot 6 \cdot 5!}{3 \cdot 2 \cdot 1 \cdot 2 \cdot 1 \cdot 5!} = 2520$

A-20. $\frac{6!}{2!2!2!} = \frac{6 \cdot 5 \cdot 4 \cdot 3 \cdot 2!}{2 \cdot 1 \cdot 2 \cdot 1 \cdot 2!} = 90$

A-2

A-21.

X	X^2	$X - \overline{X}$	$(X - \overline{X})^2$
9	81	− 3.1	9.61
17	289	4.9	24.01
32	1024	19.9	396.01
16	256	3.9	15.21
8	64	− 4.1	16.81
2	4	− 10.1	102.01
9	81	− 3.1	9.61
7	49	− 5.1	26.01
3	9	− 9.1	82.81
18	324	5.9	34.81
121	2181		716.9

$\sum X = 121 \qquad \overline{X} = \frac{121}{10} = 12.1$

$\sum X^2 = 2{,}181 \qquad (\sum X)^2 = 121^2 = 14{,}641$

$\sum(X-\overline{X})^2 = 716.9$

A-22.

X	X^2	$X - \overline{X}$	$(X - \overline{X})^2$
4	16	− 3	9
12	144	5	25
9	81	2	4
13	169	6	36
0	0	− 7	49
6	36	− 1	1
2	4	− 5	25
10	100	3	9
56	550		158

$\sum X = 56 \qquad \overline{X} = \frac{56}{8} = 7$

$\sum X^2 = 550 \qquad (\sum X)^2 = 56^2 = 3{,}136$

$\sum(X-\overline{X})^2 = 158$

A-23.

X	X^2	$X - \overline{X}$	$(X - \overline{X})^2$
5	25	− 1.4	1.96
12	144	5.6	31.36
8	64	1.6	2.56
3	9	− 3.4	11.56
4	16	− 2.4	5.76
32	258		53.20

$\sum X = 32 \qquad \overline{X} = \frac{32}{5} = 6.4$

$\sum X^2 = 258 \qquad (\sum X)^2 = 32^2 = 1{,}024$

A-23. continued

$\sum(X-\bar{X})^2 = 53.2$

A-24.

X	X^2	$X-\bar{X}$	$(X-\bar{X})^2$
6	36	– 12.75	163.5625
2	4	– 16.75	280.5625
18	324	– 0.75	0.5625
30	900	11.25	126.5625
31	961	12.25	150.0625
42	1764	23.25	540.5625
16	256	– 2.75	7.5625
5	25	– 13.75	189.0625
150	4270		1457.5000

$\sum X = 150 \qquad \bar{X} = \frac{150}{8} = 18.75$

$\sum X^2 = 4{,}270 \qquad (\sum X)^2 = 150^2 = 22{,}500$

$\sum(X-\bar{X})^2 = 1457.5$

A-25.

X	X^2	$X-\bar{X}$	$(X-\bar{X})^2$
80	6400	14.4	207.36
76	5776	10.4	108.16
42	1764	– 23.6	556.96
53	2809	– 12.6	158.76
77	5929	11.4	129.96
328	22678		1161.20

$\sum X = 328 \qquad \bar{X} = \frac{328}{5} = 65.6$

$\sum X^2 = 22{,}678 \qquad (\sum X)^2 = 328^2 = 107{,}584$

$\sum(X-\bar{X})^2 = 1161.2$

A-26.

X	X^2	$X-\bar{X}$	$(X-\bar{X})^2$
123	15129	– 15.17	230.1289
132	17424	– 6.17	38.0689
216	46656	77.83	6057.5089
98	9604	– 40.17	1613.6289
146	21316	7.83	61.3089
114	12996	– 24.17	584.1889
829	123125		8584.8334

$\sum X = 829 \qquad \bar{X} = \frac{829}{6} = 138.17$

$\sum X^2 = 123{,}125 \qquad (\sum X)^2 = 829^2 = 687{,}241$

$\sum(X-\bar{X})^2 = 8{,}584.8334$

A-27.

X	X^2	$X-\bar{X}$	$(X-\bar{X})^2$
53	2809	– 16.3	265.69
72	5184	2.7	7.29
81	6561	11.7	136.89
42	1764	– 27.3	745.29
63	3969	– 6.3	39.69
71	5041	1.7	2.89
73	5329	3.7	13.69
85	7225	15.7	246.49
98	9604	28.7	823.69
55	3025	– 14.3	204.49
693	50511		2486.10

$\sum X = 693 \qquad \bar{X} = \frac{693}{10} = 69.3$

$\sum X^2 = 50{,}511 \qquad (\sum X)^2 = 693^2 = 480{,}249$

$\sum(X-\bar{X})^2 = 2486.1$

A-28.

X	X^2	$X-\bar{X}$	$(X-\bar{X})^2$
43	1849	– 38.8	1505.44
32	1024	– 49.8	2480.04
116	13456	34.2	1169.64
98	9604	16.2	262.44
120	14400	38.2	1459.24
409	40333		6876.80

$\sum X = 409 \qquad \bar{X} = \frac{409}{5} = 81.8$

$\sum X^2 = 40{,}333 \qquad (\sum X)^2 = 409^2 = 167{,}281$

$\sum(X-\bar{X})^2 = 6876.8$

A-29.

X	X^2	$X-\bar{X}$	$(X-\bar{X})^2$
12	144	– 41	1681
52	2704	– 1	1
36	1296	– 17	289
81	6561	28	784
63	3969	10	100
74	5476	21	441
318	20150		3296

$\sum X = 318 \qquad \bar{X} = \frac{318}{6} = 53$

$\sum X^2 = 20{,}150 \qquad (\sum X)^2 = 318^2 = 101{,}124$

$\sum(X-\bar{X})^2 = 3{,}296$

A-30.

X	X^2	$X - \bar{X}$	$(X - \bar{X})^2$
−9	81	−5.67	32.1489
−12	144	−8.67	75.1689
18	324	21.33	454.9689
0	0	3.33	11.0889
−2	4	1.33	1.7689
−15	225	−11.67	136.1889
−20	778		711.3334

$\sum X = -20$ $\quad \bar{X} = \frac{-20}{6} = -3.33$

$\sum X^2 = 778$ $\quad (\sum X)^2 = -20^2 = 400$

$\sum (X - \bar{X})^2 = 711.3334$

A-3

A-31.

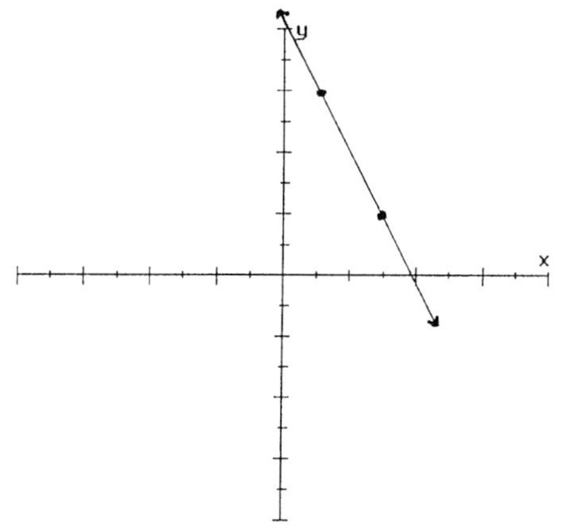

A-32.

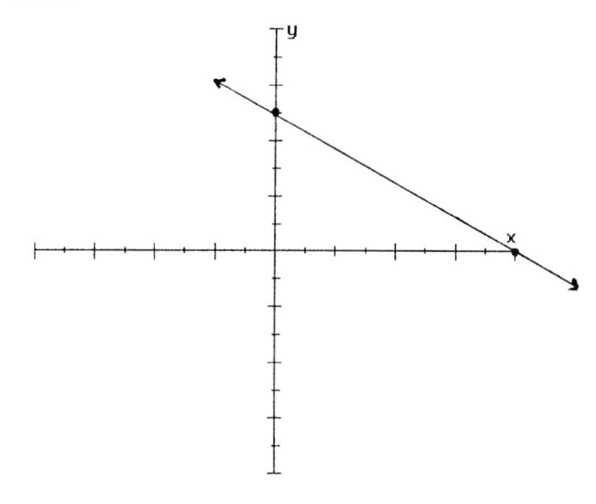

A-33.

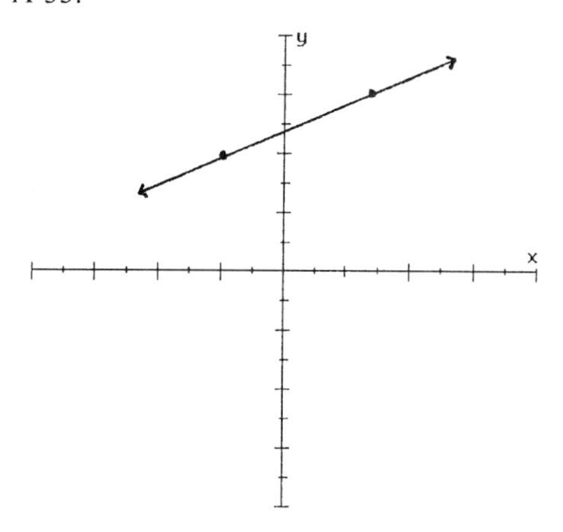

A-34.

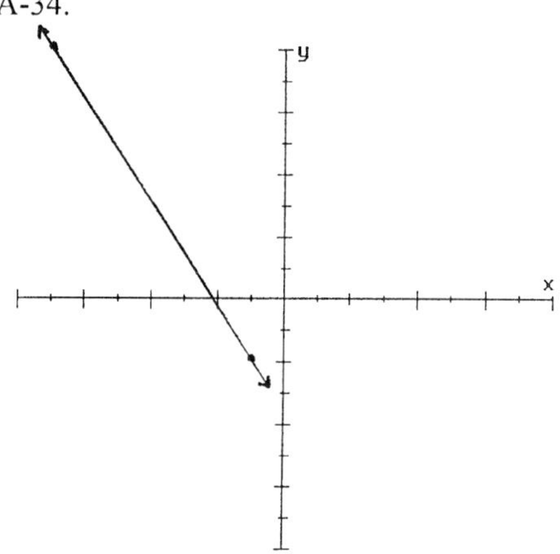

A-35.

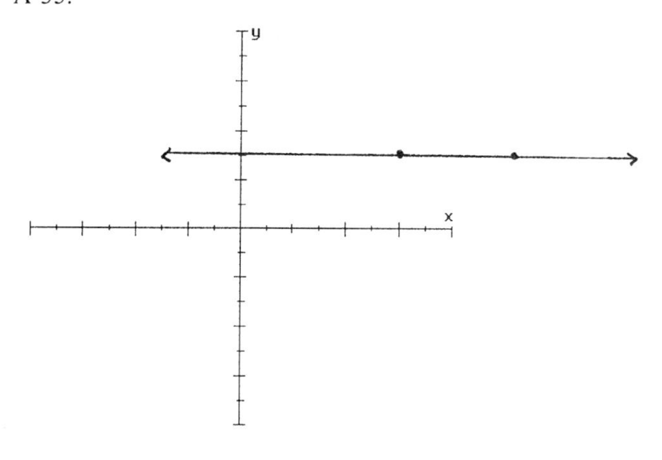

A-36.

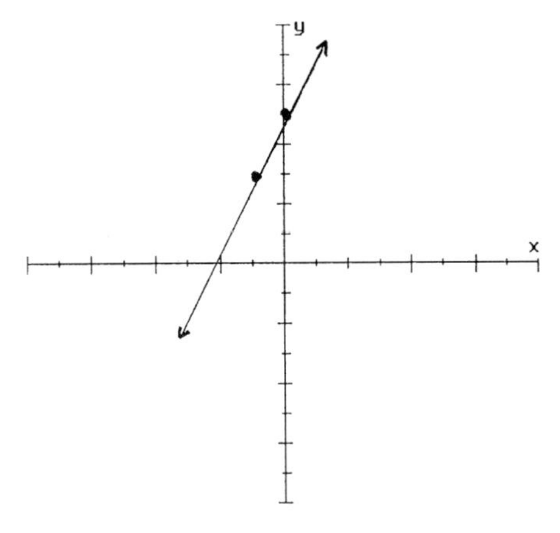

Two points are: (0, 5) and (-1, 3).

A-37.

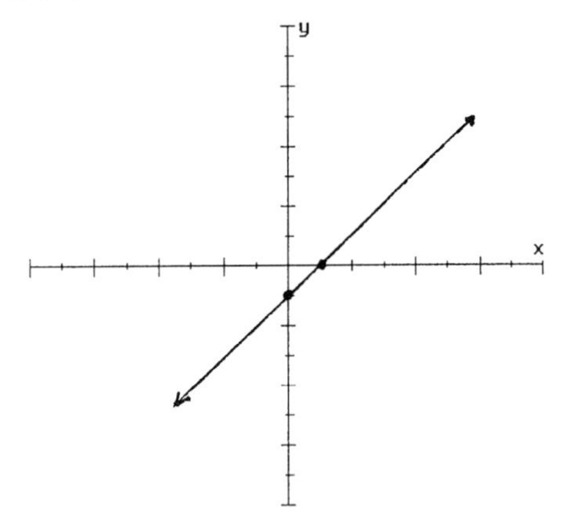

Two points are: (1, 0) and (0, -1).

A-38.

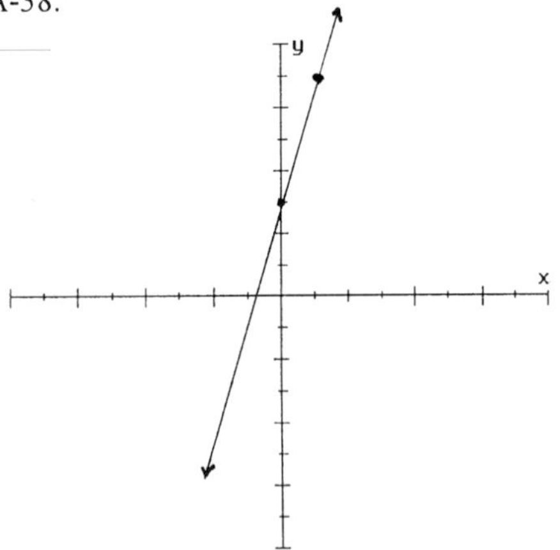

Two points are: (1, 7) and (0, 3).

A-39.

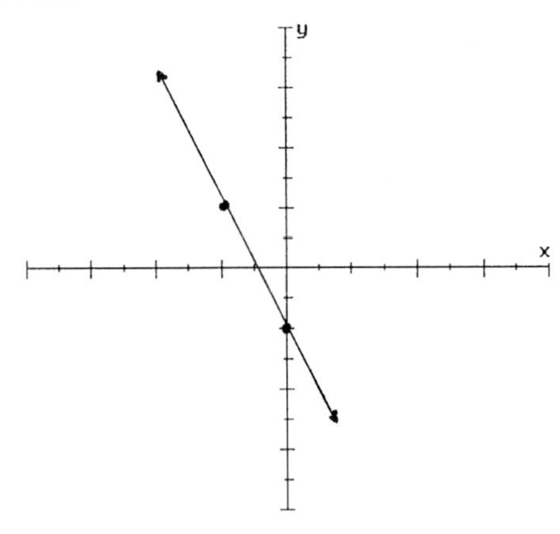

Two points are: (-2, 2) and (0, -2).

A-40.

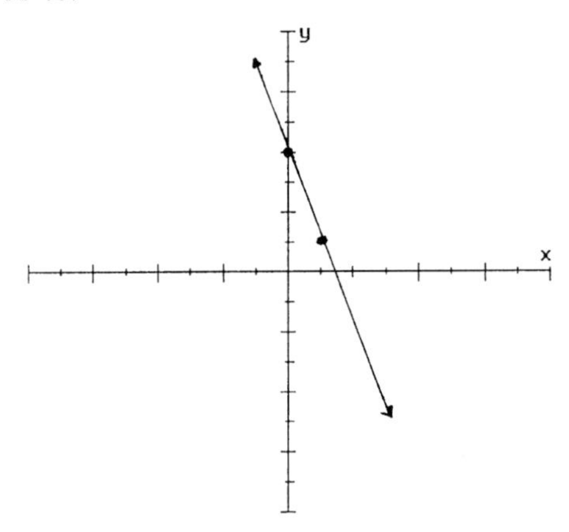

Two points are: (1, 1) and (0, 4)

B-2

B-1.

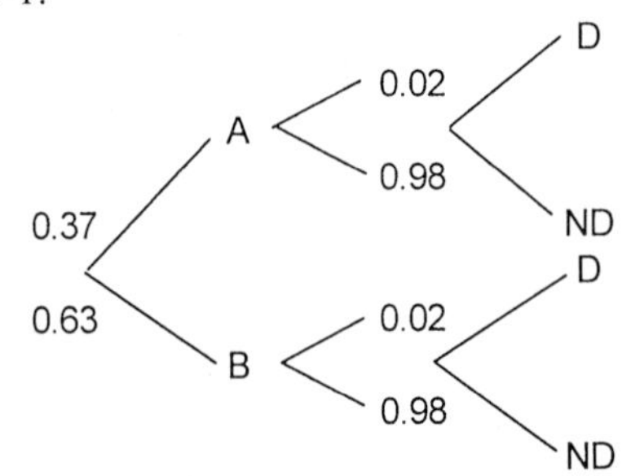

$$P(B \mid \text{defective}) = \frac{P(B)\cdot P(\text{def} \mid B)}{P(A)\cdot P(\text{def} \mid A) + P(B)\cdot P(\text{def} \mid B)}$$

$$= \frac{0.63(.02)}{0.37(0.02)+0.63(0.02)} = \frac{0.0126}{0.02} = 0.63$$

B-2.

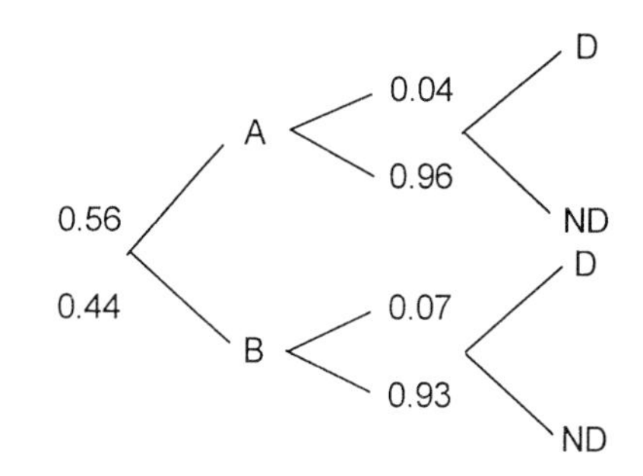

$$P(B \mid \text{defective}) = \frac{P(B)\cdot P(\text{def} \mid B)}{P(A)\cdot P(\text{def} \mid A)+P(B)\cdot P(\text{def} \mid B)}$$

$$= \frac{0.44(0.07)}{0.56(0.04)+0.44(0.07)} = \frac{0.0308}{0.0532} = 0.579$$

B-3.
Let D = person has disease
Let $\overline{D}$ = person does not have disease
Let A = positive test result
Let $\overline{A}$ = negative test result

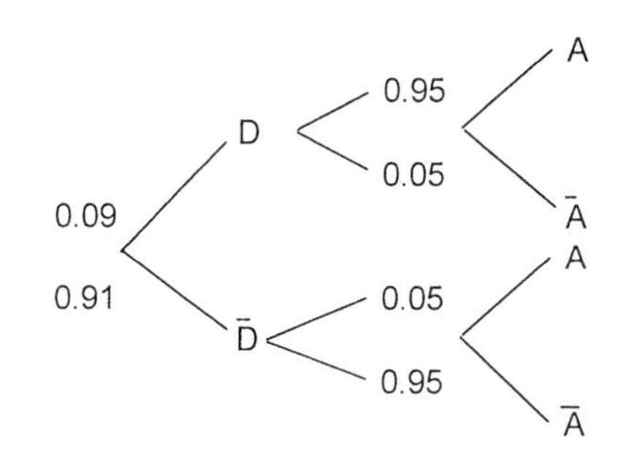

$$P(D \mid A) = \frac{P(D)\cdot P(A \mid D)}{P(D)\cdot P(A \mid D) + P(\overline{D})\cdot P(A \mid \overline{D})}$$

$$= \frac{0.09(0.95)}{0.09(0.95) + 0.91(0.05)} = 0.653$$

B-4.

$$P(D \mid \overline{A}) = \frac{P(D)\cdot P(\overline{A} \mid D)}{P(D)\cdot P(\overline{A} \mid D) + P(\overline{D})\cdot P(\overline{A} \mid \overline{D})}$$

$$= \frac{0.09(0.05)}{0.09(0.05) + 0.91(0.95)} = 0.005$$

B-5.
Let S = success
Let F = failure

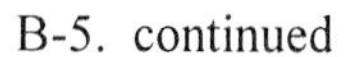
B-5. continued

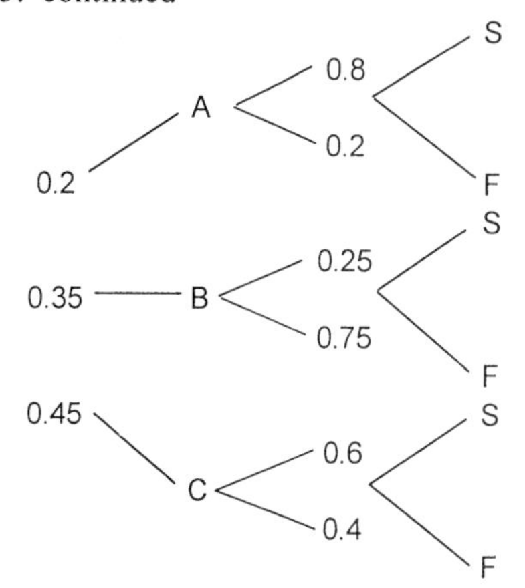

$$P(B \mid S) = \frac{P(B)\cdot P(S \mid B)}{P(A)\cdot P(S \mid A) + P(B)\cdot P(S \mid B) + P(C)\cdot P(S \mid C)}$$

$$= \frac{0.35(0.75)}{0.20(0.80) + 0.35(0.75) + 0.45(0.60)} = \frac{0.2625}{0.6925}$$

$$= 0.379$$

B-6.

$$P(C \mid F) = \frac{0.45(0.40)}{0.20(0.2) + 0.35(0.25) + 0.45(0.40)}$$

$$= 0.585$$

B-7.

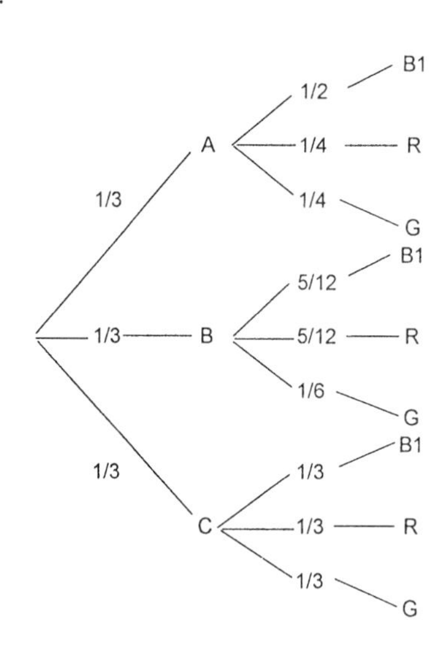

$$P(A \mid R) = \frac{P(A)\cdot P(R \mid A)}{P(A)\cdot P(R \mid A) + P(B)\cdot P(R \mid B) + P(C)\cdot P(R \mid C)}$$

$$= \frac{\frac{1}{3}\cdot\frac{1}{4}}{\frac{1}{3}\cdot\frac{1}{4} + \frac{1}{3}\cdot\frac{5}{12} + \frac{1}{3}\cdot\frac{1}{3}} = \frac{\frac{1}{12}}{\frac{1}{12} + \frac{5}{36} + \frac{1}{9}} = \frac{\frac{1}{12}}{\frac{1}{3}} = \frac{1}{4}$$

B-8.

$P(B \mid G)$

$$= \frac{P(B)\cdot P(G \mid B)}{P(A)\cdot P(G \mid A) + P(B)\cdot P(G \mid B) + P(C)\cdot P(G \mid C)}$$

$$= \frac{\frac{1}{3}\cdot\frac{1}{6}}{\frac{1}{3}\cdot\frac{1}{4}+\frac{1}{3}\cdot\frac{1}{6}+\frac{1}{3}\cdot\frac{1}{3}} = \frac{\frac{1}{18}}{\frac{9}{36}} = \frac{2}{9}$$

B-9.

Let J = traffic jam and $\bar{J}$ = no traffic jam

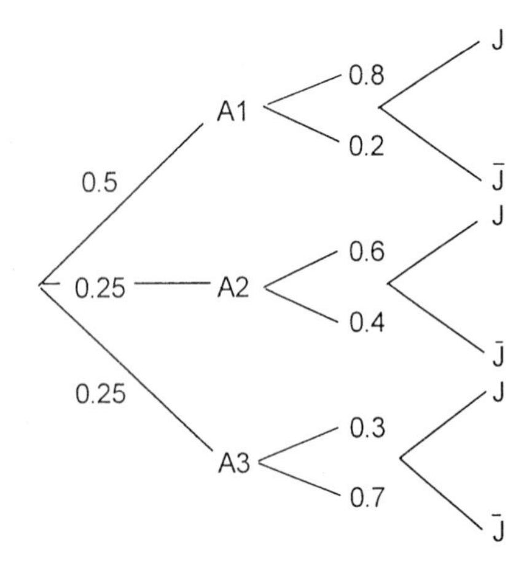

$P(A_1 \mid J) =$

$$\frac{P(A_1)\cdot P(J \mid A_1)}{P(A_1)\cdot P(J \mid A_1) + P(A_2)\cdot P(J \mid A_2) + P(A_3)\cdot P(J \mid A_3)}$$

$$\frac{0.5(0.8)}{0.5(0.8)+0.25(0.6)+0.25(0.3)} = \frac{0.4}{0.4+0.15+0.075}$$

$$= \frac{0.4}{0.625} = 0.64$$

B-10.

$$P(A_3 \mid \bar{J}) = \frac{0.25(0.7)}{0.5(0.2)+0.25(0.4)+0.25(0.7)}$$

$$= \frac{0.175}{0.1+0.1+0.175} = \frac{0.175}{0.375} = 0.467$$

B-11.

$P(A) = \frac{350}{1400} = 0.25 \quad P(B) = \frac{1050}{1400} = 0.75$

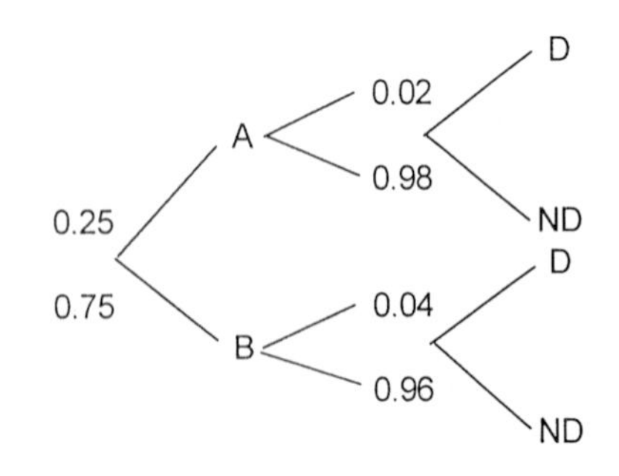

B-11. continued

$$P(B \mid D) = \frac{P(B)\cdot P(D \mid B)}{P(A)\cdot P(D \mid A) + P(B)\cdot P(D \mid B)}$$

$$= \frac{0.75(0.04)}{0.25(0.02)+0.75(0.04)} = \frac{0.03}{0.035} = 0.857$$

B-12.

Let D = a dented can and let $\overline{D}$ = a non-dented can

$P(A) = \frac{2400}{6000} = 0.4$

$P(B) = \frac{3600}{6000} = 0.6$

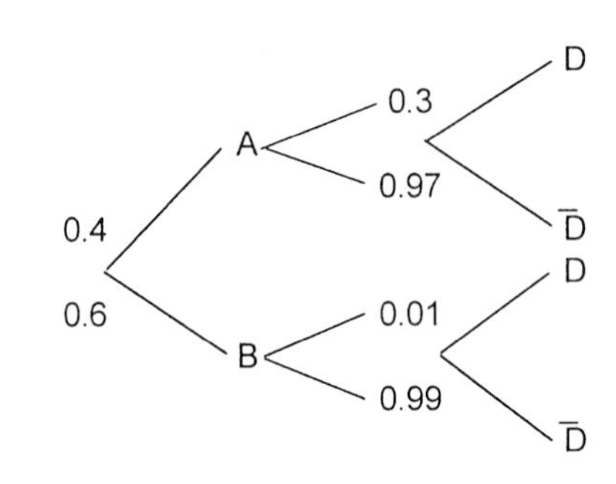

$$P(B \mid D) = \frac{P(B)\cdot P(D \mid B)}{P(A)\cdot P(D \mid A) + P(B)\cdot P(D \mid B)}$$

$$= \frac{0.6(0.01)}{0.4(0.03)+0.6(0.01)} = \frac{0.006}{0.018} = 0.33$$